James J. Kelly

Graduate Mathematical Physics

Related Titles

Trigg, G. L. (ed.)
Mathematical Tools for Physicists
686 pages with 98 figures and 29 tables
2005
Hardcover
ISBN 3-527-40548-8

Masujima, M.
Applied Mathematical Methods in Theoretical Physics
388 pages with approx. 60 figures
2005
Hardcover
ISBN 3-527-40534-8

Dubin, D.
Numerical and Analytical Methods for Scientists and Engineers Using Mathematica
633 pages
2003
Hardcover
ISBN 0-471-26610-8

Lambourne, R., Tinker, M.
Basic Mathematics for the Physical Sciences
688 pages
2000
Hardcover
ISBN 0-471-85206-6

Kusse, B., Westwig, E. A.
Mathematical Physics
Applied Mathematics for Scientists and Engineers
680 pages
1998
Hardcover
ISBN 0-471-15431-8

Courant, R., Hilbert, D.
Methods of Mathematical Physics
Volume 1
575 pages with 27 figures
1989
Softcover
ISBN 0-471-50447-5

James J. Kelly

Graduate Mathematical Physics

With MATHEMATICA Supplements

WILEY-VCH Verlag GmbH & Co. KGaA

The Author

Prof. James J. Kelly
University of Maryland
Dept. of Physics
jjkelly@umd.edu

For a Solutions Manual, lecturers should contact the editorial department at vch-college@wiley-vch.de, stating their affiliation and the course in which they wish to use the book.

Library of Congress Card No.:
applied for

British Library Cataloguing-in-Publication Data
A catalogue record for this book is available from the British Library.

Bibliographic information published by the Deutsche Nationalbibliothek
Die Deutsche Nationalbibliothek lists this publication in the Deutsche Nationalbibliografie; detailed bibliographic data are available in the Internet at <http://dnb.d-nb.de>.

Typesetting Da-TeX Gerd Blumenstein, Leipzig
Printing Strauss GmbH, Mörlenbach
Binding Litges & Dopf GmbH, Heppenheim

Printed in the Federal Republic of Germany
Printed on acid-free paper

ISBN-13: 978-3-527-40637-1
ISBN-10: 3-527-40637-9

Preface

This textbook is intended to serve a course on mathematical methods of physics that is often taken by graduate students in their first semester or by undergraduates in their senior year. I believe the most important topic for first-year graduate students in physics is the theory of analytic functions. Some students may have had a brief exposure to that subject as undergraduates, but few are adequately prepared to apply such methods to physics problems. Therefore, I start with the theory of analytic functions and practically all subsequent material is based upon it. The primary topics include: theory of analytic functions, integral transforms, generalized functions, eigenfunction expansions, Green functions, boundary-value problems, and group theory. This course is designed to prepare students for advanced treatments of electromagnetic theory and quantum mechanics, but the methods and applications are more general. Although this is a fairly standard course taught in most major universities, I was not satisfied with the available textbooks. Some popular but encyclopedic books include a broader range of topics, much too broad to cover in one semester at the depth that I thought necessary for graduate students. Others with a more manageable length appear to be targeted primarily at undergraduates and relegate to appendices some of the topics that I believe to be most important. Therefore, I soon found that preparation of lecture notes for distribution to students was evolving into a textbook-writing project.

I was not able to avoid producing too much material either. I usually chose to skip most of the chapter on Legendre and Bessel functions, assuming that graduate students already had some familiarity with them, and instead referred them to a summary of the properties that are useful for the chapter on boundary-value problems. Other instructors might choose to omit the chapter on dispersion theory instead because most of it will probably be covered in the subsequent course on electromagnetism, but I find that subject more interesting and more fun to discuss than special functions. The chapter on group theory was prepared at the request of reviewers; although I never reached that topic in one semester, I hope that it will be useful for those teaching a two-semester course or as a resource that students will use later on. It may also be useful for one-semester courses at institutions where the average student already has a sufficiently strong mastery of analytic functions that the first couple of chapters can be abbreviated or omitted. I believe that it should be possible to cover most of the remaining material well in a single semester at any mid-level university. I assume that the calculus of variations will be covered in a concurrent course on classical mechanics and that the students are already comfortable with linear algebra, differential equations, and vector calculus. Probability theory, tensor analysis, and differential geometry are omitted.

A CD containing detailed solutions to all of the problems is available to instructors. These solutions often employ *MATHEMATICA®* to perform some of the routine but tedious manipulations and to prepare figures. Some of these solutions may also be presented as additional examples of the techniques covered in this course.

Graduate Mathematical Physics. James J. Kelly

ISBN: 3-527-40637-9

Contents

Graduate Mathematical Physics. James J. Kelly

ISBN: 3-527-40637-9

Note to the Reader

I chose to prepare my lecture notes and subsequent textbook using *MATHEMATICA®* because I am very enamored of its facility for combining mathematical typesetting with symbolic manipulation, numerical computation, and graphics into notebook documents approaching publication quality. However, because students must learn the mathematical techniques in this course, not just the syntax of a program, practically all derivations in the body of the text are performed *by hand* with *MATHEMATICA®* serving primarily as a word processor. The figures were also produced using *MATHEMATICA®*, but most of the code for the figures has been removed from the main text. And, of course, I often checked my work using the symbolic manipulation tools of the program. However, I also discovered a disturbingly large number of integrals that *MATHEMATICA®* evaluated incorrectly. Some of those errors have been corrected in later versions, perhaps due in part to my error reports, but inevitably new errors emerged even for integrals that were evaluated correctly in earlier versions! The lessons that students should learn from my experience are either *caveat emptor* (let the buyer beware) and *trust but verify*. The student must understand the mathematics well enough to recognize probable errors (the *smell test*) and to check the results of any mathematical software. Software is helpful, but no software is perfect! The *wetware* between your ears must evaluate the results of the software.

I also adopted some of the notation of *MATHEMATICA®* because it is often superior to the traditional notation of mathematical literature. For example, $f[x]$ with square brackets indicates a function f whose argument is x while $f(x)$ with parentheses indicates the product $f * x$. Although the target audience would rarely confuse delimiters intended for grouping with delimiters intended for arguments, anyone who has taught lower-level courses has witnessed the havoc wrought by ambiguous notations. Therefore, I have gotten into the habit of using parentheses only for grouping terms, square brackets primarily for arguments (and commutators), and curly brackets for lists or iterators. Similarly, I use *MATHEMATICA®*'s double-struck symbols $\mathbb{i}$ for $\sqrt{-1}$, $\mathbb{e}$ for the base of the natural logarithm, $\mathbb{d}$ for differential, etc. Furthermore, I often distinguish between assignments (=) and equations (==) intended to be solved. I hope that most readers eventually agree that some of these nontraditional typesetting practices are actually preferable to traditional notation. Finally, I use several convenient acronyms: wrt for with respect to, rhs for right-hand side, lhs for left-hand side, and iff for if and only if.

I encourage students to use mathematical software to perform some of the mundane tasks encountered in homework problems and to plot their solutions. Several examples are given in the text, where *MATHEMATICA®* code is sometimes used to perform simple but tedious algebraic manipulations. However, software should not be used to circumvent the object of an exercise. For example, if the objective of a problem is to practice an integration method, then simply quoting *MATHEMATICA®*'s answer is not sufficient and sometimes would even be incorrect. It should be obvious when computer assistance is appropriate

Graduate Mathematical Physics. James J. Kelly

ISBN: 3-527-40637-9

and when it is not. Also, please take care to specify the assumptions made in the solution of problems, carefully identifying the range of validity with respect to any parameters present.

The student CD that accompanies this book includes a basic introduction to *MATHEMATICA*®, supplementary notebooks that provide solutions to selected exercises, the code used to prepare figures, and some additional material where appropriate.

1 Analytic Functions

Abstract. We introduce the theory of functions of a complex variable. Many familiar functions of real variables become multivalued when extended to complex variables, requiring branch cuts to establish single-valued definitions. The requirements for differentiability are developed and the properties of analytic functions are explored in some detail. The Cauchy integral formula facilitates development of power series and provides powerful new methods of integration.

1.1 Complex Numbers

1.1.1 Motivation and Definitions

The definition of complex numbers can be motivated by the need to find solutions to polynomial equations. The simplest example of a polynomial equation without solutions among the real numbers is $z^2 == -1$. Gauss demonstrated that by defining two solutions according to

$$z^2 == -1 \Longrightarrow z = \pm i \tag{1.1}$$

one can prove that any polynomial equation of degree n has n solutions among complex numbers of the form $z = x + iy$ where x and y are real and where $i^2 = -1$. This powerful result is now known as the *fundamental theorem of algebra*. The object i is described as an *imaginary number* because it is not a real number, just as $\sqrt{2}$ is an irrational number because it is not a rational number. A number that may have both real and imaginary components, even if either vanishes, is described as *complex* because it has two parts. Throughout this course we will discover that the rich properties of functions of complex variables provide an amazing arsenal of weapons to attack problems in mathematical physics.

The complex numbers can be represented as ordered pairs of real numbers $z = (x, y)$ that strongly resemble the Cartesian coordinates of a point in the plane. Thus, if we treat the numbers $1 = (1, 0)$ and $i = (0, 1)$ as basis vectors, the complex numbers $z = (x, y) = x \times 1 + y \times i = x + iy$ can be represented as points in the complex plane, as indicated in Fig. 1.1. A diagram of this type is often called an *Argand diagram*. It is useful to define functions Re or Im that retrieve the real part $x = \mathrm{Re}[z]$ or the imaginary part $y = \mathrm{Im}[z]$ of a complex number. Similarly, the modulus, r, and phase, θ, can be defined as the polar coordinates

$$r = \sqrt{x^2 + y^2}, \quad \theta = \mathrm{ArcTan}\left[\frac{y}{x}\right] \tag{1.2}$$

by analogy with two-dimensional vectors.

Graduate Mathematical Physics. James J. Kelly

ISBN: 3-527-40637-9

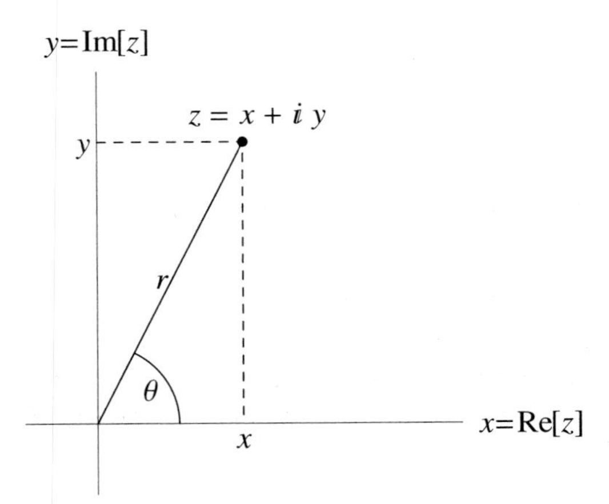

Figure 1.1. Cartesian and polar representations of complex numbers.

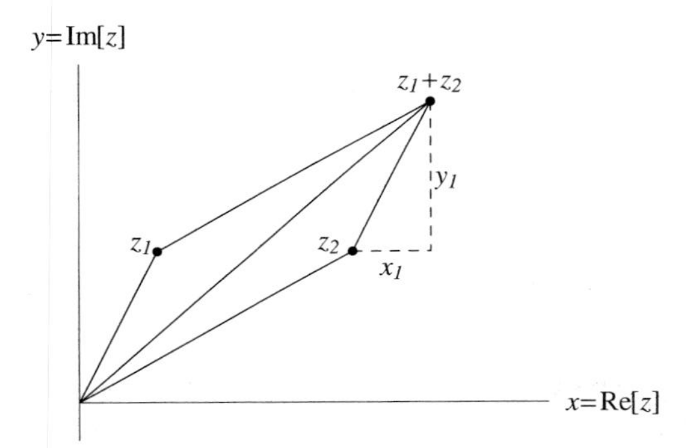

Figure 1.2. Addition of complex numbers.

Continuing this analogy, we also define the addition of complex numbers by adding their components, such that

$$z_1 + z_2 = \left(x_1 + x_2, y_1 + y_2\right) \Longleftrightarrow z_1 + z_2 = \left(x_1 + x_2\right) \times 1 + \left(y_1 + y_2\right) \times i \tag{1.3}$$

as diagrammed in Fig. 1.2. The complex numbers then form a *linear vector space* and addition of complex numbers can be performed graphically in exactly the same manner as for vectors in a plane.

However, the analogy with Cartesian coordinates is not complete and does not extend to multiplication. The multiplication of two complex numbers is based upon the distributive property of multiplication

$$\begin{aligned} z_1 z_2 &= (x_1 + i y_1)(x_2 + i y_2) \\ &= x_1 x_2 + i^2 y_1 y_2 + i(x_1 y_2 + x_2 y_1) \\ &= (x_1 x_2 - y_1 y_2) + i(x_1 y_2 + x_2 y_1) \end{aligned} \tag{1.4}$$

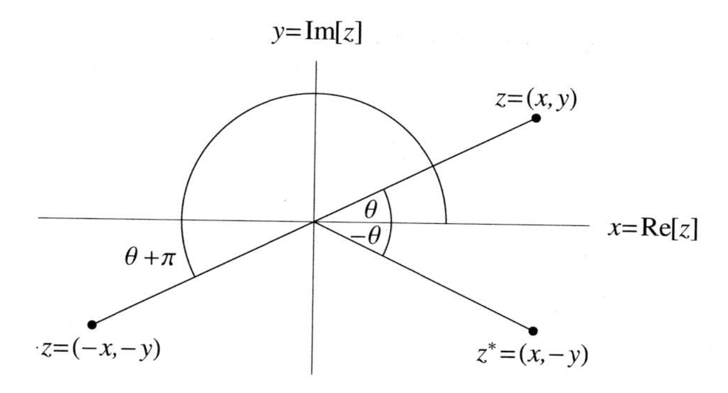

Figure 1.3. Inversion and complex conjugation of a complex number.

and the definition $i^2 = -1$. The product of two complex numbers is then another complex number with the components

$$z_1 z_2 = (x_1 x_2 - y_1 y_2, x_1 y_2 + x_2 y_1) \tag{1.5}$$

More formally, the complex numbers can be represented as ordered pairs of real numbers $z = (x, y)$ with equality, addition, and multiplication defined by:

$$z_1 = z_2 \Longrightarrow x_1 = x_2 \wedge y_1 = y_2 \tag{1.6}$$

$$z_1 + z_2 = (x_1 + x_2, y_1 + y_2) \tag{1.7}$$

$$z_1 \times z_2 = (x_1 x_2 - y_1 y_2, x_1 y_2 + x_2 y_1) \tag{1.8}$$

One can show that these definitions fulfill all the formal requirements of a *field*, and we denote the complex number field as $\mathbb{C}$. Thus, the field of real numbers is contained as a subset, $\mathbb{R} \subset \mathbb{C}$.

It will also be useful to define complex conjugation

$$\text{complex conjugation: } z = (x, y) \Longrightarrow z^* = (x, -y) \tag{1.9}$$

and absolute value functions

$$\text{absolute value: } |z| = \sqrt{x^2 + y^2} \tag{1.10}$$

with conventional notations. Geometrically, complex conjugation represents reflection across the real axis, as sketched in Fig. 1.3.

The Re, Im, and Abs functions can now be expressed as

$$\text{Re}[z] = \frac{z + z^*}{2}, \quad \text{Im}[z] = \frac{z - z^*}{2i}, \quad |z|^2 = zz^* \tag{1.11}$$

Thus, we quickly obtain the following arithmetic facts:

$$
\begin{aligned}
&i = (0, 1) \quad i^2 = -1 \quad i^3 = -i \quad i^4 = 1 \\
&\text{scalar multiplication: } c \in \mathbb{R} \Longrightarrow cz = (cx, cy) \\
&\text{additive inverse: } z = (x, y) \Longrightarrow -z = (-x, -y) \Longrightarrow z + (-z) = 0 \\
&\text{multiplicative inverse: } z^{-1} = \frac{1}{x + iy} = \frac{x - iy}{x^2 + y^2} = \frac{z^*}{|z|^2}
\end{aligned}
\tag{1.12}
$$

1.1.2 Triangle Inequalities

Distances between points in the complex plane are calculated using a metric function. A *metric* $d[a, b]$ is a real-valued function such that

1. $d[a, b] > 0$ for all $a \neq b$
2. $d[a, b] = 0$ for all $a = b$
3. $d[a, b] = d[b, a]$
4. $d[a, b] \leq d[a, c] + d[c, b]$ for any c.

Thus, the Euclidean metric $d[z_1, z_2] = |z_1 - z_2| = \sqrt{(x_1 - x_2)^2 + (y_1 - y_2)^2}$ is suitable for $\mathbb{C}$. Then with geometric reasoning one easily obtains the triangle inequalities:

$$\text{triangle inequalities : } ||z_1| - |z_2|| \leq |z_1 \pm z_2| \leq |z_1| + |z_2| \tag{1.13}$$

Note that $\mathbb{C}$ cannot be ordered (it is not possible to define $<$ properly).

1.1.3 Polar Representation

The function $e^{i\theta}$ can be evaluated using the power series

$$e^{i\theta} = \sum_{n=0}^{\infty} \frac{(i\theta)^n}{n!} = \sum_{n=0}^{\infty} (-)^n \frac{\theta^{2n}}{(2n)!} + i \sum_{n=0}^{\infty} (-)^n \frac{\theta^{2n+1}}{(2n+1)!} = \mathrm{Cos}[\theta] + i\,\mathrm{Sin}[\theta] \tag{1.14}$$

giving a result known as *Euler's formula*. Thus, we can represent complex numbers in *polar form* according to

$$z = re^{i\theta} \Longrightarrow x = r\,\mathrm{Cos}[\theta], \tag{1.15}$$

$$y = r\,\mathrm{Sin}[\theta] \quad \text{with} \quad r = |z| = \sqrt{x^2 + y^2} \quad \text{and} \quad \theta = \arg[z] \tag{1.16}$$

where r is the *modulus* or *magnitude* and θ is the *phase* or *argument* of z. Although addition of complex numbers is easier with the Cartesian representation, multiplication is usually easier using polar notation where the product of two complex numbers becomes

$$\begin{aligned} z_1 z_2 &= r_1 r_2 (\mathrm{Cos}[\theta_1] + i\,\mathrm{Sin}[\theta_1])(\mathrm{Cos}[\theta_2] + i\,\mathrm{Sin}[\theta_2]) \\ &= r_1 r_2 (\mathrm{Cos}[\theta_1]\mathrm{Cos}[\theta_2] - \mathrm{Sin}[\theta_1]\mathrm{Sin}[\theta_2] + i(\mathrm{Sin}[\theta_1]\mathrm{Cos}[\theta_2] + \mathrm{Cos}[\theta_1]\mathrm{Sin}[\theta_2])) \\ &= r_1 r_2 (\mathrm{Cos}[\theta_1 + \theta_2] + i\,\mathrm{Sin}[\theta_1 + \theta_2]) \\ &= r_1 r_2 e^{i(\theta_1+\theta_2)} \end{aligned} \tag{1.17}$$

Thus, the moduli multiply while the phases add. Note that in this derivation we did not assume that $e^{z_1} e^{z_2} = e^{z_1+z_2}$, which we have not yet proven for complex arguments, relying instead upon the Euler formula and established properties for trigonometric functions of real variables.

Using the polar representation, it also becomes trivial to prove *de Moivre's theorem*

$$\left(e^{i\theta}\right)^n = e^{in\theta} \Longrightarrow \left(\mathrm{Cos}[\theta] + i\,\mathrm{Sin}[\theta]\right)^n = \mathrm{Cos}[n\theta] + i\,\mathrm{Sin}[n\theta] \quad \text{for integer } n. \tag{1.18}$$

However, one must be careful in performing calculations of this type. For example, one cannot simply replace $(e^{in\theta})^{1/n}$ by $e^{i\theta}$ because the equation, $z^n == w$ has n solutions $\{z_k, k = 1, n\}$ while $e^{i\theta}$ is a unique complex number. Thus, there are n, n^{th}-roots of unity, obtained as follows.

$$z = re^{i\theta} \Longrightarrow z^n = r^n e^{in\theta} \tag{1.19}$$

$$z^n == 1 \Longrightarrow r = 1, \quad n\theta = 2k\pi \tag{1.20}$$

$$\therefore z = \mathrm{Exp}\left[i\frac{2\pi k}{n}\right] = \mathrm{Cos}\left[\frac{2\pi k}{n}\right] + i\,\mathrm{Sin}\left[\frac{2\pi k}{n}\right] \quad \text{for} \quad k = 0, 1, 2, \ldots, n-1 \tag{1.21}$$

In the Argand plane, these roots are found at the vertices of a regular n-sided polygon inscribed within the unit circle with the principal root at $z = 1$. More generally, the roots

$$z^n = w = \rho e^{i\phi} \Longrightarrow z_k = \rho^{1/n}\,\mathrm{Exp}\left[i\frac{\phi + 2\pi k}{n}\right] \quad \text{for} \quad k = 0, 1, 2, \ldots, n-1 \tag{1.22}$$

of $e^{i\phi}$ are found at the vertices of a rotated polygon inscribed within the unit circle, as illustrated in Fig. 1.4.

1.1.4 Argument Function

The graphical representation of complex numbers suggests that we should obtain the phase using

$$\theta \stackrel{?}{=} \arctan\left[\frac{y}{x}\right] \tag{1.23}$$

but this definition is unsatisfactory because the ratio y/x is not sensitive to the quadrant, being positive in both first and third and negative in both second and fourth quadrants. Consequently, computer programs using $\arctan[\frac{y}{x}]$ return values limited to the range $(-\frac{\pi}{2}, \frac{\pi}{2})$.

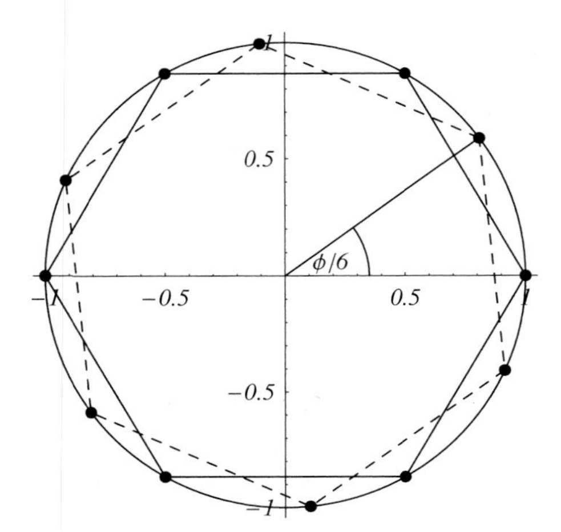

Figure 1.4. Solid: 6th roots of 1, dashed: 6th roots of $e^{i\phi}$.

A better definition is provided by a quadrant-sensitive extension of the usual arctangent function

$$\mathrm{ArcTan}[x, y] = \mathrm{ArcTan}\left[\frac{y}{x}\right] + \frac{\pi}{2}(1 - \mathrm{Sign}[x])\,\mathrm{Sign}[y] \tag{1.24}$$

that returns values in the range $(-\pi, \pi)$. (Unfortunately, the order of the arguments is reversed between Fortran and *Mathematica*.) Therefore, we define the *principal branch* of the argument function by

$$\mathrm{Arg}[z] = \mathrm{Arg}[x + iy] = \mathrm{ArcTan}[x, y] \tag{1.25}$$

where $-\pi < \mathrm{Arg}[z] \leq \pi$.

However, the polar representation of complex numbers is not unique because the phase θ is only defined modulo 2π. Thus,

$$\arg[z] = \mathrm{Arg}[z] + 2\pi n \tag{1.26}$$

is a multivalued function where n is an arbitrary integer. Note that some authors distinguish between these functions by using lower case for the multivalued and upper case for the single-valued version while others rely on context. Consider two points on opposite sides of the negative real axis, with $y \to 0^+$ infinitesimally above and $y \to 0^-$ infinitesimally below. Although these points are very close together, $\mathrm{Arg}[z]$ changes by 2π across the negative real axis.

A discontinuity of this type is usually represented by a *branch cut*. Imagine that the complex plane is a sheet of paper upon which axes are drawn. Starting at the point $(r, 0)$ one can reach any point $z = (x, y)$ by drawing a continuous circular arc of radius $r = \sqrt{x^2 + y^2}$ and we define $\arg[z]$ as the angle subtended by that arc. This function is multivalued because the circular arc can be traversed in either direction or can wind around the origin an

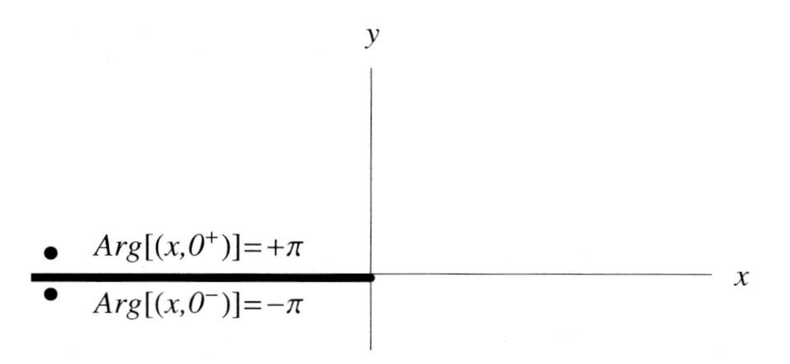

Figure 1.5. Branch cut for Arg[z].

arbitrary number of times before stopping at its destination. A single-valued version can be created by making a cut infinitesimally below the negative real axis, as sketched in Fig. 1.5, that prevents a continuous arc from subtending more than $\pm\pi$ radians. Points on the negative real axis are reached by positive (counterclockwise) arcs with $\text{Arg}[(-|z|, 0)] = \pi$ while points infinitesimally below the negative real axis can only be reached by negative arcs with $\text{Arg}[(-|z|, 0^-)] \to -\pi$. Thus, Arg[$z$] is single-valued and is continuous on any path that does not cross its branch cut, but is discontinuous across the cut.

The principal branch of the argument function is defined by the restriction $-\pi < \text{Arg}[z] \leq \pi$. Notice that one side of this range is open, represented by $<$, while the other side is closed, represented by $\leq$. This notation indicates that the cut is infinitesimally below the negative real axis, such that the argument for negative real numbers is π, not $-\pi$. This choice is not unique, but is the nearly universal convention for the argument and many related functions. The distinction between $<$ and $\leq$ many seem to be nitpicking, but attention to such details is often important in performing accurate derivations and calculations with functions of complex variables.

Many functions require one or more branch cuts to establish single-valued definitions; in fact, handling either the multivaluedness of functions of complex variables or the discontinuities associated with their single-valued manifestations is often the most difficult problem encountered in complex analysis. Although our choice of branch cut for Arg[z] is not unique (any radial cut from the origin to ∞ would serve the same purpose), it is consistent with the customary definitions of ArcTan, Log, and other elementary functions to be discussed in more detail later. The single-valued version of a function that is most common is described as its *principal branch*. For many functions there is considerable flexibility in the choice of branch cut and we are free to make the most convenient choice, provided that we maintain that choice throughout the problem. For example, in some applications it might prove convenient to define an argument function with the range $-\frac{3\pi}{4} < \text{MyArg}[z] \leq \frac{5\pi}{4}$ using the branch cut shown in Fig. 1.6. Consider the point $z_1 = (-1, -1)$ for which the standard argument function gives $\text{Arg}[z_1] = -3\pi/4$ while our new argument function gives $\text{MyArg}[z_1] = 5\pi/4$. These functions are obviously different because the same input gives different output, but both represent precisely the same ray in the complex plane. Therefore, we should consider the specification of the branch cuts as an important part of the definition of a single-valued function and recognize that different

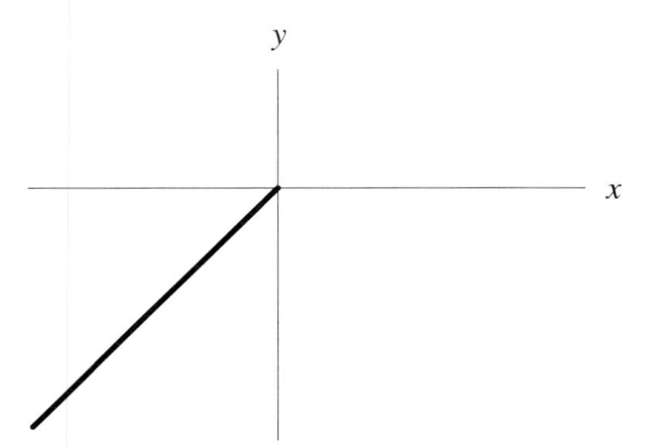

Figure 1.6. Branch cut for MyArg.

choices of cuts lead to related but different functions.

It is important to recognize that, because of discontinuities across branch cuts, simple algebraic relationships that apply to multivalent functions of complex variables, often do not pertain to their monovalent cousins. For example, using the polar representation of the product of two complex numbers we find

$$z_1 z_2 = r_1 r_2 \,\mathrm{Exp}\big[i(\theta_1 + \theta_2)\big] \Longrightarrow |z_1 z_2| = |z_1||z_2| \; \arg[z_1 z_2] = \arg[z_1] + \arg[z_2] \tag{1.27}$$

but this relationship for the phase does not necessarily apply to the principal branch because

$$\begin{aligned} \mathrm{Arg}[z_1 z_2] &= \arg[z_1 z_2] + 2\pi n \\ &= \arg[z_1] + \arg[z_2] - 2\pi n \\ &= \mathrm{Arg}[z_1] + \mathrm{Arg}[z_2] + 2\pi(n - n_1 - n_2) \end{aligned} \tag{1.28}$$

where n must be chosen to ensure that $-\pi < \mathrm{Arg}[z_1 z_2] \leq \pi$. Often the price of single-valuedness is the awkwardness of discontinuities.

1.2 Take Care with Multivalued Functions

Ambiguities in the definitions of many seemingly innocuous functions require considerable care. For example, consider the common replacement

$$\sqrt{\frac{1}{z}} \overset{?}{\longleftrightarrow} z^{-1/2} \tag{1.29}$$

that one often makes without thinking. Is this apparent equivalence correct? Compare the following two methods for evaluating these quantities when $z \to -1$.

$$z = -1 \Longrightarrow z^{-1} = -1 \Longrightarrow \sqrt{\frac{1}{z}} = i \tag{1.30}$$

$$z = e^{i\pi} \Longrightarrow z^{-1/2} = e^{-i\pi/2} = -i \tag{1.31}$$

Both calculations look correct, but their results differ in sign. These expressions are not always interchangeable! One must take more care with multivalued functions.

If we represent the complex number z in Cartesian form $z = x + iy$ where x, y are real, then

$$\frac{1}{z} = \frac{1}{x + iy} = \frac{x - iy}{x^2 + y^2} \tag{1.32}$$

If $x < 0$ and $y \to \varepsilon$ where ε is a positive infinitesimal, then z is just above and z^{-1} is just below the usual cut in the square-root function (below the negative real axis). Consequently, Arg$[z]$ and Arg$[z^{-1}]$ differ by 2π and the arguments of $\sqrt{1/z}$ and $z^{-1/2}$ differ by π, a negative sign, in the immediate vicinity of the negative real axis. It is usually not a good idea to use the surd (square-root) symbol for complex variables – for real numbers that symbol is usually interpreted as the positive square root, but for negative or complex numbers we should employ a fractional power and define the branch cut explicitly. Then, if we define $-\pi < \text{Arg}[z] \leq \pi$ with a cut infinitesimally below the negative real axis the same cut would be implied for fractional powers and the value of $z^{-1/2}$ determined using polar notation would be unambiguous on the negative real axis. Furthermore, $1/z^{1/2} = z^{-1/2}$ applies everywhere in the cut z-plane without the sign ambiguity encountered above. Of course, the sign discontinuity across the cut is still present – it is an essential feature of such functions.

Let us examine the square-root function, $w = f[z] = z^{1/2}$, in more detail. When z is a positive real number, the square-root function maps one z onto two values of $w = \pm\sqrt{x}$. Similar behavior is expected for complex z because there are always two solutions to the quadratic equation $w^2 == z$. In polar notation

$$z = re^{i\theta}, \quad z == w^2 \Longrightarrow w = \sqrt{r}\,\text{Exp}\left[i\frac{\theta}{2} + n\pi i\right] \tag{1.33}$$

where, by convention, $\sqrt{r}$ represents the positive square root for real numbers and where $n = 0, 1$ yields two distinct possibilities. Thus, the image of one point in the z-plane is two points in the w-plane. If we define $w = u + iv$, the component functions $u[x, y]$ and $v[x, y]$ can be obtained by solving the equations

$$x == u^2 - v^2 \quad y == 2uv \tag{1.34}$$

Substituting $v \to y/2u$ and solving the quadratic equation for u^2, we find

$$4u^4 - 4u^2x - y^2 == 0 \Longrightarrow u^2 = \frac{x \pm \sqrt{x^2 + y^2}}{2} \Longrightarrow u = \pm\sqrt{\frac{\sqrt{x^2 + y^2} + x}{2}} \tag{1.35}$$

where the positive root is required in u^2 to ensure real u. Then, solving for v and rationalizing the expression under the square root, we obtain

$$v = \frac{y}{2u} = \pm\frac{y}{2}\sqrt{\frac{2}{\sqrt{x^2 + y^2} + x}} = \pm y\sqrt{\frac{\sqrt{x^2 + y^2} - x}{2y^2}} \tag{1.36}$$

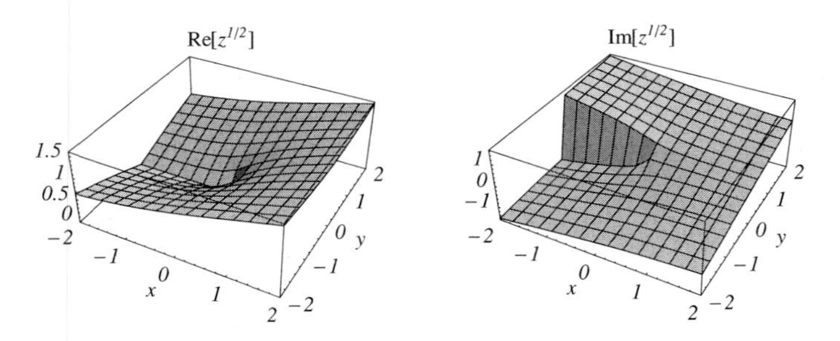

Figure 1.7. Real and imaginary components of $z^{1/2}$.

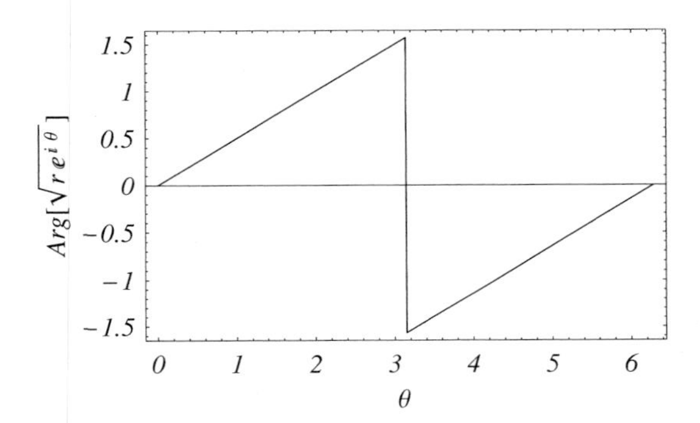

Figure 1.8. Dependence of Arg[$z^{1/2}$] upon polar angle.

Finally, taking the positive root of y^2, we obtain

$$u = \pm\sqrt{\frac{\sqrt{x^2+y^2}+x}{2}} \qquad v = \pm\,\mathrm{Sign}[y]\sqrt{\frac{\sqrt{x^2+y^2}-x}{2}} \tag{1.37}$$

where the relative sign between u and v is determined by the sign of y. Note that there are only two, not four, solutions. The principal branches of the component functions are plotted in Fig. 1.7, where it is customary, though arbitrary, to select the positive branch of u so that positive square roots are obtained on the positive real axis. Similar figures are obtained for other positive nonintegral powers, rational or irrational.

Notice that $v = \mathrm{Im}[\sqrt{z}]$ is discontinuous on the negative real axis. The real part is continuous, but its derivative with respect to y is discontinuous on the negative real axis. Consider the image of a circular path $z = re^{i\theta}$, $0 \le \theta \le 2\pi$ under the mapping $w = \sqrt{z}$. The argument of w changes abruptly from π to $-\pi$ as the negative real axis is crossed from above, as sketched in Fig. 1.8.

In order to define a well-behaved monovalent function, we must include in the definition of f a rule for selecting the appropriate output value when the mapping $z \to w$ is multivalent. The customary solution is to introduce a branch cut along the negative real axis by restricting the range of the argument of z to $-\pi < \theta \leq \pi$ and agreeing not to cross the cut in the z-plane. Thus, Sqrt and Arg employ the same branch cut, shown in Fig. 1.5. We imagine that the cut is infinitesimally below the negative real axis so that the argument of negative real numbers is π and $x < 0 \Longrightarrow \sqrt{x} = i\sqrt{|x|}$. The discontinuity in $\mathrm{Arg}[z^\alpha]$ across the branch cut depends upon α. The end points of the branch cut are known as *branch points* at which discontinuities first open. Here, the most important branch point is at $z = 0$, but one often says that there is also a branch point at ∞. This somewhat sloppy language means that for large $|z|$ the branch cut is parametrized by $z = Re^{i\theta}$ with $R \to \infty$, but the choice of θ remains arbitrary; here we happened to choose $\theta = \pi$.

Next consider the slightly more complicated function

$$f[z] = \left(z^2 - 1\right)^{1/2} = \left((z-1)(z+1)\right)^{1/2} \tag{1.38}$$

Our experience with the square root suggests that we must pay close attention to the points $z = \pm 1$ that are the branch points for $(z \mp 1)^{1/2}$. By factoring the argument of the square root and choosing ranges for the phase of each factor according to

$$z_1 = z - 1 = r_1 e^{i\theta_1}, \quad -\pi < \theta_1 \leq \pi \tag{1.39}$$

$$z_2 = z + 1 = r_2 e^{i\theta_2}, \quad -\pi < \theta_2 \leq \pi \tag{1.40}$$

we obtain a single-valued version defined by

$$f_1[z] = \sqrt{r_1 r_2}\,\mathrm{Exp}\left[i\left(\frac{\theta_1 + \theta_2}{2} + k\pi\right)\right], \quad k \in \{0, 1\} \tag{1.41}$$

and the branch cuts indicated in Fig. 1.9. The principal branch is defined, somewhat arbitrarily, by $k = 0$ because that gives a positive root for z on the real axis with $x > 1$. The two heavy points show the branch points in the two factors $(z \pm 1)^{1/2}$ and the lines anchored by those points show the associated branch cuts. Here we decided to draw both branch cuts to the left, as is customary for $\mathrm{Arg}[z]$ or $z^{1/2}$, such that both polar angles are defined in the range $-\pi < \theta_{1,2} \leq \pi$. Since it is clear that any discontinuities will be found along the real axis where the two phases may be discontinuous, the structure of the function can be investigated using strategically chosen points on both sides of the real axis labeled a–f in the figure and its accompanying Table 1.1. This table shows that $f_1[z]$ is discontinuous across the portion of real axis between the two branch points, namely $-1 < x < 1$, but is continuous elsewhere. In effect, the overlapping branch cuts cancel each other in the region $x < -1$ because, at least for this function, the discontinuity is simply a sign change; however, the behavior of overlapping cuts is not always this simple.

Alternatively, if we define the phases according to

$$z_1 = z - 1 = r_1 e^{i\theta_1}, \quad 0 \leq \theta_1 < 2\pi \tag{1.42}$$

$$z_2 = z + 1 = r_2 e^{i\theta_2}, \quad -\pi < \theta_2 \leq \pi \tag{1.43}$$

we obtain another version

$$f_2[z] = \sqrt{r_1 r_2}\,\mathrm{Exp}\left[i\left(\frac{\theta_1 + \theta_2}{2} + k\pi\right)\right], \quad k \in \{0, 1\} \tag{1.44}$$

in which the cut in $z_1^{1/2}$ is now directed toward the right while the cut in $z_2^{1/2}$ remains to the left; these cuts are illustrated in Fig. 1.10. The same selection of trial points now produces

Table 1.1. Selected values of $f_1[z]$ defined by Eq. (1.41).

Point	x	y	θ_1	θ_2	$f_1[z]$
a	$x > 1$	ε	0	0	$\sqrt{x^2-1}$
b	$x > 1$	$-\varepsilon$	0	0	$\sqrt{x^2-1}$
c	$-1 < x < 1$	ε	π	0	$i\sqrt{1-x^2}$
d	$-1 < x < 1$	$-\varepsilon$	$-\pi$	0	$-i\sqrt{1-x^2}$
e	$x < -1$	ε	π	π	$-\sqrt{x^2-1}$
f	$x < -1$	$-\varepsilon$	$-\pi$	$-\pi$	$-\sqrt{x^2-1}$

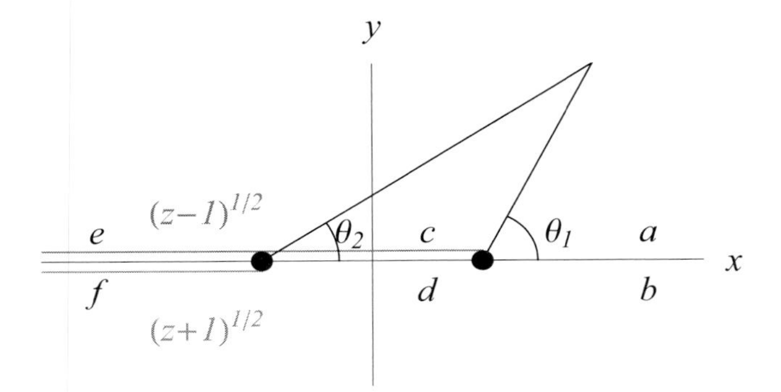

Figure 1.9. Branch cuts for $f_1[z]$ defined by Eq. (1.41).

Table 1.2 that shows that $f_2[z]$ is continuous for $-1 < x < 1$ but is discontinuous everywhere else on the real axis. Although their algebraic definitions are the same, $f_1[z]$ and $f_2[z]$ are clearly different functions, fraternal twins that are distinguished by their branch cuts. The choice of cuts is a fundamental aspect of the definition of a single-valued version of an inherently multivalued function; the definition is not complete until the cuts are specified. For any particular application the most appropriate version may depend upon other aspects of the problem, such as physical boundary conditions, or may be chosen for convenience. If one is most interested in small values of $|z|$ it will probably be more convenient to choose $f_2[z]$, but for large values $f_1[z]$ is probably preferable, but consistency must be maintained through any particular problem. Furthermore, although the two versions suggested here are the most common, they do not exhaust the possibilities.

The moral of this somewhat belabored exercise is that one must be very careful in manipulating expressions involving complex variables and avoid making unintentional assumptions about the phases of various subexpressions. Most people tend to be very careless with phases, automatically replacing $\sqrt{s^2}$ by s without knowing whether s is positive or negative or complex. Similarly, many people complain that *MATHEMATICA®* often does not perform simplifications that are perceived to be obvious. The reason for this is that *MATHEMATICA®* hates to make mistakes and will not make unjustified assumptions about

Table 1.2. Selected values of $f_2[z]$ defined by Eq. (1.44).

Point	x	y	θ_1	θ_2	$f_2[z]$
a	$x > 1$	ε	0	0	$\sqrt{x^2 - 1}$
b	$x > 1$	$-\varepsilon$	2π	0	$-\sqrt{x^2 - 1}$
c	$-1 < x < 1$	ε	π	0	$i\sqrt{1 - x^2}$
d	$-1 < x < 1$	$-\varepsilon$	π	0	$i\sqrt{1 - x^2}$
e	$x < -1$	ε	π	π	$-\sqrt{x^2 - 1}$
f	$x < -1$	$-\varepsilon$	π	$-\pi$	$\sqrt{x^2 - 1}$

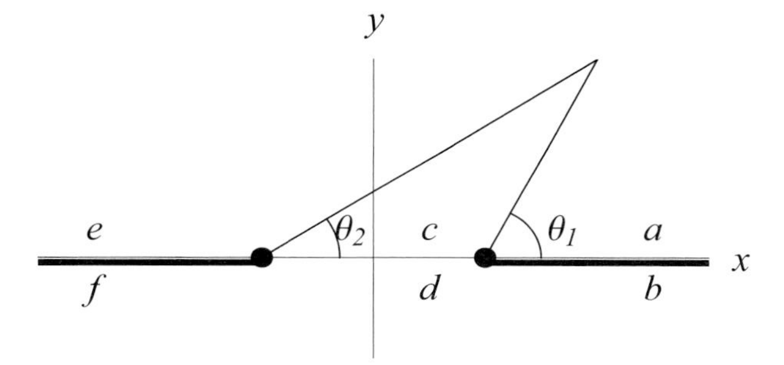

Figure 1.10. Branch cuts for $f_2[z]$ defined by Eq. (1.44).

whether a variable is real or, if it is, about its sign – it assumes that all variables are complex unless told otherwise. The Simplify function has an option that permits the user to specify permissible assumptions, such as one variable is real, another positive, a third a negative integer, etc. When you take responsibility for these assumptions, MATHEMATICA® will usually go much further in simplifying your expressions. Often it still will not reach the elegant representation that one might find in a textbook, but its manipulations will be correct and that is what matters most.

1.3 Functions as Mappings

A function f maps the complex variable $z = (x, y)$ into a complex *image* $w = (u, v)$ according to rules specified in the definition $w = f[z]$. Thus, f maps points in the complex z-plane onto points in the complex w-plane, a mapping of $\mathbb{C} \to \mathbb{C}$. For a single-valued function the image of a point is a point, but for a multiple-valued function the image of a point may be a set of points. For continuous functions, the image of a line segment (arc) in the input plane will be one or more arcs in the output plane. Often considerable insight into the properties of a function may be obtained by examining the images of *coordinate lines* (lines of constant x or constant y). Below we analyze the mappings produced by some familiar functions.

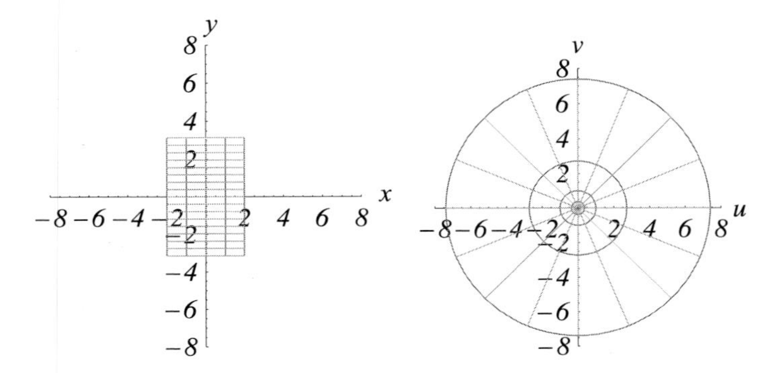

Figure 1.11. Mapping: $u + iv = e^{x+iy}$. Lines of constant y are mapped onto radial lines while lines lines of constant x are mapped onto circles in the w-plane.

1.3.1 Mapping: $w = e^z$

The exponential function is defined by the mapping

$$w = e^z = e^x(\text{Cos}[y] + i\,\text{Sin}[y]) = u + iv \Longrightarrow u[x, y] = e^x\,\text{Cos}[y] \quad v[x, y] = e^x\,\text{Sin}[y] \tag{1.45}$$

The image of a grid of coordinate lines is sketched in Fig. 1.11. Lines of constant y are mapped into radial lines, while lines of constant x are mapped into circles. The origin, with $x = 0$, is mapped onto the unit circle. Increasingly positive $x = x_0 > 0$, mapped onto circles of exponentially increasing radius e^{x_0}, and increasingly negative $x = x_0 < 0$ mapped onto exponentially tighter circles, practically indiscernible in this figure. Thus, the images of the coordinate lines remain orthogonal, but the mapping severely distorts distances.

It is important to recognize that the mapping produced by the exponential function is *many-to-one* because

$$\text{Exp}[z + 2\pi i k] = \text{Exp}[z] \quad \text{for integer } k \tag{1.46}$$

is periodic, so that infinitely many input points $z = z_0 + 2\pi i k$ are mapped onto the same image point. Thus, any strip $|y - y_0| \leq \pi$ in the z-plane is mapped onto the entire w-plane and neighboring strips would replicate the covering of the w-plane. Consequently, the inverse function $z = \log[w]$ is many-valued because it represents a *one-to-many* mapping. By convention, we define the principal branch of the logarithm function

$$\text{Log}[z] = \text{Log}[|z|] + i\,\text{Arg}[z] \quad \text{with} \quad -\pi < \text{Arg}[z] \leq \pi \tag{1.47}$$

by limiting the phase $y = \text{Arg}[w]$ to the strip $-\pi < y \leq \pi$ by means of a branch cut along the negative real axis, as indicated by the thick line in Fig. 1.12. With this convention one obtains the following principal values:

$$\text{Log}[1] = 0 \quad \text{Log}[i] = \frac{i\pi}{2} \quad \text{Log}[-1] = i\pi \quad \text{Log}[-i] = -\frac{i\pi}{2} \tag{1.48}$$

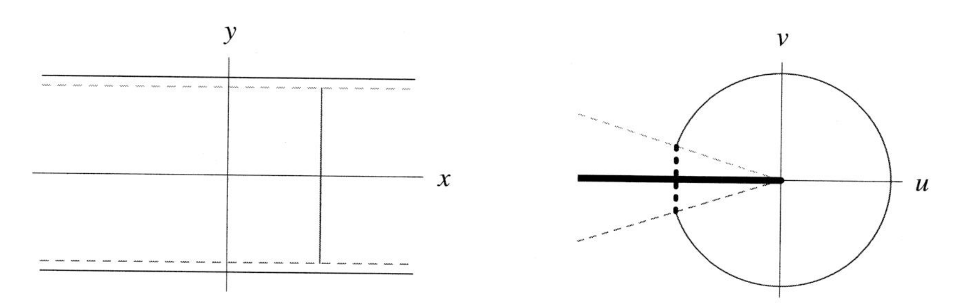

Figure 1.12. Mapping: $u + \imath v = \text{Exp}[x + \imath y]$. The mapping of any strip $|y - y_0| \leq \pi$ covers the w-plane. The branch cut for the inverse mapping is shown in the w-plane.

One might imagine the w-plane as a circular paper disk with a cut from its edge to the center, which prevents a continuous curve to be drawn crossing the cut. A vertical line segment in the z-plane between $y = -\pi + \varepsilon$ and $y = \pi - \varepsilon$ is mapped onto a circular arc between $\text{Arg}[w] = -\pi + \varepsilon$ and $\text{Arg}[w] = \pi - \varepsilon$. Although the two ends of the arc approach other from opposite sides of the branch cut as $\varepsilon \to 0^+$, they differ in phase by 2π. Therefore, although the branch cut permits the inverse function $z = \text{Log}[w]$ to be defined as a single-valued mapping $w \to z$, that function is discontinuous across the branch cut. Often single-valuedness comes only at the expense of discontinuities.

Neighboring strips $y_n = y_0 + n\pi$ simply remap the entire w-plane. One might imagine an infinite collection of w-planes, called *Riemann sheets*, stacked on top of each other such that curves which cross the branch cut move from one Riemann sheet to the next. The index n then identifies a particular Riemann sheet, with the principal branch represented by $n = 0$. Thus, it is useful to distinguish between a multivalued log and a single-valued Log defined by

$$\begin{aligned} w == e^z \Longrightarrow z = \log[w] &= \log[|w|] + \imath \arg[w] \\ &= \log[|w|] + \imath\, \text{Arg}[w] + 2\pi \imath n = \text{Log}[w] + 2\pi \imath n \end{aligned} \tag{1.49}$$

As a curve winds around the origin of the w-plane in a counterclockwise sense, the argument increases continuously and each time one crosses the branch cut one moves from one sheet to the next and increments the *winding number n* by one unit. Clockwise winding decrements n, which is permitted to be negative also. Furthermore, the choice $y_0 = 0$ is not unique and other choices would rotate the branch cut in the w-plane. The single-valued function produced by the most common choice of branch cut is described as the *principal branch*, but for some problems it may become convenient to make a different choice. However, the branch cuts used for Arg, Log, and related functions are correlated and must be chosen consistently throughout a particular calculation.

Physics calculations normally must produce a unique answer that can be compared with a measurable quantity, such that physical functions must be based upon single-valued functions. Similarly, if one is to compute the value of an expression using a computer program, there must be a unique result. Numerical methods cannot tolerate multivalued expressions – the programmer must provide an unambiguous prescription for selecting the appropriate branches of multivalued functions; a machine cannot perform that job for you. It is useful to visualize a function as a machine. When you supply appropriate input, it produces a definite and predictable output. A function is not really defined until its branch cuts and its discontinuities across those cuts are completely specified. Furthermore, there is often considerable flexibility in the selection of cuts that can be exploited to simplify the problem at hand, one selection for one problem and another for the next. Therefore, one must always be aware of the branch cuts used to regularize an inherently multivalued function.

1.3.2 Mapping: $w = \mathrm{Sin}[z]$

The sine function is extended to complex variables by the definition

$$\mathrm{Sin}[z] = \frac{e^{iz} - e^{-iz}}{2i} \tag{1.50}$$

Using

$$\begin{aligned} z = x + iy \Longrightarrow \mathrm{Sin}[z] &= \frac{e^{-y}\left(\mathrm{Cos}[x] + i\,\mathrm{Sin}[x]\right) - e^{y}\left(\mathrm{Cos}[x] - i\,\mathrm{Sin}[x]\right)}{2i} \\ &= \frac{e^{y} + e^{-y}}{2}\mathrm{Sin}[x] + i\frac{e^{y} - e^{-y}}{2}\mathrm{Cos}[x] \end{aligned} \tag{1.51}$$

and the familiar definitions

$$\mathrm{Cosh}[y] = \frac{e^{y} + e^{-y}}{2} \qquad \mathrm{Sinh}[y] = \frac{e^{y} - e^{-y}}{2} \tag{1.52}$$

for real variables, the components of the sine function become

$$u + iv = \mathrm{Sin}[x + iy] \Longrightarrow u = \mathrm{Sin}[x]\,\mathrm{Cosh}[y] \quad v = \mathrm{Cos}[x]\,\mathrm{Sinh}[y] \tag{1.53}$$

The mapping of (x, y) coordinate lines is illustrated in Fig. 1.13. Lines of constant x are mapped into hyperbolae while lines of constant y are mapped into confocal ellipses with foci at $(u, v) = (\pm 1, 0)$. The definition of the inverse mapping $z = \mathrm{ArcSin}[w]$ requires branch cuts because any strip $|x - x_0| \leq \pi$ is mapped onto the entire w-plane. It is customary to map the principal branch of ArcSin onto the strip $-\frac{\pi}{2} < x < \frac{\pi}{2}$, but two choices remain for the branch cuts. Recognizing that as $|x| \to \frac{\pi}{2}$ the hyperbolic images of the vertical lines collapse upon the real axis, the most common choice is to place the branch cuts along the open interval $|u| > 1$. This choice reduces to the standard definition of ArcSin$[x]$ for real arguments in the range $|x| \leq 1$.

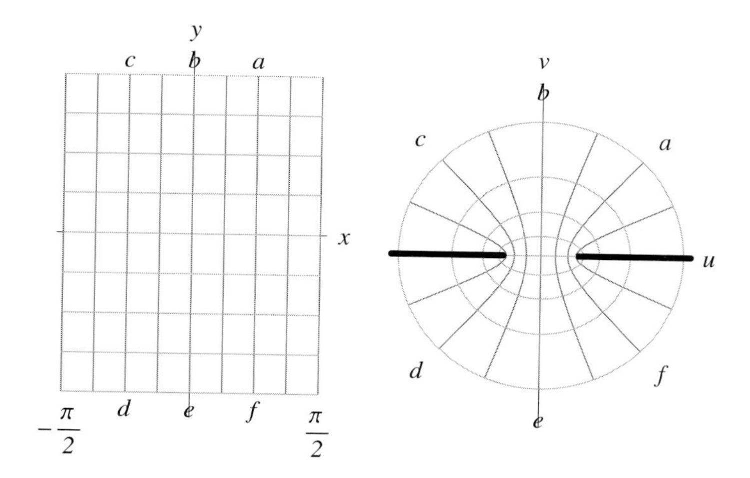

Figure 1.13. Mapping: $u + iv = \mathrm{Sin}[x + iy]$. Lines of constant x are mapped onto hyperbolas while lines of constant y are mapped onto confocal ellipses.

1.4 Elementary Functions and Their Inverses

1.4.1 Exponential and Logarithm

Some properties of the exponential are preserved by extension from the real axis to the complex plane. For example, using

$$\begin{aligned} e^{z_1} e^{z_2} &= e^{x_1} e^{x_2} \left(\mathrm{Cos}[y_1] + i\,\mathrm{Sin}[y_1]\right)\left(\mathrm{Cos}[y_2] + i\,\mathrm{Sin}[y_2]\right) \\ &= e^{x_1+x_2} \left(\mathrm{Cos}[y_1]\,\mathrm{Cos}[y_2] - \mathrm{Sin}[y_1]\,\mathrm{Sin}[y_2]\right. \\ &\quad \left. +i\left(\mathrm{Sin}[y_1]\,\mathrm{Cos}[y_2] + \mathrm{Cos}[y_1]\,\mathrm{Sin}[y_2]\right)\right) \\ &= e^{x_1+x_2} \left(\mathrm{Cos}[y_1 + y_2] + i\,\mathrm{Sin}[y_1 + y_2]\right) \end{aligned} \tag{1.54}$$

we find

$$e^{z_1} e^{z_2} = e^{z_1+z_2} \tag{1.55}$$

Similarly one can easily prove

$$\frac{1}{e^z} = e^{-z} \qquad \frac{e^{z_1}}{e^{z_2}} = e^{z_1-z_2} \tag{1.56}$$

and

$$(e^z)^n = e^{nz} \quad \text{for} \quad n \in \text{Integers} \tag{1.57}$$

for integer n. However, one must generally assume that

$$(e^{z_1})^{z_2} \neq e^{z_1 z_2} \tag{1.58}$$

for arbitrary powers z_2. For example, $(e^z)^{1/n}$ is multivalued, producing n values for integer n, while $e^{z/n}$ represents a unique complex number.

We define the multivalent logarithm function $w = \log[z]$ in terms of the solutions to the equation $z == e^w$, such that

$$\begin{aligned} &z == e^w , \\ &z = |z|e^{i \arg[z]} \Longrightarrow \log[z] = \log[|z|] + i \arg[z] = \mathrm{Log}[|z|] + i\, \mathrm{Arg}[z] + 2\pi i n \end{aligned} \tag{1.59}$$

where the principal value is used for the logarithm of the positive real number $|z|$ and where n is an arbitrary integer. Thus, this version of the logarithm function produces infinitely many values for any z . Consequently, we cannot simply replace $\log[e^z]$ by z during calculations because

$$\begin{aligned} z = x + iy \Longrightarrow e^z = e^x e^{iy} \Longrightarrow \log[e^z] &= \mathrm{Log}[|e^z|] + i\, \mathrm{Arg}[e^z] + 2\pi i n \\ &= x + i(y + 2\pi n) \\ &= z + 2\pi i n \end{aligned} \tag{1.60}$$

is ambiguous. By selecting $n \to 0$, we define the single-valued principal branch as

$$\mathrm{Log}[z] = \mathrm{Log}[|z|] + i\, \mathrm{Arg}[z] \tag{1.61}$$

and obtain

$$\mathrm{Log}[e^z] = z \tag{1.62}$$

as expected. On the other hand, some functional relationships that pertain to real arguments remain true for the multivalent version but are not necessarily true for the principal branch. For example, from

$$\begin{aligned} \log[z_1 z_2] &= \log[|z_1 z_2|] + i \arg[z_1 z_2] \\ &= \log[|z_1||z_2|] + i\big(\arg[z_1] + \arg[z_2]\big) \\ &= \log[|z_1|] + \log[|z_2|] + i\big(\arg[z_1] + \arg[z_2]\big) \end{aligned} \tag{1.63}$$

we find

$$\log[z_1 z_2] = \log[z_1] + \log[z_2] \tag{1.64}$$

but the corresponding relationship for the principal branch

$$\mathrm{Log}[z_1 z_2] = \mathrm{Log}[z_1] + \mathrm{Log}[z_2] + 2\pi i n \tag{1.65}$$

is more complicated because we must deduce the appropriate n from the phases of z_1 and z_2.

1.4.2 Powers

Powers of a complex number are defined by

$$z^\alpha = \mathrm{Exp}\big[\alpha \log[z]\big] = \mathrm{Exp}\big[\alpha\, \mathrm{Log}[|z|]\big]\, \mathrm{Exp}\big[i\alpha \arg[z]\big] \tag{1.66}$$

and are generally multivalued. In polar notation we may write

$$z = re^{i\theta} \Longrightarrow z^{\alpha} = \mathrm{Exp}\left[\alpha\,\mathrm{Log}[r]\right]\mathrm{Exp}\left[i\alpha(\theta + 2\pi n)\right] \tag{1.67}$$

where n is an integer. This definition conforms to simple expectations for rational exponents and preserves the algebraic relationships

$$\left(z^{\alpha}\right)^{\beta} = \mathrm{Exp}\left[\beta\log[z^{\alpha}]\right] = \mathrm{Exp}\left[\beta\alpha\log[z]\right] = z^{\alpha\beta} \tag{1.68}$$

$$\left(z_1 z_2\right)^{\alpha} = \mathrm{Exp}\left[\alpha\log[z_1 z_2]\right] = \mathrm{Exp}\left[\alpha(\log[z_1] + \log[z_2])\right] = z_1^{\alpha} z_2^{\alpha} \tag{1.69}$$

However, these algebraic relationships are generally multivalent. If we use a branch cut for the logarithm under the negative real axis, the same branch cut must be used for multivalent powers that are complex, nonintegral, or have negative real parts. The discontinuity in the argument of z^{α} across the branch cut is then

$$\Delta\,\mathrm{Arg}[z^{\alpha}] = \mathrm{Limit}[\mathrm{Arg}[x + i\varepsilon] - \mathrm{Arg}[x - i\varepsilon], \varepsilon \to 0] = \mathrm{Mod}[2\pi\alpha, 2\pi] \tag{1.70}$$

If α is rational there are a finite number of Riemann sheets, but for irrational α there are an infinite number of Riemann sheets. The principal branch is given by

$$\text{principal branch: } z^{\alpha} = \mathrm{Exp}\left[\alpha\,\mathrm{Log}[z]\right] = \mathrm{Exp}\left[\alpha\,\mathrm{Log}[|z|]\right]\mathrm{Exp}\left[i\alpha\,\mathrm{Arg}[z]\right] \tag{1.71}$$

1.4.3 Trigonometric and Hyperbolic Functions

Recognizing that

$$\mathrm{Cos}[x] = \frac{e^{ix} + e^{-ix}}{2} \qquad \mathrm{Sin}[x] = \frac{e^{ix} - e^{-ix}}{2i} \qquad \mathrm{Tan}[x] = \frac{\mathrm{Sin}[x]}{\mathrm{Cos}[x]} \tag{1.72}$$

$$\mathrm{Cosh}[x] = \frac{e^{x} + e^{-x}}{2} \qquad \mathrm{Sinh}[x] = \frac{e^{x} - e^{-x}}{2} \qquad \mathrm{Tanh}[x] = \frac{\mathrm{Sinh}[x]}{\mathrm{Cosh}[x]} \tag{1.73}$$

for $x \in \mathbb{R}$, we define

$$\mathrm{Cos}[z] = \frac{e^{iz} + e^{-iz}}{2} \qquad \mathrm{Sin}[z] = \frac{e^{iz} - e^{-iz}}{2i} \qquad \mathrm{Tan}[z] = \frac{\mathrm{Sin}[z]}{\mathrm{Cos}[z]} \tag{1.74}$$

$$\mathrm{Cosh}[z] = \frac{e^{z} + e^{-z}}{2} \qquad \mathrm{Sinh}[z] = \frac{e^{z} - e^{-z}}{2} \qquad \mathrm{Tanh}[z] = \frac{\mathrm{Sinh}[z]}{\mathrm{Cosh}[z]} \tag{1.75}$$

for $z \in \mathbb{C}$. Inverting these expressions gives

$$e^{iz} = \mathrm{Cos}[z] + i\,\mathrm{Sin}[z] \qquad e^{-iz} = \mathrm{Cos}[z] - i\,\mathrm{Sin}[z] \tag{1.76}$$

$$e^{z} = \mathrm{Cosh}[z] + \mathrm{Sinh}[z] \qquad e^{-z} = \mathrm{Cosh}[z] - \mathrm{Sinh}[z] \tag{1.77}$$

Obviously,

$$\mathrm{Cos}[-z] = \mathrm{Cos}[z] \qquad \mathrm{Sin}[-z] = -\,\mathrm{Sin}[z] \qquad \mathrm{Tan}[z] = -\,\mathrm{Tan}[z] \tag{1.78}$$

$$\mathrm{Cosh}[-z] = \mathrm{Cosh}[z] \qquad \mathrm{Sinh}[-z] = -\,\mathrm{Sinh}[z] \qquad \mathrm{Tanh}[z] = -\,\mathrm{Tanh}[z] \tag{1.79}$$

and

$$\text{Cos}[iz] = \text{Cosh}[z] \qquad \text{Sin}[iz] = i\,\text{Sinh}[z] \qquad \text{Tan}[iz] = i\,\text{Tanh}[z] \tag{1.80}$$

$$\text{Cosh}[iz] = \text{Cos}[z] \qquad \text{Sinh}[iz] = i\,\text{Sin}[z] \qquad \text{Tanh}[iz] = i\,\text{Tan}[z] \tag{1.81}$$

Multiplying out the expressions

$$\begin{aligned} \text{Exp}\left[i(z_1 + z_2)\right] &= \text{Cos}[z_1 + z_2] + i\,\text{Sin}[z_1 + z_2] \\ &= \left(\text{Cos}[z_1] + i\,\text{Sin}[z_1]\right)\left(\text{Cos}[z_2] + i\,\text{Sin}[z_2]\right) \\ &= e^{iz_1} e^{iz_2} \end{aligned} \tag{1.82}$$

$$\begin{aligned} \text{Exp}\left[i(z_1 - z_2)\right] &= \text{Cos}[z_1 - z_2] + i\,\text{Sin}[z_1 - z_2] \\ &= \left(\text{Cos}[z_1] + i\,\text{Sin}[z_1]\right)\left(\text{Cos}[z_2] - i\,\text{Sin}[z_2]\right) \\ &= e^{iz_1} e^{-iz_2} \end{aligned} \tag{1.83}$$

one quickly deduces the addition formulae

$$\text{Cos}[z_1 + z_2] = \text{Cos}[z_1]\,\text{Cos}[z_2] - \text{Sin}[z_1]\,\text{Sin}[z_2] \tag{1.84}$$

$$\text{Cosh}[z_1 + z_2] = \text{Cosh}[z_1]\,\text{Cosh}[z_2] + \text{Sinh}[z_1]\,\text{Sinh}[z_2] \tag{1.85}$$

$$\text{Sin}[z_1 + z_2] = \text{Sin}[z_1]\,\text{Cos}[z_2] + \text{Cos}[z_1]\,\text{Sin}[z_2] \tag{1.86}$$

$$\text{Sinh}[z_1 + z_2] = \text{Sinh}[z_1]\,\text{Cosh}[z_2] + \text{Cosh}[z_1]\,\text{Sinh}[z_2] \tag{1.87}$$

and

$$\text{Cos}[z]^2 + \text{Sin}[z]^2 = 1 \qquad \text{Cosh}[z]^2 - \text{Sinh}[z]^2 = 1 \tag{1.88}$$

Combining these results, we obtain

$$\text{Cos}[x + iy] = \text{Cos}[x]\,\text{Cosh}[y] - i\,\text{Sin}[x]\,\text{Sinh}[y] \tag{1.89}$$

$$\text{Cosh}[x + iy] = \text{Cosh}[x]\,\text{Cos}[y] + i\,\text{Sinh}[x]\,\text{Sin}[y] \tag{1.90}$$

$$\text{Sin}[x + iy] = \text{Sin}[x]\,\text{Cosh}[y] + i\,\text{Cos}[x]\,\text{Sinh}[y] \tag{1.91}$$

$$\text{Sinh}[x + iy] = \text{Sinh}[x]\,\text{Cos}[y] + i\,\text{Cosh}[x]\,\text{Sin}[y] \tag{1.92}$$

for $\{x, y\} \in \mathbb{R}$. Expressions for the real and imaginary components of inverse trigonometric functions are developed in the exercises.

1.4.4 Standard Branch Cuts

Although there is often some flexibility in the choice of branch cuts, the cuts for related functions are correlated. Table 1.3 lists the standard choices for elementary functions, but other choices can facilitate certain calculations. Parentheses (square brackets) indicate an open (closed) interval.

Table 1.3. Standard definitions for principal branch of elementary functions.

Function	Branch cuts
Abs	none
Arg	$(-\infty, 0)$
Sqrt	$(-\infty, 0)$
z^s, nonintegral s with $\mathrm{Re}[s] > 0$	$(-\infty, 0)$
z^s, nonintegral s with $\mathrm{Re}[s] \le 0$	$(-\infty, 0]$
Exp	none
Log	$(-\infty, 0]$
trigonometric functions	none
ArcSin, ArcCos	$(-\infty, -1)$ and $(1, \infty)$
ArcTan	$(-i\infty, -i]$ and $[i, i\infty)$
ArcCsc and ArcSec	$(-1, 1)$
ArcCot	$[-i, i]$
hyperbolic functions	none
ArcSinh	$(-i\infty, -i)$ and $(i, i\infty)$
ArcCosh	$(-\infty, 1)$
ArcTanh	$(-i\infty, -i]$ and $[i, i\infty)$
ArcCsch	$(-i, i)$
ArcSech	$(-\infty, 0]$ and $(1, \infty)$
ArcCoth	$[-1, 1]$

1.5 Sets, Curves, Regions and Domains

The basic concept used to characterize sets, curves, and regions in the complex plane is *neighborhood.* A neighborhood of z_0 consists of the set of all points that satisfy the inequality $|z - z_0| < \varepsilon$; the radius ε is usually assumed to be small. A point z is an *interior point* of the set S if there exists a neighborhood containing only points belonging to S. Conversely, a point is *exterior* to S if there exists a neighborhood that does not contain any points belonging to S. Finally, a *boundary point* is neither interior nor exterior to S because any neighborhood, no matter how small, contains both points which belong to S and points which do not. An *open set* is a set for which every point is an interior point; in other words, an open set contains none of its boundary points. A *closed set*, on the other hand, contains all of its boundary points. The *closure* of S consists of S plus all of its boundary points and is denoted $\bar{S}$. Note that some sets, such as $0 < |z| \le 1$, are neither open nor closed because they contain some but not all of their boundary points, while $\mathbb{C}$ is both open and closed because there are no boundary points. A set is *bounded* if all points lie within a disk $|z| < R$ for some finite R and is *unbounded* otherwise. Finally, a point z_0 is an *accumulation point* of S if every neighborhood contains at least one other point that also belongs to S. Thus, a closed set contains all of its accumulation points and, conversely, any set which contains all of its accumulation points is closed. For example, the origin is the only accumulation point of the set $\{z_n = \frac{1}{n}, n = 1, \infty\}$.

Any set of points that consists only of boundary points constitutes a *curve*. For example, the set of points that satisfy the equation $|z - z_0| == R$ describes a circle of radius R

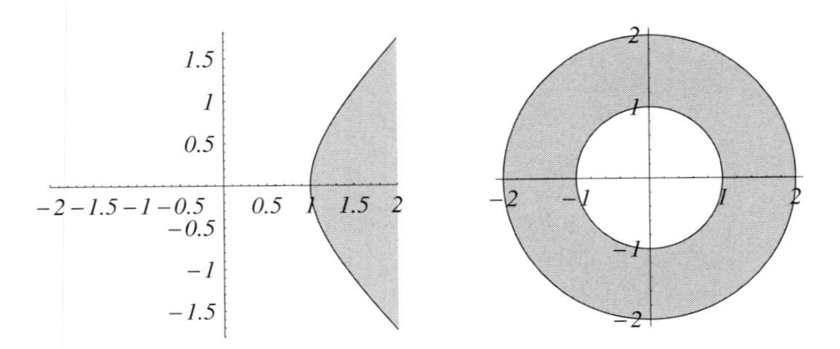

Figure 1.14. Left: A simply connected domain; right: a mutiply connected domain.

centered on (x_0, y_0) and is an example of a curve. An *arc* is a curve described by the parametric equation $z = (x[t], y[t])$ where $x[t]$ and $y[t]$ are continuous real functions of the real variable $t_{min} \leq t \leq t_{max}$. An arc is *simple* if it does not intersect itself, in other words if $t_1 \neq t_2 \Longrightarrow z[t_1] \neq z[t_2]$ for $t_{min} < t_1, t_2 < t_{max}$. A *simple closed curve* does not intersect itself except at the endpoints where $z[t_{min}] = z[t_{max}]$.

An open set is *connected* if any pair of points can be joined by a polygonal path that lies entirely within the set. An open connected set is called a *domain*. For example, the annulus $1 < |z| < 2$ is a domain because it is open and connected. Any neighborhood is also a domain. A domain D is described as *simply connected* if all simple closed curves within D enclose only points that are also within D and is described as multiply connected otherwise. A domain together with a subset of its boundary points (none, some, or all) is called a *region*. For example, $\{z \ni \text{Re}[z^2] > 1 \wedge \text{Re}[z] > 0\}$ describes a simply connected domain while $\{z \ni 1 \leq |z| \leq 2\}$ describes a multiply connected region.

1.6 Limits and Continuity

The limit of $f[z]$ as $z \to z_0$ is defined to be the complex number w_0 if for each arbitrarily small positive number ε there exists a positive number δ for which $0 < |f[z] - w_0| < \varepsilon$ whenever $0 < |z - z_0| < \delta$. Geometrically, this definition requires that the image $w = f[z]$ for any point z in a δ-neighborhood of z_0, with the possible exception of z_0 itself, should lie within an ε-neighborhood of w_0. Note that this definition requires all points in the neighborhood of z_0 to be mapped within the neighborhood of w_0 but does not require the mapping to constitute a domain because the mapping need not produce a connected set. Furthermore, the limit $z \to z_0$ may be approached in an arbitrary manner. However, the present definition does not apply to points z_0 which lie on the boundary of the domain on which $f[z]$ is defined because in that case the δ-neighborhood contains points at which $f[z]$ may be undefined. Nevertheless, we can extend the definition of limit by limiting the requirements on the inequalities to those points in the neighborhood of z_0 that lie within the domain of f.

Direct application of the definition of limits can be quite cumbersome, but a few almost self-evident theorems are quite helpful.

Theorem 1. *Let $f[z] = u[x, y] + iv[x, y]$, $z_0 = x_0 + iy_0$, and $w_0 = u_0 + iv_0$. Then*

$$\operatorname{Lim}_{z\to z_0} f[z] = w_0 \tag{1.93}$$

if and only if

$$\operatorname{Lim}_{(x,y)\to(x_0,y_0)} u[x, y] = u_0 \quad \text{and} \quad \operatorname{Lim}_{(x,y)\to(x_0,y_0)} v[x, y] = v_0 \tag{1.94}$$

Theorem 2. *Let $f_0 = \operatorname{Lim}_{z\to z_0} f[z]$ and $g_0 = \operatorname{Lim}_{z\to z_0} g[z]$. Then*

1. $\operatorname{Lim}_{z\to z_0}(f[z] + g[z]) = f_0 + g_0$
2. $\operatorname{Lim}_{z\to z_0} f[z]g[z] = f_0 g_0$
3. $\operatorname{Lim}_{z\to z_0} \frac{f[z]}{g[z]} = \frac{f_0}{g_0}$ *if* $g_0 \neq 0$
4. $\operatorname{Lim}_{z\to z_0} \operatorname{Abs}[f[z]] = \operatorname{Abs}[f_0]$

A function $f[z]$ is *continuous* at z_0 if $\lim_{z\to z_0} f[z] = f[z_0]$ and is continuous in a region R if it is continuous at all points within that region. Note that this definition implicitly requires $f[z]$ and its limit at z_0 to exist.

Theorem 3. *Let $f[z]$ be defined in a neighborhood of z_0 and suppose that for all points in that neighborhood $f[z]$ lies within the domain of $g[z]$. Then if $f[z]$ is continuous at z_0 and $g[z]$ is continuous at $f[z_0]$, it follows that $g[f[z]]$ is continuous at z_0.*

Consider a sequence of complex numbers $\{z_n\}$. The limit of a sequence $z_n \to w$ requires that $|z_n - w| < \varepsilon$ whenever $n > N[\varepsilon]$. A *Cauchy sequence* requires $|z_n - z_m| \to 0$ as $n, m \to \infty$. A sequence converges if and only if it is a Cauchy sequence.

1.7 Differentiability

1.7.1 Cauchy–Riemann Equations

Let $w = f[z] = u[x, y] + iv[x, y]$ be a function of the complex variable $z = x + iy$ and define its derivative by

$$f'[z] = \frac{dw}{dz} = \lim_{\Delta z\to 0} \frac{\Delta w}{\Delta z} = \lim_{\Delta z\to 0} \frac{f[z + \Delta z] - f[z]}{\Delta z} \tag{1.95}$$

Although this definition is simply the obvious generalization of the derivative of a real-valued function of a real variable, the higher dimensionality of complex variables imposes nontrivial requirements upon differentiable complex functions. The existence of such a derivative requires

1. $f[z]$ be defined at z
2. $f[z] \neq \infty$
3. the limit must be independent of the direction in which $\Delta z \to 0$.

The independence of direction is a strong condition which leads to the *Cauchy–Riemann equations*, henceforth denoted CR. Approaching the limit using variations along coordinate directions, one finds

$$\begin{aligned}\Delta z = \Delta x \Longrightarrow & \lim_{\Delta x \to 0, \Delta y = 0} \frac{u[x + \Delta x, y] + i v[x + \Delta x, y] - u[x, y] - i v[x, y]}{\Delta x} \\ &= \frac{\partial u}{\partial x} + i \frac{\partial v}{\partial x}\end{aligned} \tag{1.96}$$

$$\begin{aligned}\Delta z = i\Delta y \Longrightarrow & \lim_{\Delta x = 0, \Delta y \to 0} \frac{u[x, y + \Delta y] + i v[x, y + \Delta y] - u[x, y] - i v[x, y]}{i \Delta y} \\ &= -i \frac{\partial u}{\partial y} + \frac{\partial v}{\partial y}\end{aligned} \tag{1.97}$$

Equating the real and imaginary parts separately, then requires

$$\frac{\partial u}{\partial x} = \frac{\partial v}{\partial y} \qquad \frac{\partial u}{\partial y} = -\frac{\partial v}{\partial x} \tag{1.98}$$

The CR equations are necessary but not quite sufficient to ensure differentiability. To obtain sufficient conditions, we also require continuity of the partial derivatives of component functions.

Theorem 4. *Let $f[z] = u[x, y] + i v[x, y]$ be defined throughout a neighborhood $|z - z_0| < \varepsilon$ and suppose that the first partial derivatives of u and v wrt x and y exist in that neighborhood and are continuous at $z_0 = (x_0, y_0)$. Then $f'[z]$ exists if those partial derivatives satisfy the Cauchy–Riemann equations*

$$\frac{\partial u}{\partial x} = \frac{\partial v}{\partial y} \qquad \frac{\partial u}{\partial y} = -\frac{\partial v}{\partial x} \tag{1.99}$$

Conversely, if $f'[z]$ exists, then the CR equations are satisfied.

If $f[z]$ is differentiable at z_0 and throughout a neighborhood of z_0, then $f[z]$ is described as *analytic* (or regular or holomorphic) at z_0. If $f[z]$ is analytic everywhere in the finite complex plane, it is described as *entire*. Examples of entire functions include Exp, Sin, Cos, Sinh, and Cosh. Functions which are analytic except on branch cuts include Log, ArcSin, ArcCos, ArcSinh, and ArcCosh.

Recognizing that the CR equations are linear, it is trivial to demonstrate that if $f_1[z]$ and $f_2[z]$ are analytic functions in domains D_1 and D_2, then any linear combination $af_1[z] + bf_2[z]$ is also analytic in the overlapping domain $D = D_1 \cap D_2$. Similarly, it is straightforward, though tedious, to demonstrate that the product $f_1[z]f_2[z]$ also satisfies the CR equations and, hence, is analytic in D. Furthermore, one can show that $1/f_2[z]$ is analytic in D_2 where $f_2[z] \neq 0$ such that $f_1[z]/f_2[z]$ is analytic in D except possibly at the zeros of the denominator. Finally, if $f_1[w]$ is analytic at $w = f_2[z]$, then $f_1[f_2[z]]$ is analytic. Formal demonstration that these familiar properties of derivatives also apply to analytic functions of a complex variable is left to the student.

Example

$$f[z] = z^2 \Longrightarrow u = x^2 - y^2, \quad v = 2xy \Longrightarrow \frac{\partial u}{\partial x} = 2x = \frac{\partial v}{\partial y}, \quad \frac{\partial u}{\partial y} = -2y = -\frac{\partial v}{\partial x} \tag{1.100}$$

The partial derivatives are continuous throughout the complex plane and satisfy the CR equation; hence, z^2 is entire. In fact, one can show that any polynomial in z is entire.

Example

$$f[z] = z^* \Longrightarrow u = x, \quad v = -y \Longrightarrow \frac{\partial u}{\partial x} = 1 \neq \frac{\partial v}{\partial y} = -1 \tag{1.101}$$

The partial derivatives are continuous, but do not satisfy CR; hence, z^* is nowhere differentiable and is not analytic anywhere. It is important to recognize that functions of a complex variable can be smooth and continuous without being differentiable. The requirements for differentiability are stricter for complex variables than for real variables because independence from direction imposes correlations between the dependencies upon the real and imaginary parts of the independent variable. Analytic functions of one complex variable are not simply functions of two real variables.

1.7.2 Differentiation Rules

Many of the familiar differentiation rules for real functions can be applied to complex functions. Suppose that $f[z]$ and $g[z]$ are differentiable within overlapping regions. Within the intersection of those regions, we can derive differentiation rules using the definition in terms of limits. Alternatively, by separating each function into real and imaginary components, one could also employ the CR relations.

For example, one quickly finds that the derivative of a sum

$$\begin{aligned} F[z] &= f[z] + g[z] \\ \Longrightarrow \lim_{\Delta z \to 0} \frac{F[z+\Delta z] - F[z]}{\Delta z} &= \lim_{\Delta z \to 0} \frac{f[z+\Delta z] - f[z]}{\Delta z} + \lim_{\Delta z \to 0} \frac{g[z+\Delta z] - g[z]}{\Delta z} \end{aligned} \tag{1.102}$$

reduces to the sum of derivatives

$$F[z] = f[z] + g[z] \Longrightarrow F'[z] = f'[z] + g'[z] \tag{1.103}$$

if both functions are differentiable. Similarly, the familiar rule for a differentiation of a product

$$F[z] = f[z]g[z] \Longrightarrow \lim_{\Delta z \to 0} \frac{F[z+\Delta z] - F[z]}{\Delta z} = \lim_{\Delta z \to 0} \frac{f[z+\Delta z]g[z+\Delta z] - f[z]g[z]}{\Delta z} \tag{1.104}$$

is obtained using $f[z+\Delta z] \approx f[z] + f'[z]\Delta z$ and $g[z+\Delta z] \approx g[z] + g'[z]\Delta z$ for differentiable functions and retaining only first-order terms,

$$\lim_{\Delta z \to 0} \frac{f[z+\Delta z]g[z+\Delta z] - f[z]g[z]}{\Delta z} = \lim_{\Delta z \to 0} \frac{f[z]g'[z]\Delta z + f'[z]g[z]\Delta z}{\Delta z} \tag{1.105}$$

such that

$$F[z] = f[z]g[z] \Longrightarrow F'[z] = f[z]g'[z] + f'[z]g[z] \tag{1.106}$$

By similar reasoning one can verify all standard differentiation rules, subject to obvious conditions on differentiability of the various parts. Perhaps the most important is the *chain rule*

$$F[z] = g[f[z]] \Longrightarrow F'[z] = \left(g'[w]f'[z]\right)_{w=f[z]} \tag{1.107}$$

provided that f is differentiable at z and that g is differentiable at $w = f[z]$.

1.8 Properties of Analytic Functions

Suppose that $f[z] = u[x, y] + i v[x, y]$ is analytic in domain D and suppose that the second partial derivatives of the component functions u and v are continuous in D also. (We will soon prove that analytic functions are infinitely differentiable so that the component functions u and v must have continuous partial derivatives of all orders within D.) Differentiation of the CR equations then gives

$$\frac{\partial u}{\partial x} = \frac{\partial v}{\partial y} \Longrightarrow \frac{\partial^2 u}{\partial x^2} = \frac{\partial^2 v}{\partial x \partial y} = \frac{\partial^2 v}{\partial y \partial x} = -\frac{\partial^2 u}{\partial y^2} \Longrightarrow \frac{\partial^2 u}{\partial x^2} + \frac{\partial^2 u}{\partial y^2} = 0 \tag{1.108}$$

$$\frac{\partial v}{\partial x} = -\frac{\partial u}{\partial y} \Longrightarrow \frac{\partial^2 v}{\partial x^2} = -\frac{\partial^2 u}{\partial x \partial y} = -\frac{\partial^2 u}{\partial y \partial x} = -\frac{\partial^2 v}{\partial y^2} \Longrightarrow \frac{\partial^2 v}{\partial x^2} + \frac{\partial^2 v}{\partial y^2} = 0 \tag{1.109}$$

Therefore, both the real and imaginary components of f are *harmonic functions* that satisfy Laplace's equation. Furthermore, comparing the two-dimensional gradients

$$\vec{\nabla} u = \hat{x}\frac{\partial u}{\partial x} + \hat{y}\frac{\partial u}{\partial y} = \hat{x}\frac{\partial v}{\partial y} - \hat{y}\frac{\partial v}{\partial x} = \hat{n} \times \vec{\nabla} v \tag{1.110}$$

$$\vec{\nabla} v = \hat{x}\frac{\partial v}{\partial x} + \hat{y}\frac{\partial v}{\partial y} = -\hat{x}\frac{\partial u}{\partial y} + \hat{y}\frac{\partial u}{\partial x} = -\hat{n} \times \vec{\nabla} u \tag{1.111}$$

$$\therefore \vec{\nabla} u \cdot \vec{\nabla} v = 0 \tag{1.112}$$

we find that lines of constant u (*level curves*) are orthogonal to lines of constant v anywhere that $f'[z] \neq 0$. (Here $\hat{n}$ represents the outward normal to the xy-plane.) If u represents a potential function, then v represents the corresponding stream function (lines of force), or vice versa.

Consider, for example, $f[z] = z^2$ with $u = x^2 - y^2$ and $v = 2xy$. If we interpret v as an electrostatic potential, then u represents lines of force. Figure 1.15 shows equipotentials as solid lines, positive in the first quadrant and alternating sign by quadrant, and lines of force as dashed lines. The arrows indicate the direction of the force, as prescribed by $-\vec{\nabla} v$. If electrodes were shaped with surfaces parallel to equipotentials, the interior field would act as an electrostatic quadrupole lens, focussing a beam of positively-charged particles along the 45° and 225° directions and defocussing along the 135° and 315° directions.

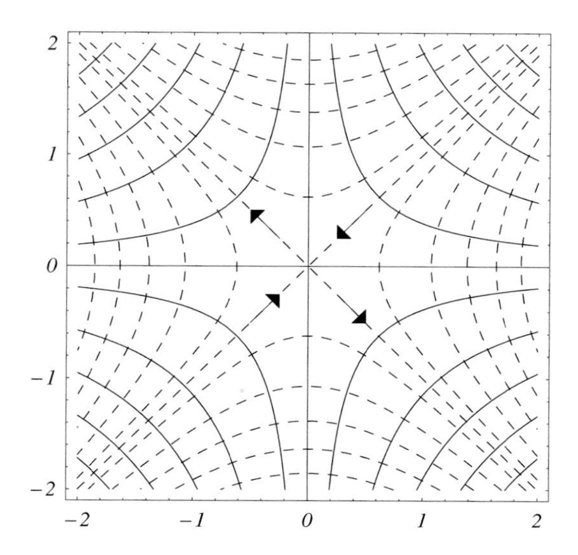

Figure 1.15. Level curves for $f[z] = z^2 = u + iv$ are shown as solid for v and dashed for u. If the solid lines are interpreted as equipotentials, the dashed lines with directions given by $-\vec{\nabla}v$ represent lines of force.

Alternatively, if v represents a magnetostatic potential, then u would represent magnetic field lines. A beam of positively-charged particles moving into the page would be vertically focussed and horizontally defocussed by a magnetic quadrupole lens whose iron pole pieces have surfaces shaped by $v \propto xy$.

It is also easy to demonstrate that, although harmonic functions may have saddle points, they cannot have extrema in the finite plane. Hence, neither component of an analytic function may have an extremum within the domain of analyticity. Figure 1.16 illustrates the typical saddle shape for components of an analytic function. Furthermore, the average value of a harmonic function on a circle is equal to the value of that function of the center of the circle. Proofs of these hopefully familiar properties of Laplace's equation are left to the exercises.

Suppose that Z_1 is a curve in the z-plane represented by the parametric equations $z_1[t] = \{x_1[t], y_1[t]\}$ and that $f[z]$ is analytic in a domain containing Z_1, such that the image W_1 of that curve in the w-plane is represented by $w_1[t] = f[z_1[t]]$. The slopes of tangent lines at a point z_0 and its image w_0 are related by the chain rule, such that

$$w_1'[t] = f'[z]z_1'[t] \Longrightarrow \arg\left[w_1'[t]\right] = \arg\left[z_1'[t]\right] + \arg\left[f'[z_0]\right] \tag{1.113}$$

Thus, the mapping $f[z]$ rotates the tangent line through an angle $\arg[f'[z_0]]$. The tangent to a second curve which passes through the same point z_0 is rotated by the same amount,

$$w_2'[t] = f'[z]z_2'[t] \Longrightarrow \arg\left[w_2'[t]\right] = \arg\left[z_2'[t]\right] + \arg\left[f'[z_0]\right] \tag{1.114}$$

such that angle between the two curves

$$\arg[w_2'] - \arg[w_1'] = \arg[z_2'] - \arg[z_1'] \tag{1.115}$$

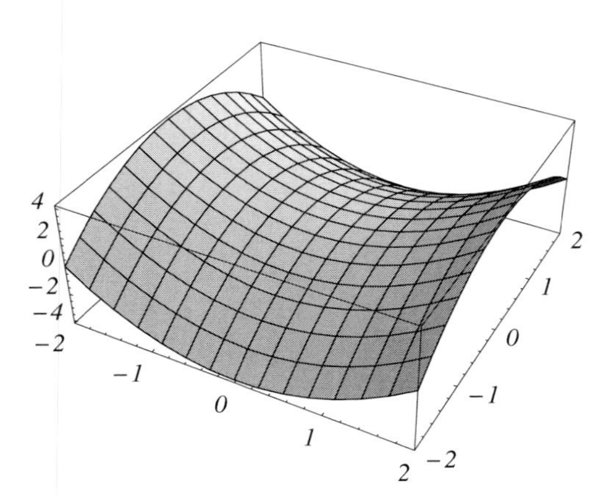

Figure 1.16. Typical saddle: $u = x^2 - y^2$.

is unchanged by the *conformal transformation* specified by an analytic function $f[z]$. Similarly, distances in the immediate vicinity of z_0 are scaled by the factor $|f'[z_0]|$, such that

$$|w - w_0| = |f'[z_0]||z - z_0| \tag{1.116}$$

Therefore, the image of a small triangle in the z-plane is a similar triangle in the w-plane that is generally rotated and scaled in size.

1.9 Cauchy–Goursat Theorem

1.9.1 Simply Connected Regions

We have seen that the components of analytic functions are harmonic and might be stimulated to pursue analogies with potential theory as far as possible. Remembering that the line integral about a closed path vanishes for a potential derived from a conservative force, we seek to evaluate

$$\oint_C f[z]\,dz = \oint_C (u\,dx - v\,dy) + i\oint_C (u\,dy + v\,dx) \tag{1.117}$$

for an analytic function $f = u + iv$ of $z = x + iy$ where $u[x, y]$ and $v[x, y]$ are real. If we require $P[x, y]$ and $Q[x, y]$ to be differentiable within the simply connected region R enclosed by the simple closed contour C, we can apply Stoke's theorem to prove

$$\oint_C (P\,dx + Q\,dy) = \int_R \left(\frac{\partial Q}{\partial x} - \frac{\partial P}{\partial y}\right) dx\,dy \tag{1.118}$$

Let

$$\vec{V} = (P, Q, 0) \Longrightarrow \hat{n} \cdot \vec{\nabla} \times \vec{V} = \frac{\partial Q}{\partial x} - \frac{\partial P}{\partial y} \tag{1.119}$$

where $\hat{n}$ is normal to the xy-plane and use $d\vec{\lambda} = (dx, dy, 0)$ as the line element and $d\vec{\sigma} = \hat{n}\, dx\, dy$ as the area element to obtain

$$\oint_C d\vec{\lambda} \cdot \vec{V} = \int_R d\vec{\sigma} \cdot \vec{\nabla} \times \vec{V} \Longrightarrow \oint_C (P\, dx + Q\, dy) = \int_R \left(\frac{\partial Q}{\partial x} - \frac{\partial P}{\partial y} \right) dx\, dy \tag{1.120}$$

Applying this result, known as Green's theorem, to the real and imaginary parts of the line integral separately, and using the CR conditions for analytic functions, we find

$$\oint_C (u\, dx - v\, dy) = \int_R \left(-\frac{\partial v}{\partial x} - \frac{\partial u}{\partial y} \right) dx\, dy = 0 \tag{1.121}$$

$$\oint_C (u\, dy + v\, dx) = \int_R \left(\frac{\partial u}{\partial x} - \frac{\partial v}{\partial y} \right) dx\, dy = 0 \tag{1.122}$$

and conclude that

$$f \text{ analytic for } z \text{ within } C \Longrightarrow \oint_C f[z]\, dz = 0 \tag{1.123}$$

This result was first obtained by Cauchy, but was later generalized by Goursat. The derivation above requires not only that $f'[z]$ exist throughout R, but also that it be continuous therein. The latter restriction can be removed.

Theorem 5. *Cauchy–Goursat theorem: If a function $f[z]$ is analytic at all points on and within a simple closed contour C, then $\oint_C f[z]\, dz = 0$.*

1.9.2 Proof

Consider the closed contour C sketched in Fig. 1.17. Divide the enclosed region R into a grid of squares and partial squares, whereby

$$\oint_C f[z]\, dz = \sum_{j=1}^{n} \oint_{C_j} f[z]\, dz \tag{1.124}$$

where the contributions made by shared interior boundaries cancel such that the net contour integral is the sum of the exterior borders of outer partial squares. For each of these cells, we construct the function

$$\delta[z, z_j] = \frac{f[z] - f[z_j]}{z - z_j} - f'[z_j] \tag{1.125}$$

where z and z_j are distinct points within or on C_j and evaluate its largest modulus

$$\delta_j = \text{Max}\left[\left| \frac{f[z] - f[z_j]}{z - z_j} - f'[z_j] \right| \right] \tag{1.126}$$

For any positive value of ε, a finite number of subdivisions is sufficient to ensure that all $\delta_j < \varepsilon$ because $f[z]$ is differentiable. Thus, we can now write

$$f[z] = f[z_j] + \left(f'[z_j] + \delta[z, z_j] \right)(z - z_j) \tag{1.127}$$

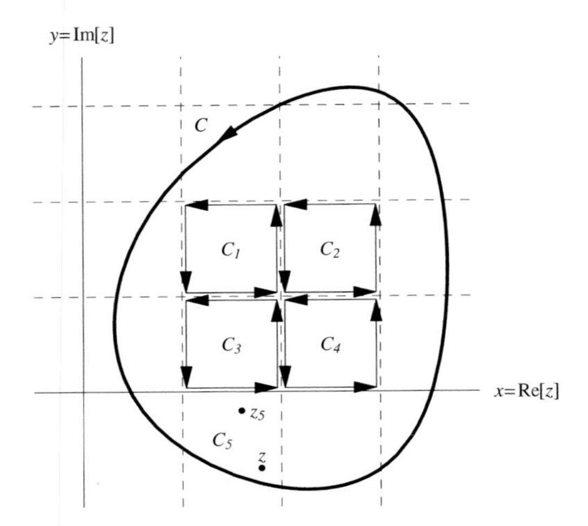

Figure 1.17. Proof of the Cauchy–Goursat theorem. Contours about four of the interior squares are labeled C_{1-4}. If $f[z]$ is continuous, contributions to the contour integral from shared sides cancel, leaving only the outer border C that passes through partial squares. In the partial square, labeled C_5, we identify two distinct points labeled z and z_5.

for any $z \subset C_j$, such that

$$\oint_{C_j} f[z]\, dz = f[z_j] \oint_{C_j} dz + f'[z_j] \oint_{C_j} (z - z_j)\, dz + \oint_{C_j} \delta[z, z_j](z - z_j)\, dz \tag{1.128}$$

The first two terms obviously vanish, leaving

$$\oint_C f[z]\, dz = \sum_{j=1}^{n} \oint_{C_j} \delta[z, z_j](z - z_j)\, dz \tag{1.129}$$

which can be bounded by

$$\left| \oint_C f[z]\, dz \right| \leq \sum_{j=1}^{n} \left| \oint_{C_j} \delta[z, z_j](z - z_j)\, dz \right| \tag{1.130}$$

If s_j is the length of the longest side of partial square C_j, then $|z - z_j| \leq \sqrt{2} s_j$. Furthermore, $|\delta_j| < \varepsilon$, such that

$$\left| \oint_{C_j} \delta[z, z_j](z - z_j)\, dz \right| \leq \sqrt{2} s_j \varepsilon (4 s_j + L_j) \tag{1.131}$$

where L_j is the length of that part of C_j that coincides with C. Because each factor is bounded and $\varepsilon \to 0$ may be taken arbitrarily small, we find that $|\oint_C f[z]\, dz|$ is also arbitrarily small and, hence, must vanish. Therefore, the Cauchy–Goursat theorem is established without assuming that f' is continuous.

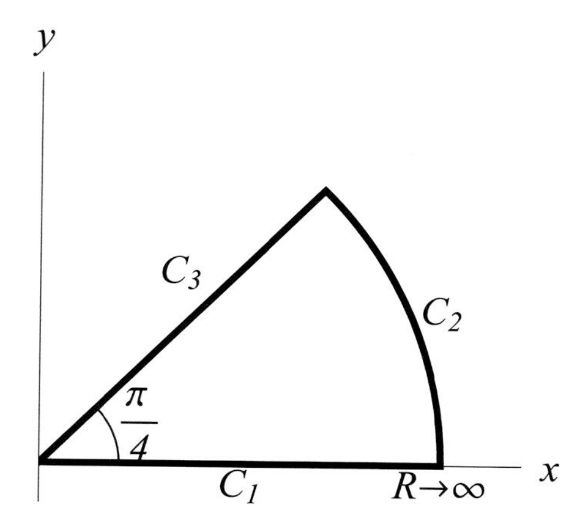

Figure 1.18. Wedge contour used for $\int_0^\infty \mathrm{Cos}[x^2]\, dx$.

1.9.3 Example

Contour integration of analytic functions provides powerful new methods for evaluation of otherwise intractable definite integrals. Although we will consider a wider variety later, for now consider the integral

$$\int_0^\infty \mathrm{Cos}[x^2]\, dx \tag{1.132}$$

which arises in the Fresnel theory of diffraction. It appears to be difficult to evaluate this integral using standard methods for real variables; nor is it obvious that this integral even converges. On the other hand, the Cauchy–Goursat theorem ensures that

$$I = \oint_C \mathrm{Exp}[iz^2]\, dz = 0 = I_1 + I_2 + I_3 \tag{1.133}$$

for a contour C consisting of a wedge of opening angle $\theta = \frac{\pi}{4}$ closed by a circular arc at $R \to \infty$; this contour is shown in Fig. 1.18. Consider first the circular arc where

$$z = Re^{i\theta} \Longrightarrow e^{iz^2} = \mathrm{Exp}\left[iR^2\,\mathrm{Cos}[2\theta]\right]\mathrm{Exp}\left[-R^2\,\mathrm{Sin}[2\theta]\right] \tag{1.134}$$

Recognizing that $0 < \mathrm{Sin}[2\theta] < 1$ is positive on the arc, the integrand is damped by a factor of order e^{-R^2} such that

$$R \to \infty \Longrightarrow I_2 = 0 \Longrightarrow I_1 = -I_3 \tag{1.135}$$

where

$$I_1 = \int_0^\infty \mathrm{Cos}[x^2]\, dx + i\int_0^\infty \mathrm{Sin}[x^2]\, dx \tag{1.136}$$

The return line is represented by

$$z = \frac{1+i}{\sqrt{2}}t \Longrightarrow dz = \frac{1+i}{\sqrt{2}}\, dt\; e^{iz^2} = e^{-t^2} \tag{1.137}$$

such that

$$I_3 = \frac{1+i}{\sqrt{2}} \int_{\infty}^{0} e^{-t^2}\, dt = -\frac{1+i}{\sqrt{2}} \frac{\sqrt{\pi}}{2} \tag{1.138}$$

Therefore, equating real and imaginary parts, we find

$$\int_0^{\infty} \text{Cos}[x^2]\, dx = \int_0^{\infty} \text{Sin}[x^2]\, dx = \sqrt{\frac{\pi}{8}} \tag{1.139}$$

rather easily. By representing the integrand in terms of analytic functions and choosing a clever contour, one can perform a surprisingly diverse variety of integrals relatively painlessly. In this case we even obtain two results for the price of one. (What a deal!)

1.10 Cauchy Integral Formula

1.10.1 Integration Around Nonanalytic Regions

Suppose that the region $R = R_1 + R_2$ enclosed by the simple closed contour C includes a localized region R_2 where the function f is nonanalytic, but that f is analytic everywhere else within C. The Cauchy–Goursat theorem can be applied to such a region by deforming the contour in a manner that encapsulates the problematic region. Figure 1.19 illustrates this technique. The colored region represents the nonanalytic region R_2 and the outer circle, when closed, represents the contour C and is traversed in a positive, counterclockwise, sense. Note that C need not actually be circular, but it is easier to draw that way. We imagine drawing line A from C to a point just outside the nonanalytic region. The contour C_2 goes around this region in a negative, clockwise sense, remaining within the analytic region R_1, ending close to its starting point. We then return along B to the contour C_1. The common path AB traversed in opposite directions between inner and outer contours is sometimes called a *contour wall* and serves to create a simply connected region R_1 for which the Cauchy–Goursat theorem requires

$$\int_{C_1} f[z]\, dz + \int_A f[z]\, dz + \int_B f[z]\, dz + \int_{C_2} f[z]\, dz = 0 \tag{1.140}$$

Recognizing that, for a continuous integrand, the contributions of A and B must become equal and opposite as the separation between those paths becomes infinitesimal, we find

$$\int_A f[z]\, dz + \int_B f[z]\, dz = 0 \Longrightarrow \oint_C f[z]\, dz = -\oint_{C_2} f[z]\, dz \tag{1.141}$$

Here the negative sign occurs because the inner contour is traversed in the opposite direction when reached by means of the contour wall. Therefore, the original contour can be *shrink-wrapped* about the nonanalytic region without changing the value of the contour integral.

If the path C encloses several localized nonanalytic regions, we simply construct several contour walls. The net contour integral is then just the sum of the contributions

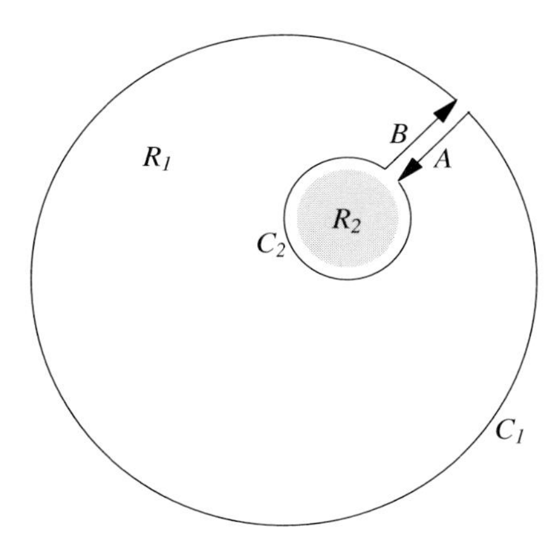

Figure 1.19. Construction of a contour wall and demonstration that a contour within an analytic region may be shrink-wrapped around and enclosed nonanalytic region.

from shrink-wrapped contours around each nonanalytic region. Take care with the signs though – if the original contour is traversed in a positive sense, the nonanalytic regions are enclosed in a negative sense by the continuous deformed contour that circumvents nonanalytic regions. However, recognizing that the entire contour integral vanishes and that the contour walls cancel, the net integral for a simple contour that encloses nonanalytic regions reduces to the sum of the contributions made by shrink-wrapped contours enclosing the nonanalytic regions in a positive sense. Therefore, if there are N isolated nonanalytic regions within the simple closed contour C, we find

$$\oint_C f[z]\, dz = \sum_{k=1}^{N} \oint_{C_k} f[z]\, dz \tag{1.142}$$

where each simple closed contour C_k encloses one of the nonanalytic regions and is traversed with the same sense as the original contour C.

We postpone consideration of extended nonanalytic regions to the next chapter, but in the next few sections consider the important special case of an isolated singularity within the contour.

1.10.2 Cauchy Integral Formula

Suppose that the contour C lies within a region R in which $f[z]$ is analytic, but that it surrounds another region R' in which f is not analytic. We demonstrated above that the contour can be deformed, such that $C \to C'$ where C' is immediately outside R', without changing the value of the contour integral

$$\oint_C f[z]\, dz = \oint_{C'} f[z]\, dz \tag{1.143}$$

Thus, a contour integral that encloses a single localized nonanalytic region can be shrink-wrapped about the border of that region. This result is particularly useful for the case of an isolated singularity for which the region of nonanalyticity consists of a single point z_0. Consider the integral

$$\oint_C ds \frac{f[s]}{s-z} \tag{1.144}$$

where f is analytic throughout the region enclosed by C while the integrand it singular at z. If z is outside C the integral vanishes because the integrand is analytic at all points within C. Alternatively, if z lies within C, we can reduce C to a small circle surrounding z, such that

$$s - z = re^{i\theta} \Longrightarrow ds = ire^{i\theta}\, d\theta \tag{1.145}$$

Thus, the integral can be approximated

$$\oint_C ds \frac{f[s]}{s-z} \approx f[z] \oint_C \frac{ire^{i\theta}\, d\theta}{re^{i\theta}} = 2\pi i f[z] \tag{1.146}$$

to arbitrary accuracy as $r \to 0$. Therefore, we obtain the *Cauchy integral formula*:

Theorem 6. *Cauchy integral formula: If a function $f[z]$ is analytic at all points on and within a simple closed contour C, then $f[z] = \frac{1}{2\pi i} \oint_C \frac{f[s]}{s-z}\, ds$ for any interior point z.*

This remarkably powerful theorem requires that the value of an analytic function at any interior point is uniquely determined by its values on any surrounding closed curve and is analogous to the two-dimensional form of Gauss' theorem. The behavior of an analytic function is severely constrained.

1.10.3 Example: Yukawa Field

Using elementary field theory, the virtual pion field surrounding a nucleon is represented in momentum space by

$$\tilde{\phi}[q] = \frac{\Lambda^2}{q^2 + \Lambda^2} \tag{1.147}$$

The spatial distribution is then obtained from the three-dimensional Fourier transform

$$\phi[r] = \int \frac{d^3q}{(2\pi)^3} e^{iq\cdot r} \tilde{\phi}[q] = \frac{4\pi}{(2\pi)^3} \frac{\Lambda^2}{r} \int_0^\infty \frac{q \operatorname{Sin}[qr]}{q^2 + \Lambda^2}\, dq \tag{1.148}$$

where spherical symmetry and the multipole expansion of the plane wave have been used to reduce the integral to one dimension. (Alternatively, the angular integrals can be evaluated directly.) Recognizing that the integrand is even, we can write

$$\int_0^\infty \frac{q \operatorname{Sin}[qr]}{q^2 + \Lambda^2}\, dq = \frac{1}{2} \int_{-\infty}^\infty \frac{q \operatorname{Sin}[qr]}{q^2 + \Lambda^2}\, dq = \frac{1}{2i} \int_{-\infty}^\infty \frac{q \operatorname{Exp}[iqr]}{q^2 + \Lambda^2}\, dq \tag{1.149}$$

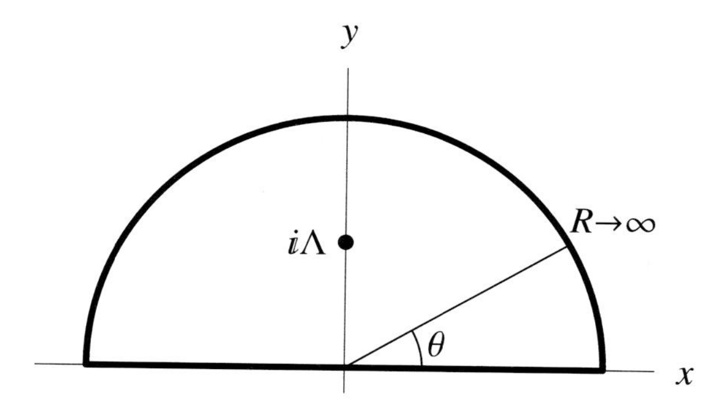

Figure 1.20. Great semicircle with enclosed pole at $z = i\Lambda$.

because the contribution from Cos[qr] vanishes by symmetry. Now consider the contour integral

$$I[\Lambda] = \oint_C \frac{g[z]}{z - i\Lambda} dz \tag{1.150}$$

where

$$g[z] = \frac{z e^{izr}}{(z + i\Lambda)} \tag{1.151}$$

is analytic in the upper half-plane. If we choose a contour C, shown in Fig. 1.20, consisting of the real axis and a semicircle in the upper half-plane with $R \to \infty$, affectionately called a *great semicircle*, this integral can be expressed as

$$I[\Lambda] = \int_{-\infty}^{\infty} \frac{q \,\mathrm{Exp}[iqr]}{q^2 + \Lambda^2} dq + iR^2 \int_0^{\pi} \frac{\mathrm{Exp}\left[irRe^{i\theta}\right]}{R^2 e^{2i\theta} + \Lambda^2} d\theta \tag{1.152}$$

Using

$$\mathrm{Exp}\left[irRe^{i\theta}\right] = \mathrm{Exp}[irR\,\mathrm{Cos}[\theta]]\,\mathrm{Exp}[-rR\,\mathrm{Sin}[\theta]] \tag{1.153}$$

and recognizing that Sin[θ] > 0 in the upper half-plane, we realize that the contribution of the circular arc decreases exponentially with R and vanishes in the limit $R \to \infty$. Therefore, with the aid of the Cauchy integral formula

$$I[\Lambda] = 2\pi i g[i\Lambda] = i\pi e^{-\Lambda r} \tag{1.154}$$

we obtain the Yukawa field

$$\phi[r] = \frac{\Lambda^2}{4\pi} \frac{e^{-\Lambda r}}{r} \tag{1.155}$$

that is central to the meson exchange model of the nucleon–nucleon interaction that binds atomic nuclei together.

1.10.4 Derivatives of Analytic Functions

Let

$$f[z] = \frac{1}{2\pi i} \oint_C dw \frac{f[w]}{w - z} \tag{1.156}$$

represent a function that is analytic in the domain D containing the simple closed contour C. First we demonstrate that differentiation can be performed under the integral sign. Using

$$\begin{aligned} f[z + \Delta z] - f[z] &= \frac{1}{2\pi i} \oint_C dw \left(\frac{f[w]}{w - z - \Delta z} - \frac{f[w]}{w - z} \right) \\ &= \frac{\Delta z}{2\pi i} \oint_C dw \frac{f[w]}{(w - z)(w - z - \Delta z)} \end{aligned} \tag{1.157}$$

we recognize that the left-hand side vanishes in the limit $\Delta z \to 0$ because $f[z]$ is continuous and must demonstrate that the right-hand side shares this property. Recognizing that the integrand is analytic everywhere within C except at z, we may reduce the contour to a small circle of radius r around z. Let $M = \max[|f[w]|]$ on the reduced contour, and use the triangle inequalities to evaluate the maximum modulus of the integrand, such that

$$\begin{aligned} \left| \Delta z \oint_C dw \frac{f[w]}{(w - z)(w - z - \Delta z)} \right| &\leq |\Delta z| \oint_C dw \frac{M}{|(w - z)(w - z - \Delta z)|} \\ &\leq 2\pi r \frac{M|\Delta z|}{r(r - |\Delta z|)} \end{aligned} \tag{1.158}$$

vanishes in the limit $\Delta z \to 0$ for finite r. Thus, we find that

$$\begin{aligned} \lim_{\Delta z \to 0} \frac{f[z + \Delta z] - f[z]}{\Delta z} &= \frac{1}{2\pi i} \oint_C dw \frac{f[w]}{(w - z)^2} \\ &\Longrightarrow f'[z] = \frac{1}{2\pi i} \oint_C dw f[w] \frac{d}{dz} (w - z)^{-1} \end{aligned} \tag{1.159}$$

is also analytic within D. Repeating this process, we obtain

$$f^{(n)}[z] = \frac{d^n f[z]}{dz^n} = \frac{n!}{2\pi i} \oint_C dw \frac{f[w]}{(w - z)^{n+1}} \tag{1.160}$$

by induction. Therefore, we have demonstrated by construction the remarkably powerful theorem that analytic functions have derivatives of all orders. This also requires all partial derivatives of its component functions to be continuous in D. This theorem will soon be used to derive series representations of analytic functions.

Theorem 7. *If a function f is analytic at a point, then derivatives of all orders exist and are analytic at that point.*

1.10.5 Morera's Theorem

The converse of Cauchy–Goursat theorem is known as *Morera's theorem*:

Theorem 8. *Morera's theorem: If a function $f[z]$ is continuous in a simply connected region R and $\oint_C f[z]\,dz = 0$ for every simple closed contour C within R, then $f[z]$ is analytic throughout R.*

If every closed path integral vanishes, the path integral between two points in the domain of analyticity D depends only upon the end points and is independent of the path, provided that the path lies entirely within D. Hence, we define the function $F[z]$ by means of the definite integral

$$F[z_2] - F[z_1] = \int_{z_1}^{z_2} f[z]\,dz \tag{1.161}$$

Clearly,

$$\int_{z_1}^{z_2} \left(f[z] - f[z_1]\right) dz = F[z_2] - F[z_1] - (z_2 - z_1) f[z_1] \tag{1.162}$$

such that the limit as $z_2 \to z_1$

$$\lim_{z_2 \to z_1} \frac{\int_{z_1}^{z_2} \left(f[z] - f[z_1]\right) dz}{z_2 - z_1} = \lim_{z_2 \to z_1} \frac{F[z_2] - F[z_1]}{z_2 - z_1} - f[z_1] = F'[z_1] - f[z_1] \tag{1.163}$$

compares $F'[z_1]$ with $f[z_1]$. However, the integral vanishes in the limit of vanishing range of integration because $f[z]$ is continuous in D, such that

$$\lim_{z_2 \to z_1} \frac{\int_{z_1}^{z_2} \left(f[z] - f[z_1]\right) dz}{z_2 - z_1} = 0 \Longrightarrow F'[z_1] = f[z_1] \tag{1.164}$$

Thus, $F[z]$ is analytic in D with $F'[z] = f[z]$. Therefore, because the derivative of an analytic function is also analytic, we conclude that $f[z]$ must also be analytic, proving Morera's theorem.

Morera's theorem is sometimes useful for proving general properties for analytic functions of various types, but is rarely of practical value to more detailed calculations.

1.11 Complex Sequences and Series

1.11.1 Convergence Tests

An infinite sequence of complex numbers $\{z_n, n = 1, 2, \ldots\}$ can be represented by combining two sequences of real numbers $\{x_n, n = 1, 2, \ldots\}$ and $\{y_n, n = 1, 2, \ldots\}$ such that $z_n = x_n + i y_n$. The sequence z_n converges to z if for any small positive ϵ there exists an integer N such that $|z_n - z| < \epsilon$ for $n > N$. Convergence of a complex series z_n to $z = x + iy$ then requires convergence of both x_n to x and y_n to y. Many of the properties of real sequences can be adapted to complex sequences with only minor and obvious changes. Therefore, we state without proof the *Cauchy convergence principle*:

Theorem 9. *The sequence $\{z_n\}$ converges if and only if for every small positive ϵ there exists an integer N_ϵ such that $|z_m - z_n| < \epsilon$ for any $m, n > N_\epsilon$.*

If $\{z_n\}$ and $\{w_n\}$ are two convergent sequences with limits z and w, then $\{az_n + bw_n\}$ and $\{z_n w_n\}$ are also convergent sequences with limits $az + bw$ and zw.

An infinite series of complex numbers z_k converges if the sequence of partial sums

$$S_n = \sum_{k=1}^{n} z_k \tag{1.165}$$

converges to S, such that

$$\lim_{n\to\infty} S_n = S \Longrightarrow S = \sum_{k=1}^{\infty} z_k \tag{1.166}$$

If the sequence of partial sums does not converge, the corresponding series diverges. A series is *absolutely convergent* if the series of moduli

$$\sum_{k=1}^{n} |z_k| \tag{1.167}$$

converges. An absolutely convergent series converges, but a convergent series need not converge absolutely. A convergent series that is not absolutely convergent is described as *conditionally convergent*. For example, the alternating harmonic series $\sum_{k=1}^{\infty}(-)^k/k$ converges conditionally but not absolutely because $\sum_{k=1}^{\infty} k^{-1}$ diverges. Term-by-term addition of convergent series yields another convergent series, but convergence of a series formed by termwise multiplication requires absolute convergence of the individual series.

If $\{z_k\}$ does not converge to zero, the corresponding series diverges because the sequence of partial sums will not satisfy the Cauchy convergence condition. However, convergence of the sequence of terms to zero does not ensure convergence of the series. The most general analysis of a series separates its terms into real and imaginary parts and then applies one of the many tests developed for series of real numbers to the real and imaginary subseries separately; the complex series then converges if both its real and imaginary subseries converge. However, it is usually simpler and often sufficient to test for absolute convergence instead. The following convergence tests familiar for real series can be generalized to complex series.

Comparison test: If $0 \le |z_k| \le a_k$ for sufficiently large k and $\sum_k a_k$ converges, then $\sum_k z_k$ converges absolutely.

Ratio test: If $|z_{k+1}/z_k| \le r$ for all $k > N$, then $\sum_k z_k$ converges absolutely if $r < 1$. Alternatively, if $r = \lim_{k\to\infty} |z_{k+1}/z_k|$ the series converges absolutely if $r < 1$ but diverges if $r > 1$. This test is inconclusive if $r = 1$.

Root test: If $|z_k|^{1/k} \le r$ for all $k > N$, then $\sum_k z_k$ converges absolutely if $r < 1$. Alternatively, if $r = \lim_{k\to\infty} |z_k|^{1/k}$ the series converges absolutely if $r < 1$ but diverges if $r > 1$. This test is also inconclusive if $r = 1$.

Integral test: Suppose that $f[k] = |z_k|$ where $f[x]$ is defined for $x \geq n \geq 1$. The series then converges absolutely if the integral $\int_n^\infty f[x]\, dx$ converges.

Note that the ratio test is indeterminate when $\lim_{k\to\infty} |z_{k+1}/z_k| = 1$. For example, the harmonic series $z_k = k^{-1}$ diverges while the alternating harmonic series converges. A "sharpened" version, established in the exercises, shows that a series converges absolutely if the ratio of successive terms takes the form

$$\left|\frac{a_{n+1}}{a_n}\right| \simeq 1 - \frac{s}{n} \tag{1.168}$$

for large n with $s > 1$.

Often the terms of a series will themselves be functions of a complex variable, z, such that

$$f_n[z] = \sum_{k=1}^{n} g_k[z] \tag{1.169}$$

represents a sequence $\{f_n[z]\}$ of partial sums. If such a sequence converges for all z in a region R, such that

$$z \in R \Longrightarrow f[z] = \lim_{n\to\infty} f_n[z] = \lim_{n\to\infty} \sum_{k=1}^{n} g_k[z] \tag{1.170}$$

then $f_n[z]$ is described as a series *representation* of the function $f[z]$ valid within the *convergence region* R. Often the convergence region takes the form of a disk, $|z - z_0| \leq R$, with center z_0 and *radius of convergence* R. If a series converges for all z within $|z - z_0| < R$ but diverges for some points on the circle $|z - z_0| = R$, one still reports a radius of convergence R. The problem then is to determine the radius of convergence.

Example

What is the radius of convergence for a geometric series, $\sum_{k=0}^{\infty} z^k$, extended to the complex plane? According to the ratio test,

$$\left|\frac{z_{k+1}}{z_k}\right| = \left|\frac{z^{k+1}}{z^k}\right| = |z| \tag{1.171}$$

this series converges absolutely for any $|z| < 1$ and diverges for $|z| > 1$. Thus, the radius of convergence is 1. Notice that even though the ratio test is inconclusive for $|z| = 1$, this series clearly diverges on the unit circle because the terms do not approach zero. Alternatively, by the ratio test

$$\lim_{k\to\infty} |z_k|^{1/k} = |z| \tag{1.172}$$

one finds convergence for $|z| > 1$ and divergence for $|z| \geq 1$. Furthermore, one can demonstrate that

$$|z| < 1 \Longrightarrow \lim_{n\to\infty} \sum_{k=0}^{n} z^k = \frac{1}{1 - z} \tag{1.173}$$

within the radius of convergence. Let

$$f_n[z] = \sum_{k=0}^{n} z^k = \frac{1 - z^{n+1}}{1 - z} \tag{1.174}$$

represent a sequence of complex numbers, where the last step is verified by direct multiplication

$$\begin{aligned}(1 - z)(1 + z + z^2 + \ldots + z^n) &= (1 + z + z^2 + \ldots + z^n) - (z + z^2 + \ldots + z^{n+1}) \\ &= 1 - z^{n+1}\end{aligned} \tag{1.175}$$

Then, separating the constant term (for fixed z) from the variable part of the sequence

$$f_n[z] = \frac{1}{1 - z} - \frac{z^{n+1}}{1 - z} \tag{1.176}$$

and recognizing

$$|z| < 1 \Longrightarrow \lim_{n\to\infty} \frac{z^{n+1}}{1 - z} = 0 \tag{1.177}$$

one finds that

$$|z| < 1 \Longrightarrow \lim_{n\to\infty} f_n[z] = \frac{1}{1 - z} \tag{1.178}$$

Therefore, the geometric series

$$|z| < 1 \Longrightarrow \sum_{k=0}^{\infty} z^k = \frac{1}{1 - z} \tag{1.179}$$

converges to a simple analytic function within the unit circle, thereby extending a familiar result from the real axis to the complex plane.

1.11.2 Uniform Convergence

A sequence of functions $\{f_n[z]\}$ is said to *converge uniformly* to the function $f[z]$ in a region R if there exists a fixed positive integer N_ϵ such that $|f_n[z] - f[z]| < \epsilon$ for any z within R when $n > N_\epsilon$. Consequently, a uniformly convergent series $f_n[z] = \sum_{k=1}^{n} g_k[z]$ provides an approximation to $f[z]$ within R with controllable accuracy – there exists a finite number of terms, even if large, that guarantees a specified degree of accuracy anywhere within the region of uniform convergence. The region of uniform convergence is always a subset of the region of convergence. For example, although the geometric series $\sum_{k=0}^{\infty} z^k$ converges uniformly to $(1 - z)^{-1}$ within any disk $|z| \leq R < 1$ with less than unit radius and is convergent within $|z| < 1$, one cannot properly claim uniform convergence throughout the open region $|z| < 1$ because the convergence becomes so slow near the circle of convergence that there will always be points within that region that require more than

N terms to achieve the desired accuracy no matter how large N is chosen. Convergence at z without uniform convergence within the region of interest is described as *pointwise*.

The most common test for uniform convergence is offered by the *Weierstrass M-test*: The series $\sum_k f_k[z]$ is uniformly convergent in region R if there exists a series of positive constants M_k such that $|f_k[z]| \leq M_k$ for all z in R and $\sum_k M_k$ converges. The proof follows directly from the comparison test. (For what it's worth, M stands for *majorant*.)

The follow theorems for manipulation of uniformly convergent series can be established by straightforward generalization of the corresponding results for real functions.

***Continuity theorem*:** a uniformly convergent series of continuous functions is continuous.

***Combination theorem*:** the sum or product of two uniformly convergent series is uniformly convergent within the overlap of their convergence regions.

***Integrability theorem*:** the integral of a uniformly convergent series of continuous functions is equal to the sum of the integrals of each term.

***Differentiability theorem*:** the derivative of a uniformly convergent series of continuous functions with continuous derivatives is uniformly convergent and is equal to the sum of the derivatives of each term.

Furthermore, by combining these results one can obtain the more general *Weierstrass theorem* establishing uniformly convergent series as analytic functions within their convergence regions. Thus, the property of uniform convergence is important because it makes available all theorems in the theory of analytic functions.

Theorem 10. *Weierstrass theorem: If the terms of a series $\sum_k g_k[z]$ are analytic throughout a simply-connected region R and the series converges uniformly throughout R, then its sum is an analytic function within R and the series may be integrated or differentiated termwise any number of times.*

1.12 Derivatives and Taylor Series for Analytic Functions

1.12.1 Taylor Series

It is now a simple matter to demonstrate the existence of power-series expansions for analytic functions. Suppose that f is analytic within a disk $|z - z_0| \leq R$ centered upon z_0 and assume that

$$f[z] = \sum_{n=0}^{\infty} a_n (z - z_0)^n \tag{1.180}$$

Consider the integral

$$I_k = \frac{1}{2\pi i} \oint dz \frac{1}{(z - z_0)^k} \tag{1.181}$$

evaluated on the circle $|z - z_0| = R$. Using $dz = iRe^{i\theta}\, d\theta$, we find that

$$I_k = \frac{R^{1-k}}{2\pi} \int_0^{2\pi} e^{i(1-k)\theta}\, d\theta = \delta_{k,1} \tag{1.182}$$

vanishes unless $k = 1$. Therefore, the coefficients of the power series can be evaluated according to

$$a_n = \frac{1}{2\pi i} \oint dz \frac{f[z]}{(z - z_0)^{n+1}} = \frac{f^{(n)}[z_0]}{n!} \tag{1.183}$$

Although we performed this calculation using circular contours, the same results would be obtained for arbitrary simple contours within the analytic region because the singularities in the integrands are confined to a single point, which can be excised. A power series centered upon the origin is sometimes called a *Maclaurin series* while a more general power series about arbitrary z_0 is called a *Taylor series*.

Theorem 11. *Taylor series: If a function f is analytic within a disk $|z - z_0| \leq R$, then the power series $f[z] = \sum_{n=0}^{\infty} a_n (z - z_0)^n$ with $a_n = \frac{f^{(n)}[z_0]}{n!}$ converges to $f[z]$ at all points within the disk. Conversely, if a power series converges for $|z - z_0| \leq R$, it represents an analytic function within that disk.*

It is instructive to demonstrate convergence of the power series directly. Expanding

$$(s - z)^{-1} = (s - z_0)^{-1} \left(1 - \frac{z - z_0}{s - z_0}\right)^{-1} = (s - z_0)^{-1} \left(A_n + \sum_{k=0}^{n} \left(\frac{z - z_0}{s - z_0}\right)^k\right) \tag{1.184}$$

where

$$A_n = \frac{(z - z_0)^{n+1}}{(s - z_0)^n (s - z)} \tag{1.185}$$

the Cauchy integral formula becomes

$$f[z] = \frac{1}{2\pi i} \oint_C ds \frac{f[s]}{s - z} = \sum_{k=0}^{n} a_k (z - z_0)^k + R_n (z - z_0)^{n+1} \tag{1.186}$$

where

$$a_k = \frac{1}{2\pi i} \oint ds \frac{f[s]}{(s - z_0)^{k+1}} = \frac{f^{(k)}[z_0]}{k!} \tag{1.187}$$

as before and where the remainder takes the form

$$R_n = \frac{1}{2\pi i} \oint_C ds \frac{f[s]}{(s - z)} \frac{(z - z_0)^{n+1}}{(s - z_0)^{n+1}} \tag{1.188}$$

Identifying

$$|z - z_0| = \rho, \quad M = \max[|f[s]|], \quad \delta = \min[|s - z|] \tag{1.189}$$

and choosing a circular contour with

$$|s - z_0| = r > \rho \tag{1.190}$$

we find

$$|R_n| \leq \left(\frac{\rho}{r}\right)^{n+1} \frac{r}{\delta} M \Longrightarrow \lim_{n\to\infty} R_n = 0 \tag{1.191}$$

Thus, this power series converges throughout the region of analyticity that surrounds z_0. Therefore, *the radius of convergence is the distance to the nearest singularity in the complex plane*. With more careful analysis, one may find that a Taylor series converges at some points on the circle of convergence also.

The Taylor series for $f[z]$ about a point $z_0 = (x_0, 0)$ on the real axis has the same form as the expansion of $f[x]$ interpreted as a function of the real variable x. More importantly, this extension of the Taylor theorem to the complex plane often provides the simplest method for evaluating the radius of convergence of a power series. Consider the hyperbolic tangent

$$\mathrm{Tanh}[z] = z - \frac{1}{3}z^3 + \frac{2}{15}z^5 - \frac{17}{315}z^7 + \frac{62}{2835}z^9 - \frac{1382}{155\,925}z^{11} + \frac{21\,844}{6\,081\,075}z^{13} + \cdots \tag{1.192}$$

It is difficult to evaluate the general term and to deduce the radius of convergence from the real-valued series, but from the function of a complex variable we know immediately that the radius of convergence is $\pi/2$ because the nearest roots of Cosh[z] are found at $z = \pm i\pi/2$.

Sometimes it is necessary to determine the radius of convergence directly from the terms of the power series. Then one finds

$$R = \left(\lim_{n\to\infty} \left|\frac{a_{n+1}}{a_n}\right|\right)^{-1} \tag{1.193}$$

using the ratio test, or

$$R = \lim_{n\to\infty} |a_n|^{-1/n} \tag{1.194}$$

using the root test. For example, the Maclaurin series for Log[$1 + z$] takes the form

$$\mathrm{Log}[1 + z] = \sum_{n=1}^{\infty} (-)^{n+1} \frac{z^n}{n} \tag{1.195}$$

and one obtains a convergence radius $R = 1$ using the ratio test. In this case the convergence radius is limited by the branch point at $z = -1$. Notice that at $z = 1$ this power series reduces to the alternating harmonic series

$$\sum_{n=1}^{\infty} \frac{(-)^{n+1}}{n} = \mathrm{Log}[2] \tag{1.196}$$

and thus converges for at least one point on the convergence circle, while at $z = -1$ the resulting harmonic series diverges. Applying the root test instead suggests a limit

$$\lim_{n\to\infty} n^{1/n} = 1 \tag{1.197}$$

that might not be obvious otherwise.

1.12.2 Cauchy Inequality

Let

$$M[r] = \max_{|z-z_0|=r} |f[z]| \tag{1.198}$$

represent the maximum modulus of an analytic function on a circle of radius r surrounding z_0. We then find that

$$|a_n| \leq \frac{1}{2\pi} \oint dz \frac{M}{r^{n+1}} = M r^{-n} \Longrightarrow |a_n| r^n \leq M[r] \tag{1.199}$$

constrains the coefficients of the Taylor series.

Theorem 12. *Cauchy inequality: If a function* $f[z] = \sum_{n=0}^{\infty} a_n (z - z_0)^n$ *is analytic and bounded in D and* $|f[z]| \leq M$ *on a circle* $|z - z_0| = r$*, then* $|a_n| r^n \leq M$.

1.12.3 Liouville's Theorem

Theorem 13. *Liouville's theorem: If a function* $f[z]$ *is analytic and bounded everywhere in the complex plane, then* $f[z]$ *is constant.*

According to the Cauchy inequality, if $|f[z]| < M$ for $|z| < R$, then $|a_n| R^n < M$. If this inequality applies in the limit $R \to \infty$, then we must require $a_n \to 0$ for $n > 0$. Therefore, if f is not constant, it must have a singularity somewhere. The behavior of functions of a complex variable is largely determined by the nature and locations of their singularities.

1.12.4 Fundamental Theorem of Algebra

Theorem 14. *Fundamental theorem of algebra: Any polynomial* $P_n[z] = \sum_{k=0}^{n} a_n z^n$ *of order* $n \geq 1$ *must have at least one zero* $z_0 \ni P_n[z_0] = 0$ *in the finite complex plane.*

Although it is difficult to prove the fundamental theorem of algebra using purely algebraic means, it is an almost trivial consequence of Liouville's theorem. If $P_n[z]$ has no zeros, then the function $f[z] = 1/P_n[z]$ would be analytic throughout the entire complex plane. Recognizing that

$$|z| = R \to \infty \Longrightarrow |P_n[z]| \to |a_n| R^n \Longrightarrow f[z] \to \frac{1}{|a_n| R^n} \tag{1.200}$$

it is clear that $f[z]$ is bounded. Thus, Liouville's theorem requires f to be constant, but $f[z]$ vanishes in the limit $R \to \infty$, which contradicts the absence of zeros in P_n. Therefore, P_n must have at least one zero. Factoring out this root, we write $P_n[z] = (z - z_1) P_{n-1}[z]$ and apply the theorem to P_{n-1}, concluding that it must also have a root if $n > 2$. Repeating this process, we determine that a polynomial of degree n must have n roots, although some might be repeated.

1.12.5 Zeros of Analytic Functions

If a function $f[z]$ is analytic at z_0 there exists a disk $|z - z_0| < R$ wherein the Taylor series converges, such that

$$|z - z_0| < R \Longrightarrow f[z] = \sum_{n=0}^{\infty} a_n (z - z_0)^n \tag{1.201}$$

Suppose that z_0 was chosen to be a root of $f[z_0] == 0$, such that $a_0 = 0$. If $a_1 \neq 0$, we describe z_0 as a simple zero of f, but if all $a_n = 0$ with $n < m$ while $a_m \neq 0$, we describe z_0 as a zero of order m. It is then useful to express the Taylor series in the form

$$f[z] = (z - z_0)^m \phi[z] \tag{1.202}$$

where the auxiliary function

$$\phi[z] = \sum_{n=0}^{\infty} a_{m+n} (z - z_0)^n \tag{1.203}$$

employing the coefficients a_n with $n \geq m$ has the nonzero value $\phi[z_0] = a_m$ at z_0. Clearly ϕ is continuous at z_0 and is analytic within the radius of convergence. Therefore, for any small positive number ε there exists a corresponding radius δ such that

$$|\phi[z] - a_m| < \varepsilon \quad \text{whenever} \quad |z - z_0| < \delta \tag{1.204}$$

Suppose there were another point z_1 in a neighborhood of z_0 where $\phi[z_1] = 0$, such that

$$\phi[z_1] = 0 \Longrightarrow |a_m| < \varepsilon \quad \text{whenever} \quad |z_1 - z_0| < \delta \tag{1.205}$$

can only be satisfied if $a_m = 0$, contrary to our assumption that z_0 is a zero of order m. Therefore, we conclude that if f is analytic and does not vanish identically, there must exist a neighborhood around any root in which no other root is found; in other words, the roots of analytic functions are isolated.

Theorem 15. *Suppose that a function $f[z]$ is analytic at z_0 and that $f[z_0] = 0$. Then there must exist a neighborhood of z_0 containing no other zeros of f unless f vanishes identically.*

1.13 Laurent Series

1.13.1 Derivation

A more general expansion which is useful in an analytic region that surrounds a nonanalytic region is provided by the Laurent series.

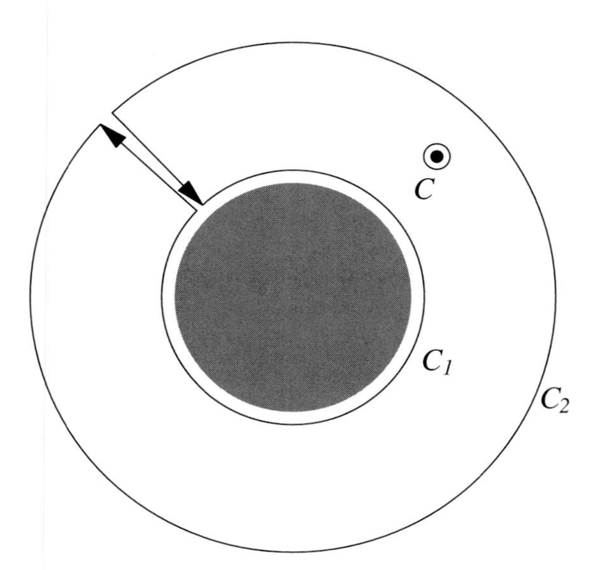

Figure 1.21. To develop the Laurent series, a small contour C within an analytic region is stretched toward the limits of the region of analyticity, indicated by C_1 and C_2, with the aid of contour wall.

Theorem 16. *Laurent series: If $f[z]$ is analytic throughout the region $R_1 < |z - z_0| < R_2$, it can be represented by an expansion*

$$f[z] = \sum_{n=-\infty}^{\infty} a_n (z - z_0)^n \tag{1.206}$$

with coefficients

$$a_n = \frac{1}{2\pi i} \oint_C \frac{f[z]\, dz}{(z - z_0)^{n+1}} \tag{1.207}$$

computed using any simple counterclockwise contour C within the analytic region.

If $R_1 \to 0$ and coefficients with $n < 0$ vanish, then the Laurent series reduces to the Taylor series.

Suppose that C is a small contour surrounding an interior point z, such that

$$f[z] = \frac{1}{2\pi i} \oint_C ds \frac{f[s]}{s - z} \tag{1.208}$$

according to the Cauchy integral formula. As shown in Fig. 1.21, we can stretch C to the limits of the annulus without changing the integral because the integrand is analytic throughout that region. Recognizing that the opposing segments of the contour wall cancel, we obtain

$$f[z] = \frac{1}{2\pi i} \oint_{R_2} ds \frac{f[s]}{s - z} - \frac{1}{2\pi i} \oint_{R_1} ds \frac{f[s]}{s - z} \tag{1.209}$$

where R_1 and R_2 denote counterclockwise circles at the inner and outer borders of the analytic annulus. For the outer integral we employ the expansion

$$(s-z)^{-1} = (s-z_0)^{-1}\left(1-\frac{z-z_0}{s-z_0}\right)^{-1} = \left(s-z_0\right)^{-1}\sum_{n=0}^{\infty}\left(\frac{z-z_0}{s-z_0}\right)^n \tag{1.210}$$

while for the inner integral

$$(s-z)^{-1} = -(z-z_0)^{-1}\left(1-\frac{s-z_0}{z-z_0}\right)^{-1} = -\left(z-z_0\right)^{-1}\sum_{n=0}^{\infty}\left(\frac{s-z_0}{z-z_0}\right)^n \tag{1.211}$$

such that

$$2\pi i f[z] = \sum_{n=0}^{\infty}(z-z_0)^n \oint_{R_2} ds \frac{f[s]}{(s-z_0)^{1+n}} + \sum_{n=0}^{\infty}(z-z_0)^{-1-n} \oint_{R_1} ds f[s](s-z_0)^n \tag{1.212}$$

Both integrands are analytic throughout the annulus and are independent of z. Hence, these integrals can be evaluated using any simple closed path within the analytic region. Therefore, we may combine the two terms into a single expression

$$f[z] = \sum_{n=-\infty}^{\infty} a_n(z-z_0)^n \tag{1.213}$$

$$a_n = \frac{1}{2\pi i}\oint_C dz \frac{f[z]}{(z-z_0)^{n+1}} \tag{1.214}$$

representing the Laurent expansion. One can also show that the Laurent expansion about a specific z_0 is unique within its analytic annulus.

1.13.2 Example

The function

$$f[z] = \frac{1}{z^2(1-z)} \tag{1.215}$$

has singular points at $z = 0, 1$. Suppose that we evaluate the Laurent coefficients using contour integration

$$a_n = \frac{1}{2\pi i}\oint_C ds \frac{f[s]}{s^{n+1}} = \frac{R^{-n-2}}{2\pi}\int_0^{2\pi} \frac{e^{-i\theta(n+2)}}{1-Re^{i\theta}}\, d\theta \tag{1.216}$$

on a circle with $s = Re^{i\theta}$ and $ds = is\, d\theta$. For $R < 1$ we can expand the integrand to obtain

$$R < 1 \Longrightarrow a_n = \frac{R^{-n-2}}{2\pi}\sum_{k=0}^{\infty} R^k \int_0^{2\pi} \text{Exp}[i(k-n-2)\theta]\, d\theta \tag{1.217}$$

Nonvanishing coefficients then require $k = n+2$ and $k \geq 0 \Longrightarrow n \geq -2$, such that

$$0 < |z| < 1 \Longrightarrow f[z] = \sum_{n=0}^{\infty} z^{n-2} \tag{1.218}$$

Alternatively, for $R > 1$ we use $(1-z)^{-1} = -z^{-1}(1-z^{-1})^{-1}$ to obtain

$$R > 1 \Longrightarrow a_n = -\frac{R^{-n-3}}{2\pi}\sum_{k=0}^{\infty} R^{-k}\int_0^{2\pi} \mathrm{Exp}[-i(k+n+3)\theta]\, d\theta \tag{1.219}$$

for which nonvanishing coefficients require $k = -n-3$ and $k \geq 0 \Longrightarrow n \leq -3$, such that

$$|z| > 1 \Longrightarrow f[z] = -\sum_{n=0}^{\infty} z^{-n-3} \tag{1.220}$$

Although the Laurent theorem provides an explicit formula for the coefficients, evaluation of the contour integrals is often difficult and one seeks simpler alternative methods. In this case we can use the partial fractions

$$\frac{1}{z^2(1-z)} = \frac{1}{z^2} + \frac{1}{z} + \frac{1}{1-z} \tag{1.221}$$

and

$$|z| < 1 \Longrightarrow \frac{1}{1-z} = \sum_{n=0}^{\infty} z^n \tag{1.222}$$

$$|z| > 1 \Longrightarrow \frac{1}{1-z} = -\frac{1}{z(1-z^{-1})} = -\sum_{n=1}^{\infty} z^{-n} \tag{1.223}$$

to obtain the same results without integration. In other cases we may be able to convert a known Taylor series into a Laurent series. For example,

$$\mathrm{Log}[1+z] = \sum_{n=1}^{\infty} (-)^{n+1}\frac{z^n}{n} \quad \text{for} \quad |z| < 1 \tag{1.224}$$

$$\Longrightarrow \mathrm{Log}\left[1+z^{-1}\right] = \sum_{n=1}^{\infty} (-)^{n+1}\frac{z^{-n}}{n} \quad \text{for} \quad |z| > 1 \tag{1.225}$$

where the latter is valid in the largest annulus that excludes the branch cut $-1 \leq x \leq 1$ on the real axis.

1.13.3 Classification of Singularities

Suppose that $f[z]$ is singular at z_0 but analytic at all other points in a neighborhood of z_0; f is then said to have an *isolated singularity* at z_0. A function that is analytic throughout the finite complex plane except for isolated singularities is described as *meromorphic*. Meromorphic functions include entire functions, such as Exp, that have no singularities in the finite plane and rational functions that have a finite number of poles. Functions, such as Log, that require branch cuts are not meromorphic.

The Laurent expansion about an isolated singularity takes the form

$$f[z] = \sum_{n=-\infty}^{\infty} a_n (z - z_0)^n \tag{1.226}$$

If $a_m \neq 0$ for some $m < 0$ while all $a_{n<m} = 0$, then z_0 is classified as a *pole of order* $-m$ and the coefficient a_{-1} is called the *residue* of the pole. A *simple pole* has $m = -1$. If the function appears to have a singularity at z_0 but all a_m vanish for $m < 0$, z_0 is described as a *removable singularity* because the function can be made analytic simply by assigning a suitable value to $f[z_0]$. For example, $z = 0$ is a removable singularity of

$$f[z] = \frac{\text{Sin}[z]}{z} = \sum_{n=0}^{\infty} \frac{(-)^n}{(2n+1)!} z^{2n} \tag{1.227}$$

because with the assignment $f[0] = 1$ the function is continuous and its Laurent series reduces to a simple Taylor series.

If the Laurent expansion has nonvanishing coefficients for arbitrarily large negative $n \to -\infty$ and the inner radius vanishes, then it has an *essential singularity* at z_0. According to *Picard's theorem*, essential singularities have the nasty property that $f[z]$ takes any, hence all, values in any arbitrarily small neighborhood infinitely often with possibly one exception. For example,

$$e^{1/z} = \sum_{n=0}^{\infty} \frac{z^{-n}}{n!} \tag{1.228}$$

has an essential singularity at the origin. The equation $w == e^{1/z}$ is satisfied by

$$w == e^{1/z} \Longrightarrow z = \frac{1}{\text{Log}[w]} = (\text{Log}\,|w| + i\,\text{Arg}[w] + 2n\pi i)^{-1} \tag{1.229}$$

for any integer n. By choosing n sufficiently large, one can make $|z|$ as small as desired. Thus, although $e^{1/z} \neq 0$, the one exception, all other values of w are obtained infinitely often in a neighborhood of $z \to 0$, as expected from Picard's theorem.

Singularities in $f[z]$ at $z = \infty$ are classified according to the behavior of $f[\frac{1}{z}]$ at $z = 0$. Thus, e^z has an essential singularity at ∞, while z^{-n} is analytic at ∞ if n is a positive integer.

1.13.4 Poles and Residues

Although the Laurent coefficients are defined in terms of an integral, it is usually easier to compute the coefficients using a derivative formula similar to that for the Taylor series. If $f[z]$ is analytic near z_0 except for an isolated m-pole at z_0, we define an auxiliary function

$$\phi[z] = (z - z_0)^m f[z] = \sum_{n=-m}^{\infty} a_n (z - z_0)^{m+n} \tag{1.230}$$

that is analytic within $|z - z_0| < R$ where R is the radius of convergence for the Laurent series. The coefficients can then be obtained by differentiation, whereby

$$a_n = \frac{1}{(m+n)!}\left(\frac{d^{m+n}}{dz^{m+n}}\phi[z]\right)_{z=z_0} \tag{1.231}$$

This result can be written more succinctly as

$$a_n = \frac{\phi^{(m+n)}[z_0]}{(m+n)!} \tag{1.232}$$

where $\phi^{(k)}[z_0]$ denotes the kth derivative of $\phi[z]$ evaluated at z_0. This formula is similar to that for the Taylor series, except that f is replaced by ϕ and the index is shifted, and reduces to the Taylor coefficients for an analytic function with $m = 0$. However, this method is not useful at an essential singularity where $m = \infty$.

Often we require only the residue of f at z_0. For a simple pole we identify the residue as

$$m = 1 \Longrightarrow a_{-1} = \phi[z_0] = \lim_{z \to z_0}(z - z_0) f[z] \tag{1.233}$$

while for an m-pole one obtains

$$m > 1 \Longrightarrow a_{-1} = \frac{\phi^{(m-1)}[z_0]}{(m-1)!} \tag{1.234}$$

For example, consider

$$f[z] = \frac{z^n}{q[z]} = \frac{z^n}{az^2 + bz + c} \tag{1.235}$$

where we assume that $n \geq 0$, $a \neq 0$, and that a, b, c are real. (Other cases can be treated separately.) The two poles at the roots of the denominator

$$z_{\pm} = \frac{-b \pm \sqrt{b^2 - 4ac}}{2a} \tag{1.236}$$

are distinct unless the discriminant $b^2 - 4ac$ happens to vanish. We can then write

$$f[z] = \frac{a^{-1}z^n}{(z - z_1)(z - z_2)} \Longrightarrow \rho_i = \frac{a^{-1}z_i^n}{z_i - z_j} \tag{1.237}$$

where ρ_i is the residue at pole z_i and z_j is the other pole. If there happens to be a double pole, we find

$$z_1 = z_2 = z_0 \Longrightarrow a_{-1} = \left(\frac{d}{dz}(a^{-1}z^n)\right)_{z_0} = a^{-1}nz_0^{n-1} = a^{-1}n\left(\frac{-b}{2a}\right)^{n-1} \tag{1.238}$$

An important special case is provided by functions of the form

$$f[z] = \frac{p[z]}{q[z]} \quad \text{with} \quad q[z_0] = 0 \quad q'[z_0] \neq 0 \quad p[z_0] \neq 0 \tag{1.239}$$

where the simple pole at z_0 is a zero of $q[z]$. The Taylor expansion of $q[z]$ then takes the form

$$q[z] \approx q'[z_0](z - z_0) \tag{1.240}$$

such that the residue of $f[z]$ at z_0 becomes

$$a_{-1} = \frac{p[z_0]}{q'[z_0]} \tag{1.241}$$

For example, the function

$$f[z] = \frac{e^{az}}{e^z + 1} \Longrightarrow z_k = (2k + 1)\pi i, \quad R_k = -e^{az_k} \tag{1.242}$$

has poles at odd-integer multiples of πi with residues easily determined using $q'[z_k] = -1$.

1.14 Meromorphic Functions

1.14.1 Pole Expansion

If a function $f[z]$ only has isolated singularities, it is described as *meromorphic*. For simplicity suppose that these singularities are simple poles at z_n where the index lists the poles in order of increasing distance from the origin. The behavior near a simple pole can be represented by $z \approx z_n \Longrightarrow f[z] \approx \frac{b_n}{z - z_n}$. Thus, the function

$$g_n[z] = f[z] - \sum_{k=1}^{n} \frac{b_k}{z - z_k} \tag{1.243}$$

is analytic in a disk $|z| \le R_n$ where the radius R_n encloses n poles. According to the Cauchy integral formula, we may write

$$g_n[z] = \frac{1}{2\pi i} \oint_{C_n} \frac{g_n[s]}{s - z} ds = \frac{1}{2\pi i} \oint_{C_n} \frac{f[s]}{s - z} ds - \frac{1}{2\pi i} \sum_{k=1}^{n} b_k \oint_{C_n} \frac{ds}{(s - z)\left(s - z_k\right)} \tag{1.244}$$

where C_n is a circle, $|z| = R_n$, that encloses n poles without any poles being on the contour itself. For any $z \ne z_k$ we can use partial fractions to express the second contribution in a form

$$\oint_{C_n} \frac{ds}{(s - z)(s - z_k)} = \oint_{C_n} \frac{ds}{s - z_k} - \oint_{C_n} \frac{ds}{s - z} = 0 \tag{1.245}$$

where cancellation between equal residues is apparent. Furthermore, if $z = z_k$ we also find

$$\oint_{C_n} \frac{ds}{(s - z_k)^2} = 0 \tag{1.246}$$

and conclude that

$$g_n[z] = \frac{1}{2\pi i} \oint_{C_n} \frac{f[s]}{s - z} ds \tag{1.247}$$

for z within C_n.

Next let

$$M_n = \max\left[\left|f\left[R_n e^{i\theta}\right]\right|\right] \tag{1.248}$$

represent the largest modulus found on the circle C_n, such that

$$|g_n[z]| \le \frac{M_n R_n}{R_n - |z|} \tag{1.249}$$

bounds g_n. If f is bounded such that $R_n \to \infty$ with finite M_n, we can construct a sequence of g_n functions which are also bounded as $|z| \to \infty$. Thus, the function

$$g[z] = \lim_{n\to\infty} g_n[z] \tag{1.250}$$

is analytic and bounded in the entire complex plane. According to Liouville's theorem, such a function must be constant! Hence, we can write

$$f[z] = g_\infty + \sum_{k=1}^{\infty} \frac{b_k}{z - z_k} \tag{1.251}$$

and all that remains is to determine the value of the constant g_∞. Using

$$f[0] = g_\infty - \sum_{k=1}^{\infty} \frac{b_k}{z_k} \Longrightarrow g_\infty = f[0] + \sum_{k=1}^{\infty} \frac{b_k}{z_k} \tag{1.252}$$

we finally obtain the *Mittag–Leffler theorem.*

Theorem 17. *Mittag–Leffler theorem: Suppose that the function $f[z]$ is analytic everywhere except for isolated simple poles, is analytic at the origin, and that there exists a sequence of circles $\{C_k : |z| = R_k, k = 1, n\}$ where each C_k encloses k poles within radius R_k. Furthermore, assume that on these circles $|f|$ is bounded as $R_n \to \infty$. The function can then be expanded in the form*

$$f[z] = f[0] + \sum_{n=1}^{\infty} \left(\frac{b_n}{z - z_n} + \frac{b_n}{z_n}\right) \tag{1.253}$$

where b_n is the residue for pole z_n.

Unlike Laurent expansions for which the choice of z_0 can be somewhat arbitrary, the *pole expansion* for meromorphic functions depends only upon intrinsic properties of the function itself. Although the present version places significant restrictions on the function, generalizations can often be made fairly easily. For example, if $f[z]$ has a pole at the origin, one can apply the theorem to the closely related function $g[z] = f[z + z_0]$ where z_0 is any convenient point where $f[z]$ is analytic. Similarly, if $M_n \propto R_n^{m+1}$ for large R_n, one can employ an expansion of the form

$$f[z] = \sum_{k=1}^{m} f^{(k)}[0]\frac{z^k}{k!} + \sum_{n=1}^{\infty} \frac{b_n}{z - z_n}\left(\frac{z}{z_n}\right)^{m+1} \tag{1.254}$$

Poles of higher order can be accommodated also, but we forego detailed analysis here.

Pole expansions appear in many branches of physics. If $f[z]$ represents the response of a dynamical system to some driving force, the poles generally represent resonances or normal modes of vibration while the residues represent the coupling of the driving force to those normal modes. Pole expansions can also be used to sum infinite series.

1.14.2 Example: Tan[z]

The function Tan[z] has simple poles at $z_n = (n + \frac{1}{2})\pi$ with residue $b_n = -1$ for integer n, both positive and negative. Thus, circles C_n of radius $R_n = n\pi$ enclose $2n$ poles without singularities on the contours. One can show that $M_n \to 1$ as $n \to \infty$. Hence, Tan[z] fulfills all requirements for application of the Mittag–Leffler theorem. The pole expansion can now be expressed as

$$\text{Tan}[z] = -\sum_{n=-\infty}^{\infty} \left(\frac{1}{z - \left(n + \frac{1}{2}\right)\pi} - \frac{1}{\left(n + \frac{1}{2}\right)\pi} \right) \tag{1.255}$$

$$= -\sum_{n=0}^{\infty} \left(\frac{1}{z - \left(n + \frac{1}{2}\right)\pi} - \frac{1}{\left(n + \frac{1}{2}\right)\pi} + \frac{1}{z + \left(n + \frac{1}{2}\right)\pi} + \frac{1}{\left(n + \frac{1}{2}\right)\pi} \right) \tag{1.256}$$

such that

$$\text{Tan}[z] = \sum_{n=0}^{\infty} \frac{2z}{\left(\left(n + \frac{1}{2}\right)\pi\right)^2 - z^2} \tag{1.257}$$

With the substitution $z \to s\pi/2$, we obtain the *partial fraction representation*

$$\frac{\pi}{4s} \text{Tan}\left[\frac{s\pi}{2}\right] = \frac{1}{1 - s^2} + \frac{1}{9 - s^2} + \frac{1}{25 - s^2} + \cdots \tag{1.258}$$

Expressions of this type can often be used to sum infinite series. For example, from

$$\lim_{s \to 0} \frac{\pi}{4s} \text{Tan}\left[\frac{s\pi}{2}\right] = \frac{\pi^2}{8} \tag{1.259}$$

one immediately obtains

$$\sum_{k=0}^{\infty} \left(\frac{1}{2k + 1} \right)^2 = \frac{\pi^2}{8} \tag{1.260}$$

Then using

$$\sum_{k=1}^{\infty} \frac{1}{k^2} = 1 + \sum_{k=1}^{\infty} \left(\frac{1}{2k} \right)^2 + \sum_{k=1}^{\infty} \left(\frac{1}{2k + 1} \right)^2 \Longrightarrow \frac{3}{4} \sum_{k=1}^{\infty} \frac{1}{k^2} = \sum_{k=0}^{\infty} \left(\frac{1}{2k + 1} \right)^2 \tag{1.261}$$

we find

$$\sum_{k=1}^{\infty} \frac{1}{k^2} = \frac{\pi^2}{6} \tag{1.262}$$

1.14.3 Product Expansion

If $f[z]$ is an entire function, its logarithmic derivative $\phi[z] = f'[z]/f[z]$ is a meromorphic function with poles at the roots of $f[z]$. If these roots are simple, the corresponding poles in ϕ will also be simple. Near a simple root we express $f[z]$ in the form

$$z \approx z_n \Longrightarrow f[z] \approx (z - z_n) p_n[z] \Longrightarrow \phi[z] \approx \frac{1}{z - z_n} \tag{1.263}$$

where $p_n[z]$ is smooth and nonvanishing near z_n. Hence, the poles of the logarithmic derivative all have residue $b_n = 1$. Provided that ϕ is suitably bounded at ∞, we can now use the pole expansion of ϕ to write

$$\frac{d\mathrm{Log}[f]}{dz} = \phi_0 + \sum_{n=1}^{\infty} \left(\frac{1}{z - z_n} + \frac{1}{z_n} \right) \tag{1.264}$$

$$\Longrightarrow \mathrm{Log}[f[z]] - \mathrm{Log}[f[0]] = z\phi_0 + \sum_{n=1}^{\infty} \left(\mathrm{Log}[z - z_n] - \mathrm{Log}[-z_n] + \frac{z}{z_n} \right) \tag{1.265}$$

where $\phi_0 = f'[0]/f[0]$. Exponentiating and simplifying this expression, we obtain the *product expansion*

$$\frac{f[z]}{f[0]} = \mathrm{Exp}\left[z \frac{f'[0]}{f[0]} \right] \prod_{n=1}^{\infty} \left(1 - \frac{z}{z_n} \right) e^{z/z_n} \tag{1.266}$$

where the z_n are the roots of $f[z]$. Like the pole expansion of meromorphic functions, the product expansion of entire functions depends only upon intrinsic properties of the function. Expansions of this type are often useful in symbolic manipulations, but generally converge too slowly to be useful for numerical evaluations.

1.14.4 Example: Sin[z]

Although Sin[z] is an entire function with simple poles at $z_n = \pm n\pi$, we cannot employ the product expansion directly because ϕ_0 is not finite. However, this difficulty is easily circumvented by considering instead the function

$$f[z] = \frac{\mathrm{Sin}[z]}{z} \Longrightarrow \phi[z] = \mathrm{Cot}[z] - \frac{1}{z}, \quad \phi[0] = 0 \tag{1.267}$$

The positive and negative roots can be accommodated by using two products

$$\frac{\mathrm{Sin}[z]}{z} = \prod_{n=1}^{\infty} \left(1 - \frac{z}{n\pi} \right) e^{\frac{z}{n\pi}} \prod_{n=1}^{\infty} \left(1 + \frac{z}{n\pi} \right) e^{-\frac{z}{n\pi}} \tag{1.268}$$

and combining factors pairwise to obtain

$$\frac{\mathrm{Sin}[z]}{z} = \prod_{n=1}^{\infty} \left(1 - \left(\frac{z}{n\pi} \right)^2 \right) \Longrightarrow \mathrm{Sin}[z] = z \prod_{n=1}^{\infty} \left(1 - \left(\frac{z}{n\pi} \right)^2 \right) \tag{1.269}$$

This form displays all the roots of Sin[z] and is, in effect, a completely factored representation of its Taylor series.

Problems for Chapter 1

1. Complex number field

In mathematics, a *field* $\mathbb{F}$ is defined as a set containing at least two elements on which two binary operations, denoted addition (+) and multiplication (×) satisfy the following conditions:

a) completeness and uniqueness of addition: $\forall a, b \in \mathbb{F}, c = a + b \in \mathbb{F}$ is unique

b) commutative law of addition: $a + b = b + a$

c) associative law of addition: $(a + b) + c = a + (b + c)$

d) $a + c = b + c \Longrightarrow a = b$

e) existence of identity element for addition: $\forall a, b \in \mathbb{F}, \exists x \ni a + x = b \Longrightarrow \exists 0 \ni a + 0 = a$

f) completeness and uniqueness of multiplication: $\forall a, b \in \mathbb{F}, c = a \times b \in \mathbb{F}$ is unique

g) commutative law of multiplication: $a \times b = b \times a$

h) associative law of multiplication: $(a \times b) \times c = a \times (b \times c)$

i) $a \times c = b \times c \wedge c \neq 0 \Longrightarrow a = b$

j) existence of identity element for multiplication: $\forall a, b \in \mathbb{F}, \exists x \neq 0 \ni a \times x = b \Longrightarrow \exists 1 \ni a \times 1 = a$

k) distributive law: $a \times (b + c) = a \times b + a \times c$

The real numbers $\mathbb{R}$ obviously form a field with respect to ordinary addition and multiplication, but it is not immediately obvious that the complex numbers $\mathbb{C}$ form a field with respect to the extended definitions of addition and multiplication. To demonstrate that $\mathbb{C}$ is a field, you must identify the identity elements for addition and multiplication and must verify that each of the 11 conditions set forth above is satisfied.

2. Triangle inequalities

Prove the triangle inequalities: $||z_1| - |z_2|| \leq |z_1 \pm z_2| \leq |z_1| + |z_2|$.

3. Applications of de Moivre's theorem

Show that

$$\text{Cos}[n\theta] = \text{Cos}[\theta]^n - \binom{n}{2} \text{Cos}[\theta]^{n-2} \text{Sin}[\theta]^2 + \binom{n}{4} \text{Cos}[\theta]^{n-4} \text{Sin}[\theta]^4 + \cdots \tag{1.270}$$

$$\text{Sin}[n\theta] = \binom{n}{1} \text{Cos}[\theta]^{n-1} \text{Sin}[\theta] - \binom{n}{3} \text{Cos}[\theta]^{n-3} \text{Sin}[\theta]^3 + \cdots \tag{1.271}$$

4. Lagrange's trigonometric identity

Prove:

$$\sum_{k=0}^{n} \text{Cos}[k\theta] = \frac{1}{2} + \frac{\text{Sin}\left[\left(n + \frac{1}{2}\right)\theta\right]}{2\,\text{Sin}\left[\frac{\theta}{2}\right]} \tag{1.272}$$

Hint: first prove

$$\sum_{k=0}^{n} z^k = \frac{1 - z^{n+1}}{1 - z} \tag{1.273}$$

5. Quadratic formula

a) Prove that

$$az^2 + bz + c == 0 \Longrightarrow z = \frac{(b^2 - 4ac)^{1/2} - b}{2a} \tag{1.274}$$

applies even when a, b, c are complex. Why did we not use a $\pm$ sign in front of the square root?

b) Use the quadratic formula to determine all roots of the equation $\text{Sin}[z] == 2$. (Hint: $\text{Sin}[z] = \frac{1}{2i}(w - \frac{1}{w})$ where $w = e^{iz}$.)

6. Assorted trigonometric equations with complex solutions

Find all solutions to the following equations assuming that a, b are real numbers and that $|a| > 1$, $|b| > 1$. Express your results in the form $z = x + iy$ where x, y are real-valued expressions that do not involve trigonometric functions and be sure to consider all cases.

a) $\text{Cos}[z] == a$

b) $\text{Cos}[z] == bi$

7. Series RLC circuit

A circuit contains resistance R, inductance L, and capacitance C in series with a generator of electromotive force $\mathcal{E}[t] = \mathcal{E}_0 \text{Cos}[\omega t]$. Let $I[t]$ represent the current flowing in the circuit and $Q[t]$ the charge stored in the capacitor. It is useful to express the physical quantities

$$\mathcal{E}[t] = \text{Re}\left[\hat{\mathcal{E}} e^{i\omega t}\right], \quad I[t] = \text{Re}\left[\hat{I} e^{i\omega t}\right], \quad Q[t] = \text{Re}\left[\hat{Q} e^{i\omega t}\right] \tag{1.275}$$

in terms of complex phasors $\hat{\mathcal{E}}$, $\hat{I}$, and $\hat{Q}$ that represent both the magnitudes and relative phases for sinusoidal time dependencies.

a) Use Kirchhoff's laws to derive a phasor generalization of Ohm's law, $\hat{\mathcal{E}} = \hat{I}\hat{Z}$, where the impedance $\hat{Z} = Ze^{i\phi}$ is generally complex. Express the modulus, Z, and the phase, ϕ, of the complex impedance in terms of the real parameters of the circuit.

b) Show that the power averaged over a cycle is given by $\bar{P} = \frac{1}{2}\text{Re}[\hat{I}\hat{\mathcal{E}}^*]$ and evaluate this quantity in terms of real parameters. Show that $\bar{P}[\omega]$ exhibits a resonance and determine its position and full width at half maximum (FWHM). Sketch $\bar{P}[\omega]$ and $\phi[\omega]$ together.

8. Smith chart

The complex impedance $Z = R + iX$ for an AC circuit is decomposed into resistive and reactive components, R and X, where $R > 0$ and $-\infty < X < \infty$. Smith proposed a representation

$$W = u + iv = \frac{Z-1}{Z+1} \tag{1.276}$$

that maps the right half-plane for Z onto the unit disk for W. Determine the mappings for lines of constant R and lines of constant X. Sketch illustrative samples of each.

9. Bilinear mapping

Study the bilinear mapping

$$w = \frac{az+b}{cz+d}, \qquad ad - bc \neq 0 \tag{1.277}$$

by determining the images in the w-plane of representative lines and circles in the z-plane.

10. Component functions

Develop explicit expressions for the real and imaginary components, $u[x, y]$ and $v[x, y]$ for the following functions of $z = x + iy$.

a) $f[z] = (z^2 - 1)^{1/2}$

b) $g[z] = (z - 1)^{1/2}(z + 1)^{1/2}$

11. Inverse trigonometric functions

Prove:

$$\arcsin[z] = -i \log\left[iz + \left(1 - z^2\right)^{1/2}\right] \tag{1.278}$$

$$\arccos[z] = -i \log\left[z + \left(z^2 - 1\right)^{1/2}\right] \tag{1.279}$$

$$\arctan[z] = \frac{i}{2} \log\left[\frac{i+z}{i-z}\right] \tag{1.280}$$

This can be done by expressing equations of the form $z == \mathrm{Sin}[f]$ in exponential form, substituting $w = e^{if}$, solving for w, and deducing $f[z]$. Determine the branch cuts needed to specify the principal branch of each function.

12. An identity

Prove: $\mathrm{ArcTan}\left[\frac{2z}{z^2-1}\right] = 2\,\mathrm{ArcCot}[z]$

13. Principal value for an imaginary power

Suppose that

$$\eta = \left(\frac{ia-1}{ia+1}\right)^{ib} \tag{1.281}$$

where a, b are real.

a) Show that this quantity is real and find a simple expression for its principal value.

b) Determine the position and magnitude of any discontinuities.

14. Derivative wrt z^*

Show that a function $f[x, y]$ of two real variables can be expressed as a function $g[z, z^*]$ of the complex variable $z = x + iy$ and its complex conjugate $z^* = x - iy$. Then show that the requirement $\partial g/\partial z^* == 0$ is equivalent to the Cauchy–Riemann equations for the components of f and argue that an analytic function is truly a function of a single complex variable, instead of two real variables.

15. Analyticity of conjugate functions

Suppose that $f[z]$ is analytic in some region.

a) Under what conditions is $g[z] = f[z^*]$ analytic in the same region?

b) Under what conditions is $h[z] = f[z]^*$ analytic?

c) Under what conditions is $w[z] = f[z^*]^*$ analytic?

16. Completion of analytic functions

Which of the following functions $u[x, y]$ are the real parts of an analytic function $f[z]$ with $z = x + iy$? If $u[x, y] = \text{Re}\, f[z]$, determine $f[z]$.

a) $u = x^3 - y^3$

b) $u = x^2 - y^2 + y$

17. Analyticity for the sum, product, quotient, or composition of two functions

Suppose that $f_1[z] = u_1[x, y] + i\, v_1[x, y]$ and $f_2[z] = u_2[x, y] + iv_2[x, y]$ are analytic functions of $z = x + iy$. Show that $f_1 + f_2$, $f_1 f_2$, f_1/f_2, and $f_1[f_2[z]]$ are analytic functions under appropriate conditions by demonstrating consistency with the Cauchy–Riemann equations. Be sure to specify the requisite conditions for each case.

18. Equipotentials and streamlines for exponential function

Sketch the equipotentials $u[x, y]$ and streamlines $v[x, y]$ for $w = e^z$ where $z = x + iy$ and $w = u + iv$.

19. Equipotentials and streamlines for Tanh

Evaluate and sketch the equipotentials and streamlines for the hyperbolic tangent.

20. Cauchy–Riemann equations in polar form

Suppose that $z = x + iy = re^{i\theta}$ is expressed in polar form and let $f[z] = Re^{i\Theta}$ where $R[r, \theta]$ and $\Theta[r, \theta]$ are real functions of r and θ. Derive Cauchy–Riemann equations relating $\frac{\partial R}{\partial r}$ to $\frac{\partial \Theta}{\partial \theta}$ and $\frac{\partial R}{\partial \theta}$ to $\frac{\partial \Theta}{\partial r}$ for differentiable functions. (Hint: consider infinitesimal displacements dz_r and dz_θ in the $\hat{r}$ and $\hat{\theta}$ directions.)

21. Circular average of analytic function

Demonstrate that if $f[z]$ is analytic within the disk $|z - z_0| \leq R$ then the average value of $f[z_0 + re^{i\theta}]$ on any circle $|z - z_0| = r < R$ is equal to the value at its center, $f[z_0]$.

22. Maximum modulus principle

Prove that, if $f[z]$ is analytic and not constant within a region R, then $|f[z]|$ does not have a maximum within the interior of R. Hence, if f is analytic and not constant within R, $|f|$ must reach its maximum value on the boundary of R.

23. Extrema of harmonic functions

Suppose that $u[x, y]$ is harmonic and not constant within region R. Prove that $u[x, y]$ has no extrema (neither maximum nor minimum) within R; hence, its extrema must be found on the boundary of R. (Hint: apply the maximum modulus principle to $e^{f[z]}$ where f is analytic within R.)

24. Absence of extrema in $|f|$ for analytic functions

If $f[z]$ is analytic in domain D, demonstrate that $|f[z]|$ has no extrema in D. Hint: use the Taylor series representation to show that no neighborhood $|z - z_0| < r$ contains an extremum.

25. An application of the Cauchy integral formula

Suppose that $f[z]$ is analytic on and within the simple closed positive contour C. Evaluate the following integrals.

a) $\frac{1}{2\pi i} \oint_C \frac{t f[t]}{t^2 - z^2}\, dt$

b) $\frac{1}{2\pi i} \oint_C \frac{t^2 + z^2}{t^2 - z^2} f[t]\, dt$

26. Derivatives of analytic functions

Assume that $f[z]$ is analytic on and within a positive simple closed contour C that encloses z. Use induction to prove

$$f^{(n)}[z] = \lim_{\Delta z \to 0} \frac{f^{(n-1)}[z + \Delta z] - f^{(n-1)}[z]}{\Delta z} = \frac{n!}{2\pi i} \oint_C dw \frac{f[w]}{(w - z)^{n+1}} \tag{1.282}$$

where $f^{(n)}$ is the nth derivative of f.

27. Fundamental theorem of integral calculus

Prove the fundamental theorem of integral calculus: If $f[z]$ is analytic in a simply connected domain D that includes z_0 and z, then

$$F[z] = \int_{z_0}^{z} f[t]\, dt \tag{1.283}$$

is also analytic in D and $f[z] = dF[z]/dz$.

28. Poisson integral formula

a) Suppose that $f[z]$ is analytic within the disk $|z| = r \le a$. Prove

$$f[z] = \frac{1}{2\pi i} \oint_C \left(\frac{f[s]}{s - z} - \frac{f[s]}{s - \frac{a^2}{z^*}} \right) ds \tag{1.284}$$

where C is a circle of radius a centered on the origin. Then deduce the *Poisson integral formula*

$$f\left[re^{i\theta}\right] = \frac{a^2 - r^2}{2\pi} \int_0^{2\pi} \frac{f[ae^{i\phi}]}{a^2 + r^2 - 2ar\,\mathrm{Cos}[\phi - \theta]} d\phi \tag{1.285}$$

b) Suppose that we know the electrostatic potential ψ on the surface of a long cylinder as the real part of an analytic function (consider only two spatial dimensions). Obtain a general formula for the potential at any point within the cylinder. Compare the potential at the origin with the mean-value theorem. Note that, although a formal proof is not required, the Poisson integral formula can be applied for any function that is harmonic within C except for a finite number of jump discontinuities upon C.

c) As a specific illustration, compute the interior potential given that $\psi[ae^{i\phi}]$ has the constant value V for $\phi_1 \le \phi \le \phi_2$ and is zero on the rest of the cylinder. Display the angular dependence for a representative selection of r values for some choice of $\phi_2 - \phi_1$.

29. Uniform convergence of power series

Suppose that the power series $f_n[z] = \sum_{k=0}^{n} a_k z^k$ converges absolutely such that $f[z] = \lim_{n\to\infty} f_n[z]$ for $|z| < R$. Show that $f_n[z]$ converges uniformly in any subdisk $|z| \le B < R$.

30. Convergence of series representation for e^z

Demonstrate explicitly that the series $\sum_{k=0}^{\infty} z^k/k!$ is absolutely convergent for all z and that it is uniformly convergent in $|z| \le R$ for any finite R. Can one properly claim uniform convergence for all z?

31. Sharpened ratio test

a) The integral test can be used to established absolute convergence of the series representation of the Riemann zeta function

$$\zeta[z] = \sum_{n=1}^{\infty} n^{-z} \tag{1.286}$$

when $\mathrm{Re}[z] > 1$. Use the Weierstrass theorem to prove that $\zeta[z]$ is analytic for $\mathrm{Re}[z] > 1$. (This function has an important role in number theory and often appears in theoretical physics. It can be extended to most of the complex plane by analytic continuation, but that is beyond the scope of this problem.)

b) Use this result to obtain a sharpened form of the ratio test that states when the ratio of successive terms takes the form

$$\left|\frac{a_{n+1}}{a_n}\right| \simeq 1 - \frac{s}{n} \tag{1.287}$$

for large n, the series converges absolutely if $s > 1$.

c) Prove the existence of Euler's constant

$$\gamma = \lim_{n\to\infty}\left(\sum_{k=1}^{n} \frac{1}{k} - \mathrm{Log}[n]\right) \tag{1.288}$$

32. Laurent series

For each of the following functions, construct a complete set of Laurent series about the specified point and specify their convergence regions.

a) $f[z] = \frac{1}{(z-1)(z-2)}$ about $z = 0$

b) $f[z] = \frac{2z}{z^2-1}$ about $z = 2$

c) $f[z] = \frac{1}{(z^2-1)^{1/2}}$ about $z = 0$

d) $f[z] = \text{Sin}\left[z + \frac{1}{z}\right]$ about $z = 0$

33. Laurent expansion for $z^2 \, \text{Log}\left[\frac{z}{1-z}\right]$

a) Define a single-valued branch for

$$f[z] = z^2 \, \text{Log}\left[\frac{z}{1-z}\right] \tag{1.289}$$

and specify the region where your definition is real.

b) What is the nature of the singularity at infinity?

c) Construct a Laurent expansion for $|z| > 1$.

34. Some trigonometric series based upon a Laurent series

Evaluate the Laurent series for $(z-a)^{-1}$ where $-1 < a < 1$ in the region $|z| > a$. Then use $z \to e^{i\theta}$ to compute $\sum_{m=1}^{\infty} a^m \, \text{Cos}[m\theta]$ and $\sum_{m=1}^{\infty} a^m \, \text{Sin}[m\theta]$.

35. Grounded cylinder normal to uniform external field

a) Suppose that an infinitely long conducting cylinder of radius a is grounded. The cylinder is subjected to a uniform external electric field directed perpendicular to its symmetry axis. Use an analytic function to evaluate and sketch the equipotential surfaces and the net electric field. (Hint: expand $\Phi[z] = \phi[z] + i\psi[z]$ as a Laurent series around the origin and determine the coefficients using the appropriate boundary conditions.)

b) A two-dimensional incompressible fluid flows around an infinite cylinder whose axis is normal to the plane of motion. At large distances the velocity field is uniform. Evaluate and sketch the streamlines near the cylinder using an analytic function.

36. Isolated singularities

Classify the isolated singularities for each of the following functions. Be sure to consider the point at ∞, using $z \to 1/w$ with $w \to 0$.

a) $\dfrac{z^2}{1+z}$

b) $\dfrac{1 - \text{Cos}[z]}{z}$

c) $z e^{1/z}$

d) $\dfrac{e^{iz}}{z^2 + \Lambda^2}$

37. Singularity sequence
Identify and classify the singularities of

$$f[z] = \frac{1}{\text{Sin}[1/z]} \tag{1.290}$$

Is the singularity at the origin isolated? Is it a branch point?

38. Residues
Locate the poles for each of the following functions and evaluate their residues.

a) $\dfrac{z+1}{z^2(z+2i)}$
b) $\text{Tanh}[z]$
c) $\dfrac{e^z}{z^2+\pi^2}$
d) $\dfrac{1}{z^n\,(e^z-1)}$ (integer n)

39. Pole expansions
Develop pole expansions for the following functions, being sure to verify that the necessary conditions are satisfied.

a) $\text{Cot}[z]$
b) $\text{Csc}[z]$

40. Product expansions
Develop product expansions for the following functions, being sure to verify that the necessary conditions are satisfied.

a) $\text{Cos}[z]$
b) $\text{Sinh}[z]$

41. Product expansion for even functions
Suppose that $f[z]$ is entire and is even, such that $f[-z] = f[z]$, and that its roots are all simple. Also assume that, except for simple poles, its logarithmic derivative is bounded at infinity such that the product expansion of $f[z]$ converges.

a) Show that the product expansion can be expressed in the form

$$f[z] = f[0]\prod_{n=1}^{\infty}\left(1-\left(\frac{z}{z_n}\right)^2\right) \tag{1.291}$$

where $f[\pm z_n] = 0$ and where the product includes only one member of each pair of roots.

b) Show that

$$\frac{f''[0]}{f[0]} = -2\sum_{n=1}^{\infty}\frac{1}{z_n^2} \tag{1.292}$$

$$\frac{f^{(4)}[0]}{f[0]} = 3\left(\frac{f''[0]}{f[0]}\right)^2 - 12\sum_{n=1}^{\infty}\frac{1}{z_n^4} \tag{1.293}$$

c) Apply these results to $f[z] = \mathrm{Sin}[z]/z$ and evaluate the following sums:

$$\sum_{n=1}^{\infty} n^{-2} \quad \sum_{n=1}^{\infty} n^{-4} \tag{1.294}$$

42. Contour integration of logarithmic derivative

Suppose that $f[z]$ is analytic within a domain D containing the positive simple closed contour C. The function $\phi[z] = f'[z]/f[z]$ is known as the logarithmic derivative of f. Let

$$I = \frac{1}{2\pi i}\oint_C \phi[z]\,dz \tag{1.295}$$

a) Suppose that z_0 is the only zero of f within D and is of order m. Show that $I = m$ if C encloses z_0.

b) Evaluate I assuming that $f'[z] \neq 0$ in D and that C encloses N roots of f but that $f[z] \neq 0$ on C.

43. Argument principle

a) Suppose that $f[z]$ is analytic and nonzero on the positive simple closed contour C and that it is meromorphic in the domain D contained within C. The function $\phi[z] = f'[z]/f[z]$ is known as the logarithmic derivative of f. Prove that

$$\frac{1}{2\pi i}\oint_C \phi[z]\,dz = N_0 - N_p \tag{1.296}$$

where N_0 is the number of zeros and N_p is the number of poles in D where each accounts for multiplicity (e.g., a double root or double pole counts twice).

b) Show that

$$\oint_C \phi[z]\,dz = i\Delta_C \arg[f] = 2\pi i(N_0 - N_p) \tag{1.297}$$

measures the change in the argument of $f[z]$ as z moves around C. (Hint: consider the image $C \to \Gamma$ under the mapping $w = f[z]$.)

44. Rouchés theorem

Prove that if $f[z]$ and $g[z]$ are both analytic on and within the simple closed contour C and $|g[z]| < |f[z]|$ on C, then $f[z]$ and $f[z] + g[z]$ have the same number of zeros within C.

2 Integration

Abstract. Contour integration provides very powerful methods for evaluating integrals. We also consider several useful tricks that are more elementary but sometimes unfamiliar. Finally, integral representations are introduced for a variety of functions.

2.1 Introduction

Integration in closed form is rapidly becoming a lost art. Unlike differentiation, whose clear rules permit direct if tedious evaluation, integration often relies upon trial and error with many dead ends. In the heroic era of theoretical physics, cleverness in changing variables or choosing contours was revered, but now practically any integral that can be done in closed form may be found in standard compilations, such as Gradshteyn and Ryzhik, or in mathematical software, such as *MATHEMATICA*®. Although some skill in symbolic integration is still needed for traditional examinations, for most physicists its usefulness beyond the PhD qualifier examination is rather limited, unless you happen to be teaching a course that still relies on the methodology of the nineteenth century. Nevertheless, compilations are not entirely complete and software packages are not perfect. Furthermore, traditional integration methods remain useful for developing insightful approximations to integrals that cannot be evaluated fully in closed form. Numerical methods provide answers but limited insight. Therefore, in this chapter we briefly present some of the most useful symbolic techniques. We also discuss integral representations of analytic functions.

2.2 Good Tricks

Presumably integration by parts and variable transformations are too familiar to merit discussion here, but there are several other elementary methods that are quite valuable but with which students are generally less familiar.

2.2.1 Parametric Differentiation

Often when one integral is known, an entire family of related integrals can be developed by differentiating with respect to a parameter in the integrand. For example, given

$$I_0[\lambda] = \int_0^\infty e^{-\lambda x}\, dx = \frac{1}{\lambda} \tag{2.1}$$

the entire family

$$I_n[\lambda] = \int_0^\infty x^n e^{-\lambda x}\, dx = \left(-\frac{\partial}{\partial\lambda}\right)^n I_0[\lambda] = \frac{n!}{\lambda^{n+1}} \tag{2.2}$$

Graduate Mathematical Physics. James J. Kelly

ISBN: 3-527-40637-9

becomes available. If one needs to evaluate an integral, like $\int_{-\infty}^{\infty} x^{2n} e^{-x^2} \, dx$, that is presented without a parameter, simply insert one to obtain

$$\int_{-\infty}^{\infty} x^{2n} e^{-\lambda x^2} \, dx = \left(-\frac{\partial}{\partial \lambda} \right)^n \sqrt{\frac{\pi}{\lambda}} = \frac{(2n-1)!!}{2^n \lambda^{n+\frac{1}{2}}} \tag{2.3}$$

and then set $\lambda \to 1$. It might surprise you how often this trick is helpful.

Notice that if a parameter appears in the limits of integration, one must also include the variation of these limits using

$$\frac{\partial}{\partial \lambda} \int_{a[\lambda]}^{b[\lambda]} f[x, \lambda] \, d\lambda = \int_a^b \frac{\partial f[x, \lambda]}{\partial \lambda} \, dx + f[b, \lambda] \frac{\partial b}{\partial \lambda} - f[a, \lambda] \frac{\partial a}{\partial \lambda} \tag{2.4}$$

with the rhs evaluated for the appropriate λ.

2.2.2 Convergence Factors

Sometimes when it is not obvious whether the integral of an oscillatory function over an infinite range will converge to a definite value, application of a convergence factor may help resolve the question. For example, it is not obvious, at least to this author, whether $\int_0^{\infty} \mathrm{Sin}[kx] \, dx$ converges. Consider instead

$$\int_0^{\infty} e^{-\lambda x} \, \mathrm{Sin}[kx] \, dx = \mathrm{Im}\left[\int_0^{\infty} e^{-\lambda x} e^{ikx} \, dx \right] = \mathrm{Im}\left[\frac{1}{\lambda - ik} \right] = \frac{k}{\lambda^2 + k^2} \tag{2.5}$$

which does converge for $\lambda > 0$. The desired integral is then obtained from the limit $\lambda \to 0$, whereby

$$\int_0^{\infty} \mathrm{Sin}[kx] \, dx = \lim_{\lambda \to 0} \int_0^{\infty} e^{-\lambda x} \, \mathrm{Sin}[kx] \, dx = \frac{1}{k} \tag{2.6}$$

Admittedly, this result does appear somewhat arbitrary and some skepticism is justified. However, if this integral were encountered in a physics problem, it probably would arise from a limiting process anyway. Either a spatial or temporal variable should be limited to a finite range or a damping mechanism should be present that ensures convergence. One should then retreat a few steps in the derivation, identify the appropriate convergence factor, and evaluate the integral before that convergence factor is lost from view.

2.3 Contour Integration

2.3.1 Residue Theorem

Suppose that $f[z]$ is analytic throughout a domain D except for isolated singularities (poles) and that a simple closed contour C within D encircles poles $\{z_k, k = 1, N\}$ with residues R_k. By deforming the contour to encircle each pole, we obtain

$$\oint_C f[z] \, dz = \sum_{k=1}^{N} \oint_{C_k} f[z] \, dz \tag{2.7}$$

where each C_k is a small circle encompassing pole k. Near each z_k we can employ a Laurent expansion

$$f[z] = \sum_{n=-m_k}^{\infty} a_{n,k} (z - z_k)^n \Longrightarrow \oint_{C_k} f[z]\, dz = \sum_{n=-m_k}^{\infty} a_{n,k} \oint_{C_k} (z - z_k)^n\, dz \tag{2.8}$$

to evaluate its contribution to the contour integral. Using the now familiar circular contour integration with $z - z_k = \rho e^{i\theta} \Longrightarrow dz = \rho e^{i\theta} i\, d\theta$, we obtain

$$\begin{aligned} \oint_{C_k} (z - z_k)^n\, dz &= i\rho^{n+1} \int_0^{2\pi} e^{i(n+1)\theta}\, d\theta = 2\pi i \delta_{n,-1} \\ &\Longrightarrow \oint_{C_k} f[z]\, dz = 2\pi i a_{-1,k} = 2\pi i R_k \end{aligned} \tag{2.9}$$

Therefore, we obtain the *residue theorem*:

Theorem 18. *Residue theorem: If $f[z]$ is analytic on and within a simple closed counter-clockwise contour C, except for interior poles $\{z_k, k = 1, N\}$ with residues R_k, then*

$$\oint_C f[z]\, dz = 2\pi i \sum_{k=1}^{N} R_k$$

This is an amazingly powerful theorem that can be used, with clever choices of contour, to evaluate a wide variety of definite integrals which might be very difficult by means of familiar antidifferentiation methods. The trick is to find a simple closed contour that contains the desired integral on one portion of the path with easier integrals on the remainder of the path (if any). The examples in following subsections will demonstrate that contour integration using the residue theorem provides some of the most versatile methods for evaluating definite integrals.

2.3.2 Definite Integrals of the Form $\int_0^{2\pi} f[\sin\theta, \cos\theta]\, d\theta$

We assume that $f[\sin\theta, \cos\theta]$ can be represented by a single-valued function of $z = e^{i\theta}$ in the relevant region of the complex plane. Often f is a rational function of $\sin\theta$ and $\cos\theta$. Then we use

$$z = e^{i\theta} \Longrightarrow d\theta = -i\frac{dz}{z}, \quad \text{Sin}[\theta] = \frac{z - z^{-1}}{2i}, \quad \text{Cos}[\theta] = \frac{z + z^{-1}}{2} \tag{2.10}$$

such that

$$\begin{aligned} \int_0^{2\pi} f[\text{Sin}[\theta], \text{Cos}[\theta]]\, d\theta &= -i \oint f\left[\frac{z - z^{-1}}{2i}, \frac{z + z^{-1}}{2}\right] \frac{dz}{z} \\ &= 2\pi \sum \text{residues within unit circle} \end{aligned} \tag{2.11}$$

where the contour is the unit circle about the origin. Special handling is needed if any of the singularities of f are on the unit circle.

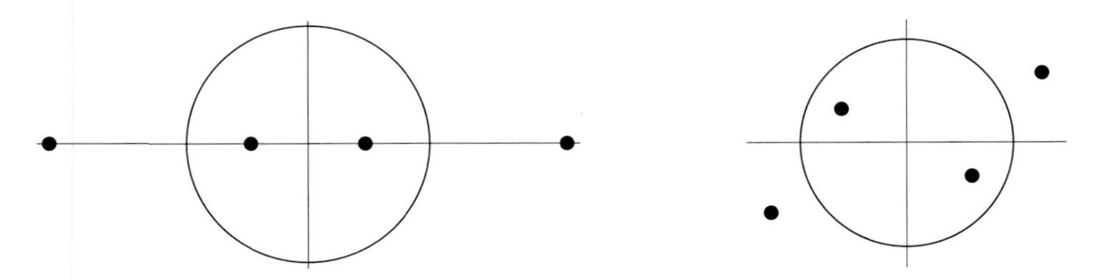

Figure 2.1. Poles in the integrand from Eq. (2.12). Left: Im[a] = 0. Right: Im[a] > 0.

2.3.2.1 Example: $I[a] = \int_0^\pi \frac{a}{a^2+\text{Sin}[\theta]^2}\, d\theta$

Although this integral is not stated in the desired form, the integrand is even. Hence, by direct application of the recipe above, we find

$$I[a] = \frac{1}{2}\int_0^{2\pi} \frac{a}{a^2 + \text{Sin}[\theta]^2}\, d\theta = -4ia \oint \frac{z}{4a^2z^2 - \left(z^2-1\right)^2}\, dz \tag{2.12}$$

The denominator is a quadratic in z^2, so that one obtains four poles at

```
denominator = 4a^2z^2 - (z^2 - 1)^2;
poles = z/. Solve[denominator == 0, z]
```

$$\left\{-a-\sqrt{1+a^2},\ a-\sqrt{1+a^2},\ -a+\sqrt{1+a^2},\ a+\sqrt{1+a^2}\right\}$$

Evaluation of the residues

```
Map[Residue[z/denominator, {z, #}] &, poles]
```

$$\left\{-\frac{1}{8a\sqrt{1+a^2}},\ \frac{1}{8a\sqrt{1+a^2}},\ \frac{1}{8a\sqrt{1+a^2}},\ -\frac{1}{8a\sqrt{1+a^2}}\right\}$$

is straightforward, even by hand. If $a > 0$ the pair of poles at $\pm\left(a-\sqrt{1+a^2}\right)$ is inside the unit circle while the other pair is outside, while if $a < 0$ the reverse is true. In either case we have two equal contributions, such that

$$I[a] = \frac{\pi}{\sqrt{1+a^2}} \tag{2.13}$$

Figure 2.1 illustrates the positions of the poles relative to the unit circle. The left figure uses a real value for a, while the right figure uses a complex value. Although one typically assumes that parameters in definite integrals are real, the method is more general and does not require that assumption.

Alternatively, if we use a trigonometric identity to express the integrand as

```
a/(a^2 + Sin[θ]^2) /. Sin[θ]^2 → (1 - Cos[2θ])/2 //Simplify
```

$$\frac{2a}{1+2a^2-\text{Cos}[2\theta]}$$

we obtain a contour integral with a quadratic denominator

$$I[a] = \int_0^{2\pi} \frac{a}{1 + 2a^2 - \text{Cos}[\theta]} \, d\theta = 2ia \oint \frac{1}{z^2 - 2(2a^2 + 1)z + 1} \, dz \tag{2.14}$$

for which evaluation of the residues is easier by hand.

```
denominator = z^2 - 2 (2a^2 + 1) z + 1;
poles = z/. Solve[denominator == 0, z]
```

$\left\{1 + 2a^2 - 2a\sqrt{1 + a^2}, 1 + 2a^2 + 2a\sqrt{1 + a^2}\right\}$

```
Map[Residue[1/denominator, {z, #}] &, poles]
```

$\left\{-\frac{1}{4a\sqrt{1 + a^2}}, \frac{1}{4a\sqrt{1 + a^2}}\right\}$

Notice that there are only two poles in the transformed function because the angular variable was replaced by $\theta \to \theta/2$. Recognizing that for either sign of a just one of the poles is within the unit circle, one obtains the same final result.

2.3.3 Definite Integrals of the Form $\int_{-\infty}^{\infty} f[x] \, dx$

We assume that $f[z]$ is analytic except for isolated singularities and vanishes faster than z^{-1} for $r \to \infty$ in either half-plane. With these conditions we can employ a semicircular contour of radius $R \to \infty$ closed in the appropriate half-plane to obtain

$$\int_{-\infty}^{\infty} f[x] \, dx = \oint f[z] \, dz = 2\pi i \sum \text{residues in half-plane} \tag{2.15}$$

To prove this result, suppose that $f[z]$ is bounded in the upper half-plane such that

$$\left|f\left[Re^{i\theta}\right]\right| \leq MR^{-\alpha} \Longrightarrow \left|\int_0^{\pi} f\left[Re^{i\theta}\right] d\theta\right| \leq MR^{-\alpha}\pi R \tag{2.16}$$

where M is a positive real number. Then

$$\begin{aligned} \alpha > 1 \Longrightarrow \lim_{R\to\infty} \left|\int_0^{\pi} f\left[Re^{i\theta}\right] d\theta\right| &\leq \lim_{R\to\infty} \pi M R^{1-\alpha} = 0 \\ \Longrightarrow \lim_{R\to\infty} \int_{-R}^{R} f[x] \, dx &= 2\pi i \sum \text{residues in half-plane} \end{aligned} \tag{2.17}$$

ensures that if f falls fast enough we need only evaluate the residues at isolated singularities of the analytic function $f[z]$ in the appropriate half-plane. Therefore, we find

$$\lim_{R\to\infty} R\left|f\left[Re^{i\theta}\right]\right| = 0 \Longrightarrow \lim_{R\to\infty} \int_{-R}^{R} f[x] \, dx = 2\pi i \sum \text{residues in half-plane} \tag{2.18}$$

using a *great semicircle*.

2.3.3.1 Example: $I[a] = \int_{-\infty}^{\infty} \frac{1}{1+x^4}\, dx$

The integral

$$\int_{-\infty}^{\infty} \frac{1}{1+x^4}\, dx = \oint_C \frac{1}{1+z^4}\, dz \tag{2.19}$$

can be evaluated using a semicircular contour in either half-plane. The integrand has poles at $z_k = \text{Exp}[i\pi \frac{2k+1}{4}]$ for $k = 1, 4$ of which two are found in each half-plane. The residues can be evaluated using

$$f[z] = \frac{1}{q[z]} \Longrightarrow R_k = \frac{1}{q'\left[z_k\right]} = \frac{1}{4z_k^3} = \frac{1}{4} e^{-3i\pi/4} i^k = -\frac{1+i}{4\sqrt{2}} i^k \tag{2.20}$$

Thus, we obtain

$$\int_{-\infty}^{\infty} \frac{1}{1+x^4}\, dx = \oint_C \frac{1}{1+z^4}\, dz = 2\pi i \left(-\frac{1+i}{4\sqrt{2}}\right)(1+i) = \frac{\pi}{\sqrt{2}} \tag{2.21}$$

2.3.4 Fourier Integrals

Consider a Fourier integral of the form

$$\tilde{f}[k] = \int_{-\infty}^{\infty} f[x] e^{ikx}\, dx \tag{2.22}$$

where k is a positive real number. Integrals of this type can often be evaluated by extending f to the complex plane and using a great semicircle in the upper half-plane, provided that the contribution of the return path

$$I_R[k] = iR \int_0^{\pi} f\left[Re^{i\theta}\right] \text{Exp}[-kR(\text{Sin}[\theta] - i\,\text{Cos}[\theta])]\, d\theta \tag{2.23}$$

vanishes in the limit $R \to \infty$. Suppose that M_R is the maximum modulus of f on this arc, such that

$$\left|I_R\right| \le RM_R \int_0^{\pi} \text{Exp}[-kR\,\text{Sin}[\theta]]\, d\theta \tag{2.24}$$

Dividing this latter result into two equal contributions now gives

$$\left|I_R\right| \le 2RM_R \int_0^{\pi/2} \text{Exp}[-kR\,\text{Sin}[\theta]]\, d\theta \tag{2.25}$$

Figure 2.2 illustrates that $\text{Sin}[\theta] > 2\theta/\pi$ on this interval. Thus, the integral on the great semicircle is limited by

$$\left|I_R\right| \le 2RM_R \int_0^{\pi/2} \text{Exp}[-2kR\theta/\pi]\, d\theta \tag{2.26}$$

or

$$\left|I_R\right| \le \pi M_R \frac{1 - e^{-kR}}{k} \tag{2.27}$$

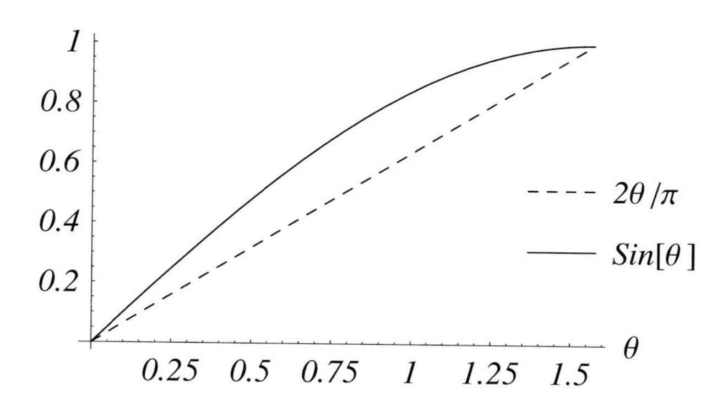

Figure 2.2.

This result is known as *Jordan's lemma*. Therefore, if $k > 0$ and if $f[z]$ vanishes on an infinite semicircle, such that

$$\lim_{R\to\infty} M_R = 0 \Longrightarrow |I_R| = 0 \tag{2.28}$$

the contribution of the return path vanishes and we may evaluate the Fourier integral using

$$\lim_{R\to\infty} \left|f\left[Re^{i\theta}\right]\right| = 0 \Longrightarrow \int_{-\infty}^{\infty} f[x]e^{ikx}\,dx = 2\pi i \sum \text{residues of integrand in half-plane} \tag{2.29}$$

If k happens to be a negative real number, we close in the lower half-plane instead and obtain the same result. Notice that this condition upon f is less restrictive than in the previous section due to the presence of the exponential factor, which is damping in the appropriate half-plane. However, convergence for $k = 0$ requires $\lim_{R\to\infty} R|f[Re^{i\theta}]| = 0$ as before.

2.3.4.1 Example: $\int_0^\infty \frac{\mathrm{Cos}[kx]}{x^2+a^2}\,dx$

Consider the integral

$$\tilde{f}[k] = \int_0^\infty \frac{\mathrm{Cos}[kx]}{x^2+a^2}\,dx \tag{2.30}$$

where k and a are positive real numbers. Although this integral is not presented in the desired form, it is simply half the real part of

$$\tilde{g}[k] = \int_{-\infty}^{\infty} \frac{\mathrm{Exp}[ikx]}{x^2+a^2}\,dx = \oint \frac{\mathrm{Exp}[ikz]}{z^2+a^2}\,dz \Longrightarrow \tilde{f}[k] = \frac{1}{2}\,\mathrm{Re}[\tilde{g}[k]] \tag{2.31}$$

and may be evaluated using a great semicircle in the upper half-plane, wherein lies one simple pole at $z = ia$. Therefore, we obtain

$$\tilde{g}[k] = 2\pi i \frac{\mathrm{Exp}[-ka]}{2ia} \Longrightarrow \tilde{f}[k] = \frac{\pi e^{-ka}}{2a} \tag{2.32}$$

without further ado. The result is actually more general than this derivation – it applies equally well for complex a provided only that $\mathrm{Re}[a] \neq 0$ to ensure that the poles are not on the real axis. Be alert to generalizations! Having expended some effort to obtain a result, it is good practice to extend it to the most general conditions possible. Also notice that one cannot use $\mathrm{Cos}[az]/(z^2+a^2)$ directly because $\mathrm{Cos}[az]$ is not bounded on the great semicircle – it diverges exponentially for large $\mathrm{Im}[z]$ in both upper and lower half-planes.

2.3.5 Custom Contours

Sometimes it is necessary to design a contour which exploits specific characteristics of the integrand. For example, when previously evaluating the integral $\int_0^\infty \mathrm{Cos}[x^2]\,dx$ we employed an arc subtending $\pi/4$ radians. Unfortunately, there are no general rules to guide one toward the optimum contour for an arbitrary integrand; one must rely on intuition and experience to minimize the amount of trial-and-error in choosing such contours. Below we give just one more example of a custom contour.

Consider the integral

$$I=\int_{-\infty}^{\infty}\frac{e^{ax}}{e^x+1}\,dx \quad 0<a<1 \tag{2.33}$$

The integral on the real axis converges because the integrand is of order e^{ax} for $x\to-\infty$ or $\mathrm{Exp}[(a-1)x]$ for $x\to\infty$ and decreases exponentially in either limit when $0<a<1$. However, the function

$$f[z]=\frac{e^{az}}{e^z+1}\Longrightarrow z_k=(2k+1)\pi i \tag{2.34}$$

with simple poles on the imaginary axis at odd-integer multiples of π does not satisfy the conditions needed to employ a great semicircle. Fortunately, the contour integral

$$\oint f[z]\,dz=\lim_{R\to\infty}\left(\int_{-R}^{R}f[(x,0)]\,dx+\int_0^{2\pi}f[(R,y)]\,dy+\int_R^{-R}f[(x,2\pi)]\,dx\right.$$
$$\left.+\int_{2\pi}^{0}f[(-R,y)]\,dy\right) \tag{2.35}$$

can be evaluated fairly easily using the rectangular strip, $(-\infty<x<\infty, 0\le y\le 2\pi)$, shown in Fig. 2.3. This contour encloses a single pole at $z_0=i\pi$ with residue $-e^{i\pi a}$, such that

$$\oint f[z]\,dz=-2\pi i e^{i\pi a} \tag{2.36}$$

The contributions from vertical segments

$$\lim_{R\to\infty}f[(R,y)]=\lim_{R\to\infty}\frac{\mathrm{Exp}[a(R+iy)]}{\mathrm{Exp}[R+iy]+1}=\lim_{R\to\infty}\mathrm{Exp}[(a-1)R]=0 \tag{2.37}$$

$$\lim_{R\to\infty}f[(-R,y)]=\lim_{R\to\infty}\frac{\mathrm{Exp}[a(-R+iy)]}{\mathrm{Exp}[-R+iy]+1}=\lim_{R\to\infty}\mathrm{Exp}[-aR]=0 \tag{2.38}$$

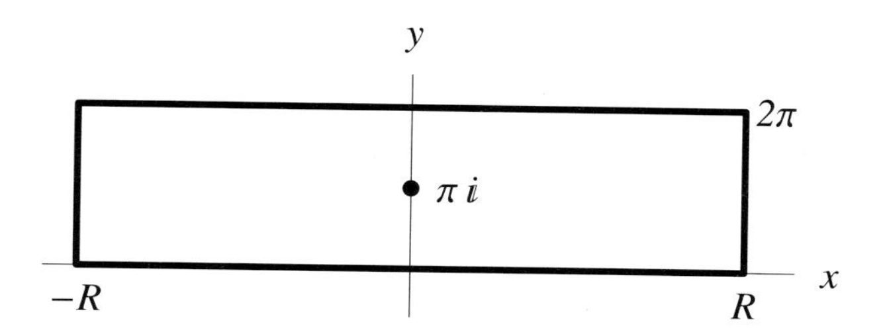

Figure 2.3. Rectangular contour used for Eq. (2.33).

vanish in the limit $R \to \infty$, while the horizontal segments are related by

$$f[(x, 2\pi)] = \frac{\text{Exp}[a(x + 2\pi i)]}{\text{Exp}[x + 2\pi i] + 1} = \text{Exp}[2\pi i a] f[(x, 0)] \tag{2.39}$$

such that

$$\oint f[z]\, dz = \left(1 - e^{2\pi i a}\right) I \Longrightarrow I = \frac{\pi}{\text{Sin}[\pi a]} \tag{2.40}$$

This result actually finds somewhat broader applicability because the contributions from the vertical segments vanish provided only that $0 < \text{Re}[a] < 1$. Therefore, we can allow a to be complex and find

$$\int_{-\infty}^{\infty} \frac{e^{ax}}{e^x + 1}\, dx = \frac{\pi}{\text{Sin}[\pi a]} \quad \text{for} \quad 0 < \text{Re}[a] < 1 \tag{2.41}$$

As always, be alert for possible generalizations.

2.4 Isolated Singularities on the Contour

2.4.1 Removable Singularity

Often one encounters isolated singularities on the integration path. For example, the integral

$$I = \int_{-\infty}^{\infty} \frac{\text{Sin}[x]}{x}\, dx \tag{2.42}$$

is important in Fourier analysis. The integrand has a removable singularity at the origin, but that would not cause any difficulty if the integral remained in this form. However, because $\text{Sin}[z]$ is divergent as $y \to \infty$, we would prefer to evaluate

$$I = \text{Im} \int_{-\infty}^{\infty} \frac{e^{ix}}{x}\, dx \tag{2.43}$$

by closing the contour in the upper half-plane. The penalty for this transformation is that the singularity at the origin is no longer removable. Fortunately, that problem can

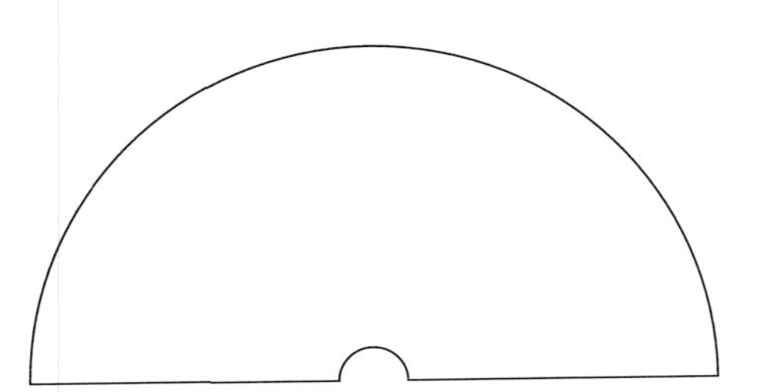

Figure 2.4. Great semicircle with small detour that excludes a removable singularity at the origin.

be avoided by also making a small semicircular detour around the origin as sketched in Fig. 2.4. According to Cauchy's theorem, the integral

$$\oint \frac{e^{iz}}{z}\,dz = 0 \tag{2.44}$$

vanishes on this contour because no singularities are enclosed. The contribution of the great semicircle vanishes because the integrand satisfies the requirements of Jordan's lemma. The linear segments

$$\lim_{\varepsilon\to 0}\left(\int_{-\infty}^{-\varepsilon} f[x]\,dx + \int_{\varepsilon}^{\infty} f[x]\,dx\right) \tag{2.45}$$

converge to the desired integral in the limit $\varepsilon \to 0$ because the integrand is well-behaved on the real axis. The contribution of the small semicircle of radius $\varepsilon \to 0$ is evaluated using

$$dz = iz\,d\theta \Longrightarrow i\int_{\pi}^{0} f[z]z\,d\theta = i\lim_{\varepsilon\to 0}\int_{\pi}^{0} \mathrm{Exp}\left[i\varepsilon e^{i\theta}\right]d\theta = -i\pi \tag{2.46}$$

Therefore, we find

$$\int_{-\infty}^{\infty} \frac{\mathrm{Sin}[x]}{x}\,dx = \pi \tag{2.47}$$

Notice that the contribution of the semicircular detour is πi times the residue of the integrand, half the value we would have obtained from the residue theorem for a complete circle around the singularity. More generally, if $f[z]$ is analytic at z_0, the detour integral

$$z - z_0 = \varepsilon e^{i\theta} \Longrightarrow \lim_{\varepsilon\to 0}\int_{\theta}^{\theta+\Delta\theta} \frac{f\left[z_0 + \varepsilon e^{i\theta}\right]}{z - z_0}\,dz = f\left[z_0\right] i\Delta\theta \tag{2.48}$$

is proportional to the angle subtended. This result is easily proven by expanding $f[z]$ around z_0. Also, notice that $\Delta\theta$ is positive for counterclockwise or negative for clockwise detours.

2.4.2 Cauchy Principal Value

An improper integral whose integrand is singular at one of the endpoints of the integration range is defined by one of the limits

$$\int_a^b f[x]\,dx = \lim_{\varepsilon\to 0}\int_{a+\varepsilon}^b f[x]\,dx \quad \text{or} \quad \int_a^b f[x]\,dx = \lim_{\varepsilon\to 0}\int_a^{b-\varepsilon} f[x]\,dx \tag{2.49}$$

However, if an isolated singularity lies within the range of integration then two limits are needed

$$\int_a^c f[x]\,dx = \lim_{\varepsilon_1\to 0}\int_a^{b-\varepsilon_1} f[x]\,dx + \lim_{\varepsilon_2\to 0}\int_{b+\varepsilon_2}^c f[x]\,dx \tag{2.50}$$

and often there will be no unique value if the two limits are taken independently. For example, applying this method to

$$\int_{-1}^1 \frac{dx}{x} = \lim_{\varepsilon_1\to 0}\int_{-1}^{-\varepsilon_1}\frac{dx}{x} + \lim_{\varepsilon_2\to 0}\int_{\varepsilon_2}^1\frac{dx}{x} = \lim_{\varepsilon_1\to 0,\varepsilon_2\to 0}\log\left[\frac{\varepsilon_1}{\varepsilon_2}\right] \tag{2.51}$$

does not provide an unambiguous result unless one decides to approach the singularity in a symmetric manner, such that $\varepsilon_1 = \varepsilon_2$. This particular value is designated the *Cauchy principal value* of the integral and denoted by $\mathcal{P}\int$, such that

$$\mathcal{P}\int_a^c f[x]\,dx = \lim_{\varepsilon\to 0}\left(\int_a^{b-\varepsilon} f[x]\,dx + \int_{b+\varepsilon}^c f[x]\,dx\right) \tag{2.52}$$

Hence, for the example above we find

$$\mathcal{P}\int_{-1}^1 \frac{dx}{x} = 0 \tag{2.53}$$

If there are several isolated singularities on the contour, the Cauchy principal value treats each symmetrically.

The Cauchy principal value is based upon antisymmetric behavior near a simple pole, where we can write

$$f[z] = \frac{a_{-1}}{z - z_0} + g[z] \tag{2.54}$$

with $g[z]$ analytic near z_0. As illustrated by Fig. 2.5, the two divergent contributions on either side cancel, leaving behind the background contribution g. Of course, this cancellation does not work for a double pole with symmetric divergence.

2.4.2.1 Example: $\mathcal{P}\int_{-\infty}^{\infty}\frac{\text{Cos}[kx]}{a^2-x^2}\,dx$

Consider the integral

$$I = \mathcal{P}\int_{-\infty}^{\infty}\frac{\text{Cos}[kx]}{a^2-x^2}\,dx = \text{Re}\left[\mathcal{P}\int_{-\infty}^{\infty}\frac{\text{Exp}[ikx]}{a^2-x^2}\,dx\right] \tag{2.55}$$

where a and k are positive real numbers. By replacing Cos with Exp we are able to close the contour in the upper half-plane and use Jordan's lemma to discard the contribution

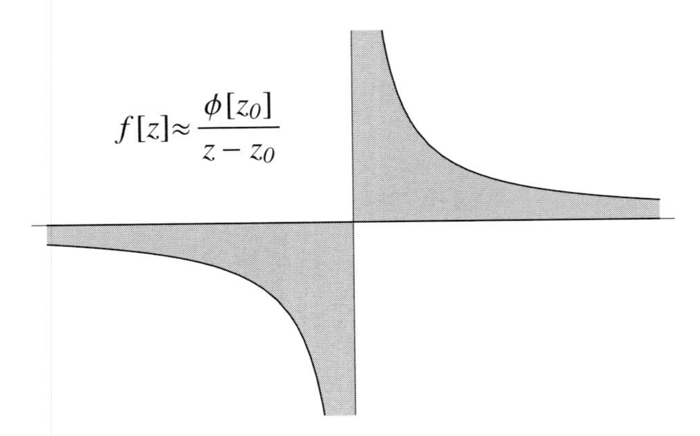

Figure 2.5. Asymmetric behavior of the divergent parts of an integrand with a simple pole.

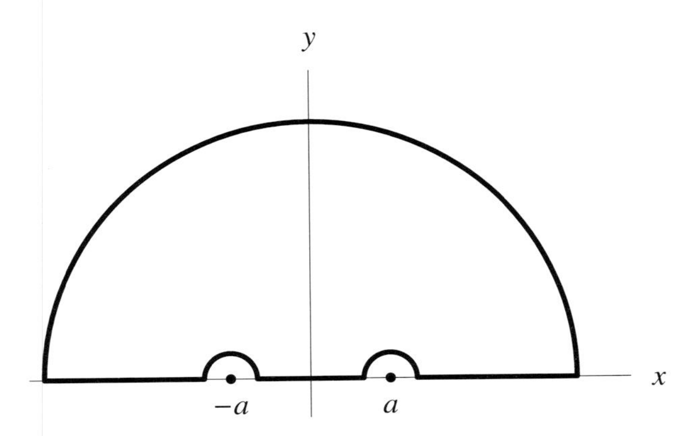

Figure 2.6. Kilroy contour: indentations that exclude poles at $\pm a$ resemble the eyes of someone peeking over a fence.

of the great semicircle. The singularities at $x = \pm a$ are avoided using small semicircular indentations, as sketched in Fig. 2.6. (Those familiar with World War II iconography might call this the *Kilroy contour*.)

The contribution from the segments along the real axis sum to the principal value, while the indentations contribute $-i\pi$ times the sum of the two residues because the semicircles are traversed in a negative sense. Thus,

$$0 = \oint_C \frac{\text{Exp}[ikz]}{a^2 - z^2}\, dz = \mathcal{P}\int_{-\infty}^{\infty} \frac{\text{Exp}[ikx]}{a^2 - x^2}\, dx - i\pi\left(\frac{e^{ika}}{-2a} + \frac{e^{-ika}}{2a}\right) \tag{2.56}$$

gives

$$\mathcal{P}\int_{-\infty}^{\infty} \frac{\text{Exp}[ikx]}{a^2 - x^2}\, dx = \frac{\pi}{a}\,\text{Sin}[ka] \Longrightarrow \mathcal{P}\int_{-\infty}^{\infty} \frac{\text{Cos}[kx]}{a^2 - x^2}\, dx = \frac{\pi}{a}\,\text{Sin}[ka] \tag{2.57}$$

Notice that the principal value is the same whether we choose indentations to exclude or to include either or both poles (four possibilities) – check this for yourself! Thus, the contribution to a principal value integral made by a simple pole on the contour is half the value it would have made if enclosed.

2.5 Integration Around a Branch Point

If the integrand involves a multivalued function, care must be taken with any branch cuts in the integration region. Such functions often differ in phase on opposite sides of the cut, so that contours with segments on opposite sides will include contributions of equal magnitude but different phase. Often the presence of a branch cut actually assists in evaluation of an integral – branch cuts are not always monsters to be feared! While it is difficult to formulate general rules, a couple of examples should suffice to illustrate the method.

Suppose that we wish to evaluate the integral

$$\int_0^\infty \frac{x^a}{x+b}\,dx \quad \text{with} \quad -1 < a < 0, \quad b > 0 \tag{2.58}$$

using contour integration. The function

$$f[z] = \frac{z^a}{z+b} \tag{2.59}$$

has a pole at $z = -b$ and requires a branch cut to define the phase of the numerator when a is nonintegral. This branch cut will connect singularities at $z = 0$ and $z \to \infty$, but we have some freedom in choosing its orientation. Although one usually cuts such a function along the negative real axis, for this purpose it is more convenient to put the cut on the positive real axis so that the desired integral is found on one segment of the contour sketched in Fig. 2.7. For now we assume that $b > 0$ so that the pole is on the negative real axis. Thus, for this problem we will cut just below the positive real axis and define the principal branch as

$$z^a = |z|^a \operatorname{Exp}[ia \arg[z]] \quad \text{with} \quad 0 \le \arg[z] < 2\pi \tag{2.60}$$

by restricting the phase of z to the range $(0, 2\pi)$. The residue of the pole is then $b^a \operatorname{Exp}[i\pi a]$ such that

$$\oint_C f[z]\,dz = 2\pi i b^a \operatorname{Exp}[i\pi a] \tag{2.61}$$

on the positive contour below, which encloses the pole without crossing the branch cut.

The contribution of the outer circle vanishes in the limit $R \to \infty$ because

$$|z| \to \infty \Longrightarrow \left|f[z]\right| \longrightarrow |z|^{a-1} \tag{2.62}$$

falls more rapidly than $|z|^{-1}$ when $a < 0$. Similarly, the integral on the inner circle with radius ε vanishes in the limit $\varepsilon \to 0$. The contributions on either side of the real axis are

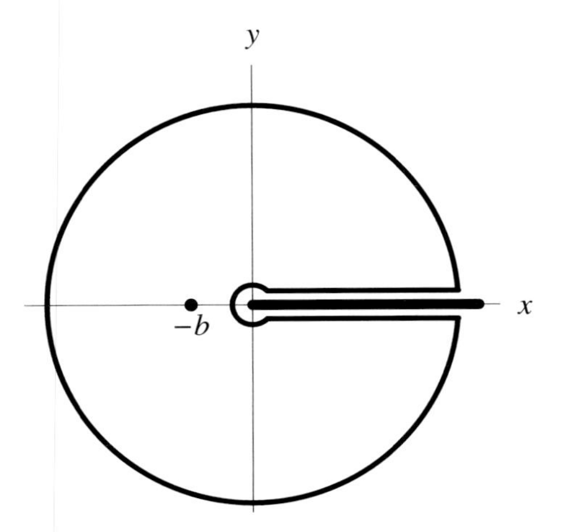

Figure 2.7. Contour used to evaluate $\int_0^\infty \frac{x^a}{x+b}\, dx$ for positive real b.

both proportional to the desired integral but differ in phase. On the top side of the cut $z^a \to x^a$ while on the bottom side $z^a \to x^a \,\mathrm{Exp}[2\pi i a]$, such that

$$(1 - \mathrm{Exp}[2\pi i a]) \int_0^\infty \frac{x^a}{x+b}\, dx = 2\pi i b^a \,\mathrm{Exp}[i\pi a] \tag{2.63}$$

Therefore, we obtain the integral

$$\int_0^\infty \frac{x^a}{x+b}\, dx = -\frac{b^a \pi}{\mathrm{Sin}[\pi a]} \quad \text{for} \quad -1 < a < 0, \quad b > 0 \tag{2.64}$$

This result can be generalized by recognizing that we need only require $-1 < \mathrm{Re}[a] < 0$ to ensure that the contribution of the outer circle vanishes as its radius becomes infinite. The definition of $b^a = \mathrm{Exp}[a\mathrm{Log}[b]] = \mathrm{Exp}[a(\mathrm{Log}[|b|] + i\mathrm{Arg}[b])]$ can be extended to complex powers. Therefore, we obtain the more general result

$$\int_0^\infty \frac{x^a}{x+b}\, dx = -\frac{b^a \pi}{\mathrm{Sin}[\pi a]} \quad \text{for} \quad -1 < \mathrm{Re}[a] < 0 \quad \text{and} \quad \mathrm{Arg}[b] \neq \pi \tag{2.65}$$

without extra work. However, this result does not apply if $\mathrm{Arg}[b] = \pi$, because then there would be a pole on the contour. To handle that situation, we employ the contour shown in Fig. 2.8 for which the segments on either side of the positive real axis are proportional to the principal value of the desired integral.

This contour encloses no poles, so that

$$b < 0 \Longrightarrow \oint_C \frac{z^a}{z+b}\, dz = 0 \tag{2.66}$$

The integral around the great circle vanishes, as before, but we now have two small semi-circular indentations to evaluate. Using $z + b = \varepsilon e^{i\theta} \Longrightarrow dz = i\varepsilon e^{i\theta}\, d\theta$, both contributions

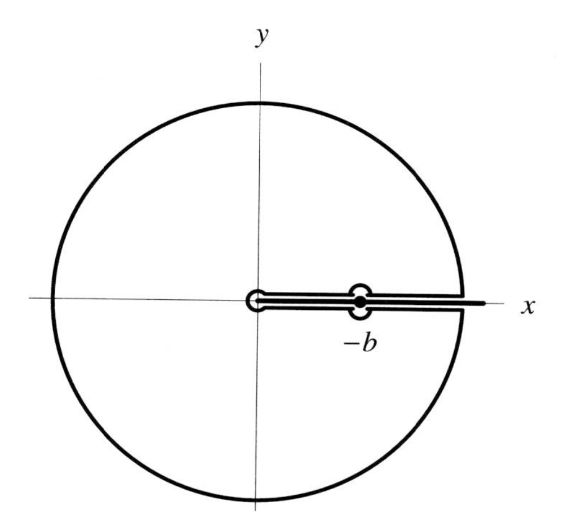

Figure 2.8. Contour used to evaluate $\int_0^\infty \frac{x^a}{x+b}\,dx$ for negative real b.

take the form

$$\lim_{\varepsilon\to 0}\int_{\theta_1}^{\theta_2} \frac{\left(-b+\varepsilon e^{i\theta}\right)^a}{\varepsilon e^{i\theta}} i\varepsilon e^{i\theta}\, d\theta = i(-b)^a e^{ia\phi}\left(\theta_2-\theta_1\right) = i(-b)^a e^{ia\phi}\Delta\theta \tag{2.67}$$

where on the upper semicircle $\phi = 0$ and $\theta_1 = \pi$, $\theta_2 = 0 \Longrightarrow \Delta\theta = -\pi$ while on the lower semicircle we must choose $\phi = 2\pi$ and $\theta_1 = 2\pi$, $\theta_2 = \pi \Longrightarrow \Delta\theta = -\pi$ for the selected branch of z^a. The integrals along the real axis on either side of the branch cut are both principal-value integrals with different phase factors because on the upper side $z^a = x^a$ while on the lower side $z^a = x^a e^{2\pi i a}$. Thus, we obtain

$$\left(1 - e^{2\pi i a}\right)\mathcal{P}\int_0^\infty \frac{x^a}{x+b}\,dx = i\pi(-b)^a\left(1+e^{2\pi i a}\right) \tag{2.68}$$

and conclude

$$b<0 \Longrightarrow \mathcal{P}\int_0^\infty \frac{x^a}{x+b}\,dx = -(-b)^a\pi\mathrm{Cot}[\pi a] \tag{2.69}$$

Notice that the cut does not diminish the influence of the pole – choosing the cut to run over the pole does not mask it; we could have placed the cut somewhere else if that had been more convenient.

2.6 Reduction to Tabulated Integrals

Some integrals that cannot be evaluated in terms of elementary functions occur sufficiently frequently to merit naming as special functions. Among the most useful are the *gamma function*

$$\Gamma[x] = \int_0^\infty t^{x-1}e^{-t}\,dt\,, \quad x>0 \tag{2.70}$$

and its cousins the *incomplete gamma function*

$$\Gamma[x, a] = \int_a^\infty t^{x-1} e^{-t}\, dt\,, \quad x > 0\,, \quad a \geq 0 \tag{2.71}$$

and the *beta function*

$$B[r, s] = \frac{\Gamma[r]\Gamma[s]}{\Gamma[r+s]} = \int_0^1 t^{r-1}(1-t)^{s-1}\, dt \tag{2.72}$$

Also important are the *sine* and *cosine integrals*

$$\mathrm{Si}[x] = \int_0^x \frac{dt}{t} \mathrm{Sin}[t] \quad \mathrm{Ci}[x] = -\int_x^\infty \frac{dt}{t} \mathrm{Cos}[t] \tag{2.73}$$

and the *exponential integrals*

$$E_n[x] = \int_1^\infty \frac{dt}{t^n} e^{-xt} \quad \mathrm{Ei}[x] = \mathcal{P} \int_{-\infty}^x \frac{dt}{t} e^t \tag{2.74}$$

The *error function* Erf[x] and *complementary error function* Erfc[x] = 1 − Erf[x] are defined by

$$\mathrm{Erf}[x] = \frac{2}{\sqrt{\pi}} \int_0^x dt\, e^{-t^2} \quad \mathrm{Erfc}[x] = \frac{2}{\sqrt{\pi}} \int_x^\infty dt\, e^{-t^2} \tag{2.75}$$

while their trigonometric cousins are the *Fresnel sine* and *cosine functions*

$$S[x] = \int_0^x dt\, \mathrm{Sin}\left[\frac{\pi t^2}{2}\right] \quad C[x] = \int_0^x dt\, \mathrm{Cos}\left[\frac{\pi t^2}{2}\right] \tag{2.76}$$

Once an integral has been reduced to one of these special functions, it can be considered done because these functions have been studied and tabulated extensively and are also available in many mathematical software packages.

2.6.1 Example: $\int_{-\infty}^{\infty} e^{-x^4}\, dx$

The integral

$$I = \int_{-\infty}^\infty e^{-x^4}\, dx = 2\int_0^\infty e^{-x^4}\, dx \tag{2.77}$$

can be expressed in terms of a gamma function using the variable transformation

$$y = x^4 \quad dy = 4x^3\, dx \Longrightarrow dx = \frac{1}{4} y^{-3/4}\, dy \tag{2.78}$$

whereby

$$I = \frac{1}{2} \int_0^\infty y^{-3/4} e^{-y}\, dy = \frac{1}{2} \Gamma\left[\frac{1}{4}\right] \tag{2.79}$$

is obtained directly.

2.6.2 Example: The Beta Function

A surprisingly wide variety of integrals can be expressed in terms of the *beta function*

$$B[p, q] = \frac{\Gamma[p]\Gamma[q]}{\Gamma[p+q]} \tag{2.80}$$

where p, q are generally complex. For the present purposes it will be sufficient to assume that p, q are nonnegative real numbers so that we can employ the simplest integral representation

$$\Gamma[p] = \int_0^\infty e^{-u} u^{p-1}\, du \tag{2.81}$$

for the more familiar gamma function. The substitution $u \to x^2$ provides

$$\Gamma[p] = \int_0^\infty e^{-u} u^{p-1}\, du = 2\int_0^\infty e^{-x^2} x^{2p-1}\, dx \tag{2.82}$$

such that

$$\begin{aligned} \Gamma[p]\Gamma[q] &= 4\int_0^\infty \int_0^\infty e^{-(x^2+y^2)} x^{2p-1} y^{2q-1}\, dx\, dy \\ &= 4\int_0^{\pi/2} \int_0^\infty e^{-r^2} \operatorname{Cos}[\theta]^{2p-1} \operatorname{Sin}[\theta]^{2q-1} r^{2p+2q+1}\, dr\, d\theta \end{aligned} \tag{2.83}$$

can be represented as integral over the first quadrant in polar coordinates. The radial integral can be performed in terms of the gamma function

$$\int_0^\infty e^{-r^2} r^{2p+2q+1}\, dr = \frac{1}{2}\Gamma[p+q] \tag{2.84}$$

such that

$$\int_0^{\pi/2} \operatorname{Cos}[\theta]^{2p-1} \operatorname{Sin}[\theta]^{2q-1}\, d\theta = \frac{\Gamma[p]\Gamma[q]}{2\Gamma[p+q]} = \frac{1}{2}B[p, q] \tag{2.85}$$

is expressed in terms of the beta function. The left-hand side would probably appear formidable to the uninitiated! We will find both gamma and beta functions very useful as the course progresses. Several additional variations are developed in the exercises.

2.6.3 Example: $\int_0^\infty \frac{\omega^n}{e^{\beta\omega}-1}\, d\omega$

Integrals of this form appear in the statistical mechanics of Bose systems. Here we assume that $\beta > 0$ and that n is a nonnegative integer. The integrand can be expanded as a power series in the small quantity $e^{-\beta\omega}$, such that

$$\int_0^\infty \frac{\omega^n}{e^{\beta\omega}-1}\, d\omega = \int_0^\infty \frac{\omega^n e^{-\beta\omega}}{1-e^{-\beta\omega}}\, d\omega = \sum_{m=1}^{\infty} \int_0^\infty \omega^n \operatorname{Exp}[-m\beta\omega]\, d\omega \tag{2.86}$$

We evaluated the integrals on the right-hand side using parametric differentiation at the beginning of this chapter. Using that result, we write

$$\int_0^\infty \frac{\omega^n}{e^{\beta\omega}-1}\,d\omega = \frac{n!}{\beta^{n+1}}\sum_{m=1}^{\infty}\frac{1}{m^{n+1}} = \frac{n!}{\beta^{n+1}}\zeta[n+1] \tag{2.87}$$

where the *Riemann zeta function* is defined by

$$\zeta[z] = \sum_{m=1}^{\infty}\frac{1}{m^z} \tag{2.88}$$

for $\mathrm{Re}[z] > 1$. Special values for integer arguments can be expressed in terms of the Bernoulli numbers and evaluated in closed form, but for our purposes we can consider the problem solved because the properties of the Riemann zeta function are well established and one can find numerical values in standard tables or mathematical software. Alternatively, the series can be evaluated numerically in a straightforward manner, if necessary.

More generally, we combine the same expansion with variable transformations to write

$$\begin{aligned}\int_0^\infty \frac{\omega^\alpha}{e^{\beta\omega}-1}\,d\omega &= \beta^{-\alpha-1}\int_0^\infty \frac{t^\alpha e^{-t}}{1-e^{-t}}\,dt = \beta^{-\alpha-1}\sum_{m=1}^{\infty}\int_0^\infty t^\alpha\,\mathrm{Exp}[-mt]\,dt \\ &= \beta^{-\alpha-1}\int_0^\infty s^\alpha\,\mathrm{Exp}[-s]\,ds\sum_{m=1}^{\infty}m^{-\alpha-1}\end{aligned} \tag{2.89}$$

Thus, we identify

$$\mathrm{Re}[\alpha] > 0,\ \mathrm{Re}[\beta] > 0 \Longrightarrow \int_0^\infty \frac{\omega^\alpha}{e^{\beta\omega}-1}\,d\omega = \beta^{-\alpha-1}\Gamma[\alpha+1]\zeta[\alpha+1] \tag{2.90}$$

with much less restrictive conditions upon the parameters.

2.7 Integral Representations for Analytic Functions

Special functions motivated by integrals for real variables can usually be extended into a portion of the complex plane simply by replacing the argument by a complex variable and employing a contour through the domain of analyticity of the integrand. Thus, one can define the complex sine integral by

$$\mathrm{Si}[z] = \int_0^z \frac{dt}{t}\,\mathrm{Sin}[t] \tag{2.91}$$

where the contour is any path between the indicated endpoints in the complex t-plane. This case is particularly simple because the integrand is entire – there is no danger of encountering singularities or branch cuts during integration. Similarly, the error function can be extended to the entire complex plane simply by using a complex argument

$$\mathrm{Erf}[z] = \frac{2}{\sqrt{\pi}}\int_0^z dt\,e^{-t^2} \quad \mathrm{Erfc}[z] = \frac{2}{\sqrt{\pi}}\int_z^\infty dt\,e^{-t^2} \tag{2.92}$$

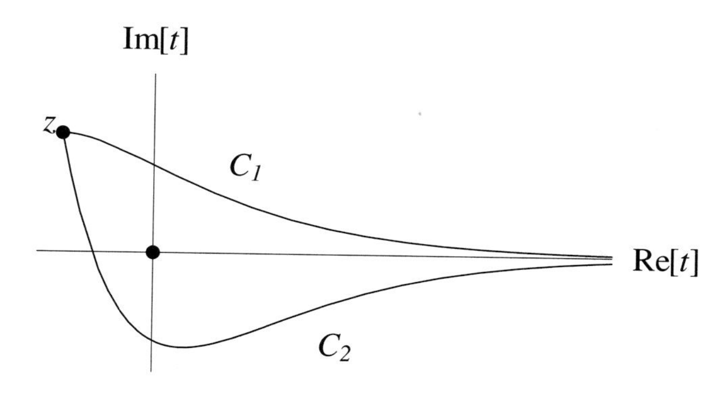

Figure 2.9. Contours for Ci[z].

where an upper limit of ∞ is interpreted as a path that asymptotically approaches the positive real axis such that $t \to (R, 0)$ where $R \to \infty$. It is also useful to define the *imaginary error function* $\text{Erfi}[z] = -i\,\text{Erf}[iz]$, such that

$$\text{Erfi}[z] = \frac{2}{\sqrt{\pi}} \int_0^{iz} dt\, e^{t^2} \tag{2.93}$$

Conversely, if the integrand is multivalued or has singularities, the definition of a function by means of an *integral representation* must constrain the contour enough to produce a unique value for the integral. For example, the integrand of the complex cosine integral

$$\text{Ci}[z] = -\int_z^{\infty} \frac{dt}{t} \text{Cos}[t] \tag{2.94}$$

has a simple pole at the origin, with unit residue. Suppose that z is found in the second quadrant, as shown in Fig. 2.9. The integral along contour C_2, which dips below the real axis for $\text{Re}[t] < 0$ and then approaches the positive real axis from below, differs by $2\pi i$ from the integral for contour C_1, which remains in the upper half-plane and approaches the positive real axis from above. (Imagine closing the contour for large Re[t] where the integrand is vanishingly small.) Contours which circle the origin several times differ by multiples of $2\pi i$, representing various branches of a multivalued function. The principal branch for this function is defined by the requirement that the positive real axis is approached from above without encircling the origin; this requirement is sufficient to provide a single-valued definition while still leaving considerable flexibility in the choice of contour.

The domain of analyticity will often be limited by requirements for the convergence of the defining integral. For example, the present definition for $\Gamma[x]$ is limited to $x > 0$, suggesting that straightforward extension to the complex plane would be limited to $\text{Re}[z] > 0$. However, we must still be wary of the branch cut needed to establish a unique value for the integrand when z is not an integer. Thus, for $\text{Re}[z] > 0$ we could use

$$\Gamma[z] = \int_0^{\infty} t^{z-1} e^{-t}\, dt\,, \quad \text{Re}[z] > 0 \tag{2.95}$$

with the contour in the left side of Fig. 2.10, where the positive real axis is approached from above. A more general definition can be made by using a theorem of *analytic continuation* that we will derive and discuss more thoroughly in a later chapter, which states that if the functions $f_1[z]$ and $f_2[z]$ are analytic in domains D_1 and D_2 and if $f_1[z] = f_2[z]$ for all $z \in D_1 \cap D_2$, then f_1 and f_2 are representations of the same analytic function $f[z]$ within their respective domains and $f[z]$ is analytic throughout $D_1 \cup D_2$. Therefore, if we can design an integral representation that provides identical values for $\mathrm{Re}[z] > 0$ while avoiding the singularity in the integrand, we would be able to extend the definition of $\Gamma[z]$ to the entire complex plane. Consider the function

$$f[z] = \int_C t^{z-1} e^{-t}\, dt = \int_C \mathrm{Exp}[-t + (z-1)\mathrm{Log}[t]]\, dt \tag{2.96}$$

for the inner keyhole contour around a branch cut on the positive real axis that is shown on the right side of Fig. 2.10 and is navigated in a counterclockwise sense. The small circle about the origin does not contribute when $\mathrm{Re}[z] > 0$. The contributions from either side of the branch cut differ in phase according to

$$t = x + i\varepsilon \Longrightarrow \mathrm{Log}[t] = \mathrm{Log}[x] \tag{2.97}$$

$$t = x - i\varepsilon \Longrightarrow \mathrm{Log}[t] = \mathrm{Log}[x] + 2\pi i \tag{2.98}$$

such that

$$f[z] = (\mathrm{Exp}[2\pi i z] - 1) \int_0^\infty t^{x-1} e^{-t}\, dt \tag{2.99}$$

Thus, we obtain a definition of the gamma function

$$\Gamma[z] = \frac{1}{\mathrm{Exp}[2\pi i z] - 1} \int_C t^{z-1} e^{-t}\, dt \tag{2.100}$$

that can be used for complex variables with positive real parts. The keyhole contour can now be deformed into the outer contour in the same figure without encountering any singularities and without altering the value of the integral. We simply require that C enters from $+\infty$ just above the real axis and exits toward $+\infty$ just below the real axis without crossing the positive real axis. Therefore, the proposed integral representation extends the definition of the gamma function to the entire complex plane. When $\mathrm{Re}[z] < 0$, one simply avoids the immediate vicinity of the origin.

It might appear that this integral representation for $\Gamma[z]$ has singularities for any integer value of z, but the singularities for positive integers are illusory (removable). When $z = n$ is an integer, t^{n-1} does not require a branch cut such that the contributions from the segments of the keyhole contour above and below the real axis cancel, leaving only the small circle about the origin. Alternatively, in the absence of a branch cut the contour can be deformed into a closed circle about the origin. (We imagine that the original contour is closed across the real axis so far out that the integrand is vanishingly small.) When $n > 0$ the integrand is analytic and the integral vanishes, leaving a 0/0 situation that suggests a removable singularity whose value should be determined by a limiting process that ensures

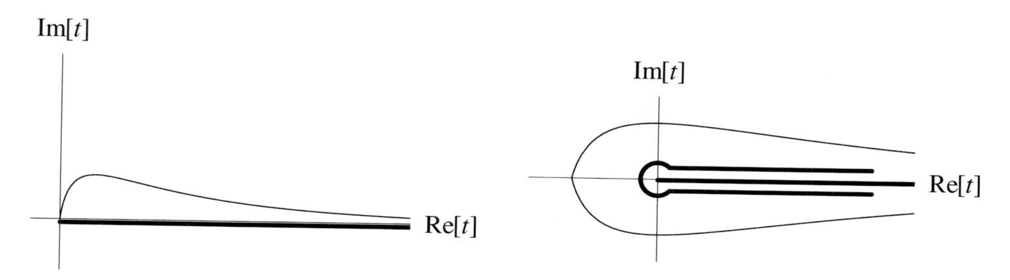

Figure 2.10. Left: contour for $\Gamma[z]$ with $\mathrm{Re}[z] > 0$. Right: more general contours for $\Gamma[z]$.

continuity. Although we expect that the appropriate value is $\Gamma[n] = (n-1)!$, it is instructive to demonstrate that this result emerges from a suitable limiting process. Performing Taylor series expansions around $z = n$,

$$t^{z-1} \approx t^{n-1}\mathrm{Log}[t](z-n), \quad \mathrm{Exp}[2\pi i z] - 1 \approx 2\pi i(z-n) \tag{2.101}$$

we find

$$z \approx n \Longrightarrow \Gamma[z] \approx \frac{1}{2\pi i}\int_C t^{n-1}e^{-t}\mathrm{Log}[t]\,dt \tag{2.102}$$

and can employ the keyhole contour for $n > 0$ without obtaining any contribution from the small circle around the origin because

$$n > 0 \Longrightarrow \lim_{\varepsilon\to 0} \varepsilon^n \mathrm{Log}[\varepsilon] = 0 \tag{2.103}$$

Placing the branch cut for $\mathrm{Log}[t]$ immediately below the positive real axis, the contributions from $\mathrm{Log}[|t|]$ on opposite sides cancel, leaving the contribution from the phase of $\mathrm{Log}[t - i\varepsilon] = \mathrm{Log}[|t|] + 2\pi i$ to give

$$\Gamma[z] \approx \frac{1}{2\pi i}\int_0^\infty t^{n-1}e^{-t}(2\pi i)\,dt = (n-1)! \tag{2.104}$$

as expected. When $n < 0$, we use

$$\oint dt\,\frac{e^{-t}}{t^{n-1}} = (-)^n \frac{2\pi i}{n!} \tag{2.105}$$

for a circle about the origin to discover that the residue for a simple pole at $z = -n$ is $(-)^n/n!$. Therefore, the integral representation teaches us that $\Gamma[z]$ is a meromorphic function with simple poles at negative integers and determines its residues. It should be clear that the new integral representation offers much more detailed information than the original definition on the real axis.

Many special functions have several different integral representations, with one or another being most useful under differing circumstances or for various purposes. These integral representations often provide the simplest or most general derivations for the properties of a function. For example, in the chapter on *Asymptotic Series* we will use integral representations to study the properties of several useful functions when $|z|$ is large. There we take advantage of the flexibility of the contour to concentrate upon a segment which dominates the integral. However, one must be careful when deforming contours of infinite length because the Cauchy–Goursat theorem on the path-independence of integrals of analytic functions was derived for finite contours. When one or both of the endpoints is at infinity, it can matter whether the approach toward infinity is along the real axis, the imaginary axis, or some intermediate direction. Penalty-free movement of a "free end" requires that the contribution of an arc of radius R subtending the angle through which the free end is moved must vanish as $R \to \infty$. Although this requirement should be obvious by now, from the care with which we studied the return paths for closing a contour of infinite length, it still bears some repetition. The contours used for integral representations offer considerable but not unlimited flexibility.

2.8 Using *MATHEMATICA®* to Evaluate Integrals

2.8.1 Symbolic Integration

Integrals can be entered in typeset form using the BasicInput palette, but if you wish to modify any options you will need to use the command line. The basic syntax for an indefinite integral is

```
Integrate[f[x], x]
```

Indefinite integrals like

$$\texttt{ans1} = \int x^n a^x \, dx$$

$$-x^{1+n} \,\texttt{Gamma}[1+n, -x\,\texttt{Log}[a]](-x\,\texttt{Log}[a])^{-1-n}$$

are returned without constants of integration and can be checked easily by differentiation

```
D[ans1, x] // Simplify
```

$$a^x x^n$$

The integral above is expressed in terms of the incomplete gamma function. *MATHEMATICA®* makes no *a priori* assumptions about the variable of integration or the constants, so that the result applies to either real or complex variables and parameters. However, it also makes no assumption regarding the values of parameters, returning results for generic values that are not always valid for special values of the parameters. Thus, the simple integral

$$\int x^a \, dx$$

$$\frac{x^{1+a}}{1+a}$$

is not valid if a happens to have the value -1.

The basic syntax for a definite integral is

```
Integrate[f[x], {x, a, b}]
```

The limits of integration may be numerical, symbolic, or infinite. In early versions any symbolic parameters or limits would also be interpreted in a generic sense. Thus, one would obtain a result for

$$\int_a^b x^n\,dx = \frac{b^{n+1} - a^{n+1}}{n+1} \tag{2.106}$$

that would valid for most choices of parameters, but fails for $n = -1$, while

$$\int_a^b x^{-1}\,dx = \mathrm{Log}[b] - \mathrm{Log}[a] \tag{2.107}$$

would fail for negative limits of integration. In more recent versions MATHEMATICA® usually returns conditional results

$$\int_0^1 \mathbf{x^n\,dx}$$

$$\mathtt{If}\left[\mathtt{Re[n]} > -1,\ \frac{1}{1+n},\ \mathtt{Integrate}\left[x^n,\ \{x, 0, 1\},\ \mathtt{Assumptions} \to \mathtt{Re[n]} \le -1\right]\right]$$

$$\int_a^b \mathbf{x^n\,dx}$$

$$(-a+b)\ \mathtt{If}\left[\mathrm{Re}\left[\frac{a}{a-b}\right] \ge 1 \,\|\, \mathrm{Re}\left[\frac{a}{-a+b}\right] \ge 0 \,\|\, \mathrm{Im}\left[\frac{a}{-a+b}\right] \ne 0,\ \frac{a^{1+n} - b^{1+n}}{(a-b)(1+n)},\ \mathtt{Integrate}\left[(a + (-a+b)x)^n,\ \{x, 0, 1\},\ \mathtt{Assumptions} \to !\left(\mathrm{Re}\left[\frac{a}{a-b}\right] \ge 1 \,\|\, \mathrm{Re}\left[\frac{a}{-a+b}\right] \ge 0 \,\|\, \mathrm{Im}\left[\frac{a}{-a+b}\right] \ne 0\right)\right]\right]$$

$$\int_a^b \mathbf{x^{-1}\,dx}$$

$$(-a+b)\ \mathtt{If}\left[\mathrm{Re}\left[\frac{a}{a-b}\right] \ge 1 \,\|\, \mathrm{Re}\left[\frac{a}{-a+b}\right] \ge 0 \,\|\, \mathrm{Im}\left[\frac{a}{-a+b}\right] \ne 0,\ \frac{\mathrm{Log}[a] - \mathrm{Log}[b]}{a-b},\ \mathtt{Integrate}\left[\frac{1}{a + (-a+b)x},\ \{x, 0, 1\},\ \mathtt{Assumptions} \to !\left(\mathrm{Re}\left[\frac{a}{a-b}\right] \ge 1 \,\|\, \mathrm{Re}\left[\frac{a}{-a+b}\right] \ge 0 \,\|\, \mathrm{Im}\left[\frac{a}{-a+b}\right] \ne 0\right)\right]\right];$$

that test any parameters that appear either in the integrand or the limits of integration. Although such results are less attractive and the conditions are often expressed in unnecessarily complicated forms, there is less chance of obtaining incorrect answers due to careless unstated assumptions about the parameters. Although one can use the option GenerateConditions → False

```
Integrate[x^n, {x, a, b}, GenerateConditions → False]
```

$$\frac{-a^{1+n} + b^{1+n}}{1+n}$$

to obtain a simple expression instead of a conditional statement, we strongly discourage that reckless practice and recommend instead that one supply the assumptions that apply to the parameters in your integrand directly, as follows.

```
Integrate[x^n, {x, a, b}, Assumptions → {Re[n] > -1, 0 < a < b}]
```

$$\frac{-a^{1+n} + b^{1+n}}{1+n}$$

MATHEMATICA® claims to be able to produce practically any integral in standard compilations, such as the massive compilation by Gradshteyn and Ryzhik. Thus, one finds that many seemingly unpromising integrals can be evaluated in terms of recognizable functions even if the output appears complicated. However, despite its impressive versatility, there remains a nontrivial error rate in the symbolic integration package. We decline to present specific examples here because each revision of the program seems to correct some errors while introducing new ones. Nevertheless, investigation of most (but not all) disagreements you find with *MATHEMATICA®* will eventually show that you have made a mistake. *The software is good, but not perfect!* Therefore, it remains useful to be able to perform such integrals independently. Furthermore, any result that is important should be checked. A very useful method for checking a symbolic integral is to compare with numerical evaluation for representative choices of the parameters. This technique cannot prove that a result is valid for all parameters satisfying the requisite conditions, but it will sometimes find errors.

Moral: trust but verify!

2.8.2 Numerical Integration

The basic syntax for numerical integration is

```
NIntegrate[f[x], {x, a, b}]
```

where the limits must be numerical and the integrand must evaluate to a number when given a numerical value for x. Numerical integration is generally more reliable than symbolic integration. Symbolic integration requires a vast library of pattern-matching rules for which it is difficult to ensure that all special cases are handled properly, while numerical integration is much more mechanical. The basic technique for numerical integration is to sample the integrand at strategically located positions, construct an interpolating polynomial, and then integrate the polynomial. The accuracy of the integral can be tested by subdividing the interval and applying the method to smaller parts. One can also sample more points where the integrand changes most rapidly. Many reliable algorithms have been developed and the one used by *MATHEMATICA®* is quite good. If it does encounter trouble with a particular integrand, there are options that can often be used to overcome those difficulties. If the difficulties persist, one should examine the integrand and handle its singularities more carefully.

2.8.3 Further Information

More information about integration using *MATHEMATICA*®, including multiple integrals and symbolic and numerical contour integration, can be found in *calculus.nb* at my Essential-Mathematica website.

Problems for Chapter 2

You may use *MATHEMATICA*® to check your work, but do not trust its symbolic integration too much. When evaluating integrals, you must specify the contour, convincingly justify neglect of any vanishing portions, and define phases near any branch cuts carefully. Be sure to consider special cases and be alert to possible generalizations.

1. Some related trigonometric integrals
Evaluate

$$\int_0^{2\pi} \frac{d\theta}{a + b\,\mathrm{Cos}[\theta]} \quad \text{for} \quad a > b > 0 \tag{2.108}$$

using contour integration and then deduce

$$\int_0^{2\pi} \frac{d\theta}{(a + b\,\mathrm{Cos}[\theta])^2} \quad \text{and} \quad \int_0^{2\pi} \frac{\mathrm{Cos}[\theta]\,d\theta}{(a + b\,\mathrm{Cos}[\theta])^2} \tag{2.109}$$

by more elementary means.

2. Trigonometric integrals on unit circle

a) $$\int_0^{2\pi} \frac{\mathrm{Sin}[\theta]^2}{1 + a\,\mathrm{Cos}[\theta]}\, d\theta \quad \text{for} \quad -1 < a < 1 \tag{2.110}$$

b) $$\int_0^{2\pi} \mathrm{Exp}[\mathrm{Cos}[\theta]]\,\mathrm{Cos}[n\theta - \mathrm{Sin}[\theta]]\, d\theta \quad \text{for integer } n \tag{2.111}$$

c) $$\int_0^{\pi} \mathrm{Sin}[\theta]^{2n}\, d\theta \quad \text{for nonnegative integer } n \tag{2.112}$$

d) $$\int_0^{2\pi} \frac{d\theta}{a + b\,\mathrm{Cos}[\theta]} \quad \text{for arbitrary complex } a, b \tag{2.113}$$

3. Magnetic flux through circle from coplanar wire
An infinitely long wire carries current $\mathcal{I}$. Compute the magnetic flux through a coplanar circle of radius a that is at a distance d from the wire, where $d \geq a$. Compare the limits $d \gg a$ and $d \gtrsim a$.

4. Average power radiated by charged harmonic oscillator
The angular distribution for the power radiated by a charge in simple harmonic motion with amplitude a and frequency ω is given by

$$\frac{dP}{d\Omega} = K\,\mathrm{Sin}[\theta]^2 \frac{\mathrm{Cos}[\omega t]^2}{(1+\beta\,\mathrm{Cos}[\theta]\,\mathrm{Sin}[\omega t])^5} \tag{2.114}$$

where $\beta = a\omega/c$ is the amplitude for the velocity oscillation (relative to light speed) and $K = e^2a^2\omega^4/4\pi c^3$. Evaluate the angular distribution of radiated power averaged over a period. Finally, evaluate the total average power integrated over angles.

5. Assorted integrals on infinite range
Use contour integration to evaluate the following integrals.

a) $$\int_0^\infty \frac{1}{1+x^n}\,dx \quad \text{where} \quad n>1 \text{ is a positive integer.} \tag{2.115}$$

(Hint: try an arc subtending $2\pi/n$ radians.)

b) $$\int_{-\infty}^\infty \frac{\mathrm{Cos}[px]-\mathrm{Cos}[qx]}{x^2}\,dx \quad \text{with } p, q \text{ real} \tag{2.116}$$

c) $$\mathcal{P}\int_{-\infty}^\infty \frac{\mathrm{Cos}[kx]}{1+x^3}\,dx \quad \text{with } k \text{ real} \tag{2.117}$$

d) $$\int_0^\infty \frac{\mathrm{Cos}[ax]}{\left(x^2+b^2\right)^2}\,dx \quad \text{for real } a, b, |b|>0 \tag{2.118}$$

e) $$\int_0^\infty \frac{\mathrm{Sin}[ax]}{x\left(x^2+b^2\right)}\,dx \quad \text{for real } a, b, |b|>0 \tag{2.119}$$

f) $$\int_{-\infty}^\infty \frac{x\,\mathrm{Sin}[2\theta]}{\left(1+x^2\right)\left(1-2x\,\mathrm{Cos}[\theta]+x^2\right)}\,dx \quad \text{for } 0\le\theta\le 2\pi \tag{2.120}$$

g) $$\int_{-\infty}^\infty \frac{dx}{\mathrm{Cosh}[kx]} \quad \text{with } k \text{ real and nonzero} \tag{2.121}$$

6. An integral from diffraction theory
Evaluate

$$\int_{-\infty}^\infty \frac{\mathrm{Sin}^2[x]}{x^2}\,dx \tag{2.122}$$

7. Integrals used in Fourier transform of oscillator wave functions

a) Evaluate

$$\int_0^\infty e^{-ax^2} \operatorname{Cos}[bx]\, dx \tag{2.123}$$

for $a, b > 0$.

b) Produce a general method for evaluating integrals of the form

$$\int_0^\infty e^{-ax^2} x^{2n+1} \operatorname{Sin}[bx]\, dx \tag{2.124}$$

with $a, b > 0$ and integer $n \geq 0$. Display explicit results for $n = 0, 1$.

Integrals of this type arise in the Fourier transform of oscillator wave functions.

8. Integration around branch cuts

a) Evaluate

$$\int_{-1}^{1} \frac{dx}{(a+bx)\sqrt{1-x^2}} \quad \text{with } a > b > 0 \tag{2.125}$$

using a suitable contour. (Hint: put the branch cut on the integration interval and define phases carefully.)

b) Evaluate

$$\int_0^\infty \frac{\ln x}{1+x^2}\, dx \tag{2.126}$$

using a suitable branch of Log[z].

c) Evaluate

$$\int_0^\infty \frac{(\ln x)^2}{1+x^2}\, dx \tag{2.127}$$

using a suitable branch of Log[z].

d) Evaluate

$$\int_0^1 \frac{x^{1/4}(1-x)^{3/4}}{(1+x)^3}\, dx \tag{2.128}$$

using a suitable contour. (Hint: put the branch cut on the integration interval and define phases carefully.)

e) Evaluate

$$\int_0^\infty \frac{x^{-a}}{1+x^4}\, dx \tag{2.129}$$

for complex a and specify the conditions required for convergence.

f) Evaluate

$$\int_0^\infty \frac{x^{-a}}{1+x^n}\,dx \tag{2.130}$$

for complex a and integer n and specify the conditions required for convergence.

g) Evaluate

$$\int_0^1 x^{a-1}(1-x)^{-a}\,dx \tag{2.131}$$

for $0 < \text{Re}[a] < 1$. (Hint: try a large circular contour and then deform the contour to snugly embrace the branch cut.)

h) Evaluate

$$\mathcal{P}\int_0^\infty \frac{x^{-\lambda}}{x-a}\,dx \tag{2.132}$$

for real $a > 0$ and real $0 < \lambda < 1$.

i) $$\mathcal{P}\int_0^\infty \frac{x^{1/4}}{x^2-x-2}\,dx \tag{2.133}$$

9. Using $\int_0^\infty g[x]\,dx \to \oint_C \text{Log}[z]g[z]\,dz$ for meromorphic integrands
Consider an integral of the form

$$\int_0^\infty g[x]\,dx \tag{2.134}$$

where $g[x]$ is well-behaved on the positive real axis but is not symmetric with respect to the sign of x, such that the integration interval cannot be extended for use with a great semicircle. If the corresponding $g[z]$ is a meromorphic function that decreases sufficiently rapidly for $z \to \infty$ and is sufficiently small at the origin, one can often use $f[z] = g[z]\text{Log}[z]$ with a branch cut on the positive real axis and a *PacMan* contour of the form shown in Fig. 2.11. Apply this method for the following integrals.

a) $$\int_0^\infty \frac{dx}{(x+1)\left(x^2+2x+2\right)} \tag{2.135}$$

b) $$\int_0^\infty \frac{dx}{x^3+1} \tag{2.136}$$

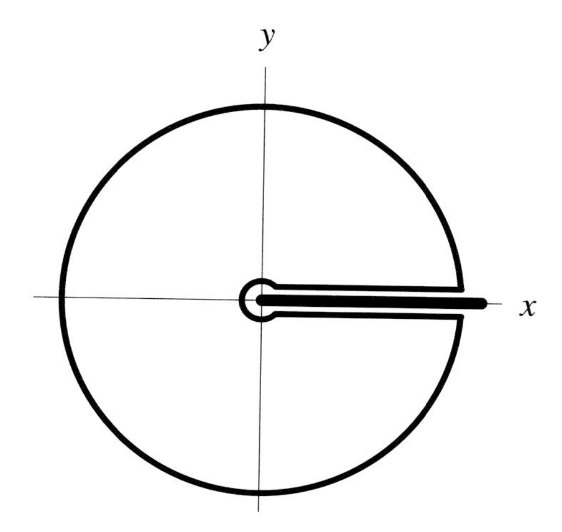

Figure 2.11. Contour used for $\int_0^\infty g[x]\,dx \to \oint_C \mathrm{Log}[z]g[z]\,dz$.

10. Reduction to standard functions

Express the following integrals in terms of standard functions.

a) $$\int_0^\infty x^m e^{-x^n}\,dx \quad \text{for } n > 0 \tag{2.137}$$

b) $$\int_0^\infty e^{-ax^2}\,\mathrm{Sin}[bx]\,dx \quad \text{with } a, b > 0 \tag{2.138}$$

c) $$\int_0^\infty \frac{e^\omega \omega^\alpha}{(e^\omega - 1)^2}\,d\omega \quad \text{with } \alpha > 1 \tag{2.139}$$

11. Using the beta function

Express each of the following integrals in terms of the beta function and then in terms of more familiar gamma functions. Be sure to identify the conditions that must be satisfied by the parameters in each integral.

a) $$\int_0^1 t^{r-1}(1-t)^{s-1}\,dt \tag{2.140}$$

b) $$\int_0^1 u^r\left(1-u^2\right)^s\,du \tag{2.141}$$

c) $$\int_0^\infty \frac{u^r}{(1+u)^s}\,du \tag{2.142}$$

d) $$\int_{-1}^{1} (1-x)^r (1+x)^s \, dx \tag{2.143}$$

e) $$\int_0^{\infty} \frac{(\mathrm{Sinh}[x])^a}{(\mathrm{Cosh}[x])^b} \, dx \quad \text{(Hint: use } \mathrm{Sinh}[x]^2 \to u \text{ and the result of c.)} \tag{2.144}$$

12. Beta probability distribution

The probability density for the so-called beta distribution is proportional to

$$P[x] \propto x^{a-1}(1-x)^{b-1} \tag{2.145}$$

for $0 \leq x \leq 1$ where a, b are nonnegative constants. Evaluate the normalization, mean, and variance for this distribution.

13. oscillation period in potential $V = kx^{2n}/2$

A classical particle with mass m and total energy E oscillates in a potential of the form $V = kx^{2n}/2$ where n is a positive integer. Determine its oscillation period. Can you generalize this result?

3 Asymptotic Series

Abstract. Asymptotic series describe the behavior of a function for large values of an argument. The leading asymptotic behavior can often be deduced from an integral representation using the method of steepest descent. Repeated partial integration or expansion of the integrand provide series that improve upon the leading approximation.

3.1 Introduction

Often we must characterize the behavior of a function of z when $|z|$ becomes very large. For example, in scattering theory we require the asymptotic behavior of Bessel functions. An *asymptotic expansion* for $f[z]$ may take the form

$$f[z] \simeq \varphi[z]g[z] \tag{3.1}$$

where $\varphi[z]$ represents the leading asymptotic behavior in terms of familiar functions, often exponential or logarithmic, while the *asymptotic series*

$$g[z] = \sum_{k=0}^{\infty} a_k z^{-k} \tag{3.2}$$

represents a correction factor expanded in inverse powers of z. Ordinarily we normalize the series such that $a_0 = 1$. However, in the common situation that the ratio $f[z]/\varphi[z]$ has an essential singularity at $z \to \infty$, the series for $g[z]$ diverges. Nevertheless, for any finite value of z there may be a truncated series

$$g_n[z] = \sum_{k=0}^{n} a_k z^{-k} \tag{3.3}$$

which provides a very good approximation to $f[z]/\varphi[z]$ with an accuracy that improves as $|z|$ increases. The sequence of analytic functions $f_n[z] = \varphi[z]g_n[z]$ is then said to *converge asymptotically* to $f[z]$. More formally, the function f_n is described as *asymptotically equal* to $f[z]$ if

$$\lim_{|z|\to\infty} (f[z] - f_n[z]) = 0 \Longrightarrow f_n[z] \simeq f[z] \tag{3.4}$$

such that their difference becomes arbitrarily small if $|z|$ is sufficiently large. This relationship is represented by the operator $\simeq$ or sometimes $\sim$. A series is asymptotically convergent if

$$\lim_{|z|\to\infty} z^n \left(\frac{f[z]}{\varphi[z]} - g_n[z] \right) = 0 \Longrightarrow f_n[z] = \varphi[z]g_n[z] \simeq f[z] \tag{3.5}$$

Graduate Mathematical Physics. James J. Kelly

ISBN: 3-527-40637-9

for any positive n. This relationship requires that the difference between f_n and f can be made arbitrarily small if $|z|$ is sufficiently large for any choice of n. Of course, the smaller n is, the larger $|z|$ would have to be to achieve a specified accuracy.

Unlike more familiar convergent series, $z^n g_n[z]$ usually diverges for fixed z as n increases such that

$$\lim_{n\to\infty} z^n g_n[z] \to \infty \tag{3.6}$$

Thus, the accuracy of $f_n[z]$ as an approximation to $f[z]$ actually deteriorates if too many terms are included in the series. Given that the error in summing a series is smaller than the first neglected term, the best approximation to $f[z]$ for finite z is then obtained by truncating the series at the smallest term (or, perhaps, the one before). Occasionally $f_n[z]$ is actually convergent with respect to n even though there remains a slight difference from $f[z]$ for finite z that decreases as z increases. We consider a series of this type also to be an asymptotic approximation to f even though some authors limit that term to divergent series only.

If $f_n[z]$ and $g_n[z]$ are asymptotic series for $f[z]$ and $g[z]$, one can show that

$$f_n[z] + g_n[z] \simeq f[z] + g[z] \tag{3.7}$$

$$f_n[z] \times g_n[z] \simeq f[z] \times g[z] \tag{3.8}$$

Asymptotic series may integrated, but there is no guarantee that $f_n'[z]$ is asymptotically equal to $f'[z]$. Finally, the asymptotic series for a given function is unique, but the same asymptotic series can represent more than one function in an asymptotic sense. For example, the asymptotic series for $f[z] + e^{-z}$ for $\mathrm{Re}[z] > 0$ is the same as that for $f[z]$.

In this chapter we explore a few of the methods that can be used to obtain asymptotic approximations to some of the functions relevant to theoretical physics. We start with saddle-point methods because they often provide the simplest methods for obtaining the leading asymptotic behavior. In particular, we discuss the method of steepest descent in some detail because it is the most versatile while the closely related but more limited method of stationary phase is developed in the problems at the end of the chapter. These methods are often suitable for analyzing integral representations that are obtained as approximations to the physical behavior of a system under specific conditions, whereas some of the other methods discussed later in the chapter are more suitable for analysis of mathematical expressions that are, in principle, exact.

3.2 Method of Steepest Descent

Often one encounters functions of the form

$$f[z] = \int_C F[t, z]\, dt \tag{3.9}$$

where C is a specified contour in the complex t-plane and F is analytic in a domain including C. We assume that, for the relevant choices of the parameter z, the contribution made by the immediate vicinities of the endpoints are negligible and seek an approximation scheme that takes advantage of the topography of analytic functions.

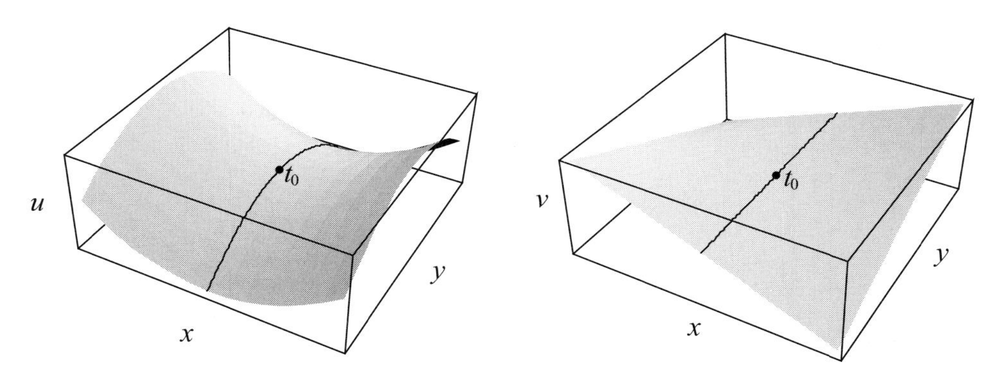

Figure 3.1. Typical saddle with path of steepest descent.

Suppose that the integrand can be represented in the exponential form

$$F[t, z] \approx \text{Exp}[zg[t, z]] \tag{3.10}$$

where $g[t, z]$ is an analytic function of the complex variable t. It is useful to express $z = re^{i\theta}$ in polar form and to absorb the phase into that analytic function by writing

$$z = re^{i\theta} \Longrightarrow zg[t, z] = r(u[t] + iv[t]) \Longrightarrow F[t, z] \approx \text{Exp}[ru[t]]\,\text{Exp}[irv[t]] \tag{3.11}$$

where the real functions $u[x, y]$ and $v[x, y]$ are harmonic and $t = x + iy$. Recall that neither u nor v can have an extremum, so that any point t_0 where $g'[t_0] == 0$ must represent a simultaneous saddle point of both $u[x, y]$ and $v[x, y]$, and that level curves of these two functions are orthogonal to each other. Therefore, v is constant on the path where u changes most rapidly. Figure 3.1 illustrates these relationships. The point t_0 is a saddle point for both u and v. The path of steepest descent for u is a path of constant v. By integrating along a path of constant v, we avoid the delicate cancellations in the integrand due to the rapid oscillation of $\text{Exp}[irv]$ for large r. By selecting the path of steepest descent in u, where u can be parametrized in the form $u \approx u_0 - \alpha s^2$ with positive α, we ensure rapid convergence of the integral of $\int e^{ru[t]}\,dt \propto \int e^{-\alpha s^2}\,ds$. Thus, the *method of steepest descent* can be used whenever it is possible to deform the contour C so that it passes through a saddle point along this special path. The contribution made by the immediate vicinity of the saddle point will often provide a very good approximation when r is sufficiently large and thus represents the lowest term of an asymptotic expansion. The method can also be generalized for application to more complicated surfaces featuring several saddle points between the endpoints of the contour. One simply applies this analysis to each saddle point and adds their contributions.

Near the saddle point we can expand

$$g[t] = g[t_0] + g''[t_0]\frac{(t - t_0)^2}{2} + \cdots \tag{3.12}$$

in parabolic form. It is useful to define

$$z = re^{i\theta}, \quad t - t_0 = se^{i\phi}, \quad g''[t_0] = g_2 e^{i\gamma} \tag{3.13}$$

where r and g_2 are positive real numbers representing the magnitude of $|z|$ and the curvature of u at the saddle point, respectively. The angle γ represents the phase of $g''[t]$ near the saddle point. A straight path through the saddle point is represented by a constant value of ϕ, describing the orientation of the path, and a real number s varying continuously from negative to positive values measuring the distance along the path relative to t_0. Thus, the argument of the exponential on a path through the saddle point is parametrized by

$$zg[t] \approx zg_0 + \frac{g_2 r s^2}{2} e^{i(\theta+\gamma+2\phi)} \tag{3.14}$$

We are free to choose ϕ in a manner that optimizes our approximation to the contour integral. The path of steepest descent is selected by choosing ϕ according to

$$\theta + \gamma + 2\phi = \pi \Longrightarrow \phi = \frac{\pi - \theta - \gamma}{2}, \quad dt = s e^{i\phi}\, ds \tag{3.15}$$

such that

$$zg[t] \approx zg_0 - \alpha s^2 \tag{3.16}$$

where

$$\alpha = \frac{g_2 r}{2} \tag{3.17}$$

is large and positive for large r, provided that g_2 is not abnormally small (if it is small, we may need to carry another term). The integrand is then so sharply peaked around $s = 0$ that we can extend the limits of integration to $\pm\infty$ with negligible error and it is easy to evaluate the resulting Gaussian integral. Hence, we obtain

$$\begin{aligned} f[z] &\simeq \mathrm{Exp}[zg_0 + i\phi] \int_{-\infty}^{\infty} \mathrm{Exp}[-\alpha s^2]\, ds = \mathrm{Exp}[zg_0 + i\phi]\sqrt{\frac{\pi}{\alpha}} \\ &= \mathrm{Exp}[zg_0 + i\phi]\sqrt{\frac{2\pi}{g_2 r}} \end{aligned} \tag{3.18}$$

Therefore,

$$f[z] \simeq \mathrm{Exp}[zg_0 + i\phi]\sqrt{\frac{2\pi}{g_2 |z|}} \tag{3.19}$$

provides the first term in an asymptotic expansion for $f[z]$ applicable for large $|z|$.

To apply the method we must express the integrand in exponential form and select an appropriate saddle point of the argument of that exponential through which the contour can be deformed in a manner that ensures that the integral is dominated by the immediate vicinity of the saddle point. By identifying the proper contour phase, the leading asymptotic behavior can be obtained quite easily. Often one can also develop a systematic series of corrections, thereby obtaining an asymptotic expansion in negative powers of z.

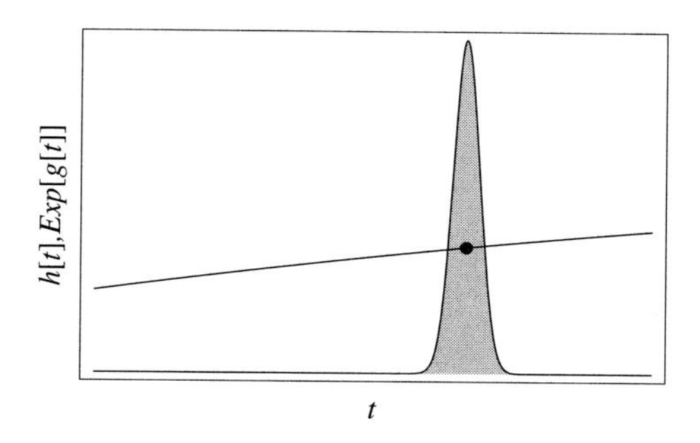

Figure 3.2. Integration of slow × fast functions.

The method can be generalized to functions expressible in the form

$$f[z] = \int_C h[t] \,\mathrm{Exp}[zg[t, z]] \, dt \tag{3.20}$$

where $h[t]$ varies much more slowly on the integration path than does $\mathrm{Exp}[zg[t, z]]$. Figure 3.2 illustrates that if $h[t]$ changes little over the width of the Gaussian approximation to the exponential factor near the saddle point of $g[t, z]$, then $h[t] \approx h[t_0]$ can be extracted from the integral such that

$$f[z] \simeq h[t_0] \,\mathrm{Exp}[zg_0 + i\phi] \sqrt{\frac{2\pi}{g_2 |z|}} \tag{3.21}$$

simply has an additional factor.

Approximation is as much art as science. The method of steepest descent provides one strategy for obtaining a useful approximation to a particular type of function, but its application depends upon the nature of the particular problem. Rather than memorize Eq. (3.21), with its definitions of $\{t_0, g_0, g_2, \phi\}$, it is better to understand how the strategy works and then, if applicable, adapt it to the problem at hand.

3.2.1 Example: Gamma Function

The gamma function for $\mathrm{Re}[z] > 0$ can be defined in terms of the integral

$$\Gamma[z+1] = \int_0^\infty t^z e^{-t} \, dt = \int_0^\infty \mathrm{Exp}[z \,\mathrm{Log}[t] - t] \, dt = \int_0^\infty \mathrm{Exp}[zg[t]] \, dt \tag{3.22}$$

where $z = re^{i\theta}$ and where

$$g[t] = \mathrm{Log}[t] - \frac{t}{z} \tag{3.23}$$

$$g'[t] = \frac{1}{t} - \frac{1}{z} \tag{3.24}$$

$$g''[t] = -\frac{1}{t^2} \tag{3.25}$$

is an analytic function of t with a parametric dependence on z. The saddle point is defined by

$$g'[t_0] == 0 \Longrightarrow t_0 = z \tag{3.26}$$

such that

$$g[t_0] = \text{Log}[z] - 1, \quad g''[t_0] = -\frac{1}{z^2} = -\frac{e^{-2i\theta}}{r^2} \tag{3.27}$$

Thus, we identify

$$g_2 = \frac{1}{r^2}, \quad \gamma = \pi - 2\theta \tag{3.28}$$

and solve

$$\pi = \theta + \gamma + 2\phi = \pi - \theta + 2\phi \Longrightarrow \phi = \frac{\theta}{2} \tag{3.29}$$

to determine that the path of steepest descent is along $\phi = \theta/2$. Substituting these values, we obtain

$$\Gamma[z+1] \simeq \text{Exp}\left[z(\text{Log}[z] - 1) + i\frac{\theta}{2}\right]\sqrt{2\pi r} = \sqrt{2\pi z}\, z^z e^{-z} \tag{3.30}$$

Finally, recognizing that $\Gamma[z+1] = z\Gamma[z]$, we obtain an approximation

$$\Gamma[z] \simeq \sqrt{2\pi}\, z^{z-\frac{1}{2}} e^{-z} \tag{3.31}$$

that is useful for large $|z|$ and $\text{Re}[z] > 0$.

Figure 3.3 illustrates the path of steepest descent for the particular case $z = 10(1+i)$, for which $\theta = \pi/4 \Longrightarrow \phi = \pi/8$. Contours for the real and imaginary parts of $zg[t,z]$ are indicated by solid and dashed lines, the saddle point is labeled by t_0, and the path of steepest descent is shown as a thick line. The complete contour in the t plane begins at the origin and is approximated by this line in a region surrounding t_0 that is large enough to accumulate most of the integral, and then approaches the positive real axis, $t \to \infty + i\varepsilon$, from above. In fact, by using the more general integral representation of the gamma function, given by

$$\Gamma[z] = \frac{1}{\text{Exp}[2\pi i z] - 1} \int_C t^{z-1} e^{-t}\, dt \tag{3.32}$$

for the positive contour indicated in Fig. 3.4 (recall Eq. 2.100)), one can demonstrate that the same asymptotic approximation also applies when $\text{Re}[z] < 0$. One simply deforms the contour to pass through the saddle point $t_0 = z$ with the proper slope. This generalization is left as an exercise for the reader.

The accuracy of this approximation is illustrated in Fig. 3.5, which was produced using the *MATHEMATICA*® code below. Curves are shown for $0 \leq \theta \leq 2\pi$ in steps of $\pi/8$. Interestingly, the approximation remains successful even for rather small r, except in the immediate vicinity of $\theta = \pi$ where the poles in $\Gamma[z]$ are found; this is the curve which stands out.

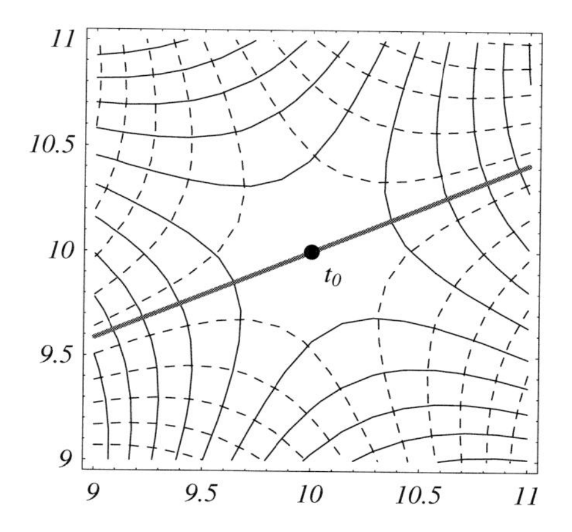

Figure 3.3. Portion of a contour for the integral representation of $\Gamma[10(1 + i)]$ near the saddle point for t_0. The heavy curve shows the path of steepest descent while level curves of u and v are shown as thin solid and dashed curves, respectively.

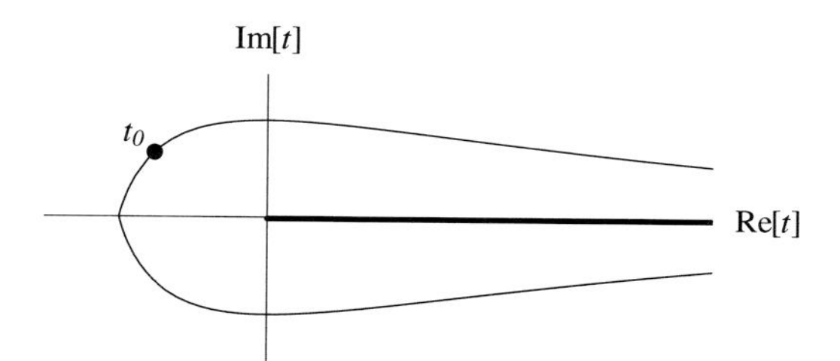

Figure 3.4. A contour for Eq. (3.32) that passes through a saddle point t_0 for $\text{Re}[z] < 0$.

Sometimes results of this type have broader applicability than suggested by the assumptions employed in their derivation.

```
Plot[Evaluate[Table[Abs[1 - √(2π) z^(z-1/2) ⅇ^(-z) / Gamma[z]] /.z → r ⅇ^(iθ)},
  {θ, 0, 2π, π/8}]], {r, 1, 10}, PlotRange → {Automatic, {0, 0.08}},
  Frame → True, FrameLabel → {"r", "relative error"}];
```

3.3 Partial Integration

Suppose that a function is defined by an integral of the form

$$f[z] = \int_z^\infty g[t, z]\, dt \quad \text{or} \quad f[z] = \int_0^z g[t, z]\, dt \tag{3.33}$$

where the independent variable appears in one of the limits of integration. Repeated integration by parts can then generate a series in z leaving a remainder term in the form of an

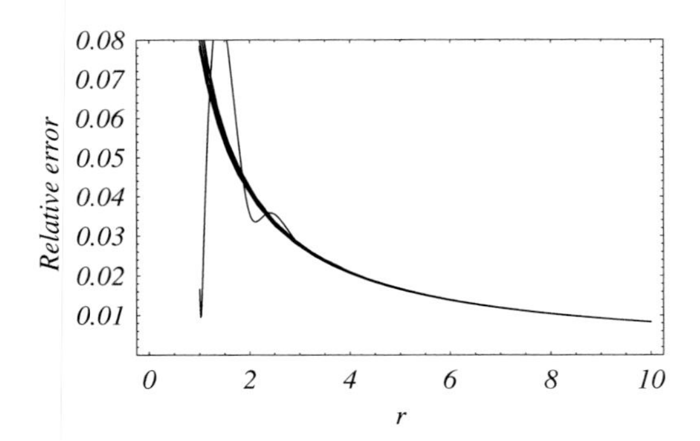

Figure 3.5. Accuracy of asymptotic approximation to $\Gamma[re^{i\Theta}]$. Curves are shown for $0 \leq \theta \leq 2\pi$ in steps of $\pi/8$; only the curve for $\theta = \pi$ is clearly separated for small r.

integral that is hopefully much smaller in the limit $|z| \to \infty$. One example should suffice to illustrate the method.

3.3.1 Example: Complementary Error Function

The complementary error function

$$\text{erfc}[x] = \frac{2}{\sqrt{\pi}} \int_x^\infty e^{-t^2}\, dt \tag{3.34}$$

representing the fraction of the area in the wings of a Gaussian probability distribution is important in statistics. It is useful to re-express this function as

$$\text{erfc}[x] = \frac{2}{\sqrt{\pi}} \int_x^\infty e^{-t^2}\, dt = \frac{1}{\sqrt{\pi}} \int_{x^2}^\infty \frac{dw}{\sqrt{w}} e^{-w} \tag{3.35}$$

and then integrate by parts repeatedly to obtain

$$\begin{aligned}
\text{erfc}[x] &= \frac{1}{\sqrt{\pi}} \left(\left(-w^{-1/2} e^{-w}\right)_{x^2}^{\infty} - \frac{1}{2} \int_x^\infty dw e^{-w} w^{-3/2} \right) \\
&= \frac{1}{\sqrt{\pi}} \left(\frac{e^{-x^2}}{x} - \frac{1}{2} \left(\left(-w^{-3/2} e^{-w}\right)_{x^2}^{\infty} - \frac{3}{2} \int_x^\infty dw e^{-w} w^{-5/2} \right) \right) \\
&= \frac{1}{\sqrt{\pi}} \left(\frac{e^{-x^2}}{x} - \frac{1}{2} \frac{e^{-x^2}}{x^3} + \frac{3}{4} \left(\left(-w^{-5/2} e^{-w}\right)_{x^2}^{\infty} - \frac{5}{2} \int_x^\infty dw e^{-w} w^{-7/2} \right) \right)
\end{aligned} \tag{3.36}$$

The pattern should now be achingly clear. Thus, we write

$$\text{erfc}[x] \simeq \frac{1}{\sqrt{\pi}} \frac{e^{-x^2}}{x} \left(1 + \sum_{k=1}^{n} \frac{(2k-1)!!}{(-2x^2)^k} \right) \tag{3.37}$$

where n should be chosen to give the best approximation for specified x.

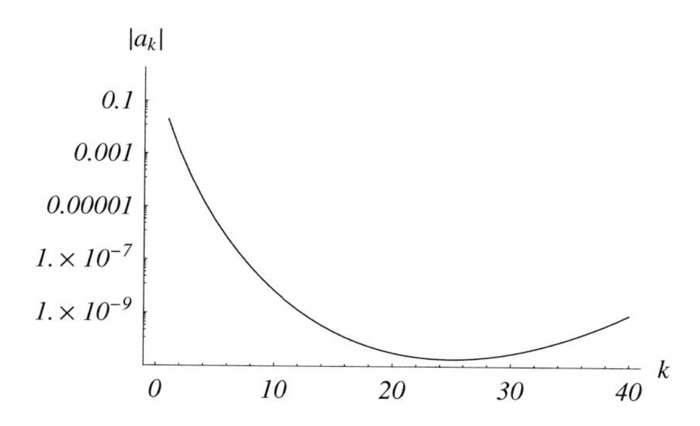

Figure 3.6. Convergence for erfc[5].

Let

$$S_n = \sum_{k=1}^{n} a_k, \quad a_k = \frac{(2k-1)!!}{(-2x^2)^k} \tag{3.38}$$

represent a sequence of partial sums. For any finite x the sequence eventually diverges because the terms a_k begin to grow for $k > k_{\max}$. The optimum number of terms is given by

$$\left|\frac{a_{k+1}}{a_k}\right| < 1 \Longrightarrow \frac{2k+1}{2x^2} < 1 \Longrightarrow k_{\max} = \text{Max}\left[0, \text{Floor}\left[x^2 - \tfrac{1}{2}\right]\right] \tag{3.39}$$

where Floor[x] is the largest integer less than or equal to x. Figure 3.6, showing $|a_k|$, demonstrates that for $x = 5$ the first term already provides 2 significant figures while the accuracy of the optimum sum is better than 10 significant figures. It is also clear that accuracy is lost if too many terms are taken.

However, Fig. 3.7 plotting terms a_k for $x = 2$, for which $k_{\max} = 3$, suggests that this approximation is limited to a few percent accuracy if x is as small as 2. These characteristics illustrate the basic nature of asymptotic expansions.

A better approximation is available for series whose terms alternate in sign. Although the term a_4 is slightly too large in the figure above, we expect that S_3 is slightly too small and we expect that a better approximation will be given by $S_3 + \frac{1}{2}a_4$. Thus, it would appear that the best asymptotic approximation is obtained by averaging partial sums for $k_{\max}$ and $k_{\max} + 1$ terms, such that

$$\text{erfc}[x] \simeq \frac{1}{\sqrt{\pi}} \frac{e^{-x^2}}{x} \left(1 + S_{k_{\max}} + \tfrac{1}{2} a_{k_{\max}+1}\right) \tag{3.40}$$

although we do not have a formal proof of that hypothesis. This technique of optimizing an asymptotic approximation is sometimes called *Richardson extrapolation*. The following *MATHEMATICA*® function evaluates our asymptotic approximation to erfc[x]. Richardson extrapolation is used by default, but can be disabled by including the option Extrapola-

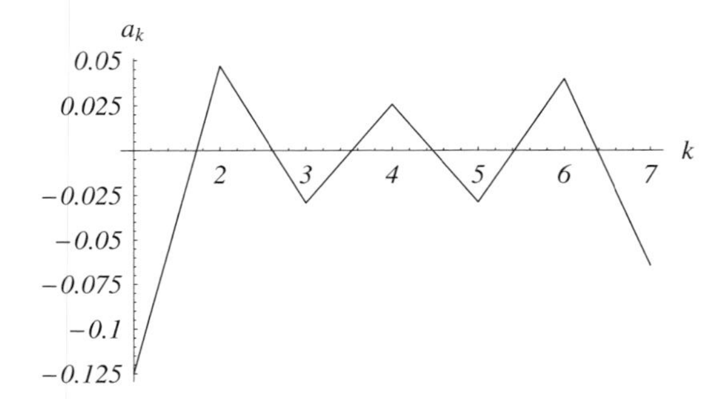

Figure 3.7. Convergence for erfc[2].

tion → False.

```
Clear[erfc];
Options[erfc] = {Extrapolate → True};
erfc[x_, opts___Rule] := Module[{kmax, sum, extrapolate},
  extrapolate = (Extrapolate/.{opts})/.Options[erfc];
  kmax = Floor[x^2 - 1/2];
  sum = If[kmax > 0, 1 + Plus@@ (2Range[kmax] - 1)!!/(-2x^2)^Range[kmax] +
    If[extrapolate, (2kmax + 1)!!/(2(-2x^2)^(kmax+1)), 0], 1]; 1/Sqrt[π] e^(-x^2)/x sum]
```

The improvement obtained by extrapolation is illustrated in Fig. 3.8, which was produced by the *MATHEMATICA®* code below. Extrapolation clearly improves the approximation, which is better than 1 % for $x > 1.3$. The approximation is poor for $x < 1$, but improves rapidly.

```
Plot[{1 - erfc[x]/Erfc[x], 1 - erfc[x, Extrapolate → False]/Erfc[x]},
  {x, 1, 2}, PlotRange → All, PlotStyle → {{}, Dashing[{0.02, 0.02}]},
  Frame → True, FrameLabel → {"x", "relative error"},
  PlotLabel → "asymptotic approximation to erfc[x]",
  PlotLegend → {"with extrapolation", "without extrapolation"},
  LegendPosition → {-0.2, -0.4},
  LegendSize → {0.7, 0.3}, LegendShadow → None];
```

3.4 Expansion of an Integrand

If the function is defined in terms of a definite integral

$$f[z] = \int_a^b g[t, z]\, dt \tag{3.41}$$

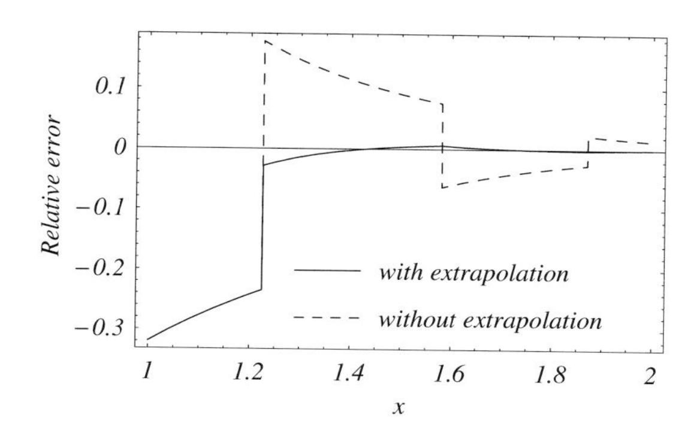

Figure 3.8. Asymptotic approximation to Erfc[x].

where a and b are fixed, while the integrand $g[t, z]$ is amenable to asymptotic expansion with respect to z, the expansion for $f[z]$ can be generated by integration term by term. This method is most useful when each term is sufficiently localized with respect to t that the limits of integration can be extended without appreciable change in the integral. Again, one example should suffice to illustrate the method.

3.4.1 Example: Modified Bessel Function

Without deriving this integral representation at this time, we can use

$$I_0\left[\frac{x^2}{2}\right] = \frac{2}{\pi} e^{x^2/2} \int_0^{\pi/2} \mathrm{Exp}[-x^2 \,\mathrm{Sin}[t]^2]\, dt \tag{3.42}$$

to derive an asymptotic approximation to the modified Bessel function of the first kind with order 0, namely $I_0[x]$. This integral representation is convenient because as x increases the integral becomes increasingly dominated by small t, which permits extension of the upper limit to ∞ with little error. Hence, we consider the integral

$$f[x] = \int_0^{\pi/2} \mathrm{Exp}[-x^2 \,\mathrm{Sin}[t]^2]\, dt \tag{3.43}$$

and use a change of variables

$$w = \mathrm{Sin}[t] \Longrightarrow dw = \mathrm{Cos}[t]\, dt = \sqrt{1 - w^2}\, dt \tag{3.44}$$

to write

$$f[x] = \int_0^1 \frac{e^{-x^2 w^2}}{\sqrt{1 - w^2}}\, dw = \int_0^1 dw\, e^{-x^2 w^2} \left(1 + \frac{w^2}{2} + \frac{3w^4}{8} + \cdots\right) \tag{3.45}$$

or

$$f[x] = \int_0^1 dw\, e^{-x^2 w^2} \left(1 + \sum_{k=1}^{\infty} \frac{(2k-1)!!}{2^k k!} w^k\right) \tag{3.46}$$

If x is large, the Gaussian factor is sharply peaked in w and we can extend the upper limit of integration to ∞ with little error such that

$$f[x] \simeq \int_0^\infty dw\, e^{-x^2 w^2}\left(1+\sum_{k=1}^{\infty}\frac{(2k-1)!!}{2^k k!}x^k\right) \tag{3.47}$$

The integrals can now be evaluated using

$$M_0[\lambda] = \int_0^\infty dw\, e^{-\lambda w^2} = \frac{1}{2}\sqrt{\frac{\pi}{\lambda}} \tag{3.48}$$

$$M_n[\lambda] = \int_0^\infty dw\, e^{-\lambda w^2} w^{2n} = \left(-\frac{\partial}{\partial\lambda}\right)^n M_0 = \frac{\sqrt{\pi}}{2}\frac{(2n-1)!!}{2^n \lambda^{n+\frac{1}{2}}} \tag{3.49}$$

and the series becomes

$$f[x] \simeq \frac{\sqrt{\pi}}{2x}\left(1+\sum_{k=1}^{\infty}\left(\frac{(2k-1)!!}{2^k k!}\right)^2 x^{-2k}\right) = \frac{\sqrt{\pi}}{2x}\left(1+\tfrac{1}{4}x^{-2}+\tfrac{9}{64}x^{-4}+\tfrac{25}{256}x^{-6}+\cdots\right) \tag{3.50}$$

Unlike the asymptotic expansion for Erfc[x], this series converges for $|x| > 1$ because

$$|x| > 1 \Longrightarrow \left|\frac{a_{k+1}}{a_k}\right| = \left(\frac{(2k+1)}{2(k+1)x}\right)^2 < 1 \tag{3.51}$$

and hence is asymptotically equal to $f[x]$ for large x without need for truncation. Thus, we obtain an asymptotic series for

$$I_0\left[\frac{x^2}{2}\right] \simeq \frac{e^{x^2/2}}{\sqrt{\pi}\,x}\left(1+\sum_{k=1}^{\infty}\left(\frac{(2k-1)!!}{2^k k!}\right)^2 x^{-2k}\right) \tag{3.52}$$

or

$$I_0[x] \simeq \frac{e^x}{\sqrt{2\pi x}}\left(1+\sum_{k=1}^{\infty}\left(\frac{(2k-1)!!}{2^k k!}\right)^2 (2x)^{-k}\right) \tag{3.53}$$

The following function and accompanying plots explore the accuracy of this asymptotic series for $I_0[z]$. Figure 3.9 demonstrates that the expansion becomes increasingly accurate as x increases, but even with just the first term of the series one obtains a few percent accuracy for $x \gtrsim 1$. The series converges quite rapidly for $x > 1$. Figure 3.10 shows that the relative error for $k_{\max} = 5$ decreases rapidly as x increases. However, this series must still be classified as an asymptotic series, even if convergent, because there is a minimum error for finite x that originates from extending the range of integration.

```
error[x_, kmax_] :=
  1 - (e^x/Sqrt[2π x] (1 + Sum[((2k - 1)!!/(2^k k!))^2 (2x)^-k, {k, 1, kmax}]))/BesselI[0, x]
  Plot[Evaluate[Table[error[x, kmax], {kmax, 0, 5}]],
  {x, 0, 10}, PlotRange → {Automatic, {-0.05, 0.05}},
  Frame → True, FrameLabel → {"x", "relative error"},
  PlotLabel → "Accuracy of Asymptotic Expansion for BesselI[0, x]"];
```

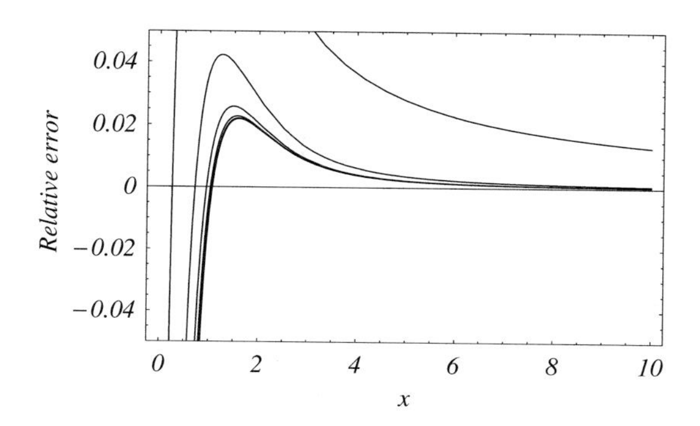

Figure 3.9. Accuracy of asymptotic expansion for $I_0[x]$ as k_{max} increases from 0 to 5.

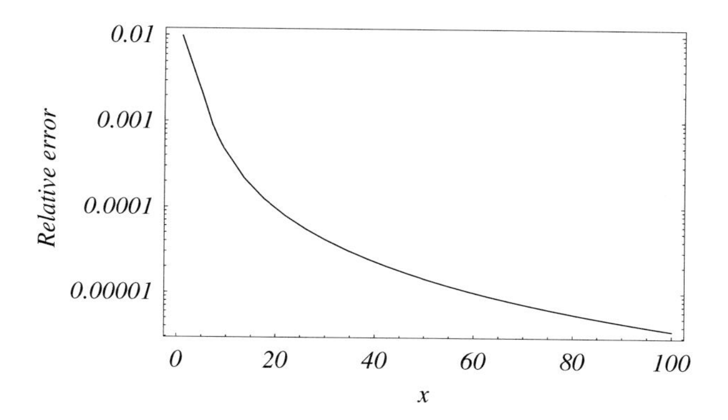

Figure 3.10. Accuracy of asymptotic expansion for $I_0[x]$ with $k_{max} = 5$.

```
LogPlot[Abs[error[x, 5]], {x, 1, 100},
  Frame → True, FrameLabel → {"x", "relative error"},
  PlotLabel → "Accuracy of Asymptotic Expansion for BesselI[0, x]"];
```

Furthermore, even though our derivation employed real variables, it is clear that the series is valid more generally for $-\frac{\pi}{2} < \text{Arg}[z] < \frac{\pi}{2}$. However, as $|\text{Arg}[z]| \to \frac{\pi}{2}$ the accuracy deteriorates for given r. Figure 3.11 shows the r-dependence of the relative error in $I_0[r e^{i\theta}]$ for $0 \le \theta \le \frac{3\pi}{8}$ in steps of $\frac{\pi}{8}$ for $k_{max} = 5$. Although the error for small r increases with θ, this expansion still provides reasonable accuracy for modest k_{max} over a substantial range of polar angles.

```
LogPlot[Evaluate[Table[Abs[error[r e^(i θ), 5]], {θ, 0, 3π/8, π/8}]],
  {r, 1, 100}, Frame → True,
  FrameLabel → {"r", "relative error"},
  PlotLabel → "Accuracy of Asymptotic Expansion for BesselI[0, x]"];
```

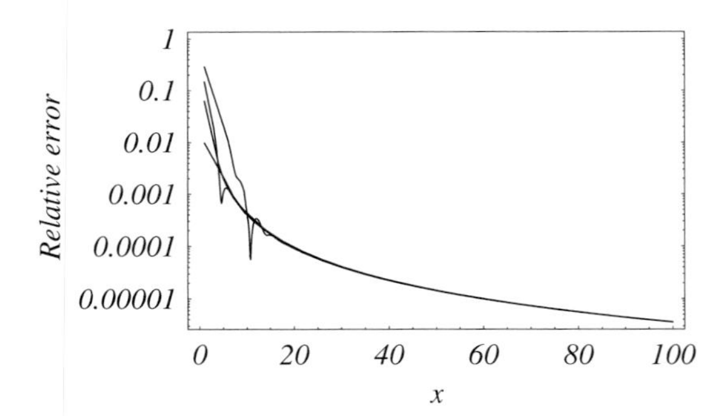

Figure 3.11. Accuracy of asymptotic expansion for $I_0[re^{i\theta}]$ with $k_{max} = 5$ and $0 \le \theta \le \frac{3\pi}{8}$ in steps of $\frac{\pi}{8}$.

Problems for Chapter 3

1. Landau straggling

If a fast charged particle passes through a thin film of material, it will lose energy by ionizing atoms via a stochastic process. Landau used a Laplace transform technique to derive the probability distribution for this energy-loss process. Without presenting the details, suffice it to say that the function

$$\varphi[\lambda] = \frac{1}{2\pi i} \int_{c-i\infty}^{c+i\infty} p^p e^{\lambda p}\, dp \tag{3.54}$$

plays an important role in the theory. The Laplace parameter c is an arbitrary positive real number. Use the method of steepest descent to derive an asymptotic approximation for this function that is useful for $\lambda \ll 0$.

2. stellar reaction rate

A star is powered by nuclear reactions in its core. Suppose that the average rate of nuclear collisions with center-of-mass energy between E and $E + dE$ is $N e^{-E/T} E\, dE$ where N is a constant and T is the core temperature in energy units. Also suppose that the probability that a collision with energy E results in a nuclear reaction is $M e^{-\alpha/\sqrt{E}}$ where M and α are constants. Use the method of steepest descent to determine the reaction rate assuming that $T \ll \alpha^2$.

3. incomplete gamma function

The incomplete gamma function, $\Gamma[a, z]$, can be defined in terms of the contour integral

$$\Gamma[a, z] = \int_z^\infty dt\, e^{-t} t^{a-1} \tag{3.55}$$

where a and z are complex and where the contour in the complex t-plane starts at z and approaches $t \to (+\infty, 0)$ smoothly. If necessary, a branch cut is made on the negative real t-axis. Use the method of steepest descent to determine the leading asymptotic behavior of the function $f[z] = \Gamma[z+1, z]$ when $|\arg[z]| < \pi$.

4. Airy function

The Airy functions Ai[z] and Bi[z] are solutions to the differential equation

$$w''[z] - zw[z] == 0 \tag{3.56}$$

and arise in boundary-value problems in quantum mechanics and electrodynamics. A useful integral representation of Ai[x] for positive real x is given by

$$\mathrm{Ai}[x] = \frac{1}{2\pi}\int_{-\infty}^{\infty} \mathrm{Exp}\left[-i\left(\omega x + \frac{\omega^3}{3}\right)\right] d\omega \tag{3.57}$$

Use the method of steepest descent to determine the leading asymptotic behavior of Ai[x] for large x. (Hint: choose the appropriate saddle point carefully.)

5. Bessel function

The *Schläfli integral* representation of the Bessel function $J_\nu[z]$ takes the form

$$J_\nu[z] = \frac{1}{2\pi i}\int_C \frac{dt}{t^{\nu+1}} \mathrm{Exp}\left[\left(t - \frac{1}{t}\right)\frac{z}{2}\right] \tag{3.58}$$

where the complex t-plane is cut on the negative real axis and where C enters from $-\infty$ just below the cut, circles the origin once in a positive sense, and then exits toward $-\infty$ just above the cut but is otherwise arbitrary. Use the method of steepest descent to obtain the leading asymptotic behavior, being careful to include both saddle points.

6. exponential integral

Develop an asymptotic series for the exponential integral

$$E_n[z] = \int_1^{\infty} e^{-zt} t^{-n}\, dt \tag{3.59}$$

for large $|z|$ with $n > 0$. You may assume that n is an integer. How many terms can be included for finite $|z|$?

7. asymptotic expansion of Fresnel functions

The Fresnel functions for real variables are defined by

$$C[x] = \int_0^x \mathrm{Cos}\left[\frac{\pi t^2}{2}\right] dt, \quad S[x] = \int_0^x \mathrm{Sin}\left[\frac{\pi t^2}{2}\right] dt \tag{3.60}$$

a) Determine the limiting values $C[\infty]$ and $S[\infty]$. (Hint: use contour integration of an exponential on a wedge of opening angle $\pi/4$.)

b) Obtain asymptotic expansions for the Fresnel functions. (Hint: evaluate corrections to the asymptotic values by integrating from x to ∞ by parts.)

8. method of stationary phase

The method of stationary phase is a somewhat simplified variation of the method of steepest descent that is useful for Fourier integrals of the form

$$f[t] = \int_{-\infty}^{\infty} g[\omega]\,\mathrm{Exp}[i\phi[\omega]t]\, d\omega \tag{3.61}$$

where the amplitude $g[\omega]$ depends relatively slowly upon frequency ω and the phase $\phi[\omega]$ is real. For large times $t \to \infty$, the phase of the exponential factor varies extremely rapidly with frequency unless $\phi[\omega]$ is practically constant. Therefore, the asymptotic value of the integral tends to be dominated by points of stationary phase.

a) Develop a general formula for the contribution of a point of stationary phase to the asymptotic behavior of $f[t]$. How does one handle several such points?

b) Suppose that $\phi'[\omega] \neq 0$ anywhere in the interval $\omega_1 < \omega < \omega_2$. Use integration by parts to show that

$$\int_{\omega_1}^{\omega_2} g[\omega]\,\mathrm{Exp}[i\phi[\omega]t]\,d\omega \tag{3.62}$$

is of order t^{-1} and, hence, that intervals lacking a point of stationary phase can be neglected in the limit $t \to \infty$.

c) Suppose that a wave function is represented by

$$\psi[x,t] = \frac{1}{2\pi}\int_{-\infty}^{\infty} \tilde{\psi}[k]\,\mathrm{Exp}[i(kx-\omega t)]\,dk, \qquad \omega = \frac{k^2}{2m} \tag{3.63}$$

where

$$\tilde{\psi}[k] = \int_{-\infty}^{\infty} dx e^{-ikx}\psi[x,0] \tag{3.64}$$

is the spectral representation of the initial wave function and varies slowly with k if $\psi[x,0]$ is well localized. Evaluate the intensity, $|\psi|^2$, for $x \to \infty$ and $t \to \infty$ with constant ratio $\xi = x/t$.

9. surface waves on deep water

One-dimensional surfaces waves on deep water in a narrow channel take the form

$$\psi[x,t] = \int_{-\infty}^{\infty} \dot{\mathcal{G}}[x-y,t]\psi[y,0]\,dy + \int_{-\infty}^{\infty} \mathcal{G}[x-y,t]\dot{\psi}[y,0]\,dy \tag{3.65}$$

where $\dot{f}[x,t] = \partial f[x,t]/\partial t$ and

$$\mathcal{G}[x,t] = \int_{-\infty}^{\infty} \frac{\mathrm{Sin}[\omega t]}{\omega} e^{ikx}\frac{dk}{2\pi} \tag{3.66}$$

with $\omega = \sqrt{g|k|}$ (g is the gravitational acceleration). Here $\psi[x,0]$ and $\dot{\psi}[x,0]$ represent the initial displacement and vertical velocity profiles and we evaluate $\psi[x,t]$ for later times, $t \geq 0$. (See A. L. Fetter and J. D. Walecka, *Theoretical Mechanics of Particles and Continua*, McGraw-Hill, 1980.) Use the method of stationary phase to evaluate $\mathcal{G}[x,t]$ for $t \to \infty$ and $x \to \infty$ simultaneously in the constant ratio $\xi = x/t$. Sketch and interpret this function.

4 Generalized Functions

Abstract. The analysis of physical systems is often simplified using idealizations that represent an impulse being delivered instantaneously or a charge distribution being confined to a mathematical point without volume. Here we develop the Dirac delta function in terms of limiting properties of nascent delta functions. Similar methods are used for related generalized functions (or distributions).

4.1 Motivation

Often it is useful to idealize an external force as instantaneous or a charge distribution as two-dimensional. For example, if the response of a system is much slower than the duration of an applied force, it becomes useful to employ impulse, defined as the time integral of the force

$$\mathcal{I} = \int_{t_0-\tau/2}^{t_0+\tau/2} F[t]\, dt = \Delta p \tag{4.1}$$

in a limiting process where the duration $\tau \to 0$ approaches zero while the peak force $F[t_0] \to \infty$ such that the product $\tau F[t_0] = \Delta p$ remains constant. Thus, we imagine that the entire change of momentum, Δp, is delivered instantaneously at time t_0. If the response of the system is much slower than the duration of the force, it becomes insensitive to the detailed shape of $F[t]$ and responds only to the net impulse Δp. There is then considerable latitude in choosing a convenient representation for $F[t]$.

Consider a Gaussian function

$$\delta[x, \sigma] = \frac{\operatorname{Exp}\left[-\frac{x^2}{2\sigma^2}\right]}{\sqrt{2\pi\sigma^2}} \tag{4.2}$$

defined with unit integral

$$\int_{-\infty}^{\infty} \delta[x, \sigma]\, dx = 1 \tag{4.3}$$

As the width σ decreases, the height increases to maintain a constant integral. Thus, we could represent an impulsive force by

$$F[t] = \lim_{\tau \to 0} \mathcal{I}\delta[t, \tau] \tag{4.4}$$

such that the integral

$$\int_{-\epsilon}^{\epsilon} F[t]\, dt = \mathcal{I} \tag{4.5}$$

Graduate Mathematical Physics. James J. Kelly

ISBN: 3-527-40637-9

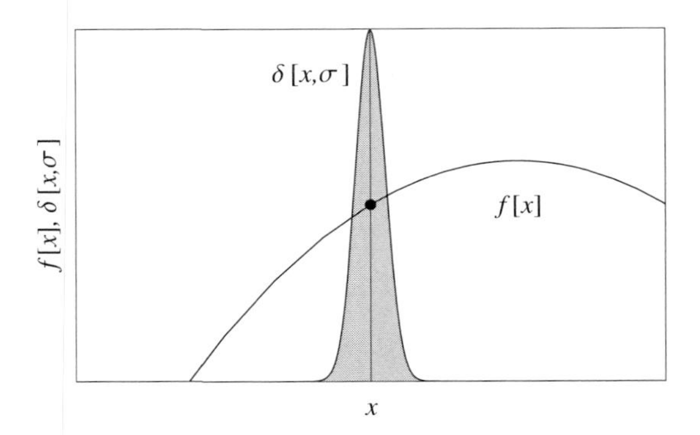

Figure 4.1. Integration of a broad function, $f[x]$, against a narrow function, $\delta[x, \sigma]$.

over any small but finite interval, $|t| < \epsilon$ with $\epsilon \to 0$, provides the same impulse $\mathcal{I}$. Therefore, one of the representations of the *Dirac delta function* $\delta[x]$ is offered by the limit

$$\delta[x] = \lim_{\sigma\to 0} \delta[x, \sigma] = \lim_{\sigma\to 0} \frac{\mathrm{Exp}\left[-\frac{x^2}{2\sigma^2}\right]}{\sqrt{2\pi\sigma^2}} \tag{4.6}$$

However, $\delta[x]$ is clearly not an ordinary function – this limit vanishes for any $x \neq 0$ and does not exist for $x = 0$.

$$x \neq 0 \Longrightarrow \lim_{\sigma\to 0} \frac{\mathrm{Exp}\left[-\frac{x^2}{2\sigma^2}\right]}{\sqrt{2\pi\sigma^2}} = 0 \tag{4.7}$$

$$x \to 0 \Longrightarrow \frac{\mathrm{Exp}\left[-\frac{x^2}{2\sigma^2}\right]}{\sqrt{2\pi\sigma^2}} \to \frac{1}{\sigma\sqrt{2\pi}} \to \infty \tag{4.8}$$

Despite the fact that the limit of the integrand vanishes everywhere but the one point where it diverges, the integral remains finite, in this representation constant, and converges to unity. We describe $\delta[x, \sigma]$ as a *nascent* delta function, a delta function in the making.

Consider an ordinary function $f[x]$ which is continuous at $x = 0$. Despite the peculiar divergence of $\delta[x]$, the integral

$$\int_{-\infty}^{\infty} f[x]\delta[x]\, dx = f[0] \tag{4.9}$$

converges simply to $f[0]$. Figure 4.1 illustrates the evaluation of this integral. We assume that $f[x]$ is continuous in some interval around $x = 0$ and choose σ small enough to ensure that $f[x]$ is well behaved throughout the domain where $\delta[x, \sigma]$ has appreciable strength. As σ is reduced further, the product $\delta[x, \sigma]f[x]$ is approximated increasingly well by $f[0]\delta[x, \sigma]$ because the variation of $f[x]$ within the width of $\delta[x, \sigma]$ can be made arbitrarily small. Consequently, we can extract the constant $f[0]$ from the integral

$$\int_{-\infty}^{\infty} f[x]\delta[x, \sigma]\, dx \approx f[0] \int_{-\infty}^{\infty} \delta[x, \sigma]\, dx = f[0] \tag{4.10}$$

to obtain

$$\int_{-\infty}^{\infty} f[x]\delta[x]\, dx = \lim_{\sigma \to 0} \int_{-\infty}^{\infty} f[x]\delta[x, \sigma]\, dx = f[0] \tag{4.11}$$

without reference to the detailed shape of $\delta[x, \sigma]$. Notice that the limits of integration need not be infinite, just much larger than σ.

Notice that the order of integration and limit operations are not interchangeable – one must evaluate the integral before the limit because the limit of the nascent delta function, $\delta[x, \sigma]$, is not a respectable function. Therefore, $\delta[x]$ is only defined within the context of integration against a continuous function. The Dirac delta function is an example of a *generalized function* defined in terms of its behavior under an integral. Mathematicians prefer to call generalized functions *distributions* instead, but to this author the term distribution hardly seems applicable to an entity that vanishes at all but one point. Other useful distributions include the Heaviside step function, the square function, and derivatives of the delta function. Familiarity with the properties of generalized functions can greatly simply the evaluation of integrals or the solutions of differential equations that involve large disparities between the spatial or temporal profile of one component and the response of another. Of course, one could employ only well-behaved narrow functions and evaluate the appropriate limits explicitly, thereby recreating a generalized function, but who has the time?

4.2 Properties of the Dirac Delta Function

If $f[x]$ is continuous at x_0, then

$$\int_{-\infty}^{\infty} f[x]\delta[x - x_0]\, dx = f[x_0] \tag{4.12}$$

selects the value at x_0. This property is easily verified using the change of variables

$$y = x - x_0 \Longrightarrow \int_{-\infty}^{\infty} f[x]\delta[x - x_0]\, dx = \int_{-\infty}^{\infty} f[y + x_0]\delta[y]\, dy \tag{4.13}$$

and the definition

$$\int_{-\infty}^{\infty} g[y]\delta[y]\, dy = g[0] \tag{4.14}$$

where

$$g[y] = f[y + x_0] \Longrightarrow g[0] = f[x_0] \tag{4.15}$$

Similarly, one can prove that

$$a \neq 0 \Longrightarrow \int_{-\infty}^{\infty} f[x]\delta[ax]\, dx = \frac{f[0]}{|a|} \tag{4.16}$$

using

$$a > 0 \Longrightarrow \int_{-\infty}^{\infty} f[x]\delta[ax]\,dx = \frac{1}{a}\int_{-\infty}^{\infty} f\left[\frac{y}{a}\right]\delta[y]\,dy = \frac{f[0]}{a} = \frac{f[0]}{|a|} \tag{4.17}$$

$$a < 0 \Longrightarrow \int_{-\infty}^{\infty} f[x]\delta[ax]\,dx = \frac{1}{a}\int_{\infty}^{-\infty} f\left[\frac{y}{a}\right]\delta[y]\,dy = -\frac{f[0]}{a} = \frac{f[0]}{|a|} \tag{4.18}$$

Thus, we write

$$a \neq 0 \Longrightarrow \delta[ax] = \frac{\delta[x]}{|a|} \tag{4.19}$$

as an equality even though neither side represents a function because the replacement of one side with the other is operationally correct within the context of an appropriate integral. Notice that this result suggests that $\delta[x]$ is even; hence, it is useful to require that property of nascent delta functions.

How should one interpret an expression of the form $\delta[x^2 - a^2]$? The interpretation is based upon the behavior of that generalized function under an integral. We assume that a is real and, without loss of generality, that $a > 0$. Recognizing that $x^2 - a^2$ has roots at $x = \pm a$, we can separate the integral into two contributions

$$\int_{-\infty}^{\infty} f[x]\delta[x^2 - a^2]\,dx = \int_{-\infty}^{0} f[x]\delta[(x+a)(x-a)]\,dx + \int_{0}^{\infty} f[x]\delta[(x+a)(x-a)]\,dx \tag{4.20}$$

where in the first integral

$$x \to -a \Longrightarrow x - a \to -2a \Longrightarrow \int_{-\infty}^{0} f[x]\delta[(x+a)(x-a)]\,dx = \frac{f[-a]}{|2a|} \tag{4.21}$$

while in the second

$$x \to a \Longrightarrow x + a \to 2a \Longrightarrow \int_{0}^{\infty} f[x]\delta[(x+a)(x-a)]\,dx = \frac{f[a]}{|2a|} \tag{4.22}$$

such that

$$a \neq 0 \Longrightarrow \int_{-\infty}^{\infty} f[x]\delta[x^2 - a^2]\,dx = \frac{f[a] + f[-a]}{|2a|} \tag{4.23}$$

provided only that $f[x]$ is continuous near both $x = \pm a$ and that $a \neq 0$. Therefore, from an operational point of view, we may write

$$a \neq 0 \Longrightarrow \delta[x^2 - a^2] = \frac{\delta[x-a] + \delta[x+a]}{|2a|} \tag{4.24}$$

because both sides produce the same result when integrated against any suitable function. (Note, however, that $\delta[x^2]$ is uninterpretable.)

Finally, we can generalize this procedure to evaluate $\delta\big[g[x]\big]$ where $g[x]$ is a function with an arbitrary number of isolated roots.

Theorem 19. *Suppose that $g[x]$ has isolated simple roots at $\{x_n, n = 1, 2, \dots\}$, such that*

$$x \approx x_n \Longrightarrow g[x] \approx (x - x_n)g'[x_n]$$

with $g'[x_n] \neq 0$. Then

$$\delta\big[g[x]\big] = \sum_n \frac{\delta[x - x_n]}{|g'[x_n]|} \tag{4.25}$$

If the roots of $g[x]$ are isolated, we may evaluate the integral

$$\int_{-\infty}^{\infty} f[x]\delta\big[g[x]\big]\,dx = \sum_n \int_{x_n-\epsilon}^{x_n+\epsilon} f[x]\delta\big[(x - x_n)g'[x_n]\big]\,dx \tag{4.26}$$

as a sum over contributions from each root of $g[x]$ where ϵ is a positive number smaller than the spacing between roots. Then using the scaling property $\delta[ax] = \delta[x]/|a|$, we obtain

$$\int_{-\infty}^{\infty} f[x]\delta\big[g[x]\big]\,dx = \sum_n \frac{f[x_n]}{|g'[x_n]|} \tag{4.27}$$

Therefore, we obtain an operational definition

$$\{g[x_n] = 0, g'[x_n] \neq 0\} \Longrightarrow \delta\big[g[x]\big] = \sum_n \frac{\delta[x - x_n]}{|g'[x_n]|} \tag{4.28}$$

that incorporates all preceding results from this section.

4.3 Other Useful Generalized Functions

4.3.1 Heaviside Step Function

A useful version of the unit step function, known as the *Heaviside* function, can be constructed by integration over a delta function, whereby

$$\Theta[x] = \int_{-\infty}^{x} \delta[y]\,dy \tag{4.29}$$

serves as an operational definition. A more controlled definition of a nascent step function is based upon a suitable nascent delta function, such that

$$\Theta[x, \sigma] = \int_{-\infty}^{x} \delta[y, \sigma]\,dy \tag{4.30}$$

$$\Theta[x] = \lim_{\sigma \to 0} \Theta[x, \sigma] \tag{4.31}$$

If we choose the Gaussian representation, then we obtain

$$\delta[x, \sigma] = \frac{\mathrm{Exp}\left[-\frac{x^2}{2\sigma^2}\right]}{\sqrt{2\pi\sigma^2}} \Longrightarrow \Theta[x, \sigma] = \frac{1}{2}\left(1 + \mathrm{Erf}\left[\frac{x}{\sqrt{2}\sigma}\right]\right) \tag{4.32}$$

where $\mathrm{Erf}[x]$ is the error function. Figure 4.2 demonstrates that the transition between values of 0 for $x \ll 0$ or 1 for $x \gg 1$ becomes very sharp as σ becomes small. Furthermore,

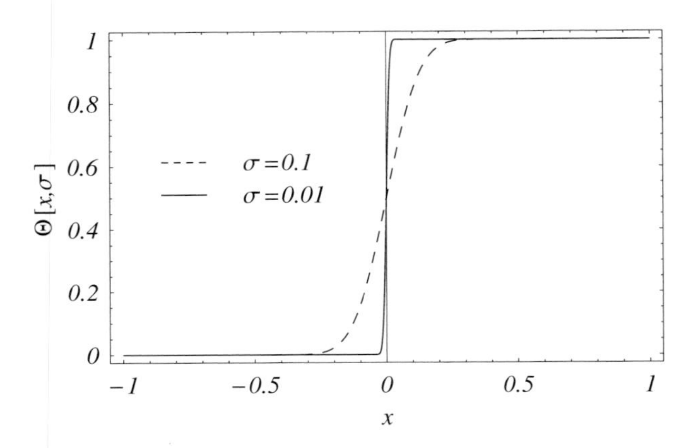

Figure 4.2. Nascent step functions.

because $\delta[x]$ is even, we obtain $\Theta[0, \sigma] = 1/2$. Therefore, the Heaviside step function assumes the values

$$\Theta[x] = \begin{cases} 0 & x < 0 \\ \frac{1}{2} & x = 0 \\ 1 & x > 0 \end{cases} \tag{4.33}$$

Notice that $\Theta[x]$ is an ordinary function despite the discontinuity at $x = 0$. There are other versions of the step function defining the value at $x = 0$ to be either 0 or 1, but $\Theta[x]$ is the version that arises naturally in calculations that employ the Dirac delta function. Also notice that once the limit $\sigma \to 0$ is taken $\Theta[x]$ becomes independent of the representation of $\delta[x, \sigma]$, depending only upon the fact that the delta function is defined with unit area and the properties $\delta[x \neq 0] = 0$ and $\delta[-x] = \delta[x]$.

Despite the discontinuity of $\Theta[x]$ at $x = 0$, its definition in terms of integration over a delta function can be used to deduce an operational definition for the derivative of $\Theta[x]$, namely

$$\Theta[x] = \int_{-\infty}^{x} \delta[y]\, dy \Longrightarrow \Theta'[x] = \delta[x] \tag{4.34}$$

where the derivative of the unit discontinuity becomes a spike of unit area. This result is consistent with the behavior of the underlying nascent functions

$$\Theta[x, \sigma] = \int_{-\infty}^{x} \delta[y, \sigma]\, dy \Longrightarrow \Theta'[x, \sigma] = \delta[x, \sigma] \tag{4.35}$$

4.3.2 Derivatives of the Dirac Delta Function

Can one interpret the derivative of a delta function, $\delta'[x]$? Once again we appeal to the Gaussian representation of a nascent delta function to evaluate

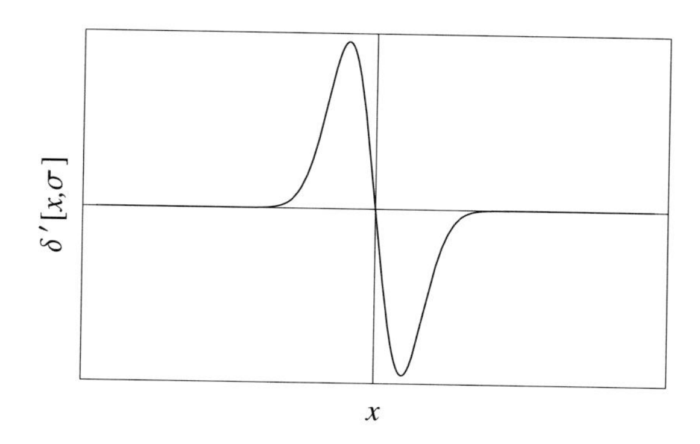

Figure 4.3. Derivative of nascent delta function.

$$\delta[x,\sigma] = \frac{\text{Exp}\left[-\frac{x^2}{2\sigma^2}\right]}{\sqrt{2\pi\sigma^2}} \Longrightarrow \delta'[x,\sigma] = \frac{\partial\delta[x,\sigma]}{\partial x} = -\frac{x\,\text{Exp}\left[-\frac{x^2}{2\sigma^2}\right]}{\sigma^3\sqrt{2\pi}} \tag{4.36}$$

Figure 4.3 demonstrates that this function has two equal and opposite lobes centered near $x \approx \pm\sigma$ with area proportional to σ^{-1}. As σ decreases, these lobes become taller and closer together. Thus, one expects integration of $\delta'[x,\sigma]$ against a continuous function to produce a result of the form

$$\int_{-\infty}^{\infty} f[x]\delta'[x,\sigma]\,dx \propto \frac{f[-\sigma]-f[\sigma]}{\sigma} \propto -f'[0] \tag{4.37}$$

wherein the ratio between the difference of the function on opposite sides of $x = 0$ and the width of the interval is proportional to the derivative of $f[x]$ and of opposite sign. Although $\delta'[x,\sigma]$ obviously does not yield an ordinary function in the limit $\sigma \to 0$, its behavior under an integral serves as the operational definition of a generalized function. Assuming that $f[x]$ is differentiable near $x = 0$, integration by parts yields

$$\int_{-\infty}^{\infty} f[x]\delta'[x,\sigma]\,dx = (f[x]\delta[x,\sigma])_{-\infty}^{\infty} - \int_{-\infty}^{\infty} f'[x]\delta[x,\sigma]\,dx \tag{4.38}$$

The surface terms vanish for any reasonably well-behaved $f[x]$, while the remaining integral reduces to $f'[0]$ in the limit $\sigma \to 0$, such that

$$\int_{-\infty}^{\infty} f[x]\delta'[x]\,dx = \lim_{\sigma\to 0}\int_{-\infty}^{\infty} f[x]\delta'[x,\sigma]\,dx = -\int_{-\infty}^{\infty} f'[x]\delta[x]\,dx = -f'[0] \tag{4.39}$$

It is now straightforward to generalize this procedure to obtain an interpretation

$$\int_{-\infty}^{\infty} f[x]\delta^{(n)}[x]\,dx = (-)^n f^{(n)}[0] \tag{4.40}$$

of arbitrary derivatives, assuming that $f[x]$ is differentiable at least n times at $x = 0$. Note that we use the notation

$$\delta^{(n)}[x] = \frac{d^n\delta[x]}{dx^n}, \quad f^{(n)}[x] = \frac{d^n f[x]}{dx^n} \tag{4.41}$$

4.4 Green Functions

Suppose that a damped harmonic oscillator is subjected to an external force $F[t]$ with finite duration and is at rest before the onset of that force. The displacement satisfies a differential equation of the form

$$m\ddot{x}[t] + 2m\gamma\dot{x}[t] + m\omega_0^2 x[t] == F[t] \tag{4.42}$$

where ω_0 is the natural frequency of undamped oscillations, γ is the damping constant, and time derivatives are indicated by the dot notation. The homogeneous equation has solutions of the form

$$m\ddot{x}[t] + 2m\gamma\dot{x}[t] + m\omega_0^2 x[t] == 0 \Longrightarrow x[t] = \text{Exp}[-\gamma t](A\,\text{Sin}[\omega t] + B\,\text{Cos}[\omega t]) \tag{4.43}$$

where A and B are constants and where we assume weak damping to obtain

$$\gamma < \omega_0 \Longrightarrow \omega = \sqrt{\omega_0^2 - \gamma^2} \tag{4.44}$$

Now suppose that the mass is struck at time t' such that there is an abrupt change in velocity. We assume that the impulse acts during an infinitesimal time interval that is too short for the position of the mass to change. Thus, the dissipative and restoring forces have no effect and we find that the instantaneous change in velocity

$$F[t] \to \mathcal{I}\delta[t - t'] \Longrightarrow \Delta\dot{x}[t'] = \frac{\mathcal{I}}{m} = \lim_{\varepsilon \to 0}(\dot{x}[t' + \varepsilon] - \dot{x}[t' - \varepsilon]) \tag{4.45}$$

reduces to simply $\mathcal{I}/m$, where $\mathcal{I}$ is the impulse delivered. Recognizing that the homogeneous equation applies both before and after the impulse, solutions for those intervals take the form

$$t < t' \Longrightarrow x[t] = \text{Exp}[-\gamma t](A_- \,\text{Sin}[\omega t] + B_- \,\text{Cos}[\omega t]) \tag{4.46}$$

$$t > t' \Longrightarrow x[t] = \text{Exp}[-\gamma t](A_+ \,\text{Sin}[\omega t] + B_+ \,\text{Cos}[\omega t]) \tag{4.47}$$

where the coefficients are chosen to satisfy the boundary conditions

$$x[t' - \varepsilon] = \dot{x}[t' - \varepsilon] = 0 \Longrightarrow A_- = B_- = 0 \tag{4.48}$$

$$x[t' + \varepsilon] = 0, \dot{x}[t' + \varepsilon] = \frac{\mathcal{I}}{m} \Longrightarrow A_+ = \frac{\mathcal{I}}{m\omega} e^{\gamma t'} \,\text{Cos}[\omega t'], B_+ = \frac{\mathcal{I}}{m\omega} e^{\gamma t'} \,\text{Sin}[\omega t'] \tag{4.49}$$

such that the *Green function*

$$G[t, t'] = \frac{\text{Exp}[-\gamma(t - t')]}{m\omega} \text{Sin}[\omega(t - t')]\Theta[t - t'] \tag{4.50}$$

represents the response of the system at time t to a unit impulse at time t'. This response takes the form of a damped sinusoidal oscillation. The step function enforces *causality* – there can be no effect before its cause. Also, notice that $G[t, t'] = G[t - t']$ depends only upon the difference between the times because the homogeneous equation has no explicit time dependence and thus is insensitive to clock offsets. We could have simplified the algebra (omitted anyway) by exploiting this symmetry from the beginning. The response of the system to a disturbance of finite duration can be evaluated by using the *superposition principle* for linear differential equations to sum the effects of a sequence of impulses that represents the time profile according to

$$F[t] = \int_{-\infty}^{\infty} F[t']\delta[t-t']\,dt' \Longrightarrow x[t] = x_0[t] + \int_{-\infty}^{\infty} G[t,t']F[t']\,dt' \tag{4.51}$$

where $x_0[t]$ is a solution to the homogeneous equation

$$\ddot{x}_0[t] + 2\gamma\dot{x}_0[t] + \omega_0^2 x_0[t] == 0 \tag{4.52}$$

Often the boundary conditions $x[-\infty] = \dot{x}[-\infty] = 0$ specifying that the system is at rest before the action of an external disturbance require $x_0[t] = 0$; if not, $x_0[t]$ is designed to match the appropriate boundary conditions. To demonstrate this solution formally, let

$$\mathcal{L} = m\frac{d^2}{dt^2} + 2m\gamma\frac{d}{dt} + m\omega_0^2 \tag{4.53}$$

represent a linear differential operator such that the equation of motion becomes

$$\mathcal{L}x[t] == F[t] \tag{4.54}$$

The homogeneous solution satisfies

$$\mathcal{L}x_0[t] == 0 \tag{4.55}$$

while the Green function satisfies

$$\mathcal{L}G[t,t'] == \delta[t-t'] \tag{4.56}$$

where the operator applies to the first argument of $G[t,t']$ with the second held fixed. Applying $\mathcal{L}$ to the proposed solution,

$$\begin{aligned} x[t] = x_0[t] + \int G[t,t']F[t']\,dt' \Longrightarrow \mathcal{L}x[t] &= \int \mathcal{L}G[t,t']F[t']\,dt' \\ &= \int \delta[t-t']F[t']\,dt' = F[t] \end{aligned} \tag{4.57}$$

one recovers the original differential equation. Therefore, the function

$$x[t] = x_0[t] + \int_{-\infty}^{\infty} G[t,t']F[t']\,dt' \tag{4.58}$$

with

$$G[t,t'] = \frac{\text{Exp}[-\gamma(t-t')]}{m\omega}\text{Sin}[\omega(t-t')]\Theta[t-t'] \tag{4.59}$$

provides a solution to the inhomogeneous problem in the form of a *convolution integral* in which the displacement at time t is determined by the net response of the system to external forces at all earlier times. The influence of a force at time t' upon a later time t is reduced by the exponential damping factor, such that recent forces tend to matter most. Notice that causality is enforced by the step function in $G[t,t']$ and that we may limit the range of integration strictly to the past, such that

$$x[t] = x_0[t] + \int_{-\infty}^{t} G[t,t']F[t']\,dt' \tag{4.60}$$

should be interpreted as the sum of effects produced by past causes for all past times up to the present. Nevertheless, one often extends the upper limit of integration to ∞ for convenience, confident that the Green function eliminates sensitivity to the future.

Note that our derivation was more schematic than rigorous; for example, we implicitly assumed uniform convergence in order to move $\mathcal{L}$ inside the convolution integral. Nevertheless, it is consistent with the spirit of generalized functions and can be made rigorous by replacing the delta and step functions with the appropriate nascent functions and evaluating the limits more carefully. Although the $G[t,t']$ obtained here is actually an ordinary function, it is described as a generalized function because the inhomogeneous term upon which it is based is idealized as a delta function. It is also important to recognize that the Green function depends not only upon $\mathcal{L}$ but also upon the boundary conditions. The solution to

$$\mathcal{L}G[t,t'] == \delta[t-t'] \tag{4.61}$$

is usually not unique without specification of the boundary conditions. The version we derived above is often described as a *retarded Green function* because the effect follows the cause (is retarded in time), and is denoted by $G^{(+)}[t,t']$. If $\mathcal{L}$ describes a wave equation, we might be interested in an advanced solution, $G^{(-)}$, or a standing-wave solution, $G^{(0)}$.

The present analysis is typical of the Green-function method for linear systems. In the next several chapters we will develop a variety of useful techniques for construction of the Green functions for a variety of systems. Often there will be several equivalent representations and the most useful representation may depend upon its anticipated application.

4.5 Multidimensional Delta Functions

A point mass at $\vec{r}\,' = (x',y',z')$ is represented by a density distribution of the form

$$\rho[x,y,z] = m\delta[\vec{r}-\vec{r}\,'] = m\delta[x-x']\delta[y-y']\delta[z-z'] \tag{4.62}$$

where in Cartesian coordinates the three-dimensional delta function factors into a product of three one-dimensional delta functions of identical form such that

$$\begin{aligned}
\int_V \rho[x,y,z]\,d^3r &= m\int_V \delta[\vec{r}-\vec{r}\,']\,d^3r \\
&= m\int \delta[x-x']\,dx \int \delta[y-y']\,dy \int \delta[z-z']\,dz \\
&= \begin{cases} m & \vec{r}\,' \in V \\ 0 & \vec{r}\,' \notin V \end{cases}
\end{aligned} \tag{4.63}$$

returns the mass for any volume that contains the particle and vanishes otherwise. If necessary, we can work with a spherically symmetric nascent delta function of the form

$$\delta[\vec{r}-\vec{r}'] = \lim_{\sigma\to 0}\delta[\vec{r}-\vec{r}',\sigma] = \lim_{\sigma\to 0}\frac{\mathrm{Exp}\left[-\frac{(\vec{r}-\vec{r}')^2}{2\sigma^2}\right]}{(2\pi\sigma^2)^{3/2}} \tag{4.64}$$

that represents a compact charge distribution and is normalized according to

$$\int_V \delta[\vec{r}-\vec{r}',\sigma]\, d^3r = \begin{cases} 1 & \vec{r}' \in V \\ 0 & \vec{r}' \notin V \end{cases} \tag{4.65}$$

Notice that a three-dimensional delta function has dimensions of inverse volume. Of course, many alternative nascent delta functions with these same basic properties are also available.

In curvilinear coordinate systems (spherical, cylindrical, etc.), one must include Jacobian factors for each one-dimensional delta function, with respect to a particular coordinate, that match the differential volume element. For example, in spherical coordinates (r, θ, ϕ) we factor the three-dimensional delta function in the form

$$\delta[\vec{r}-\vec{r}'] = \frac{\delta[r-r']}{rr'}\frac{\delta[\theta-\theta']}{\mathrm{Sin}[\theta]}\delta[\phi-\phi'] = \frac{\delta[r-r']}{rr'}\delta\big[\mathrm{Cos}[\theta]-\mathrm{Cos}[\theta']\big]\delta[\phi-\phi'] \tag{4.66}$$

such that

$$\begin{aligned}\int \delta[\vec{r}-\vec{r}']\, d^3r &= \int_0^\infty \frac{\delta[r-r']}{rr'}r^2\,dr \int_0^\pi \frac{\delta[\theta-\theta']}{\mathrm{Sin}[\theta]}\,\mathrm{Sin}[\theta]\,d\theta \int_0^{2\pi}\delta[\phi-\phi']\,d\phi \\ &= \int_0^\infty \frac{\delta[r-r']}{rr'}r^2\,dr \int_{-1}^{1}\delta\big[\mathrm{Cos}[\theta]-\mathrm{Cos}[\theta']\big]\,d\mathrm{Cos}[\theta] \int_0^{2\pi}\delta[\phi-\phi']\,d\phi \\ &= 1\end{aligned} \tag{4.67}$$

where for the polar angle it is often more convenient to employ $\zeta = \mathrm{Cos}[\theta]$ instead of θ directly. Also note for the radial delta function we chose to express the denominator in the symmetric form rr' instead of either r^2 or r'^2, but all three forms are equivalent because the delta function carries the instruction $r \to r'$. Similarly, in cylindrical coordinates (ξ, ϕ, z) we find

$$dV = \xi\, d\xi\, d\phi\, dz \Longrightarrow \delta[\vec{r}-\vec{r}'] = \frac{\delta[\xi-\xi']}{\xi}\delta[\phi-\phi']\delta[z-z'] \tag{4.68}$$

Suppose that charge q is distributed uniformly on a disk of radius a in the equatorial plane. The charge density can then be represented by

$$\rho[x,y,z] = \frac{3q}{2\pi a^3}\Theta[a-r]\delta\big[\mathrm{Cos}[\theta]\big] = \frac{3q}{2\pi a^3}\Theta[a-r]\delta\left[\theta-\frac{\pi}{2}\right] = \frac{q}{\pi a^2}\Theta[a-\xi]\delta[z] \tag{4.69}$$

where the first two forms employ variations of spherical coordinates and the last employs cylindrical coordinates. The step function, constructed as an integral over a delta function, is dimensionless. The coefficients must be chosen to ensure proper normalization of the total charge for each representation of the differential volume element. Again, physically realizable distributions could employ nascent delta and step functions but idealized functions can be used if we are not interested in details that depend upon the precise sharpness of the edges or the width of the disk.

Problems for Chapter 4

1. Nascent delta functions

Sketch the following functions for small positive σ and demonstrate that they can serve as nascent delta functions.

a) $\delta[x,\sigma] = \begin{cases} 0 & |x| > \sigma \\ \frac{1}{2\sigma} & |x| < \sigma \end{cases}$

b) $\delta[x,\sigma] = \frac{\sigma}{\pi}\frac{1}{\sigma^2+x^2}$

c) $\delta[x,\sigma] = \frac{1}{\pi}\frac{\mathrm{Sin}[x/\sigma]}{x}$

d) $\delta[x,\sigma] = \frac{\sigma}{\pi}\left(\frac{\mathrm{Sin}[x/\sigma]}{x}\right)^2$

Briefly discuss their advantages and disadvantages. Consider, for example, $\lim_{\sigma\to 0}\delta[x,\sigma]$ for $x \neq 0$ and also the behavior of

$$\lim_{\sigma\to 0}\int_{-\infty}^{\infty} f[x]\delta[x,\sigma]\,dx \tag{4.70}$$

when $f[x]$ diverges at $\pm\infty$. Contrast, where appropriate, with the behavior of the Gaussian representation.

2. Fermi distribution as nascent step function

The Fermi distribution

$$n[\varepsilon,\tau] = (e^{\varepsilon/\tau}+1)^{-1} \tag{4.71}$$

represents the occupation probability for a fermion state with energy ε relative to the Fermi level at temperature τ (in energy units). Show that as $\tau \to 0$ this function is closely related to a nascent step function and deduce the corresponding nascent delta function.

3. Higher derivatives of the delta function

Sketch $\delta''[x,\sigma]$ and $\delta'''[x,\sigma]$. Discuss these curves in terms of the finite-difference formulas for $f''[x]$ and $f'''[x]$.

4. Derivative of delta function with monomial factor

Use induction to demonstrate that

$$x^m\delta^{(n)}[x] = \begin{cases} (-)^m\frac{n!}{(n-m)!}\delta^{(n-m)}[x] & n \ge m \ge 0 \\ 0 & m > n \ge 0 \end{cases} \tag{4.72}$$

in the operational sense of a generalized function integrated against a suitable function $f[x]$. What conditions must be imposed upon $f[x]$?

5. Recoil factor for knockout reactions

Suppose that projectile a is scattered by nucleus A producing a final state containing the scattered particle b, a knocked out particle c, and residual nucleus B such that $a+A \longrightarrow b+c+B$. Let $(\varepsilon_i, \vec{p}_i)$ represent the energy and momentum of a particle labeled $i \in \{a, b, c, A, B\}$ with mass m_i and use relativistic kinematics throughout. If B is left in a discrete state of mass m_B, the reaction calculation often involves a recoil factor

$$\mathcal{R} = \int \delta[\varepsilon_b + \varepsilon_c + \varepsilon_B - \varepsilon_a - m_A]\, d\varepsilon_c \tag{4.73}$$

defined by an integral over a delta function that enforces energy conservation, where we have assumed that the target is at rest such that $\varepsilon_A = m_A$ and where

$$\varepsilon_B^2 = \vec{p}_B^{\,2} + m_B^2 = (\vec{p}_a - \vec{p}_b - \vec{p}_c)^2 + m_B^2 \quad \text{with} \quad \vec{p}_c^{\,2} = \varepsilon_c^2 - m_c^2 \tag{4.74}$$

according to momentum conservation. (Note that we use natural units with the light speed set to unity.) Find an explicit expression for $\mathcal{R}$ in terms of the energies and momenta of c and B. (Hint: the kinematics of particles a and b and the direction of $\vec{p}_c$ are fixed while ε_B and $|\vec{p}_c|$ depend upon ε_c.)

6. Green function for stretched string

Suppose that a string clamped at $x = 0$ and $x = L$ is under tension T and is subjected to a static load $F[x]$. The displacement $y[x]$ satisfies

$$Ty''[x] == F[x] \tag{4.75}$$

for sufficiently small F/T. Note that F is the force per unit length. Evaluate the Green function, $G[x, x']$, which satisfies

$$T\frac{\partial^2}{\partial x^2}G[x, x'] == \delta[x - x'] \tag{4.76}$$

and express the general solution for an arbitrary load distribution in terms of the Green function.

7. Green function for damped oscillator

A damped harmonic oscillator satisfies a differential equation of the form

$$m\ddot{x}[t] + 2m\gamma\dot{x}[t] + m\omega_0^2 x[t] == F[t] \tag{4.77}$$

where m is the mass, γ is the damping parameter, ω_0 is the frequency of free undamped oscillations, and F is an additional external force. Evaluate and graphically compare the retarded Green functions for underdamped ($\gamma < \omega_0$), critically damped ($\gamma = \omega_0$), and overdamped ($\gamma > \omega_0$) systems.

8. Response of damped oscillator to step using Green function

a) Use the Green function for a damped oscillator to solve

$$\ddot{x}[t] + 2\gamma\dot{x}[t] + \omega_0^2 x[t] == f_0\Theta[t] \tag{4.78}$$

for a step function. Find explicit solutions for the underdamped, overdamped, and critically damped cases. Compare these three solutions graphically and explain their behavior.

b) Combine two of these solutions to determine the response to a square pulse

$$\ddot{x}[t] + 2\gamma\dot{x}[t] + \omega_0^2 x[t] == f_0(\Theta[t] - \Theta[t - \tau]) \tag{4.79}$$

Plot the solutions for $\gamma/\omega_0 = 0.5, 1.0, 2.0$ with $\tau = 10/\omega_0$ as functions of $\omega_0 t$ and explain their general characteristics. If this were an RLC circuit, under what conditions would the output follow the input most closely?

5 Integral Transforms

Abstract. Integral transforms find widespread application in the solution of differential equations and the construction of Green functions that describe the response of linear systems to external influences. We illustrate these methods using the Fourier and Laplace transforms, but the technique is more general. Several practical examples of the use of Fast Fourier transforms are also included.

5.1 Introduction

The function $\tilde{f}[z]$ defined by

$$\tilde{f}[z] = \int_a^b dt K[z,t] f[t] \tag{5.1}$$

is described as an *integral transform* of $f[t]$ with respect to the *kernel* $K[z,t]$. The range of integration must also be specified in the definition of the transform. The variables t and z are then described as *conjugate* to each other. It is useful to represent the integral transformation by a linear operator $\mathcal{K}$ acting on the function f to produce a new function $\tilde{f}$ according to

$$\tilde{f} = \mathcal{K} f \Longleftrightarrow \tilde{f}[z] = \int_a^b dt K[z,t] f[t] \tag{5.2}$$

where the functional relationship on the left-hand side is uncluttered by variable names. When it is useful to indicate symbols for the pair of conjugate variables, we can write

$$\tilde{f}[z] = \mathcal{K} f[z] = \mathcal{K}[f[t]][z] \quad \text{or} \quad \tilde{f} = \mathcal{K}[f[t]] \tag{5.3}$$

depending upon which variable is of interest. If the integral transforms for two functions f and g both exist, the transform of a linear combination

$$h = af + bg \Longrightarrow \tilde{h} = a\tilde{f} + b\tilde{g} \tag{5.4}$$

is the same linear combination of their transforms because $\mathcal{K}$ is linear. Similarly, the inverse transformation can be represented by

$$f = \mathcal{K}^{-1} \tilde{f} \Longleftrightarrow f[t] = \int_C dz K^{-1}[t,z] \tilde{f}[z] \tag{5.5}$$

where K^{-1} represents the inversion kernel (not the reciprocal of K!) and C is an appropriate contour. Thus, the inversion is also represented by a linear operator $\mathcal{K}^{-1}$.

Graduate Mathematical Physics. James J. Kelly

ISBN: 3-527-40637-9

An infinite variety of integral transforms is possible, but the most useful are listed in Table 5.1. Note that we use the same tilde notation for the transformed function and trust that the intended operator will be clear from the context. The following sections discuss the most important properties of these transformations and present examples of their application to problem solving. One often finds that application of an integral transform to a linear differential, integral, or integro-differential equation produces a much simpler, often purely algebraic, equation. Therefore, this technique finds widespread application to problems involving the response of a linear system to some external influence.

Table 5.1. Common integral transforms.

Type	Forward transform	Inverse transform
Fourier	$\tilde{f}[\omega] = \int_{-\infty}^{\infty} dt e^{i\omega t} f[t]$	$f[t] = \frac{1}{2\pi} \int_{-\infty}^{\infty} d\omega e^{-i\omega t} \tilde{f}[\omega]$
Bessel	$\tilde{f}_n[k] = \int_0^{\infty} dx x J_n[kx] f[x]$	$f[x] = \int_0^{\infty} dk k J_n[kx] \tilde{f}_n[k]$
Laplace	$\tilde{f}[s] = \int_0^{\infty} dt e^{-st} f[t]$	$f[t] = \frac{1}{2\pi i} \int_{\gamma-i\infty}^{\gamma+i\infty} ds e^{st} \tilde{f}[s]$
Mellin	$\tilde{f}[z] = \int_0^{\infty} dt t^{z-1} f[t]$	$f[t] = \frac{1}{2\pi i} \int_{-i\infty}^{i\infty} dz t^{-z} \tilde{f}[z]$
Hilbert	$\tilde{f}[z] = \frac{\mathcal{P}}{\pi} \int_{-\infty}^{\infty} dt \frac{f[t]}{t-z}$	$f[t] = \frac{\mathcal{P}}{\pi} \int_{-\infty}^{\infty} dz \frac{\tilde{f}[z]}{z-t}$

5.2 Fourier Transform

5.2.1 Motivation

We assume that you are already familiar with Fourier sine and cosine series and their value in the analysis of periodic phenomena. Suppose that $f[t]$ is a piecewise smooth function for $-\frac{T}{2} \leq t \leq \frac{T}{2}$. By piecewise smooth we mean that there are, at most, a finite number of distinct discontinuities within $|t| \leq T/2$ and that $f[t]$ has continuous first and second derivatives in each interval between discontinuities. The Fourier theorem then tells us that at any point where $f[t]$ is continuous, the series

$$f[t] = \frac{a_0}{2} + \sum_{n=1}^{\infty} a_n \operatorname{Cos}[\omega_n t] + \sum_{n=1}^{\infty} b_n \operatorname{Sin}[\omega_n t], \qquad \omega_n = \frac{2\pi n}{T} \tag{5.6}$$

$$a_n = \frac{2}{T} \int_{-T/2}^{T/2} dt \operatorname{Cos}[\omega_n t] f[t] \tag{5.7}$$

$$b_n = \frac{2}{T} \int_{-T/2}^{T/2} dt \operatorname{Sin}[\omega_n t] f[t] \tag{5.8}$$

converges to the original function. We will not repeat the proof of this theorem here, but instead will use Euler's formula to combine these series using exponentials with complex arguments, which often simplifies their manipulation and facilitates generalization to the

Fourier transform in the limit $T \to \infty$. Thus,

$$\mathrm{Cos}[\omega_n t] \to \frac{1}{2}\left(\mathrm{Exp}[i\omega_n t] + \mathrm{Exp}[-i\omega_n t]\right) \tag{5.9}$$

$$\mathrm{Sin}[\omega_n t] \to \frac{1}{2i}\left(\mathrm{Exp}[i\omega_n t] - \mathrm{Exp}[-i\omega_n t]\right) \tag{5.10}$$

gives

$$\begin{aligned} f[t] &= \frac{a_0}{2} + \frac{1}{2}\sum_{n=1}^{\infty} a_n\left(\mathrm{Exp}[i\omega_n t] + \mathrm{Exp}[-i\omega_n t]\right) \\ &\quad + \frac{1}{2i}\sum_{n=1}^{\infty} b_n\left(\mathrm{Exp}[i\omega_n t] - \mathrm{Exp}[-i\omega_n t]\right) \\ &= \frac{a_0}{2} + \frac{1}{2}\sum_{n=1}^{\infty}(a_n - ib_n)\,\mathrm{Exp}[i\omega_n t] + \frac{1}{2}\sum_{n=1}^{\infty}(a_n + ib_n)\,\mathrm{Exp}[-i\omega_n t] \end{aligned} \tag{5.11}$$

Recognizing that

$$\omega_{-n} = -\omega_n \Longrightarrow b_{-n} = -b_n, \quad b_0 = 0 \tag{5.12}$$

we can combine these series by extending n to the negative integers, such that

$$f[t] = \sum_{n=-\infty}^{\infty} \frac{(a_n + ib_n)}{2}\,\mathrm{Exp}[-i\omega_n t] \tag{5.13}$$

Finally, we define

$$c_n = \frac{a_n + ib_n}{2} = \frac{1}{T}\int_{-T/2}^{T/2} dt\,\mathrm{Exp}[i\omega_n t] f[t] \tag{5.14}$$

to obtain the complex Fourier series

$$f[t] = \sum_{n=-\infty}^{\infty} c_n\,\mathrm{Exp}[-i\omega_n t] \tag{5.15}$$

$$c_n = \frac{1}{T}\int_{-T/2}^{T/2} dt\,\mathrm{Exp}[i\omega_n t] f[t] \tag{5.16}$$

subject to the same conditions as the sine and cosine series.

Note that if $f[t]$ is real, then

$$f^* = f \Longrightarrow c_n^* = c_{-n} \tag{5.17}$$

Similarly, for even or odd functions one obtains the symmetry conditions

$$f[-t] = f[t] \Longrightarrow c_{-n} = c_n \tag{5.18}$$

$$f[-t] = -f[t] \Longrightarrow c_{-n} = -c_n \tag{5.19}$$

such that the combined reality and reflection symmetries become

$$\text{real, even} \implies c_n^* = c_n \tag{5.20}$$

$$\text{real, odd} \implies c_n^* = -c_n \tag{5.21}$$

Therefore, the complex Fourier series contains the cosine series for even or sine series for odd functions as special cases. For more general functions, the coefficients are complex because both sine and cosine terms contribute in unequal proportions. The complex series is completely equivalent to the original Fourier series, but it is often easier to sum the exponential form than the trigonometric form.

It is important to recognize that the Fourier series is periodic by construction. Hence, if $f[t]$ is not actually periodic, the Fourier series constructed for the interval $|t| \leq T/2$ does not represent the original function outside that interval. To produce a good approximation to a nonperiodic function over a wider interval, one must increase T. Thus, a faithful representation of a nonperiodic function requires the limit $T \to \infty$ where the spacing between frequencies becomes infinitesimal and the summation over ω_n becomes an integration over the continuous variable ω. The Fourier series then becomes the Fourier transform described in the next section.

5.2.2 Definition and Inversion

A pair of functions related by

$$f[t] = \int_{-\infty}^{\infty} \frac{d\omega}{2\pi} e^{-i\omega t} \tilde{f}[\omega] \tag{5.22}$$

$$\tilde{f}[\omega] = \int_{-\infty}^{\infty} dt e^{i\omega t} f[t] \tag{5.23}$$

are described as *Fourier transforms* of each other. The apportionment of two factors of $(2\pi)^{-1/2}$ between these definitions is purely a matter of convention – some authors multiply both integrals by $(2\pi)^{-1/2}$, but I prefer to apply both factors to one of the integrals. It is often convenient to employ the operator notation

$$\tilde{f} = \mathcal{F} f \iff \mathcal{F}\big[f[t]\big][\omega] = \int_{-\infty}^{\infty} dt\; e^{i\omega t} f[t] \tag{5.24}$$

$$f = \mathcal{F}^{-1}\tilde{f} \iff \mathcal{F}^{-1}\left[\tilde{f}[\omega]\right][t] = \int_{-\infty}^{\infty} \frac{d\omega}{2\pi} e^{-i\omega t} \tilde{f}[\omega] \tag{5.25}$$

where the more detailed version includes the input function as an argument and indicates the output variable explicitly.

These forward and inverse Fourier transforms are related by the *orthogonality relations*

$$\int_{-\infty}^{\infty} dt\; e^{i(\omega-\omega')t} = 2\pi\, \delta[\omega - \omega'] \tag{5.26}$$

$$\int_{-\infty}^{\infty} d\omega\; e^{i\omega(t-t')} = 2\pi\, \delta[t - t'] \tag{5.27}$$

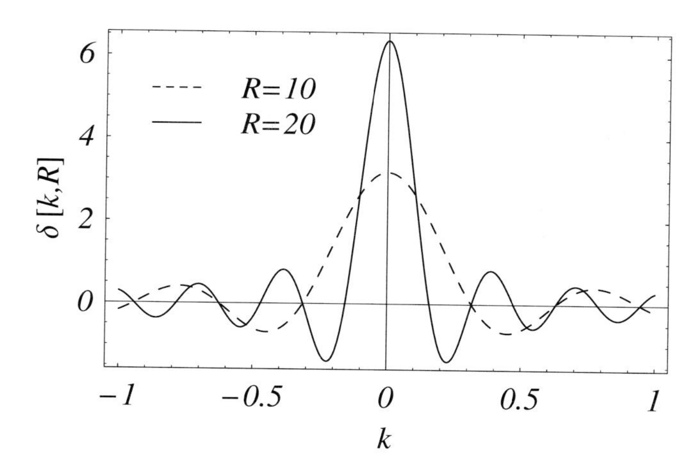

Figure 5.1. Nascent delta functions.

that express the orthogonality of Fourier components with different frequencies. Although these formulas can be derived by careful analysis of Fourier series when the interval approaches infinite length, we prefer to employ the method of generalized functions. Consider the integral

$$\int_{-R}^{R} e^{ikx}\, dx = 2\frac{\mathrm{Sin}[kR]}{k} \tag{5.28}$$

whose result is proportional to one of the nascent delta functions

$$\delta[k, R] = \frac{1}{\pi}\frac{\mathrm{Sin}[kR]}{k} \tag{5.29}$$

discussed in the previous chapter, except there we used $\sigma = R^{-1} \to 0$ and here we use $R \to \infty$ instead. This function is sketched in Fig. 5.1 for two choice of R. The area of this function is obtained from

$$\int_{-\infty}^{\infty} \frac{\mathrm{Sin}[kR]}{k}\, dk = \mathrm{Im}\left[\mathcal{P}\int_{-\infty}^{\infty} \frac{e^{ikR}}{k}\, dk\right] \tag{5.30}$$

where we use the principal value of the exponential form because the singularity in the original form is removable. Closing with a great semicircle in the upper half-plane, which does not contribute, the integral reduces to πi times the unit residue of the pole on the real axis, such that

$$\int_{-\infty}^{\infty} \frac{\mathrm{Sin}[kR]}{k}\, dk = \pi \tag{5.31}$$

Thus, the area is independent of R but becomes increasingly concentrated in a central lobe around $k = 0$ with height

$$\lim_{k\to 0} \frac{\mathrm{Sin}[kR]}{k} = R \tag{5.32}$$

and width $\pi/R \to 0$ as $R \to \infty$; furthermore, the secondary oscillations are much smaller and their areas cancel almost completely. Therefore, we identify

$$\int_{-R}^{R} e^{ikx}\, dx = 2\pi\delta[k, R] \Longrightarrow \lim_{R\to\infty} \int_{-R}^{R} e^{ikx}\, dx = 2\pi\delta[k] \tag{5.33}$$

to obtain the fundamental result

$$\int_{-\infty}^{\infty} e^{ikx}\, dx = 2\pi\, \delta[k] \tag{5.34}$$

that underlies the theory of Fourier transforms.

For the sake of completeness we should mention the conditions required for the existence of a Fourier transform – the Fourier transform $\tilde{f}[\omega]$ exists if $f[t]$ is piecewise smooth and is absolutely integrable, meaning that

$$\int_{-\infty}^{\infty} |f[t]|\, dt < \infty \tag{5.35}$$

exists and is finite. Then

$$\tilde{f}[\omega] = \int_{-\infty}^{\infty} dt\; e^{i\omega t} f[t] \tag{5.36}$$

is uniformly convergent by the comparison test and $\tilde{f}[\omega]$ satisfies the properties outlined below. Rigorous proofs of the Fourier theorem can be found in specialized texts, but are beyond the scope of ours. Furthermore, we sometimes find that it is possible to relax some of these conditions if we are willing to employ generalized functions. For example, below we will derive the Fourier transform of the step function even though it violates the condition on integrability. Thus, these conditions are sufficient, but not always strictly necessary, for use of the Fourier transform.

5.2.3 Basic Properties

5.2.3.1 Phase

If we require that ω be real, then inspection of

$$\tilde{f}[\omega] = \int_{-\infty}^{\infty} dt e^{i\omega t} f[t] \Longrightarrow \tilde{f}^{*}[\omega] = \int_{-\infty}^{\infty} dt e^{-i\omega t} f^{*}[t] \tag{5.37}$$

demonstrates that

$$f^{*}[t] = f[t] \Longrightarrow \tilde{f}^{*}[\omega] = \tilde{f}[-\omega] \tag{5.38}$$

$$f^{*}[t] = -f[t] \Longrightarrow \tilde{f}^{*}[\omega] = -\tilde{f}[-\omega] \tag{5.39}$$

such that the Fourier transform of a real function is even while the Fourier transform of an imaginary function is odd.

5.2.3.2 Reflection

If $f[t]$ has a definite reflection symmetry, such that

$$f[-t] = +f[t] \Longrightarrow \tilde{f}[-\omega] = +\tilde{f}[\omega] \tag{5.40}$$
$$f[-t] = -f[t] \Longrightarrow \tilde{f}[-\omega] = -\tilde{f}[\omega] \tag{5.41}$$

then $\tilde{f}[\omega]$ shares that symmetry. One can then use either the cosine or sine transforms for $\omega > 0$ and $t > 0$, as discussed in a later section.

5.2.3.3 Shifting and Attenuation

Using

$$\mathcal{F}\big[f[t-\tau]\big] = \int_{-\infty}^{\infty} dt e^{i\omega t} f[t-\tau] = \int_{-\infty}^{\infty} dt' e^{i\omega(t'+\tau)} f[t'] \tag{5.42}$$

one finds that the effect of a time shift on the Fourier transform is an additional phase factor, such that

$$\mathcal{F}\big[f[t-\tau]\big] = e^{i\omega\tau}\mathcal{F}\big[f[t]\big] \tag{5.43}$$

Alternatively, suppose that an exponential damping factor is applied such that

$$\mathcal{F}\big[e^{-\gamma t} f[t]\big] = \int_{-\infty}^{\infty} dt e^{i\omega t} e^{-\gamma t} f[t] = \tilde{f}[\omega + i\gamma] \tag{5.44}$$

shifts the frequency into the complex plane.

5.2.4 Parseval's Theorem

The integrated product of two functions is related to the integrated product of their Fourier transforms as follows.

$$\begin{aligned}\int_{-\infty}^{\infty} dt f[t] g^*[t] &= \int_{-\infty}^{\infty} dt \int_{-\infty}^{\infty} \frac{d\omega}{2\pi} e^{-i\omega t} \tilde{f}[\omega] \int_{-\infty}^{\infty} \frac{d\omega'}{2\pi} e^{i\omega' t} \tilde{g}^*[\omega'] \\ &= \int_{-\infty}^{\infty} \frac{d\omega}{2\pi} \int_{-\infty}^{\infty} \frac{d\omega'}{2\pi} \delta[\omega - \omega'] \tilde{f}[\omega] \tilde{g}^*[\omega'] \\ &= \int_{-\infty}^{\infty} \frac{d\omega}{2\pi} \tilde{f}[\omega] \tilde{g}^*[\omega]\end{aligned} \tag{5.45}$$

If we know that $f[t]$ and $g[t]$ are real-valued functions, we can use the phase properties of Fourier transforms to express Parseval's theorem in the form

$$f, g \text{ real} \Longrightarrow \int_{-\infty}^{\infty} dt f[t] g[t] = \int_{-\infty}^{\infty} \frac{d\omega}{2\pi} \tilde{f}[\omega] \tilde{g}[-\omega] \tag{5.46}$$

such that

$$\begin{aligned} f \text{ real} \implies \int_{-\infty}^{\infty} dt |f[t]|^2 &= \int_{-\infty}^{\infty} \frac{d\omega}{2\pi} |\tilde{f}[\omega]|^2 = \int_{-\infty}^{\infty} \frac{d\omega}{2\pi} \tilde{f}[\omega] \tilde{f}[-\omega] \\ &= 2 \int_{0}^{\infty} \frac{d\omega}{2\pi} |\tilde{f}[\omega]|^2 \end{aligned} \tag{5.47}$$

where the squared modulus is often related to the power absorbed or radiated by a system. Parseval's theorem then has a simple interpretation – the total energy (power integrated over time) is the sum of the power found in each interval of frequency. Thus, $|\tilde{f}[\omega]|^2$ is proportional to the *spectral power density* or power radiated per unit interval of frequency. Naturally, there are variations of the normalization convention.

5.2.5 Convolution Theorem

Suppose that $f = g \otimes h$ is defined by the convolution integral

$$f[t] = \int_{-\infty}^{\infty} g[t - t'] h[t'] \, dt' \tag{5.48}$$

where $g[t - t']$ represents the contribution to $f[t]$ made by a source $h[t']$. Convolution integrals are found throughout mathematical physics and are often easiest to evaluate using Fourier transforms. The Fourier transform of f can be related to the Fourier transforms of g and h, as follows.

$$\begin{aligned} \tilde{f}[\omega] &= \int_{-\infty}^{\infty} dt e^{i\omega t} \int_{-\infty}^{\infty} dt' g[t - t'] h[t'] \\ &= \int_{-\infty}^{\infty} dt \int_{-\infty}^{\infty} dt' e^{i\omega t} g[t - t'] h[t'] \\ &= \int_{-\infty}^{\infty} dt \int_{-\infty}^{\infty} dt' (e^{i\omega(t-t')} g[t - t'])(e^{i\omega t'} h[t']) \\ &= \left(\int_{-\infty}^{\infty} ds e^{i\omega s} g[s] \right) \left(\int_{-\infty}^{\infty} dt' e^{i\omega t'} h[t'] \right) \\ &= \tilde{g}[\omega] \tilde{h}[\omega] \end{aligned} \tag{5.49}$$

Notice that we inserted a factor of unity, represented as $e^{-i\omega t'} e^{i\omega t'}$, into the third line in order to factor the double integral. This useful trick often simplifies integral transforms. Thus, we find

$$\mathcal{F}[g \otimes h] = \mathcal{F}[g] \mathcal{F}[h] \implies \tilde{f}[\omega] = \tilde{g}[\omega] \tilde{h}[\omega] \tag{5.50}$$

assuming that both $\tilde{g}$ and $\tilde{h}$ exist. Therefore, the Fourier transform of a convolution of two functions is the product of the Fourier transforms of the individual functions. Be aware, though, that other conventions for normalization of the Fourier transform result in different factors of 2π in Parseval's theorem and/or the convolution theorem.

Observe that the correlation operator is commutative, associative, and distributive

$$h \otimes g = g \otimes h \tag{5.51}$$

$$g \otimes (h \otimes p) = (g \otimes h) \otimes p \tag{5.52}$$

$$g \otimes (h + p) = g \otimes h + g \otimes p \tag{5.53}$$

if all required integrals exist. These relationships can be demonstrated directly using the temporal definition, but are almost trivial using the convolution theorem because multiplication shares these properties.

5.2.6 Correlation Theorem

Suppose that the functions $g[t]$ and $h[t]$ become similar if their relative time scales are shifted to match their shapes as closely as possible. We would then describe those functions as correlated, at least over a finite range of times. A quantitative measure of the degree of correlation between two functions, $g[t]$ and $h[t]$, is provided by the *cross correlation*

$$C[g, h][t] = \int_{-\infty}^{\infty} g[t + t']h[t']^* \, dt' \tag{5.54}$$

where the complex conjugation of one factor ensures that the correlation between complex exponentials

$$g[t] = e^{-i\omega_1 t}, \quad h[t] = e^{-i\omega_2 (t+\phi)} \Longrightarrow C[g, h] = 2\pi\delta(\omega_1 - \omega_2)\,\text{Exp}[-i\omega_1(t - \phi)] \tag{5.55}$$

reduces to a delta function in frequency with a phase that expresses the time shift. Although convolution and correlation are different, their similarities suggest that the Fourier transform probably provides a simple theorem for correlation also. Thus, if we let $f[t] = C[g, h]$, we find

$$\begin{aligned}
\tilde{f}[\omega] &= \int_{-\infty}^{\infty} dt e^{i\omega t} \int_{-\infty}^{\infty} dt' g[t + t']h[t']^* \\
&= \int_{-\infty}^{\infty} dt \int_{-\infty}^{\infty} dt' e^{i\omega t} g^*[t + t']h[t']^* \\
&= \int_{-\infty}^{\infty} dt \int_{-\infty}^{\infty} dt' (e^{i\omega(t+t')} g^*[t + t'])(e^{-i\omega t'} h[t']^*) \\
&= \left(\int_{-\infty}^{\infty} ds e^{i\omega s} g[s]\right)\left(\int_{-\infty}^{\infty} dt' e^{i\omega t'} h[t']\right)^* \\
&= \tilde{g}[\omega]\tilde{h}[\omega]^*
\end{aligned} \tag{5.56}$$

Therefore, the correlation theorem takes a form

$$\tilde{C}[g, h] = \tilde{g}[\omega]\tilde{h}[\omega]^* \tag{5.57}$$

that is similar to the convolution theorem, but complex conjugation of one factor does have important consequences. Although correlation is associative and distributive, it is not commutative:

$$C[h,g][t] = C^*[g,h][-t] \Longleftrightarrow \tilde{C}[h,g] = \tilde{C}[g,h]^* \tag{5.58}$$

An important special case of correlation is the *autocorrelation function*, $C[g,g]$, that is highly sensitive to periodic or quasiperiodic behavior. The autocorrelation is obviously strongest at $t = 0$ and would also exhibit peaks at multiples of the fundamental frequency for a periodic function, similar to the Fourier series. The closer to true periodicity the narrower such peaks would be. Thus, the autocorrelation function provides an important empirical tool for the study of dynamical or statistical systems. Furthermore, the Fourier transform of the autocorrelation function

$$\tilde{C}[g,g] = \left|\tilde{g}[\omega]\right|^2 \tag{5.59}$$

is proportional to the power spectrum. This result is known as the *Wiener–Khintchine theorem* and is important to the analysis of coherence in optics or the relationship between fluctuations and dissipation in statistical physics.

5.2.7 Useful Fourier Transforms

5.2.7.1 Gaussian

Suppose that

$$f[t] = \frac{\mathrm{Exp}\left[-\frac{t^2}{2\sigma^2}\right]}{\sqrt{2\pi\sigma^2}} \tag{5.60}$$

is a Gaussian of width σ and unit area. The Fourier transform

$$\begin{aligned} \mathcal{F}f &= \frac{1}{\sqrt{2\pi\sigma^2}} \int_{-\infty}^{\infty} dt e^{i\omega t}\, \mathrm{Exp}\left[-\frac{t^2}{2\sigma^2}\right] \\ &= \frac{\mathrm{Exp}[-\omega^2\sigma^2/2]}{\sqrt{2\pi\sigma^2}} \int_{-\infty}^{\infty} dt\, \mathrm{Exp}\left[-\frac{(t-i\omega\sigma^2)^2}{2\sigma^2}\right] \end{aligned} \tag{5.61}$$

can be evaluated by completing the square; we have already demonstrated that such an integral can be treated as if the argument of the exponential were real. Thus, we obtain a Gaussian

$$\mathcal{F}f = \mathrm{Exp}[-\omega^2\sigma^2/2] \tag{5.62}$$

with width σ^{-1} and area $\sqrt{2\pi/\sigma^2}$. Notice that

$$\int_{-\infty}^{\infty} f[t]\,dt = 1 \Longrightarrow \tilde{f}[0] = 1 \tag{5.63}$$

and that the widths of the two Gaussians are reciprocals of each other. For a generic function $g[x]$ it is useful to define moments

$$M_n = \int_{-\infty}^{\infty} g[x]x^n \, dx \tag{5.64}$$

such that

$$\bar{x} = \langle x \rangle = \frac{M_1}{M_0}, \quad \sigma^2 = \left\langle (x-\bar{x})^2 \right\rangle = \frac{M_2}{M_0} \tag{5.65}$$

represent the mean and variance of g. Therefore, the variances σ_t and σ_ω of a Gaussian and its Fourier transform are related by

$$\sigma_\omega \sigma_t = 1 \tag{5.66}$$

Similar relationships are observed for other functions that can be described as a localized peak – the narrower the distribution in one variable (such as time), the broader the distribution in the conjugate variable (such as frequency). The Heisenberg uncertainty principle in quantum mechanics is closely related to this reciprocal relationship between widths of a function and its Fourier transform.

5.2.7.2 Chopped Sinusoid

Next, consider the chopped sinusoid

$$f[t] = \begin{cases} 0 & t < 0 \\ \text{Exp}[-i\omega_0 t] & 0 \le t \le T \\ 0 & T < t \end{cases} \tag{5.67}$$

for which

$$\tilde{f}[\omega] = \int_{-T}^{T} dt e^{i(\omega-\omega_0)t} = 2T \frac{\text{Sin}[(\omega-\omega_0)T]}{(\omega-\omega_0)T} \tag{5.68}$$

is the nascent delta function plotted in Fig. 5.2. Although the Fourier transform of a finite wave train is peaked at its fundamental frequency, ω_0, the limitation to a finite time interval $\pm T$ spreads the central peak and produces substantial subsidiary peaks or side-bands. These characteristics are similar to those of single-slit diffraction: instead of chopping a wave front in space we are now chopping a wave train in time, but the mathematics is the same.

An important special case is the rectangular pulse, for which $\omega_0 \to 0$, such that

$$f[t] = \Theta[T - |t|] \Longrightarrow \tilde{f}[\omega] = \int_{-T}^{T} dt e^{i\omega t} = 2\frac{\text{Sin}[\omega T]}{\omega} \tag{5.69}$$

clearly exhibits the reciprocal relationship between the widths of a function and its Fourier transform.

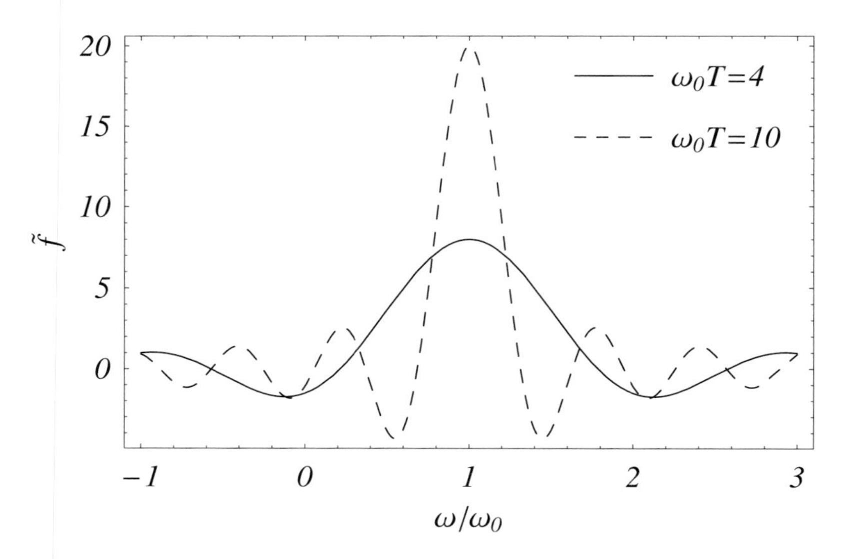

Figure 5.2. Fourier transform of chopped sinusoid: $\mathcal{F}\left[e^{-i\omega t}\Theta[-T \leq t \leq T]\right]$.

5.2.7.3 Step Function

Often it is useful to idealize a quantity, such as a force or voltage, that has a rapid onset followed by a sustained level using a step function, but the corresponding Fourier transform requires special handling as a generalized function because the step function is not absolutely integrable. First we consider a rectangular pulse of finite duration, for which

$$f[t] = \Theta[t] - \Theta[t-\tau] \Longrightarrow \tilde{f}[\omega] = \frac{e^{i\omega\tau} - 1}{i\omega} \tag{5.70}$$

In the limit that $\tau \to \infty$, the contribution of $e^{i\omega\tau}$ varies rapidly in phase and would tend to average to zero. This can be done in a more controlled fashion using a convergence factor,

$$g[t] = \lim_{\varepsilon \to 0^+} \Theta[t] e^{-\varepsilon t} \tag{5.71}$$

such that

$$\begin{aligned} \tilde{g}[\omega] &= \lim_{\varepsilon \to 0^+} \int_0^\infty dt \, \mathrm{Exp}[i(\omega + i\varepsilon)t] \\ &= \lim_{\varepsilon \to 0^+} \left(\frac{\mathrm{Exp}[i\omega t]\,\mathrm{Exp}[-\varepsilon t]}{i(\omega + i\varepsilon)} \right)_0^\infty = \lim_{\varepsilon \to 0^+} \frac{i}{\omega + i\varepsilon} \end{aligned} \tag{5.72}$$

where we leave the limit unevaluated because we recognize that this result must be treated as a generalized function. Hence, we find that the Fourier transform of $\Theta[t]$ is given by i/ω, but this is not a rigorous proof and we must verify that the inverse Fourier transform does provide the desired result. However, because the pole at the origin is on the integration path

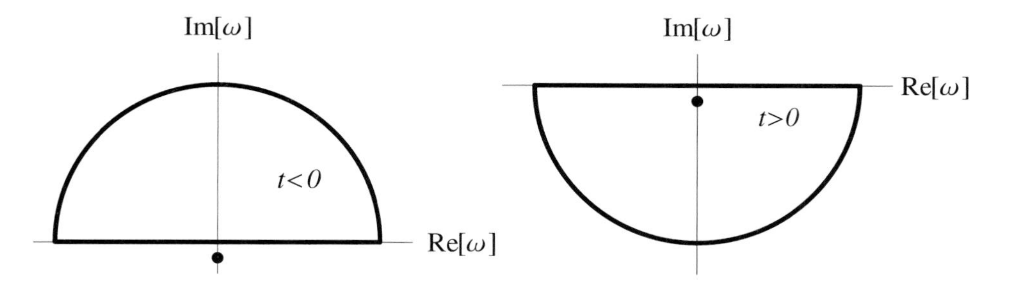

Figure 5.3. Contours used in Eq. (5.74) for Fourier representation of step function.

for the inverse transform, we must evaluate its contribution carefully. Thus, the inversion integral

$$\lim_{\varepsilon\to 0^+} \frac{i}{2\pi} \int_{-\infty}^{\infty} \frac{e^{-i\omega t}}{\omega + i\varepsilon}\, d\omega = \begin{cases} 0 & t < 0 \\ 1 & t > 0 \end{cases} \tag{5.73}$$

with a pole in the lower half-plane just below the real axis for $\varepsilon \to 0^+$, where closure is made either up or down according to the sign of t, does indeed produce a step function as sketched in Fig. 5.3. Therefore, the Fourier transform of the Heaviside step function becomes

$$\Theta[t] = \lim_{\varepsilon\to 0^+} \frac{1}{2\pi i} \int_{-\infty}^{\infty} \frac{e^{-i\omega t}}{\omega + i\varepsilon}\, d\omega \Longrightarrow \mathcal{F}\Theta = \lim_{\varepsilon\to 0^+} \frac{i}{\omega + i\varepsilon} = \frac{i}{\omega^+} \tag{5.74}$$

where the notation ω^+ implies addition to ω of an infinitesimal imaginary part, $i\varepsilon$, that has the effect of shifting the pole slightly below the real axis with the limit $\varepsilon \to 0^+$ taken later. Notice that this infinitesimal imaginary contribution is similar to the attenuation property of Fourier transforms, such that ε represents the damping coefficient that suppresses infinite times as if we had applied a convergence factor, $\mathrm{Exp}[-\varepsilon t]$, to the step function.

Alternatively, consider the principal-value integral

$$\frac{\mathcal{P}}{2\pi i} \int_{-\infty}^{\infty} \frac{e^{-i\omega t}}{\omega}\, d\omega = \frac{1}{2}(1 - 2\Theta[t]) = \begin{cases} \frac{1}{2} & t < 0 \\ -\frac{1}{2} & t > 0 \end{cases} \tag{5.75}$$

using the contours sketched in Fig. 5.4. Solving for

$$\Theta[t] = \frac{1}{2} - \frac{\mathcal{P}}{2\pi i} \int_{-\infty}^{\infty} \frac{e^{-i\omega t}}{\omega}\, d\omega \tag{5.76}$$

and evaluating the Fourier transform of both sides, we obtain

$$\int_{-\infty}^{\infty} e^{i\omega t}\Theta[t]\, dt \tag{5.77}$$

$$= \frac{1}{2} \int_{-\infty}^{\infty} e^{i\omega t}\, dt - \int_{-\infty}^{\infty} dt\, e^{i\omega t} \frac{\mathcal{P}}{2\pi i} \int_{-\infty}^{\infty} \frac{e^{-i\omega' t}}{\omega'}\, d\omega' \tag{5.78}$$

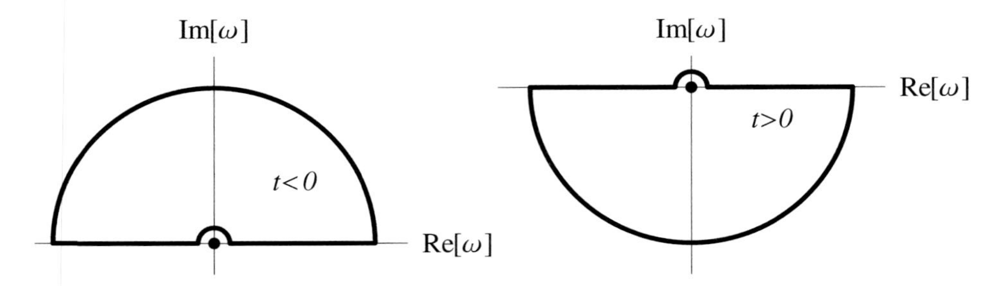

Figure 5.4. Contours used in Eq. (5.77) for Fourier representation of step function.

where we assume that the time integration can be performed first. This result can be expressed in a compact operator representation as

$$\mathcal{F}\Theta = \pi\delta[\omega] + i\frac{\mathcal{P}}{\omega} \tag{5.79}$$

in which we understand that the delta function applies under a Fourier integral and that the principal value is to be taken for the singularity at the origin. This representation is equivalent to the former, except that now we imagine that the limit $\varepsilon \to 0^+$ is taken first and that the pole pushes the contour upward, making a small semicircular indentation in the portion of the contour that is on the real axis. Consequently, we interpret the "bump in the road" as

$$\mathcal{F}\Theta = \lim_{\varepsilon\to 0^+} \frac{i}{\omega + i\varepsilon} = \pi\delta[\omega] + i\frac{\mathcal{P}}{\omega} \tag{5.80}$$

5.2.8 Fourier Transform of Derivatives

Many applications of Fourier transforms exploit the simple transformation properties of derivatives to transform a differential equation into an algebraic equation for the Fourier transform. Assuming that the Fourier transforms exist for $f[t]$ and its derivatives, the simplest method for deducing the Fourier transform of derivatives is to differentiate under the integral

$$f[t] = \int_{-\infty}^{\infty} \frac{d\omega}{2\pi} e^{-i\omega t} \tilde{f}[\omega] \Longrightarrow f'[t] = \int_{-\infty}^{\infty} \frac{d\omega}{2\pi} e^{-i\omega t} (-i\omega \tilde{f}[\omega]) \tag{5.81}$$

and thus to identify

$$\mathcal{F}\big[f'[t]\big] = -i\omega\mathcal{F}\big[f[t]\big] \tag{5.82}$$

Therefore, each derivative in a differential equation can be replaced by a power of $(-i\omega)$ according to

$$\frac{\partial}{\partial t} \longrightarrow -i\omega \tag{5.83}$$

$$\frac{\partial^n}{\partial t^n} \longrightarrow (-i\omega)^n \tag{5.84}$$

5.2.9 Summary

A brief table of Fourier transforms and their properties is given in Table 5.2. The parameters τ, ω_0, γ, σ, and λ are real. We also assume that the required derivatives exist and integrals converge. Be careful in applying theorems for derivatives and other operations to functions with discontinuities.

Table 5.2. Brief table of Fourier transforms.

Type	$f[t]$	$\tilde{f}[\omega]$
definition	$f[t] = \int_{-\infty}^{\infty} \frac{d\omega}{2\pi} e^{-i\omega t} \tilde{f}[\omega]$	$\tilde{f}[\omega] = \int_{-\infty}^{\infty} dt e^{i\omega t} f[t]$
step function	$\Theta[t]$	$\lim_{\varepsilon \to 0^+} \frac{i}{\omega + i\varepsilon} = \pi\delta[\omega] + i\frac{\mathcal{P}}{\omega}$
delta function	$\delta[t-\tau]$	$e^{i\omega\tau}$
chopped sinusoid	$e^{-i\omega_0 t}\Theta[\tau - \lvert t\rvert]$	$2\frac{\mathrm{Sin}[(\omega-\omega_0)\tau]}{(\omega-\omega_0)}$
shifting	$f[t-\tau]$	$e^{i\omega\tau}\tilde{f}[\omega]$
attenuation	$e^{-\gamma t} f[t]$	$\tilde{f}[\omega + i\gamma]$
dilatation	$f[\lambda t]$	$\lambda^{-1}\tilde{f}[\omega/\lambda]$
convolution	$f[t] = \int_{-\infty}^{\infty} d\tau g[t-\tau]h[\tau]$	$\tilde{f}[\omega] = \tilde{g}[\omega]\tilde{h}[\omega]$
derivatives	$f^{(n)}[t]$	$(-i\omega)^n \tilde{f}[\omega]$
multiplication by powers	$t f[t]$	$-i\tilde{f}'[\omega]$
Gaussian	$\mathrm{Exp}\left[-\frac{t^2}{2\sigma^2}\right] \Big/ \sqrt{2\pi\sigma^2}$	$\mathrm{Exp}[-\omega^2\sigma^2/2]$
exponential	$e^{-\gamma t}$	$2\pi\delta[\omega + i\gamma]$
	$e^{-\gamma t}\Theta[t]$	$\frac{i}{\omega + i\gamma}$
trigonometric	$\mathrm{Cos}[\omega_0 t]$	$\pi(\delta[\omega-\omega_0] + \delta[\omega+\omega_0])$
	$\mathrm{Sin}[\omega_0 t]$	$i\pi(\delta[\omega-\omega_0] - \delta[\omega+\omega_0])$

5.3 Green Functions via Fourier Transform

The Fourier transform can be very useful in analyzing the response of a linear system to external stimuli. In this section we illustrate this technique using a few relatively simple examples.

5.3.1 Example: Green Function for One-Dimensional Diffusion

Suppose that $\psi[x,t]$ represents the temperature at position x and time t relative to the background temperature for an effectively infinite homogeneous system. If the initial temperature $\psi[x,0] = f[x]$ is elevated, heat will diffuse to distant regions according to the *heat diffusion equation*

$$\frac{\partial^2 \psi}{\partial x^2} == \frac{1}{\kappa}\frac{\partial \psi}{\partial t} \tag{5.85}$$

where the diffusion constant κ is proportional to the thermal conductivity of the material and inversely proportional to its heat capacity per unit volume. Thus, we expect that the temperature for neighboring regions will initially rise as heat passes through them but will eventually return to ambient levels as the energy spreads out over an infinite volume. One method for determining the evolution of temperature throughout this sample is to employ a spatial Fourier transform to write

$$\psi[x,t] = \int_{-\infty}^{\infty} \frac{dk}{2\pi} e^{ikx} \tilde{\psi}[k,t] \tag{5.86}$$

$$\tilde{\psi}[k,t] = \int_{-\infty}^{\infty} dx e^{-ikx} \psi[x,t] \tag{5.87}$$

such that the differential equation becomes

$$-k^2 \tilde{\psi}[k,t] = \frac{1}{\kappa}\frac{\partial \tilde{\psi}[k,t]}{\partial t} \Longrightarrow \tilde{\psi}[k,t] = \tilde{\psi}[k,0]\,\mathrm{Exp}[-\kappa x k^2] \tag{5.88}$$

The initial condition requires

$$\tilde{\psi}[k,0] = \tilde{f}[k] = \int_{-\infty}^{\infty} dx e^{-ikx} f[x] \tag{5.89}$$

The inverse Fourier transform becomes

$$\begin{aligned} \psi[x,t] &= \int_{-\infty}^{\infty} \frac{dk}{2\pi} e^{ikx} \tilde{f}[k]\,\mathrm{Exp}[-\kappa t k^2] \\ &= \int_{-\infty}^{\infty} \frac{dk}{2\pi} \int_{-\infty}^{\infty} dx' f[x']\,\mathrm{Exp}\big[-\big(\kappa t k^2 - ik(x-x')\big)\big] \end{aligned} \tag{5.90}$$

The integral with respect to k can be evaluated by completing the square in the exponential, whereby

$$\psi[x,t] = (4\pi\kappa t)^{-1/2} \int_{-\infty}^{\infty} dx' f[x']\,\mathrm{Exp}\left[-\frac{(x-x')^2}{4\kappa t}\right] \tag{5.91}$$

takes the form of a convolution integral.

Suppose that the region of elevated temperature is initially restricted to a plane, such that

$$f[x] = \psi_0 \delta[x] \Longrightarrow \psi[x,t] = \frac{\psi_0}{(4\pi\kappa t)^{1/2}}\,\mathrm{Exp}\left[-\frac{x^2}{4\kappa t}\right] \tag{5.92}$$

describes both the time and spatial temperature variation produced by a localized disturbance. This special solution is an example of a *Green function* representing the response of a linear system to a localized perturbation. The Green function for a one-dimensional diffusion equation with a localized initial condition satisfies the equation

$$\left(\frac{\partial^2}{\partial x^2} - \frac{1}{\kappa}\frac{\partial}{\partial t}\right) G[x - x', t] = \delta[x - x'] \Longrightarrow$$
$$G[x - x', t] = (4\pi\kappa t)^{-1/2}\,\mathrm{Exp}\left[-\frac{(x - x')^2}{4\kappa t}\right] \tag{5.93}$$

Notice that this Green function is a nascent delta function of Gaussian form with width parameter $\sigma = \sqrt{2\kappa t}$, such that

$$\lim_{t\to 0} G[x - x', t] = \delta[x - x'] \tag{5.94}$$

Solutions for more general initial conditions are then obtained using the convolution

$$\psi[x, 0] = f[x] \Longrightarrow \psi[x, t] = \int_{-\infty}^{\infty} G[x - x', t] f[x']\, dx' \tag{5.95}$$

Therefore, the Green function describes the essential dynamics of the problem and reduces problems with specialized initial conditions to an integral which can be evaluated numerically, if not analytically. The Green function for this problem is sketched in Fig. 5.5. The delta function at $x = 0$ must be cut off for plotting. It is clear from the dimensions of the original differential equation that the spatial scale must be proportional to $\sqrt{\kappa t}$. Thus, the width of the Gaussian temperature profile spreads at a rate proportional to $t^{-1/2}$ as the disturbance propagates. The Green function is often described as a *propagator* – it governs the propagation of a disturbance (cause) at one location or time to a different location or later time (effect). The net effect of a distributed disturbance is obtained from the folding or convolution of the propagator with the profile of the disturbance.

5.3.2 Example: Three-Dimensional Green Function for Diffusion Equations

In three dimensions the initial-value problem for diffusion within an infinite uniform medium takes the form

$$\frac{1}{\kappa}\frac{\partial\psi[\vec{r}, t]}{\partial t} == \nabla^2\psi[\vec{r}, t], \quad \psi[\vec{r}, 0] = \psi_0[\vec{r}] \tag{5.96}$$

where $\psi_0[\vec{r}]$ is the initial spatial distribution and where we assume that κ is a positive constant. We assume that the initial distribution is localized and at least piecewise continuous so that its Fourier transform exists in all three dimensions. Using a three-dimensional Fourier transform

$$\psi[\vec{r}, t] = \int \tilde{\psi}[\vec{k}, t]\,\mathrm{Exp}[i\vec{k}\cdot\vec{r}]\frac{d^3k}{(2\pi)^3} \tag{5.97}$$

$$\tilde{\psi}[\vec{k}, t] = \int \psi[\vec{r}, t]\,\mathrm{Exp}[-i\vec{k}\cdot\vec{r}]\, d^3r \tag{5.98}$$

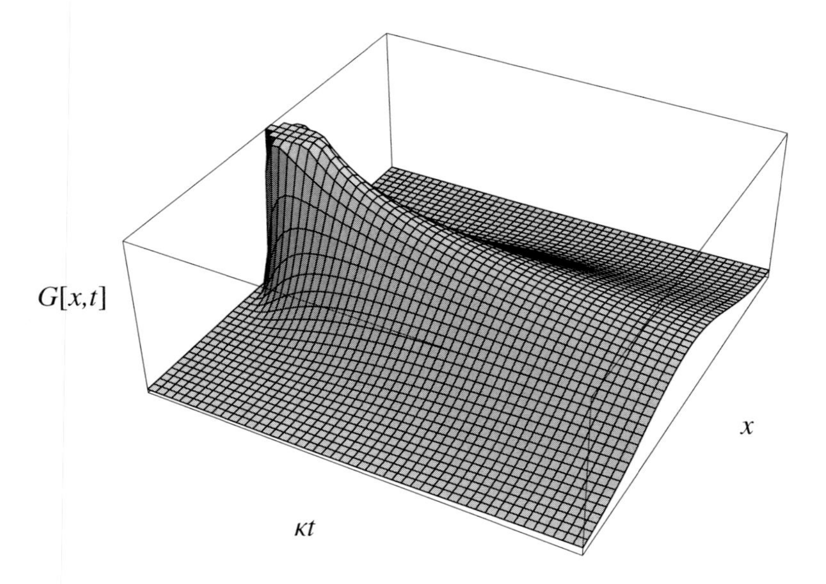

Figure 5.5. Dissipation of temperature disturbance (arb. units).

we find

$$\left(\frac{1}{\kappa}\frac{\partial}{\partial t} - \nabla^2\right)\psi == 0 \Longrightarrow \int \left(\frac{1}{\kappa}\frac{\partial}{\partial t} + k^2\right)\tilde{\psi}[\vec{k}, t]\,\mathrm{Exp}[i\vec{k}\cdot\vec{r}]\frac{d^3k}{(2\pi)^3} == 0 \tag{5.99}$$

If this equation is to be satisfied for any $\vec{r}$, we must require the integrand to vanish. Hence, we obtain

$$\left(\frac{1}{\kappa}\frac{\partial}{\partial t} + k^2\right)\tilde{\psi}[\vec{k}, t] == 0 \Longrightarrow \tilde{\psi}[\vec{k}, t] = \tilde{\psi}_0[\vec{k}]\,\mathrm{Exp}[-\kappa k^2 t] \tag{5.100}$$

where

$$\tilde{\psi}_0[\vec{k}] = \int \psi_0[\vec{r}]\,\mathrm{Exp}[-i\vec{k}\cdot\vec{r}]\,d^3r \tag{5.101}$$

is the initial spectral distribution. To obtain the Green function, we substitute

$$\psi_0[\vec{r}] = \delta[\vec{r} - \vec{r}\,'] \Longrightarrow \tilde{\psi}_0[\vec{k}] = \mathrm{Exp}[-i\vec{k}\cdot\vec{r}\,'] \tag{5.102}$$

such that

$$G[\vec{r} - \vec{r}\,', t] = \int \mathrm{Exp}[-\kappa k^2 t]\,\mathrm{Exp}[i\vec{k}\cdot(\vec{r} - \vec{r}\,')]\frac{d^3k}{(2\pi)^3} \tag{5.103}$$

can be factored according to

$$G[\vec{r} - \vec{r}\,', t] = \prod_{i=1}^{3} \int \mathrm{Exp}[-\kappa k_i^2 t]\,\mathrm{Exp}[ik_i\cdot(r_i - r_i')]\frac{dk_i}{2\pi} \tag{5.104}$$

where the index i runs over the three Cartesian components. Each factor is a one-dimensional Green function for the diffusion equation; hence, using our previous result, we find that

$$G[\vec{r}-\vec{r}',t] = \prod_{i=1}^{3}(4\pi\kappa t)^{-1/2}\,\mathrm{Exp}[-\frac{(r_i-r_i')^2}{4\kappa t}] \tag{5.105}$$

depends upon spatial coordinates through $|\vec{r}-\vec{r}'|$. Therefore, we finally obtain

$$G[|\vec{r}-\vec{r}'|,t] = (4\pi\kappa t)^{-3/2}\,\mathrm{Exp}\left[-\frac{|\vec{r}-\vec{r}'|^2}{4\kappa t}\right] \tag{5.106}$$

such that

$$\psi[\vec{r},t] = \int G[|\vec{r}-\vec{r}'|,t]\psi_0[\vec{r}']\,d^3r' \tag{5.107}$$

is the general solution for the initial-value problem for the diffusion equation in a uniform medium.

5.3.3 Example: Green Function for Damped Oscillator

The Fourier transform of the differential equation

$$\left(m\frac{\partial^2}{\partial t^2} + 2m\gamma\frac{\partial}{\partial t} + k\right)x[t] == mf[t] \tag{5.108}$$

for a driven harmonic oscillator with linear damping becomes an algebraic equation

$$\left(-m\omega^2 - 2i\omega m\gamma + k\right)\tilde{x}[\omega] == m\tilde{f}[\omega] \tag{5.109}$$

where $\tilde{f}[\omega]$ is the Fourier transform of the driving force (per unit mass). Thus, we immediately obtain a particular solution of the form

$$\tilde{x}[\omega] = \frac{\tilde{f}[\omega]}{\omega_0^2 - \omega^2 - 2i\omega\gamma} \tag{5.110}$$

where $\omega_0 = \sqrt{k/m}$ is the natural frequency for free oscillation. A complete solution can then be expressed in the form

$$x[t] = x_0[t] + \int_{-\infty}^{\infty}\frac{d\omega}{2\pi}e^{-i\omega t}\frac{\tilde{f}[\omega]}{\omega_0^2 - \omega^2 - 2i\omega\gamma} \tag{5.111}$$

where $x_0[t]$ is a solution to the homogeneous equation

$$\left(\frac{\partial^2}{\partial t^2} + 2\gamma\frac{\partial}{\partial t} + \omega_0^2\right)x_0[t] == 0 \tag{5.112}$$

designed to satisfy the appropriate boundary conditions. Often one can assume that $x_0 = 0$ if $f[t]$ is of finite duration and occurs after any earlier motions have decayed away due to the damping term.

Alternatively, we recently demonstrated that a general solution to this equation can be expressed as

$$x[t] = x_0[t] + \int_{-\infty}^{\infty} G[t - t'] f[t'] \, dt' \tag{5.113}$$

where the Green function satisfies

$$\left(\frac{\partial^2}{\partial t^2} + 2\gamma \frac{\partial}{\partial t} + \omega_0^2 \right) G[t] == \delta[t] \tag{5.114}$$

with

$$t < 0 \Longrightarrow G[t] = 0, \quad \frac{\partial G[t]}{\partial t} = 0 \tag{5.115}$$

We can demonstrate equivalence between these representations by constructing the spectral (frequency) representation of the Green function directly. Assuming that

$$f[t] = \delta[t] \Longrightarrow \tilde{f}[\omega] = 1 \tag{5.116}$$

is a unit impulse at time t' and that the mass is initially at rest, we obtain

$$\tilde{G}[\omega] = \frac{1}{\omega_0^2 - \omega^2 - 2i\omega\gamma} \tag{5.117}$$

such that

$$G[t] = \int_{-\infty}^{\infty} \frac{d\omega}{2\pi} \frac{e^{-i\omega t}}{\omega_0^2 - \omega^2 - 2i\omega\gamma} \tag{5.118}$$

is the corresponding temporal representation. Thus, we can use either of the representations

$$x[t] = x_0[t] + \frac{1}{2\pi} \int_{-\infty}^{\infty} e^{-i\omega t} \tilde{G}[\omega] \tilde{f}[\omega] \, d\omega \tag{5.119}$$

$$x[t] = x_0[t] + \int_{-\infty}^{\infty} G[t - t'] f[t'] \, dt' \tag{5.120}$$

Just to belabor the point, we substitute the spectral representation of $G[t]$ into the second equation

$$x[t] = x_0[t] + \int_{-\infty}^{\infty} \int_{-\infty}^{\infty} \frac{d\omega}{2\pi} \frac{e^{-i\omega(t-t')}}{\omega_0^2 - \omega^2 - 2i\omega\gamma} f[t'] \, dt' \tag{5.121}$$

and perform the integration over t' to obtain

$$\tilde{f}[\omega] = \int_{-\infty}^{\infty} dt' e^{i\omega t'} f[t'] \Longrightarrow x[t] = x_0[t] + \int_{-\infty}^{\infty} \frac{d\omega}{2\pi} \frac{e^{-i\omega t}}{\omega_0^2 - \omega^2 - 2i\omega\gamma} \tilde{f}[\omega] \tag{5.122}$$

as before. Alternatively, we could have substituted $\tilde{f}[\omega]$ into the first version and performed the integration over ω to obtain the second. (Try it!)

To complete the analysis, we must construct explicit expressions for $G[t]$ by inverting the Fourier transform. First, we consider the underdamped case for which the integrand has two poles

$$\gamma < \omega_0 \Longrightarrow \omega_\pm = -i\gamma \pm \omega_R \quad \text{with} \quad \omega_R = \sqrt{\omega_0^2 - \gamma^2} \tag{5.123}$$

symmetrically placed about the imaginary axis in the lower half of the complex ω-plane. In order to keep the exponential factor finite, we must close the contour with great semi-circles in the upper half-plane for $t < t'$ or the lower half-plane for $t > t'$. Thus, we find $G[t, t'] = 0$ for $t < t'$ because there are no poles in the upper half-plane while for $t > t'$

$$t > t' \Longrightarrow G[t, t'] = \frac{1}{2\pi} \oint_C \frac{e^{-i\omega t}}{\omega_0^2 - \omega^2 - 2i\omega\gamma} d\omega = -i \left(R_+ e^{-i\omega_+(t-t')} + R_- e^{-i\omega_-(t-t')} \right) \tag{5.124}$$

where the negative sign accounts for the clockwise contour and where the residues are

$$\gamma < \omega_0 \Longrightarrow R_\pm = \mp \frac{\text{Exp}[-i\omega_\pm(t - t')]}{2\omega_R} \tag{5.125}$$

such that

$$\gamma < \omega_0 \Longrightarrow G[t, t'] = \omega_R^{-1} \text{Exp}[-\gamma(t - t')] \,\text{Sin}[\omega_R(t - t')]\Theta[t - t'] \tag{5.126}$$

describes damped oscillations following a sharp blow. If the damping is small the oscillation frequency is near the natural frequency and the oscillations persist for a long time when the poles are near the real axis. Next, we consider the overdamped case

$$\gamma > \omega_0 \Longrightarrow \omega_\pm = -i(\gamma \pm \kappa) \quad \text{with} \quad \kappa = \sqrt{\gamma^2 - \omega_0^2} \tag{5.127}$$

where both poles are found on the negative imaginary axis with residues

$$\gamma > \omega_0 \Longrightarrow R_\pm = \mp \frac{\text{Exp}[-(\gamma \pm \kappa)(t - t')]}{2\kappa} \tag{5.128}$$

such that

$$\gamma > \omega_0 \Longrightarrow G[t, t'] = \kappa^{-1} \text{Exp}[-\gamma(t - t')]\text{Sinh}[\kappa(t - t')]\Theta[t - t'] \tag{5.129}$$

describes the damping of the initial velocity. (Note that $\gamma > \kappa$.) Finally, for the critically damped case the two poles coalesce into a single double pole with residue

$$\gamma = \omega_0 \Longrightarrow R = -i(t - t') \,\text{Exp}[-\gamma(t - t')] \tag{5.130}$$

such that

$$\gamma = \omega_0 \Longrightarrow G[t, t'] = (t - t') \,\text{Exp}[-\gamma(t - t')]\Theta[t - t'] \tag{5.131}$$

These results are summarized in Table 5.3. Notice that the underdamped and overdamped solutions are related by the replacement $\omega_R \to i\kappa$ as γ increases and that the critically damped solution is related to either of the other two by the limit $\omega_R \to 0$ or $\kappa \to 0$.

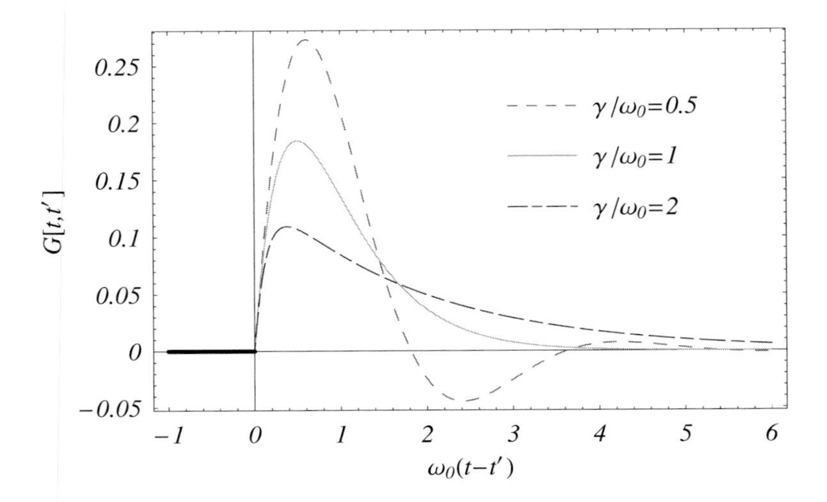

Figure 5.6. Green function for damped oscillator.

Figure 5.6 compares these functions for representative values of γ/ω_0. All are damped exponentially but the underdamped solution also oscillates while the critically damped and overdamped do not. The amplitude is suppressed for the overdamped oscillator, but the duration can be long because the response is slow. Notice that $\mathrm{Sinh}[\kappa t]$ contains exponentially growing and decaying components, but that because $\kappa < \gamma$ the net effect is damping with a relatively slow asymptotic form proportional to $\mathrm{Exp}[-(\gamma-\kappa)t]$. By contrast, the critically-damped oscillator has a fairly strong response for a more limited duration because the damping is sufficiently strong to suppress oscillations but weak enough to permit a significant response to the impulse.

Table 5.3. Green functions for damped oscillator.

Type	Condition	$G[t,t']$	$\omega_\pm$
underdamped	$\gamma < \omega_0$	$\omega_R^{-1}\,\mathrm{Exp}[-\gamma(t-t')]\,\mathrm{Sin}[\omega_R(t-t')]\Theta[t-t']$	$-i\gamma \pm \omega_R$
overdamped	$\gamma > \omega_0$	$\kappa^{-1}\,\mathrm{Exp}[-\gamma(t-t')]\,\mathrm{Sinh}[\kappa(t-t')]\Theta[t-t']$	$-i(\gamma \pm \kappa)$
critically damped	$\gamma = \omega_0$	$\mathrm{Exp}[-\gamma(t-t')](t-t')\Theta[t-t']$	$-i\gamma$

It is also of interest to consider the Green function for an ideal undamped oscillator with $\gamma = 0$. This function can be obtained by evaluating the limit $\gamma \to 0^+$ in which the poles for an underdamped oscillator approach the real axis from below, whereby

$$\gamma \to 0^+ \implies G[t,t'] = \omega_0^{-1}\,\mathrm{Sin}[\omega_0(t-t')]\Theta[t-t'] \tag{5.132}$$

exhibits undamped oscillations with amplitude ω_0^{-1} following a unit impulse. However, if we were to evaluate the Fourier integral

$$G[t,t'] = \int_{-\infty}^{\infty} \frac{d\omega}{2\pi} \frac{e^{-i\omega(t-t')}}{\omega_0^2-\omega^2} \tag{5.133}$$

directly, we would need a prescription for handling the poles on the real axis. In order to obtain a physically sensible result in which effects (oscillation) follow causes (impulse), we need to exclude the poles for $t < t'$ and include them for $t > t'$. This prescription is realized by the substitution $\omega_0 \to \omega_0 - i\gamma$ that recognizes that any real physical system should have a damping mechanism that will shift its resonances slightly below the real axis and provide convergence for the delayed response of the system. Of course, we must have $\gamma > 0$ to ensure damping rather than spontaneous growth of small perturbations.

5.3.4 Operator Method

The differential equation for the damped oscillator can be expressed as

$$\mathcal{L}x[t] == f[t] \quad \text{with} \quad \mathcal{L} = \frac{\partial^2}{\partial t^2} + 2\gamma\frac{\partial}{\partial t} + \omega_0^2 \tag{5.134}$$

where $\mathcal{L}$ is a linear differential operator whose Fourier transform becomes

$$\tilde{\mathcal{L}} = -\omega^2 - 2i\omega\gamma + \omega_0^2 \Longrightarrow \tilde{\mathcal{L}}\tilde{x} == \tilde{f} \tag{5.135}$$

The Green function represents the response to an impulse whose spectral representation is a constant, such that

$$\tilde{\mathcal{L}}\tilde{G} == 1 \Longrightarrow \tilde{G} = \tilde{\mathcal{L}}^{-1} \tag{5.136}$$

Therefore, a formal solution can be expressed in the form

$$\tilde{x}[\omega] = \tilde{x}_0[\omega] + \tilde{G}[\omega]\tilde{f}[\omega] \quad \text{with} \quad \tilde{\mathcal{L}}\tilde{x}_0 == 0, \quad \tilde{G} = \tilde{\mathcal{L}}^{-1} \tag{5.137}$$

where $\tilde{x}_0$ is in the null space of the operator $\tilde{\mathcal{L}}$. The zeros of $\tilde{\mathcal{L}}$ correspond to the poles of $\tilde{G}$ where

$$\tilde{\mathcal{L}} == 0 \Longrightarrow \omega = -i\gamma + (\omega_0^2 - \gamma^2)^{1/2} \tag{5.138}$$

Thus, the poles of $\tilde{G}$ in the complex frequency plane represent the resonances of the system. The oscillation frequency is determined by the real part and the damping by the imaginary part of the complex resonant frequency. Confinement of the poles to the lower half-plane is required by causality, such that effects follow causes, and the necessity that the effect of small perturbations decay rather than grow with time.

5.4 Cosine or Sine Transforms for Even or Odd Functions

For functions that are either even or odd with respect to reflection, such that $f[-t] = \pm f[t]$, it is sometimes convenient to employ Fourier cosine or sine transforms defined by

$$\mathcal{F}_C\big[f[t]\big][\omega] = \tilde{f}_C[\omega] = \int_0^\infty dt\, \text{Cos}[\omega t] f[t] \tag{5.139}$$

$$\mathcal{F}_S\big[f[t]\big][\omega] = \tilde{f}_S[\omega] = \int_0^\infty dt\, \text{Sin}[\omega t] f[t] \tag{5.140}$$

for $\omega > 0$. The inverse transformations are obtained with the assistance of the orthogonality relations

$$\int_0^\infty dt \,\mathrm{Cos}[\omega t]\,\mathrm{Cos}[\omega' t] = \frac{\pi}{2}(\delta[\omega - \omega'] + \delta[\omega + \omega']) \tag{5.141}$$

$$\int_0^\infty dt \,\mathrm{Sin}[\omega t]\,\mathrm{Sin}[\omega' t] = \frac{\pi}{2}(\delta[\omega - \omega'] - \delta[\omega + \omega']) \tag{5.142}$$

such that

$$f[-t] = +f[t] \Longrightarrow f[t] = \frac{2}{\pi}\int_0^\infty d\omega \,\mathrm{Cos}[\omega t]\tilde{f}_C[\omega] \tag{5.143}$$

$$f[-t] = -f[t] \Longrightarrow f[t] = \frac{2}{\pi}\int_0^\infty d\omega \,\mathrm{Sin}[\omega t]\tilde{f}_S[\omega] \tag{5.144}$$

Obviously, the relevant representation is dictated by the reflection symmetry of the original function because the inversion of the cosine (sine) transform automatically produces an even (odd) function of t.

The Fourier sine and cosine transforms can also be applied to generic functions with arbitrary reflection properties, but in such cases we need both. Decomposing a generic function into symmetric and antisymmetric components

$$f^{(\pm)}[t] = \frac{1}{2}(f[t] \pm f[-t]) \Longrightarrow f[\pm t] = f^{(+)}[t] \pm f^{(-)}[t] \tag{5.145}$$

with transforms

$$\tilde{f}_C^{(+)}[\omega] = \int_0^\infty dt \,\mathrm{Cos}[\omega t] f^{(+)}[t] \Longrightarrow f^{(+)}[t] = \frac{2}{\pi}\int_0^\infty d\omega \,\mathrm{Cos}[\omega t]\tilde{f}_C^{(+)}[\omega] \tag{5.146}$$

$$\tilde{f}_S^{(-)}[\omega] = \int_0^\infty dt \,\mathrm{Sin}[\omega t] f^{(-)}[t] \Longrightarrow f^{(-)}[t] = \frac{2}{\pi}\int_0^\infty d\omega \,\mathrm{Sin}[\omega t]\tilde{f}_S^{(-)}[\omega] \tag{5.147}$$

one obtains

$$f[t] = \frac{2}{\pi}\int_0^\infty d\omega \left(\tilde{f}_C^{(+)}[\omega]\,\mathrm{Cos}[\omega t] + \tilde{f}_S^{(-)}[\omega]\,\mathrm{Sin}[\omega t]\right) \tag{5.148}$$

The usual Fourier transform is then given by the combination

$$\tilde{f}[\omega] = 2\left(\tilde{f}_C^{(+)}[\omega] + i\tilde{f}_S^{(-)}[\omega]\right) \tag{5.149}$$

The details are left to the reader.

5.5 Discrete Fourier Transform

The continuous Fourier transform is extremely valuable for formal analysis, but often we are left with Fourier integrals that cannot be evaluated symbolically. However, numerical analysis using computer programs is usually limited to finite intervals and to discrete arrays instead of continuous functions. Alternatively, one often samples the response of a physical system at a set of equally spaced times and wishes to perform a spectral (Fourier) analysis of such data. Therefore, it is of practical interest to return to the complex Fourier series but now using a discrete time variable. The discrete Fourier transform finds innumerable applications in the physical sciences. Among them are:

- modeling processes where the response of a system is described as convolution of a driving force with a Green function;
- analysis of data where the desired signal is convoluted with an instrumental resolution;
- identification of periodic components using a power spectrum or an autocorrelation function;
- suppression of noise by digital filtering.

In this section we will survey some of the technology that facilitates practical applications of the Fourier transform to problems requiring either numerical methods or analysis of noisy data. However, this is a very broad subject and we cannot possibly study the specialized techniques that have been optimized for various types of applications in any real depth. Our intention here is to familiarize the student with some of the underlying principles, but the researcher will need to consult more specialized literature.

5.5.1 Sampling

It is useful to express times and frequencies as

$$t_j = (j-1)\Delta t\,, \qquad j = 1, N \tag{5.150}$$

$$\omega_k = (k-1)\Delta\omega\,, \qquad k = 1, N \tag{5.151}$$

where

$$T = (N-1)\Delta t\,, \qquad \Delta\omega = 2\pi/T \tag{5.152}$$

Similarly, discretized functions become

$$f_j = f[t_j]\,, \qquad \tilde{f}_k = \tilde{f}[\omega_k] \tag{5.153}$$

and one anticipates that the discrete Fourier transform would take the form

$$f_j = \sum_{k=1}^{N} \tilde{f}_k \, \mathrm{Exp}[-i\omega_k t_j] = \sum_{k=1}^{N} \tilde{f}_k \, \mathrm{Exp}\left[-\frac{2\pi i}{N}(j-1)(k-1)\right] \tag{5.154}$$

$$\tilde{f}_k = \frac{1}{N}\sum_{j=1}^{N} f_j \, \mathrm{Exp}[i\omega_k t_j] = \frac{1}{N}\sum_{j=1}^{N} f_j \, \mathrm{Exp}\left[\frac{2\pi i}{N}(j-1)(k-1)\right] \tag{5.155}$$

where we have chosen an asymmetric but convenient normalization convention for which the first element of the transform

$$\tilde{f}_1 = \frac{1}{N} \sum_{j=1}^{N} f_j \tag{5.156}$$

reduces to the average value of f. Note that some authors choose indices $0 \le j < N - 1$ or $1 - \frac{N}{2} \le j \le \frac{N}{2}$ for even N, but we chose $1 \le j \le N$ because that is usually most suitable for indexing the elements of an array within a computer program.

To verify this analogy between continuous and discrete Fourier transforms, we need to demonstrate the discrete version of the orthogonality relation. Substitution of the Fourier coefficients into the series gives

$$f_j = \frac{1}{N} \sum_{k=1}^{N} \sum_{j'=1}^{N} f_{j'} \,\mathrm{Exp}[i\omega_k (t_{j'} - t_j)] = \frac{1}{N} \sum_{k=1}^{N} \sum_{j'=1}^{N} f_{j'} \,\mathrm{Exp}\left[\frac{2\pi i}{N}(k-1)(j'-j)\right] \tag{5.157}$$

We are free to interchange the summation order because these are finite series of finite elements, such that

$$f_j = \frac{1}{N} \sum_{j'=1}^{N} f_{j'} \sum_{k=1}^{N} (z_{j'-j})^{k-1} \tag{5.158}$$

where

$$z_m = \mathrm{Exp}[2\pi i m/N] \Longrightarrow z_m^N = 1 \tag{5.159}$$

for integer m is one of the Nth roots of unity. When $j' = j \Longrightarrow z_{j'-j} = 1$, the inner summation simply consists of N terms of unit value. When $j' \neq j$, the inner sum is a finite geometric series

$$\sum_{k=1}^{N} z_m^{k-1} = \frac{1 - z_m^N}{1 - z_m} \longrightarrow 0 \tag{5.160}$$

that vanishes for any nonzero integer m that is not a multiple of N. Therefore, the summation over k yields a Kronecker delta function

$$\frac{1}{N} \sum_{k=1}^{N} \mathrm{Exp}\left[\frac{2\pi i}{N}(k-1)(j'-j)\right] = \delta_{j,j'} \tag{5.161}$$

that is the discrete analog of the Dirac delta function. The summation over j' then selects just the term f_j from the right-hand side, yielding an identity.

The time required for straightforward computation of the discrete Fourier transform using the definition above scales with N^2, but much more efficient algorithms that exploit the periodicities of complex exponentials accomplish the same task in a time that scales with $N \log_2 N$. To appreciate the enormous savings realized by the so-called Fast Fourier

Transform (FFT), consider Table 5.4. Here the array dimensions increase by factors of 2, the second and third columns represent the number of operations for straightforward and optimized Fourier transforms, while the final column of ratios shows the factor by which FFT is faster. Note that we have chosen $N = 2^n$ on purpose – the maximum savings are realized when N is a power of 2. The FFT for even N is still better than the straightforward algorithm but the advantage is smaller; one should avoid odd or, even worse, prime values of N for which FFT is no better than the standard algorithm. If your sample set is not a power of 2, simply pad it with some extra zeros – the computational savings far outweigh the cost in extra memory and, as we will see below, zero-padding is often useful anyway.

Table 5.4. Scaling of conventional versus fast Fourier transform with sample size.

N	N^2	$N \log_2 N$	Ratio
512	262 144	4 608	57
1 024	1 048 576	10 240	102
2 048	4 194 304	22 528	186
4 096	16 777 216	49 152	341
8 192	67 108 864	106 496	630
16 384	268 435 456	229 376	1 170
32 768	1 073 741 824	491 520	2 185

The first widely disseminated FFT subroutine was developed by Cooley and Tukey in the mid-60s and revolutionized numerical signal processing and related fields. Similar FFT programs are now standard tools for numerical analysis and we assume that you can find one in whatever computational environment you would use. Since FFT programs are now so widely available that hardly anyone would consider writing his/her own anymore, we will not discuss the details of those algorithms here. Useful discussions can be found in books like *Numerical Recipes* (W. H. Press, B. P. Flannery, S. A. Teukolsky, and W. T. Vetterling, Cambridge University Press).

An important feature of the discrete Fourier transform is its periodicity:

$$\tilde{f}_{N+n} = \tilde{f}_n \Longrightarrow f_{N+n} = f_n \tag{5.162}$$

However, the discrete Fourier transform is intended to be an approximation of the continuous Fourier transform that is suitable for numerical computation using computers, yet the primary motivation for the Fourier transform was its ability to describe nonperiodic functions. It seems that we have circled back to the Fourier series, but the Fourier series is capable, at least in principle, of describing continuous functions $f[t]$ with arbitrary accuracy (except at discontinuities) while the discrete Fourier transform represents sampled quantities, f_i, instead of continuous functions. On the other hand, arbitrary accuracy is not really possible numerically because an infinite number of Fourier coefficients would be needed, not to mention the inevitable truncation errors due to the finite number of bits available to the digital representation of numbers in a computer. The discrete Fourier transform represents N samples f_j faithfully, whether they come from a smooth function or not, except for truncation errors related to machine precision. The price is unwanted periodicity and its effects upon transformations, like convolution, that require knowledge of the

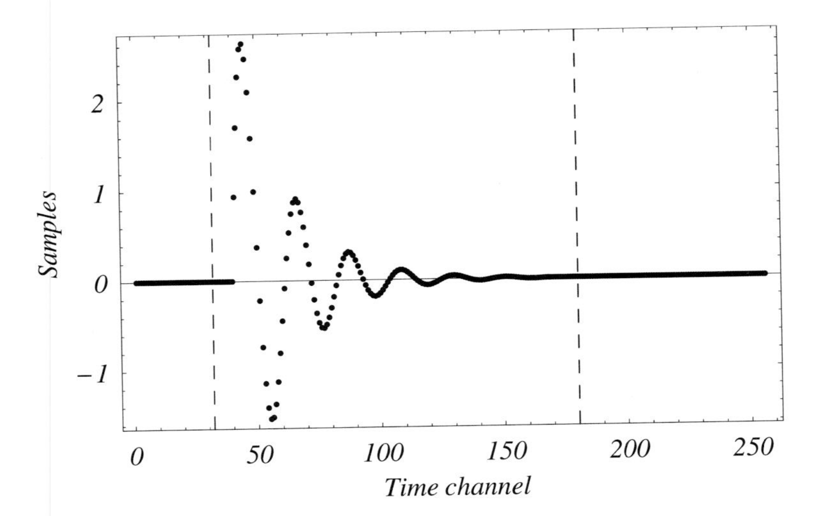

Figure 5.7. Sampling of the temporal Green function for a damped oscillator. The region within vertical dashed lines is an interior subset of the working array that contains the sampled function. The exterior buffer zones contain zero-padding.

underlying physical function $f[t]$ outside its sampled range. It is useful to think of f_j as a *working array* that contains sampled values $f[t_j]$ in an interior subset, $l \leq j \leq m$, with *zero-padding* in buffer zones on either side, such that $f_j = 0$ for $1 \leq j < l$ or $m < j \leq N$. The great speed of FFT means we need not be too concerned with increasing the array size somewhat with padding, provided that the size of the working array remains a power of 2. This arrangement is sketched in Fig. 5.7, where the data sample the Green function for an underdamped oscillator and where the interior and buffer zones are demarcated by dashed vertical lines. Note that the time axis is labeled by the sample index, often called a channel. Provided that $f[t]$ is negligibly small in the outer ranges, as in this example, the artificial periodicity induced by the discrete Fourier transform would have no adverse effects on manipulations performed in the interior range and we would simply ignore output for the buffer zones. We will show a practical example of this procedure in the convolution example later. If $f[t]$ does not decay fast enough when approaching the buffer zones, it might be necessary to impose a suitable decay by hand, but the details and consequences of such modifications vary for each case and would take us too far afield. The art of numerical computation is as interesting, important, and challenging as the more theoretical concerns of the present course but is beyond its scope.

For the remainder of this section we will assume that all f_j values are real, as befitting the sampling of a measurable quantity. The transform then exhibits the reflection symmetry

$$\begin{aligned}\tilde{f}_{N-k+2} &= \frac{1}{N}\sum_{j=1}^{N} f_j \,\mathrm{Exp}\left[\frac{2\pi i}{N}(j-1)(N-k+1)\right] \\ &= \frac{1}{N}\sum_{j=1}^{N} f_j \,\mathrm{Exp}\left[-\frac{2\pi i}{N}(j-1)(k-1)\right]\end{aligned} \tag{5.163}$$

such that

$$f = f^* \Longrightarrow \tilde{f}_{N-k+2} = \tilde{f}_k^* \tag{5.164}$$

Thus, all f_j can be reconstructed using only the first $N/2$ values of $\tilde{f}_k$; the rest are redundant. Has information been lost? Not really, because $N/2$ complex numbers still contain N real numbers so that the transform is indeed a faithful representation of the input data. At least two samples are needed to determine both the amplitude and the phase of any Fourier component. On the other hand, if we think of $\{f_j\}$ as a discretized approximation to a continuous real function $f[t]$, the limitation of the information content to the first half of the spectrum means that the discrete Fourier transform cannot reproduce any temporal variations with frequencies greater than $\omega_c = \pi/\Delta t$, where Δt is the sampling interval. This maximum frequency is known as the *Nyquist frequency* and limits the accuracy with which discretized representations can reproduce continuous functions. If the Fourier components with $\omega > \omega_c$ are known to be negligible for our target function, then sampling works well and calculations using discrete Fourier transforms should be limited only by machine precision; if not, then we should reduce the sampling interval enough to achieve the required precision. The first step in many applications is to use electronic or digital filters to suppress high-frequency components, which is especially useful when high-frequencies contain more noise than signal.

Signals whose Fourier components are limited to $\omega < \omega_{max} \leq \omega_c$ are described as *bandwidth limited* and are well-suited to the discrete Fourier transform. Unfortunately, if the input signal is not bandwidth-limited, high-frequency components can distort the lower-frequency information in the discrete Fourier transform. This phenomenon is known as *aliasing* because a component with $\omega/\omega_c = n$ will be mistaken for contribution to $\mathrm{Mod}[\omega, \omega_c]$ because the high-frequency wave oscillates n times between samples. This nature of this problem is illustrated schematically in Fig. 5.8. Both sinusoidal functions have the same values at each sampled time and are thereby indistinguishable to the discrete Fourier transform; yet they are completely different between samples.

5.5.2 Convolution

The convolution of two continuous functions is defined by

$$h = f \otimes g \Longrightarrow h[t] = \int_{-\infty}^{\infty} f[\tau]g[t-\tau]\,d\tau \tag{5.165}$$

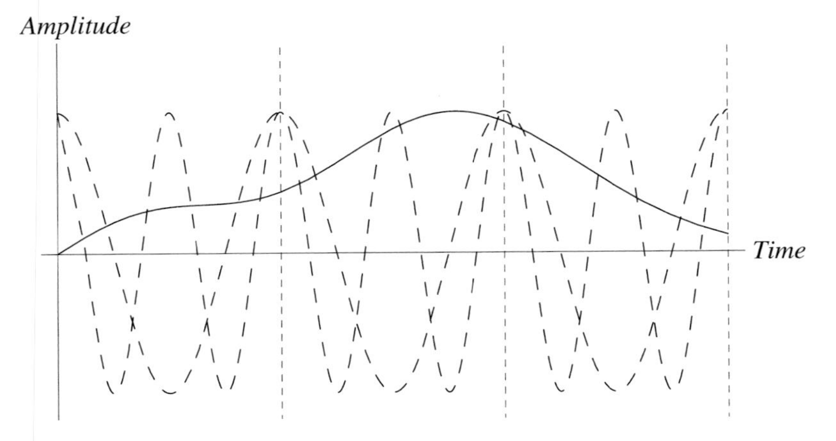

Figure 5.8. Schematic illustration of aliasing. The solid line represents an input signal and the vertical dashed lines the sampling times. Also shown are $\mathrm{Cos}[\omega_c t]$ and $\mathrm{Cos}[2\omega_c t]$.

where we assume that both functions vanish for $|t| \to \infty$. The convolution theorem then tells us that the Fourier transform of a convolution

$$\tilde{h}[\omega] = \tilde{f}[\omega]\tilde{g}[\omega] \tag{5.166}$$

is the product of the Fourier transforms of its components. Similarly, for sampled functions we define

$$h = f \otimes g \Longrightarrow h_j = \frac{1}{N}\sum_{k=1}^{N} f_k g_{j-k+1} \tag{5.167}$$

where we assume that the meaningful values of sampled quantities are confined to the interior region of suitably padded working arrays. Note the index for g_{j-k+1} accounts for the fact that k starts with 1 instead of 0. We now seek the discrete analog of the convolution theorem. By direct calculation, we write

$$\begin{aligned} \tilde{h}_k &= \frac{1}{N}\sum_{j=1}^{N} h_j \,\mathrm{Exp}\left[\frac{2\pi i}{N}(j-1)(k-1)\right] \\ &= \frac{1}{N^2}\sum_{j=1}^{N}\sum_{m=1}^{N} f_m g_{j-m+1} \,\mathrm{Exp}\left[\frac{2\pi i}{N}(j-1)(k-1)\right] \end{aligned} \tag{5.168}$$

and substitute

$$f_m = \sum_{n=1}^{N} \tilde{f}_n \,\mathrm{Exp}\left[-\frac{2\pi i}{N}(m-1)(n-1)\right] \tag{5.169}$$

$$g_{j-m+1} = \sum_{l=1}^{N} \tilde{g}_l \,\mathrm{Exp}\left[-\frac{2\pi i}{N}(j-m)(l-1)\right] \tag{5.170}$$

to obtain

$$\tilde{h}_k = \frac{1}{N^2} \sum_{j=1}^{N} \sum_{m=1}^{N} \sum_{n=1}^{N} \sum_{l=1}^{N} \tilde{f}_n \tilde{g}_l \cdot \text{Exp}\left[\frac{2\pi i}{N}\big((j-1)(k-1) - (m-1)(n-1) - (j-m)(l-1)\big)\right] \quad (5.171)$$

or

$$\tilde{h}_k = \frac{1}{N^2} \sum_{j=1}^{N} \sum_{m=1}^{N} \sum_{n=1}^{N} \sum_{l=1}^{N} \tilde{f}_n \tilde{g}_l \, \text{Exp}\left[\frac{2\pi i}{N}\big((j-1)(k-l) - (m-1)(n-l)\big)\right] \quad (5.172)$$

The summation over j reduces to $N\delta_{k,l}$, which then eliminates the summation over l also, such that

$$\tilde{h}_k = \frac{1}{N} \sum_{m=1}^{N} \sum_{n=1}^{N} \tilde{f}_n \tilde{g}_k \, \text{Exp}\left[-\frac{2\pi i}{N}(m-1)(n-k)\right] \quad (5.173)$$

Next the summation over m reduces to $N\delta_{n,k}$ and we finally obtain

$$\tilde{h}_k = \tilde{f}_k \tilde{g}_k \quad (5.174)$$

as the discrete form of the convolution theorem with the present normalization convention.

Before applying the convolution theorem to practical examples, we must consider the effects of the offsets used to place functions comfortably within the central region of the working array. Generally this means that the arrays are related to the underlying functions by

$$f_j = f[t_j + S_f], \quad g_j = g[t_j + S_g] \quad (5.175)$$

where S_f and S_g are somewhat arbitrary time shifts in f and g, chosen to make their sampling more convenient. The continuous Fourier transforms of shifted functions

$$\tilde{f}[\omega] = e^{-i\omega S_f} \int_{-\infty}^{\infty} f[t] e^{i\omega t} \, dt \quad (5.176)$$

$$\tilde{g}[\omega] = e^{-i\omega S_g} \int_{-\infty}^{\infty} g[t] e^{i\omega t} \, dt \quad (5.177)$$

include phase factors that depend upon the shifts. It is then useful to apply similar phase shifts to the corresponding discrete Fourier transforms, whereby

$$\tilde{f}_k = \frac{\text{Exp}\left[2\pi i(k-1)S_f/T\right]}{N} \sum_{j=1}^{N} f[t_j] \, \text{Exp}\left[\frac{2\pi i}{N}(j-1)(k-1)\right] \quad (5.178)$$

$$\tilde{g}_k = \frac{\text{Exp}[2\pi i(k-1)S_g/T]}{N} \sum_{j=1}^{N} g[t_j] \, \text{Exp}\left[\frac{2\pi i}{N}(j-1)(k-1)\right] \quad (5.179)$$

To illustrate the use of the convolution theorem with adjustable phase shifts, consider an underdamped harmonic oscillator with Green function $G[t]$ and driving force $F[t]$, such that the net displacement is given by

$$x[t] = \int_{-\infty}^{\infty} G[t-\tau]F[\tau]\,d\tau \Longrightarrow \tilde{x} = \tilde{f}\tilde{g} \tag{5.180}$$

We have already derived the Green function using the continuous Fourier transform and found

$$\begin{aligned} \gamma < \omega_0 \Longrightarrow G[t-\tau] &= \omega_R^{-1}\,\mathrm{Exp}[-\gamma(t-\tau)]\,\mathrm{Sin}[\omega_R(t-\tau)]\Theta[t-\tau], \\ \omega_R &= \sqrt{\omega_0^2-\gamma^2} \end{aligned} \tag{5.181}$$

Even if we know the continuous functions $f[t]$ and $g[t]$, the convolution integral is usually too difficult to perform symbolically and we need to use numerical methods. Often we do not know the underlying functions and have only measurements made at discrete times. In either case, let f_j sample the force and g_j sample the Green function. For computational reasons we shift g_j within the working array using zero padding on the left side, as shown in Fig. 5.9. Suppose that the driving force is a pulse with both positive and negative lobes centered upon $t = 0$. This time scale is also inconvenient for numerical computation, so we shift the sampled function into the working array. However, suppose that we were not too clever in our choice of shift and happened to place f somewhat too far to the right, as shown in Fig. 5.9. We then evaluate the displacement using the bare convolution theorem without compensatory phase shifts. The bulk of the resulting function is then rather far to the right of center and there appears to be a significant response for very early times before the driving force even becomes active. Does the model violate causality? No, this behavior is simply an artifact of the periodicity of the discrete Fourier transform and our injudicious sampling choices. After all, the Green function really vanishes for negative times. By placing it in the middle of the working array, the result of convolution is artificially shifted to the right. With a force that is also shifted to the right, the response goes past the end of the array and reappears, by periodicity, at the beginning. We might try placing g closer to the left edge of the array, but that could cause other unwanted wrap-around effects for functions that do not feature a sharp left edge.

The solution to these numerical problems is to multiply $\tilde{f}_k\tilde{g}_k$ by a phase $\mathrm{Exp}[-2\pi i(k-1)S_g/T]$ before computing the inverse Fourier transform of $\tilde{x}$. This phase compensates for the offset of g_j and aligns the response with the driving force, as shown in Fig. 5.10. With a larger shift, we could also compensate for the poor placement of f, but we must not use a shift so large that it wraps around the left side and back onto the right. Because the indexing of sampled functions is merely a computational issue, we are free to adjust it in any manner that ensures numerical accuracy and convenience.

5.5.3 Temporal Correlation

Suppose that two functions, $f[t]$ and $g[t]$, are qualitatively similar to each other, except that there is a time shift between them. If we had graphs of those functions we could estimate

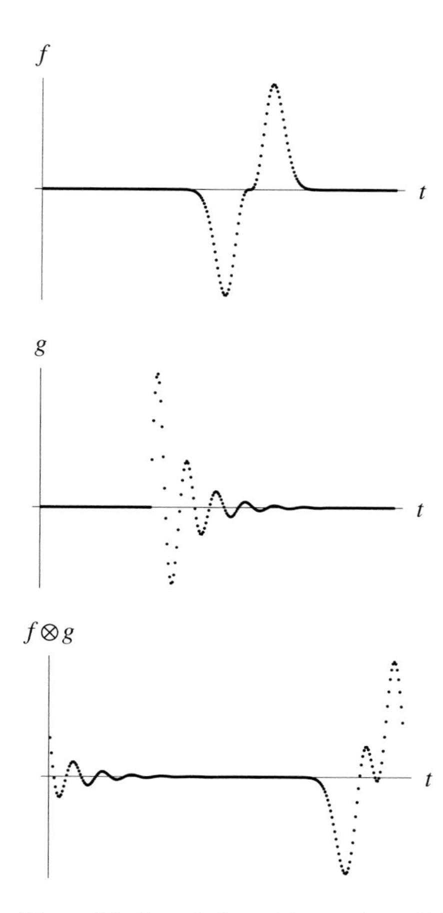

Figure 5.9. Convolution without a phase shift. If either f or g is too close to the right side, periodicity of the discrete Fourier transform produces artificial strength near the left side of their convolution, an effect described as wrap-around.

the time shift by sliding those graphs to the left or right until we obtain the best overlap between them. If these functions are quasiperiodic or include several features with different periods, there may be several shifts which result in significant overlap. The *correlation function*

$$C_j[f, g] = \frac{1}{N} \sum_{k=1}^{N} f_{j+k} g_k^* \tag{5.182}$$

provides a systematic method for evaluating the correlation between two sampled functions. The correlation function obviously closely resembles convolution and we can deduce

$$\tilde{C}_k[f, g] = \tilde{f}_k \tilde{g}_k^* \tag{5.183}$$

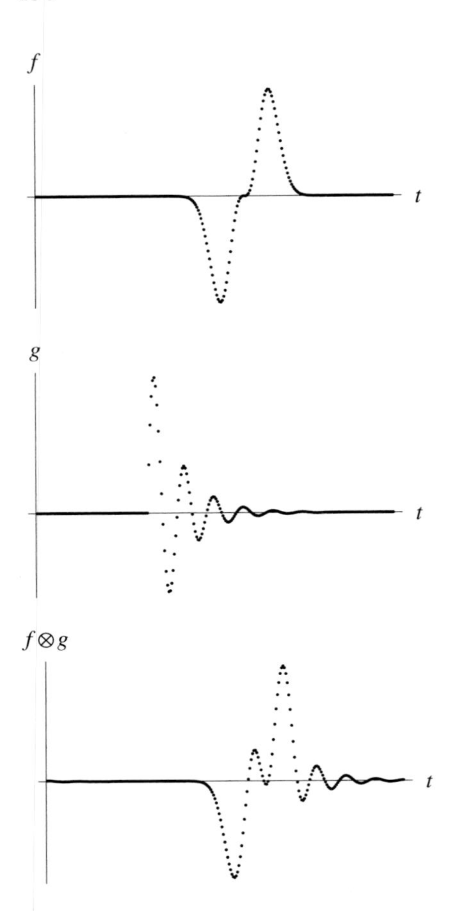

Figure 5.10. Convolution using a shift of = $-0.3125T$ aligns the response with the driving force and eliminate wrap-around.

without further ado. Often one uses the *autocorrelation function*

$$C_j[f, f] = \frac{1}{N} \sum_{k=1}^{N} f_{j+k} f_k^* \Longrightarrow \tilde{C}_k[f, f] = |\tilde{f}_k|^2 \tag{5.184}$$

to identify periodic behavior within a single time series. Thus, the Fourier transform of the autocorrelation function is simply proportional to the spectral power distribution. If we also employ ensemble averaging, spectral distributions in statistical physics are seen to be closely related to probability distributions for fluctuations about equilibrium.

Consider the data $\{x_k\}$ plotted in Fig. 5.11. There appear to be three distinct bands but, without laborious scanning of the raw data, it is not obvious to this viewer whether or not there is additional structure within those bands or whether there is a pattern to how the visible bands are visited. Perhaps the simplest method for studying data of this type is to use standard mathematical software to evaluate Fourier transforms and to form the

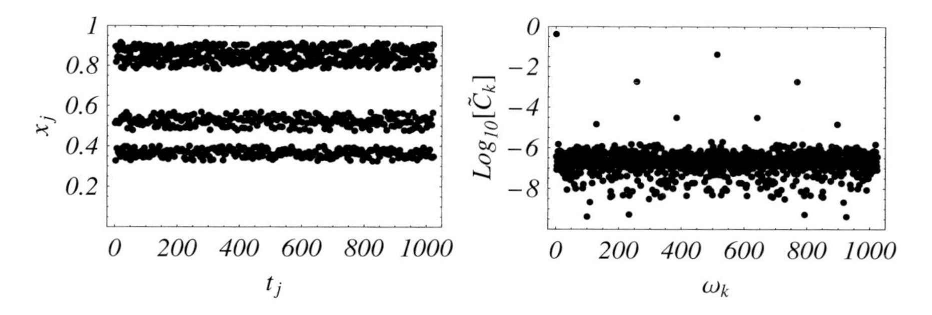

Figure 5.11. Left: noisy time series. Right: autocorrelation spectrum.

autocorrelation spectrum. This can be accomplished with only a few lines of code using MATHEMATICA®, for example. We then obtain the accompanying figure for $\{\tilde{C}_j\}$; notice the logarithmic scale. The strongest channel, $k = 1$, contains the square of the sum of $\{x_j\}$. Three other strong frequencies are clearly visible, plus four weaker frequencies. Therefore, these data actually contain an 8-cycle. (Note that alternation between two values, a 2-cycle, only corresponds to one frequency.) In addition, there is a spectrum of white noise that tends to obscure the patterns in $\{x_j\}$.

Now that we know there is an embedded 8-cycle, we can partition the data into groups of 8 and average the groups to obtain an estimated cycle $\{\bar{x}_j, j = 1, 8\}$ for which random fluctuations are suppressed by averaging; positive fluctuations tend to cancel negative fluctuations. Finally, if the underlying dynamics are periodic and deterministic, the value of x_{j+1} is predicted by x_j. Therefore, Fig. 5.12 plots $\bar{x}_{j+1}$ versus $\bar{x}_j$ for the cycle. We now see individual points that were obscured in the noisy raw data by random fluctuations. In a similar plot for the raw data without averaging, these pairs of points would smear out into indistinct blobs. These points lie within the bands in the $\{t_j, x_j\}$ plot, but the highest two pairs appeared to merge into a single band in the noisy data. The lowest band was narrowest because the underlying pair has the smallest separation. These points appear to lie along a parabola, so we also show a curve $\mu x(1-x)$ where μ is fitted to the cycle data.

The data for this simulation were constructed using the logistic map

$$x_{j+1} = \mu x_j(1 - x_j) \tag{5.185}$$

with $\mu = 3.55$, which settles onto an 8-cycle after discarding enough of the initial iterations to allow transients to decay. We then added uniformly distributed random fluctuations to simulate the noise that might be encountered in the measurement of the response of a dynamical system. The noise amplitude was chosen to be large enough to smear the bands in x but small enough not to obscure the smaller frequency peaks in $\tilde{C}$. The unperturbed cycle for this μ is actually $\{0.506, 0.887, 0.355, 0.813, 0.540, 0.882, 0.370, 0.828\}$, which agrees well with the extracted cycle. Thus, the upper band actually contains 4 values while the lower 2 bands contain 2 each, but it would take very sharp eyes, and perhaps some imagination, to discern that pattern within the noisy data. Nevertheless, with the aid of the autocorrelation spectrum we can discern the periodicities of the data and use that infor-

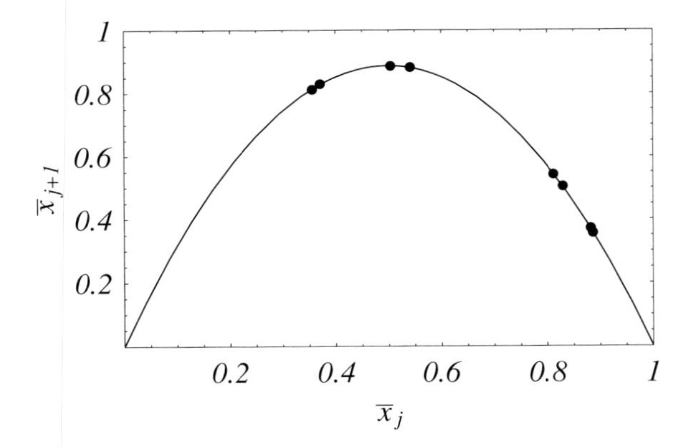

Figure 5.12. Cycle and recursion relation extracted from data in Fig. 5.11.

mation to discover the underlying dynamics despite the superimposed noise. The present example is admittedly artificial, but it should be clear that this type of analysis can be very valuable in the study of real dynamical systems.

5.5.4 Power Spectrum Estimation

Estimation of the power spectrum for a persistent function $f[t]$ that is not localized in time using discretely sampled measurements is a surprisingly tricky task for which there is much lore and literature; too much to survey here. The main problem is that the limitation of data to a finite interval necessarily sacrifices information about times outside that interval. Discrete sampling also limits the maximum meaningful frequency to $\omega \leq \omega_c = \pi/T$ for real functions or $2\omega_c$ for complex functions. Finally, there are precision and noise issues that we will not discuss but which are important in practice. Here we will present a very brief survey of some of the issues in estimating power spectra but leave more detailed discussion to specialized texts.

Suppose that $f[t] = \text{Exp}[-i\omega_0 t]$ represents a simple harmonic vibration with unique frequency ω_0. Sampling necessarily limits the Fourier transform to a finite interval T. (Who can afford to watch the vibration forever?) Thus, the continuous Fourier transform over a finite interval becomes

$$\tilde{f}[\omega] = \int_0^T \text{Exp}[i\omega t]\,\text{Exp}[-i\omega_0 t]\,dt = 2\,\text{Exp}[i(\omega - \omega_0)T/2]\frac{\text{Sin}[(\omega - \omega_0)T/2]}{\omega - \omega_0} \tag{5.186}$$

with a spectral power density

$$P[\omega] = |\tilde{f}[\omega]|^2 = \left(T\frac{\text{Sin}[(\omega - \omega_0)T/2]}{(\omega - \omega_0)T/2}\right)^2 \tag{5.187}$$

that peaks at ω_0 but which is spread over a considerable range of frequencies when $\omega_0 T$ is not large. Many periods of oscillation must be observed in order to measure ω_0 accurately, especially when the signal contains noise or measurement errors. However, actual measurements made are at discrete times $t_j = T(j-1)/(N-1)$, not continuously, so we estimate the power spectrum using the discrete Fourier transform

$$\begin{aligned}\tilde{f}_k &= \sum_{j=1}^{N} \mathrm{Exp}\left[\frac{2\pi i}{N}(j-1)(k-1)\right] f[t_j] \\ &= \sum_{j=1}^{N} \mathrm{Exp}\left[\frac{2\pi i}{N}(j-1)(k-1)\right] \mathrm{Exp}\left[-i\omega_0 T \frac{j-1}{N-1}\right]\end{aligned} \tag{5.188}$$

This expression takes the form of a finite geometric series that can be evaluated in closed form. After some tedious algebra, we obtain

$$\tilde{f}_k = \mathrm{Exp}\left[-i\pi\left(\frac{k-1}{N} + \frac{\omega_0 T}{2\pi}\right)\right] \frac{\mathrm{Sin}\left[\frac{\omega_0 T}{2}\frac{N}{N-1}\right]}{\mathrm{Sin}\left[\frac{\omega_0 T}{2(N-1)} - \pi\frac{(k-1)}{N}\right]} \tag{5.189}$$

The first factor is just a phase that does not affect the power spectrum. It is useful to define $T = (N-1)\tau$ where τ is the sampling interval, such that

$$P_k = \left(\frac{\mathrm{Sin}[N\omega_0\tau/2]}{\mathrm{Sin}[\omega_0\tau/2 - \pi(k-1)/N]}\right)^2 \tag{5.190}$$

represents the discrete power spectrum. This spectrum exhibits a strong peak where

$$k \approx \kappa = 1 + \left(\frac{\omega_0\tau}{2\pi} - m\right)N \tag{5.191}$$

is close to a root of the denominator. Here m is an integer chosen to ensure that κ is in the range $1 \le \kappa \le N$. Notice that we indicated approximate equality because k is an integer while κ usually is not. The peak of the power spectrum then has a finite width, proportional to T^{-1}, that is similar to the continuous Fourier transform of a finite wave train. When κ actually is an integer, both the numerator and the denominator have coincident roots and we use L'Hôpital's rule to determine that

$$\frac{N\omega_0\tau}{2\pi} \to \kappa + mN - 1 \Longrightarrow \tilde{f}_k \to N\delta_{k,\kappa} \tag{5.192}$$

is consistent with the normalization of the orthogonality relation for the discrete Fourier transform. In principle, such a peak is limited to only one channel, but signal noise or numerical precision will generally produce nonzero amplitudes for nearby channels.

The sensitivity of the power spectra to slight differences in frequency is illustrated in Fig. 5.13. Both figures use FFT with $N = 2^8 = 256$ channels. Notice the logarithmic scales. The peak at $\kappa = 86.33$ for $\omega_0\tau = 2\pi/3$ is not integral, resulting in a noticeable spreading about channel $k = 86$. By contrast, using the very similar frequency $\omega_0\tau = (2\pi/3)(N-1)/N$ produces a discrete delta function at channel 86; the negligible, but apparently nonzero, power in other channels is due to round-off errors for numerical calculation

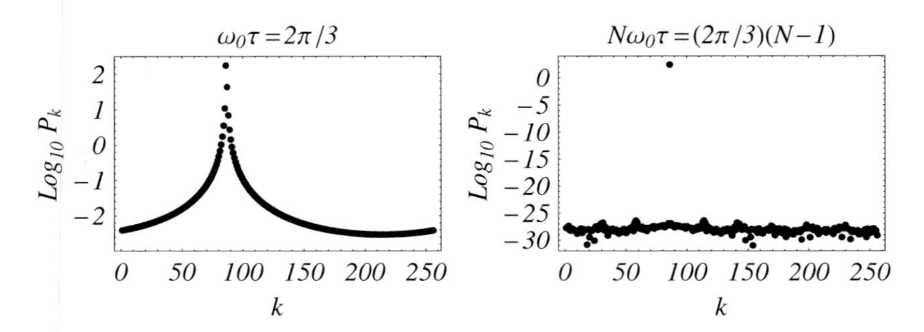

Figure 5.13. Power spectra for discrete Fourier transforms of Exp$[-i\omega_0 t]$ that use $\omega_0\tau \approx 2\pi/3$.

of the Fourier transform with a machine precision of 10^{-16}. The striking difference between these spectra is an artifact of discretization that has nothing to do with the nature of the underlying continuous function. In one case we chose, probably fortuitously, a sampling interval that divides the period perfectly. In real life our signals would not be pure sinusoids and we would not know the frequency well enough in advance to choose such a precise sampling interval – if we had such knowledge, we would not need to perform numerical analysis. Therefore, the panel on the left is the usual situation and shows that the spectrum is spread by sampling even for a pure sinusoidal oscillation. The panel on the right, on the other hand, requires precise calculations and delicate cancellations to achieve a discrete delta function. The alert reader might wonder how the discrete Fourier transform can produce a delta function using a finite observation time T while the continuous Fourier transform for the same observation time results in appreciable spread. The difference is that the discrete Fourier transform automatically assumes that the underlying function is periodic and persists forever while the continuous Fourier transform does not.

The term $-mN$ in the expression for κ is a manifestation of aliasing – even when the frequency ω_0 is very large, there is still a peak within the sampled range of frequencies because discrete sampling cannot distinguish how many times a function oscillates between samples. Consequently, very high frequency contributions to a continuous signal corrupt the discrete power spectrum for lower frequencies. In Fig. 5.14 we chose a frequency for which $\omega_0\tau = 2135$ is much larger than $2\pi N$ for $N = 256$. The peak power for the continuous spectrum should then be well beyond the end of this spectrum. Nevertheless, we observe a peak at channel 205 corresponding to $m = 339$. Our sampling is much too coarse to obtain a realistic spectrum for this high-frequency signal. Signal processing cannot be performed by blind application of numerical algorithms.

As our final example, we consider the behavior of the *van der Pol* oscillator, described by the differential equation

$$\frac{\partial^2 x[t]}{\partial t^2} == -x[t] + \varepsilon(1 - x[t]^2)\frac{\partial x[t]}{\partial t} \tag{5.193}$$

where ε is a positive constant and x will be described as a displacement. This equation reduces to a simple harmonic oscillator when $\varepsilon \to 0$, but for positive ε the velocity-dependent term pumps energy into the motion for small displacements and drains it for

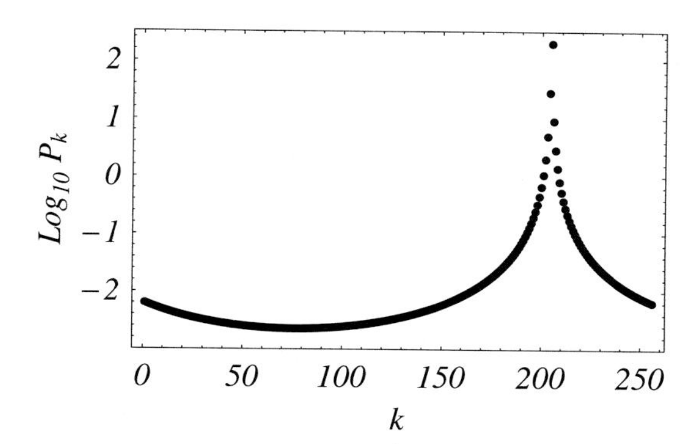

Figure 5.14. Power spectra for discrete Fourier transform of Exp$[-i\omega_0 t]$ with $\omega_0 \tau = 2135$. The peak is an artifact of aliasing; the frequency of the actual signal is much too large for this sampling interval.

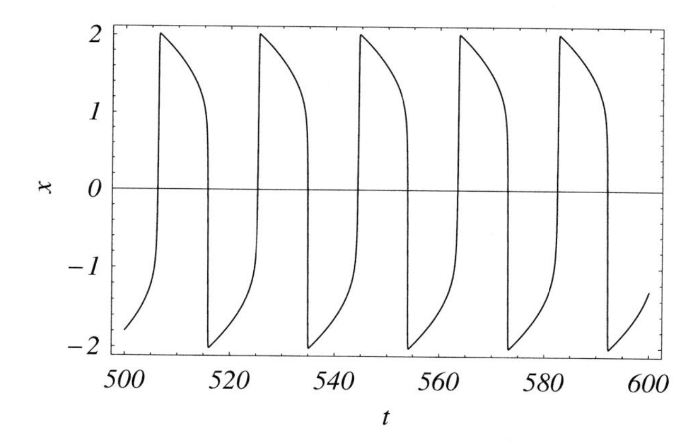

Figure 5.15. van der Pol oscillation, $\varepsilon = 10$.

large displacements. The net effect is to drive the system toward a stable "limit-cycle" oscillation that is decidedly nonsinusoidal for large ε. This nonlinear differential equation cannot be solved symbolically, so we must resort to numerical methods. The solution for large t, long after transients have decayed, is shown in Fig. 5.15 for $\varepsilon = 10$. The behavior is periodic but not simple, featuring relatively slow variations near either extreme, but with rapid transitions between those extremes. Obviously, the time sampling must be fine enough to represent the abrupt transitions.

The discrete power spectrum obtained using 4096 samples in the range $200 \leq t \leq 600$ is shown in Fig. 5.16. The peaks are found at odd multiples of a fundamental frequency, ν_1, that is approximately 21 channels. The fact that the harmonics are odd can be understood by observing that the nonlinear term responsible for those harmonics is odd. From the plot of $x[t]$ we see that the oscillation period, p, is approximately 19 time units, such that

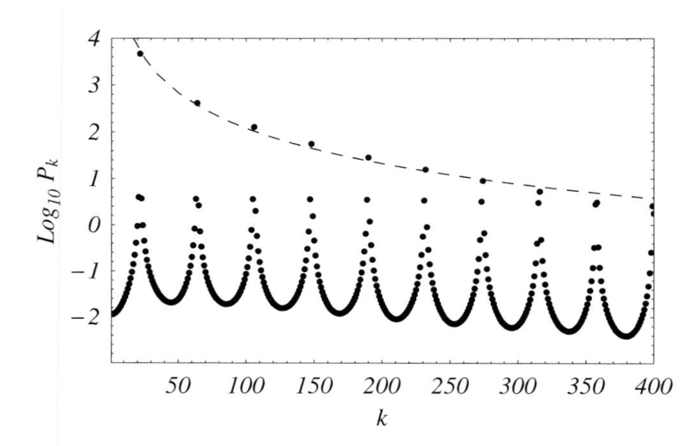

Figure 5.16. van der Pol spectrum, $\varepsilon = 10$.

$\nu_1 p \approx T$ where T is the total observation time. The dashed line shows that the intensity of higher harmonics is roughly proportional to $k^{-5/2}$, at least over this range. If the time samples represent observation of a physical system, the power spectrum can be used to study the underlying dynamics. We might, for example, attempt to guess a differential equation that has these features and fit its parameters to the observed behavior.

After transients have decayed away, the behavior of this system is periodic. If we knew its functional form and period precisely, we could use the Fourier series to deduce the power in each odd multiple of the fundamental frequency, reducing the power spectrum from a sequence of peaks to a series of discrete spikes of zero width. The spreading seen in the discrete Fourier transform is partly due to the limitations of sampling in which the period is not divided into a precisely integral number of samples. We might be able to reduce the widths of these peaks if we knew the period in advance or could interpolate within the data array to improve the sampling interval. The mismatch between the sample interval and the period is analogous to chopping the fundamental frequency. The sharp turn-on at $t = 0$ and turn-off at $t = T$ spreads each discrete frequency into a peak of finite width. A closer approximation to a Fourier series of narrow peaks can be obtained by multiplying the sampled data by a window function $w[t]$ that has a broad, relatively flat central region between smooth turn-on and turn-off regions. This technique is explored in one of the end-of-chapter problems. However, in real applications, additional spreading would be produced by measurement errors and noise. The effect of noise can be reduced by dividing the data into several subintervals, each containing an integral number of periods, and averaging the data for those subintervals to improve the signal-to-noise ratio. The discrete Fourier transform would then be taken for the averaged data.

5.6 Laplace Transform

5.6.1 Definition and Inversion

The Laplace transform $\mathcal{L}f$ is defined by

$$\mathcal{L}[f[t]][s] = \tilde{f}[s] = \int_0^\infty dt e^{-st} f[t] \tag{5.194}$$

and is useful for functions which vanish for $t < 0$ and remain finite for $t > 0$, ensuring convergence of an integral transform that uses an exponential kernel. The Laplace transform is related to the Fourier transform with respect to an imaginary variable $\omega \to is$. The primary difficulty in using the Laplace transform is defining and evaluating the inverse transformation $\mathcal{L}^{-1}$. Using the analogy with the Fourier transform, we might guess that the kernel for the inverse transform should take the form $e^{-i\omega t} \to e^{st}$, but the exponential growth for $t > 0$ clearly presents problems for convergence. Consider the Fourier transform of the function

$$g[t] = e^{-\gamma t} f[t] \Longrightarrow \tilde{g}[\omega] = \mathcal{F}g[\omega] = \int_0^\infty dt e^{i\omega t} g[t] = \int_0^\infty dt e^{-(\gamma - i\omega)t} f[t] \tag{5.195}$$

whose inverse Fourier transform satisfies

$$e^{-\gamma t} f[t] = \mathcal{F}^{-1}[\tilde{g}] = \frac{1}{2\pi} \int_{-\infty}^\infty d\omega e^{-i\omega t} \tilde{g}[\omega] \Longrightarrow f[t] = \frac{1}{2\pi} \int_{-\infty}^\infty d\omega e^{(\gamma - i\omega)t} \tilde{g}[\omega] \tag{5.196}$$

The integral for $f[t]$ will converge if γ is large enough to ensure that $e^{st} g[s] \to 0$ for $s \to \pm i\infty$. The variable change $s = \gamma - i\omega$ then gives

$$f[t] = \frac{1}{2\pi i} \int_{\gamma - i\infty}^{\gamma + i\infty} ds e^{st} \mathcal{F}g[i(s - \gamma)] \tag{5.197}$$

where the Fourier transform of g evaluated for the imaginary frequency $i(s - \gamma)$ such that

$$\mathcal{F}g[i(s - \gamma)] = \int_0^\infty dt e^{-(s-\gamma)t} g[t] = \int_0^\infty dt e^{(\gamma - s)t} e^{-\gamma t} f[t] = \mathcal{L}f[s] \tag{5.198}$$

reduces to the Laplace transform of f. Rigorous derivations may be found in more specialized literature.

Therefore, the Laplace transform and its inverse are defined by

$$\tilde{f}[s] = \mathcal{L}f = \int_0^\infty dt\, e^{-st} f[t] \tag{5.199}$$

$$f[t] = \mathcal{L}^{-1}\tilde{f} = \frac{1}{2\pi i} \int_{\gamma - i\infty}^{\gamma + i\infty} ds\, e^{st} \tilde{f}[s] \tag{5.200}$$

where γ is a real number chosen to place the integration path to the right of all singularities in $\tilde{f}[s]$. The inverse Laplace transform is often called the *Bromwich integral* or the *Mellin inversion formula*. Although the Laplace transform was originally defined for real values

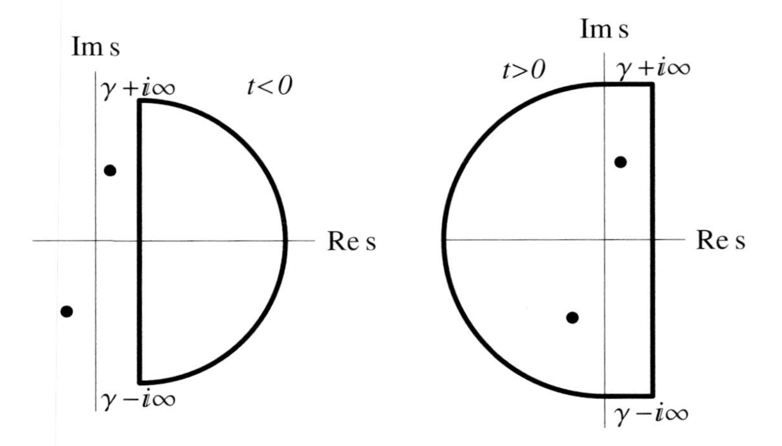

Figure 5.17. Contours for the inverse Laplace transform. We must choose γ to place the primary vertical segment to the right of any singularities. Here we show just two poles; additional detours may be needed to avoid branch cuts in the left half-plane.

of s, the inversion is performed in the complex s-plane using the domain in which $\tilde{f}[s]$ is analytic. Contours for inversion of a Laplace transform are sketched in Fig. 5.17, where the heavy dots indicate isolated singularities. The region to the left of the vertical portion of the contour may also contain branch cuts, if needed to define a single-valued function $\tilde{f}[s]$. When $t < 0$ we close the contour using a great semicircle in the right half-plane, on which the integrand vanishes in the limit of infinite radius, and conclude that $f[t < 0] = 0$ because no singularities are enclosed. This result is consistent with the requirements on f needed for application of the Laplace transform. When $t > 0$ we close the contour using a great semicircle in the left half-plane, with detours around any branch cuts. The residue theorem can be used to evaluate the contour integral, but to extract the inverse function we may need to subtract contributions of unwanted portions of the contour.

Notice that when $s \to 0$ the Laplace transform becomes the total integral

$$s \to 0 \Longrightarrow \mathcal{L}f = \int_0^\infty f[t]\, dt \tag{5.201}$$

However, it is not necessary for this integral to exist for the Laplace transform to exist. Existence of the Laplace transform requires that for any positive M there exists a real number γ such that $\left|e^{-\gamma t} f[t]\right| \le M$ is bounded for large t. A function satisfying this condition is described as *exponentially bounded.* Thus, the Laplace transform of $e^{\lambda t}$ for positive λ exists in the domain $\mathrm{Re}[s] > \lambda$ even though its total integral does not. On the other hand, $\mathrm{Exp}[t^2]$ does not have a Laplace transform because it is not bounded by an exponential of t. Nor does the Laplace transform exist for t^{-n} when $n > 1$ because the transformation integral is divergent at its lower end.

5.6.2 Laplace Transforms for Elementary Functions

Let $\Theta[t]$ represent the unit step function

$$\Theta[t < 0] = 0 \tag{5.202}$$
$$\Theta[t > 0] = 1 \tag{5.203}$$

whose Laplace transform

$$\mathcal{L}\Theta = \int_0^\infty dt e^{-st}\Theta[t] = \frac{1}{s} \tag{5.204}$$

exhibits a simple pole at the origin with unit residue. We will leave the value at $t = 0$ undefined for the moment, and evaluate it using the inverse Laplace transform. The inverse transform for $t \neq 0$ is obtained using

$$\Theta[t] = \frac{1}{2\pi i}\int_{\gamma-i\infty}^{\gamma+i\infty} ds\frac{e^{st}}{s} \tag{5.205}$$

where $\gamma > 0$ may be taken arbitrarily small. Consider the contour integral

$$\frac{1}{2\pi i}\oint_C ds\frac{e^{st}}{s} \tag{5.206}$$

where C consists of a vertical line in the right half-plane closed by a great semicircle of radius R where $R \to \infty$. The only singularity of the integrand is found at the origin. For $t < 0$ we choose a great semicircle to the right to ensure that closure does not contribute and that the contour integral reduces to the inverse Laplace transform; hence, the inverse transform for $t < 0$ vanishes because no singularities are enclosed by C. For $t > 0$ we must close in the left half-plane and include two short horizontal segments represented by $s = \gamma\beta \pm iR$ where $0 \leq \beta \leq 1$. The contributions of these segments vanish in the limit $R \to \infty$ because

$$s = \gamma\beta \pm iR \Longrightarrow \left|\int_{\text{segment}} ds\frac{e^{st}}{s}\right| \leq \gamma\int_0^1 d\beta\left|\frac{\text{Exp}[(\gamma\beta \pm iR)t]}{\gamma\beta \pm iR}\right| \leq \gamma\frac{e^{\gamma t}}{R} \to 0 \tag{5.207}$$

Thus, the contour integral again reduces to the inverse Laplace transform but this time has the value unity because a single pole with unit residue is enclosed. For $t = 0$ we must be more careful because the inversion integral takes the form

$$\Theta[0] = \frac{1}{2\pi i}\int_{\gamma-i\infty}^{\gamma+i\infty} ds\frac{1}{s} \tag{5.208}$$

and the contribution of a great semicircle does not vanish. On the contrary, if we close to the right

$$\frac{1}{2\pi i}\oint_C \frac{ds}{s} = \lim_{R\to\infty}\left(\frac{1}{2\pi i}\int_{\gamma-iR}^{\gamma+iR}\frac{ds}{s} + \frac{1}{2\pi}\int_{\pi/2}^{-\pi/2} d\theta\right) = \Theta[0] - \frac{1}{2} = 0 \Longrightarrow \Theta[0] = \frac{1}{2} \tag{5.209}$$

or if we close to the left

$$\frac{1}{2\pi i}\oint_C \frac{ds}{s} = \lim_{R\to\infty}\left(\frac{1}{2\pi i}\int_{\gamma-iR}^{\gamma+iR}\frac{ds}{s} + \frac{1}{2\pi}\int_{\pi/2}^{3\pi/2} d\theta\right) = \Theta[0] + \frac{1}{2} = 1 \Longrightarrow \Theta[0] = \frac{1}{2} \tag{5.210}$$

where the short horizontal segments can be neglected and where one choice encloses a pole, while the other does not. Either way, we find that the consistent definition of the step function in terms of the inverse Laplace transform of s^{-1} requires $\Theta[t = 0] = 1/2$.

The same result can be obtained using $\gamma \to 0$ and a contour C consisting of the imaginary axis and an infinitesimal semicircular detour around the origin that ventures into the right half-plane. For $t < 0$ we again close C to the right while for $t > 0$ we close to the left and recover the previous results. For $t = 0$ we obtain

$$t = 0 \Longrightarrow \oint_C ds\frac{e^{st}}{s} = 2\pi i = \mathcal{P}\int_{i\infty}^{i\infty}\frac{ds}{s} + i\pi \Longrightarrow \Theta[0] = \frac{1}{2} \tag{5.211}$$

and is composed of a principal-value integral that gives the inverse transform and the contribution of the semicircular detour. Therefore, combining these results we obtain

$$\frac{1}{2\pi i}\int_{\gamma-i\infty}^{\gamma+i\infty} ds\frac{e^{st}}{s} = \Theta[t] \tag{5.212}$$

and recover the original function. Obvious, perhaps, but the unit step function is an implicit factor for any function for which one applies the Laplace transform because the requirement $f[t < 0] = 0$ can be met by the replacement $f[t] \to f[t]\Theta[t]$. However, if $f[0] \neq 0$, there are sometime subtle problems in limits $t \to 0^+$ related to the value of $\Theta[0]$. These are usually harmless, but may require careful analysis.

Next consider an exponential function for $t > 0$ whose Laplace transform

$$f[t] = e^{\alpha t}\Theta[t] \Longrightarrow \tilde{f}[s] = \int_0^\infty dt e^{-st} e^{\alpha t} = \frac{1}{s-\alpha} \quad \text{for } s > \text{Re}[\alpha] \tag{5.213}$$

exhibits a simple pole with unit residue at $s = \alpha$. If the real part of α is positive, the pole is in the right half-plane. Thus, poles in the right half-plane correspond to exponentially growing functions and convergence of the inverse Laplace transform requires γ to be sufficiently large to overcome the strongest of these exponential features. Poles in the left half-plane correspond to exponentially decreasing contributions which pose no difficulties for the inversion procedure. The imaginary parts of these poles correspond to oscillatory features in the original function. The unit step function without an exponential factor is equivalent to an exponential with $\alpha = 0$ and contributes a pole at the origin. The Laplace transforms for trigonometric functions are easily deduced from that for an exponential, whereby

$$\mathcal{L}\big[\text{Sin}[\omega t]\big] = \frac{1}{2i}\left(\frac{1}{s-i\omega} - \frac{1}{s+i\omega}\right) = \frac{\omega}{s^2+\omega^2} \tag{5.214}$$

$$\mathcal{L}\big[\text{Cos}[\omega t]\big] = \frac{1}{2}\left(\frac{1}{s - i\omega} + \frac{1}{s + i\omega}\right) = \frac{s}{s^2 + \omega^2} \tag{5.215}$$

Notice that when

$$s \to 0 \Longrightarrow \mathcal{L}\big[\text{Sin}[\omega t]\big] \to \frac{1}{\omega}, \quad \mathcal{L}\big[\text{Cos}[\omega t]\big] \to 0 \tag{5.216}$$

we obtain finite values for the improper integrals

$$\int_0^\infty dt\, \text{Sin}[\omega t] = \frac{1}{\omega} \tag{5.217}$$

$$\int_0^\infty dt\, \text{Cos}[\omega t] = 0 \tag{5.218}$$

that are consistent with earlier results obtained with the aid of a convergence factor (see Sec. 2.2.2).

If the Laplace transform is a rational function, often the simplest inversion method is to expand in partial fractions and sum the resulting exponential functions. For example, given

$$\tilde{f}[s] = \frac{s}{(s-a)(s+b)} = \frac{1}{(a+b)}\left(\frac{a}{s-a} + \frac{b}{s+b}\right) \tag{5.219}$$

we can immediately determine

$$f[t] = \frac{a e^{at} + b e^{-bt}}{a+b} \tag{5.220}$$

without performing contour integration.

The Laplace transform has a simple shifting property

$$\tilde{f}[s-a] = \int_0^\infty dt e^{-(s-a)t} f[t] = \mathcal{L}\big[e^{at} f[t]\big] \tag{5.221}$$

which permits one to deduce

$$\mathcal{L}\big[e^{-\gamma t}\, \text{Sin}[\omega t]\big] = \frac{\omega}{(s+\gamma)^2 + \omega^2} \tag{5.222}$$

$$\mathcal{L}\big[e^{-\gamma t}\, \text{Cos}[\omega t]\big] = \frac{s+\gamma}{(s+\gamma)^2 + \omega^2} \tag{5.223}$$

effortlessly. Conversely, if we apply a step function to $f[t]$ to ensure that the integrand vanishes for $t < 0$, as required for the Laplace transform, multiplying the transform by an exponential

$$e^{-\alpha s}\tilde{f}[s] = \int_0^\infty dt e^{-s(t+\alpha)} f[t] = \int_\alpha^\infty dt e^{-st} f[t-\alpha] = \int_0^\infty dt e^{-st} f[t-\alpha]\Theta[t-\alpha] \tag{5.224}$$

has the effect of shifting the integrand. Hence, we find

$$\mathcal{L}\big[f[t-\alpha]\Theta[t-\alpha]\big] = e^{-\alpha s}\tilde{f}[s] \tag{5.225}$$

Finally, consider the delta function $\delta[t-t_0]$ with $t_0 > 0$, for which

$$\mathcal{L}\big[\delta[t-t_0]\big] = e^{-st_0} \Longrightarrow \mathcal{L}^{-1}[e^{-st_0}] = \delta[t-t_0] \tag{5.226}$$

In the limit $t_0 \to 0^+$, we find formal results

$$t_0 \to 0^+ \Longrightarrow \mathcal{L}\big[\delta[t]\big] = 1 \Longrightarrow \mathcal{L}^{-1}[1] = \delta[t] \tag{5.227}$$

that are often useful in initial-value problems. Although this result is similar to that for Fourier transforms, we must remember that the Laplace transform is limited to $t > 0$.

5.6.3 Laplace Transform of Derivatives

The Laplace transform finds some of its most important applications in the solution of differential equations. The Laplace transform of a derivative

$$\mathcal{L}[f'[t]] = \int_0^\infty dt e^{-st} f'[t] = (e^{-st} f[t])_0^\infty + s\int_0^\infty dt e^{-st} f[t] \tag{5.228}$$

can be integrated by parts. Assuming that $f[t]$ remains finite as $t \to \infty$ or restricting $\mathrm{Re}[s] > \gamma$, we obtain

$$\mathcal{L}\big[f'[t]\big] = s\mathcal{L}\big[f[t]\big] - f[0^+] \tag{5.229}$$

where the integration constant is evaluated in the limit $t \to 0^+$ for infinitesimal positive t. Iteration now gives

$$\mathcal{L}\big[f''[t]\big] = s^2\mathcal{L}\big[f[t]\big] - sf[0^+] - f'[0^+] \tag{5.230}$$

or, more generally,

$$\mathcal{L}\big[f^{(n)}[t]\big] = s^n\mathcal{L}\big[f[t]\big] - \sum_{k=0}^{n-1} s^{n-1-k} f^{(k)}[0^+] \tag{5.231}$$

Therefore, when we apply the Laplace transform to a linear differential equation, the initial conditions become part of the resulting algebraic equation, making this technique particularly useful for initial-value problems. The Fourier transform, on the other hand, is often more useful for boundary-value problems.

It is also useful to consider derivatives of the transform

$$\tilde{f}[s] = \int_0^\infty dt e^{-st} f[t] \Longrightarrow \tilde{f}'[s] = \int_0^\infty dt e^{-st}(-tf[t]) \tag{5.232}$$

whereby

$$\mathcal{L}\big[tf[t]\big][s] = -\frac{\partial}{\partial s}\tilde{f}[s] \tag{5.233}$$

Thus, the Laplace transform of $t^n f[t]$ is related to $\mathcal{L}f$ by the operator

$$t \longrightarrow -\frac{\partial}{\partial s} \tag{5.234}$$

$$t^n \longrightarrow \left(-\frac{\partial}{\partial s}\right)^n \tag{5.235}$$

Alternatively, using

$$\frac{e^{-st}}{t} = \int_s^\infty d\sigma e^{-\sigma t} \tag{5.236}$$

and changing the order of integration in

$$\int_0^\infty dt t^{-1} e^{-st} f[t] = \int_0^\infty dt \int_s^\infty d\sigma e^{-\sigma t} f[t] = \int_s^\infty d\sigma \int_0^\infty dt e^{-\sigma t} f[t] \tag{5.237}$$

we obtain

$$\mathcal{L}\big[t^{-1} f[t]\big] = \int_s^\infty d\sigma \tilde{f}[\sigma] \tag{5.238}$$

Thus, while positive powers of t are related to differentiation, negative powers of t are related to integration of the Laplace transform with respect to the conjugate variable s.

5.6.4 Convolution Theorem

Consider a function $f[t]$ obtained by the convolution of $g[\tau]$ and $h[t-\tau]$ over the finite interval $0 \le \tau \le t$, such that

$$f = g \otimes h = h \otimes g \Longrightarrow f[t] = \int_0^t g[\tau]h[t-\tau]\,d\tau = \int_0^t h[\tau]g[t-\tau]\,d\tau \tag{5.239}$$

We wish to demonstrate that the Laplace transform satisfies a convolution theorem

$$\mathcal{L}[g \otimes h] = \mathcal{L}[g]\mathcal{L}[h] \tag{5.240}$$

which is similar to that for Fourier transforms. We assume that both g and h are exponentially bounded so that their Laplace transforms exist. The Laplace transform of the convolution is then given by an integral

$$\tilde{f}[s] = \lim_{T\to\infty} \int_0^T dt e^{-st} \int_0^t d\tau g[\tau]h[t-\tau] \tag{5.241}$$

over a triangular area in the (t, τ) plane. Assuming that we can change the order of integration

$$\tilde{f}[s] = \lim_{T\to\infty} \int_0^T d\tau g[\tau] \int_\tau^T dt e^{-st} h[t-\tau] \tag{5.242}$$

and using the variable change $u = t - \tau$,

$$\tilde{f}[s] = \lim_{T\to\infty} \int_0^T d\tau e^{-s\tau} g[\tau] \int_0^{T-\tau} du e^{-su} h[u] \tag{5.243}$$

we obtain an integral over a triangular region of the (u, τ) plane. Recognizing that $h[u]$ is exponentially bounded, the relative contribution to the inner integral

$$\tilde{f}[s] = \lim_{T\to\infty} \int_0^T d\tau e^{-s\tau} g[\tau] \left(\int_0^T du e^{-su} h[u] - \int_{T-\tau}^T du e^{-su} h[u] \right) \tag{5.244}$$

made by the region $T - \tau < u < T$ becomes negligible as $T \to \infty$. Therefore, we obtain

$$\tilde{f}[s] = \tilde{g}[s]\tilde{h}[s] \tag{5.245}$$

as desired.

The important role of the Laplace convolution theorem in the solution of inhomogeneous differential equations, similar to the analogous theorem for the Fourier transform, will be explored soon and in the exercises. It can also be used to evaluate Laplace transforms for functions defined by definite integrals. By interpreting a definite integral as a convolution with the step function

$$\int_0^t f[t']\, dt' \longrightarrow \int_0^t \Theta[t-t'] f[t']\, dt' \tag{5.246}$$

we obtain

$$\mathcal{L}\left[\int_0^t f[t']\, dt'\right] = \frac{1}{s}\mathcal{L}[f[t]] \tag{5.247}$$

whenever $\mathcal{L}f$ exists. For example, using

$$\mathcal{L}\left[\frac{\text{Sin}[t]}{t}\right] = \int_s^\infty d\sigma \frac{1}{\sigma^2+1} = \int_0^{s^{-1}} dx \frac{1}{1+x^2} = \text{ArcTan}[s^{-1}] \tag{5.248}$$

the Laplace transform of the sine integral function becomes

$$\text{Si}[t] = \int_0^t \frac{\text{Sin}[x]}{x} dx \Longrightarrow \mathcal{L}[\text{Si}] = s^{-1}\text{ArcTan}[s^{-1}] \tag{5.249}$$

The convolution theorem can also be used as an alternative to the partial-fraction decomposition. For example, we can use

$$\begin{aligned} \mathcal{L}^{-1}\left[(s+\alpha)^{-1}(s^2+\omega^2)^{-1}\right] &= \omega^{-1} \int_0^t e^{-\alpha(t-t')}\, \text{Sin}[\omega t']\, dt' \\ &= \frac{1}{2i\omega}\left(\frac{\text{Exp}[-\alpha(t-t') + i\omega t']}{\alpha + i\omega} - \frac{\text{Exp}[-\alpha(t-t') - i\omega t']}{\alpha - i\omega} \right)_0^t \end{aligned} \tag{5.250}$$

to find

$$\mathcal{L}^{-1}\left[(s+\alpha)^{-1}(s^2+\omega^2)^{-1}\right] = \frac{e^{-\alpha t} - \mathrm{Cos}[\omega t] + \frac{\alpha}{\omega}\mathrm{Sin}[\omega t]}{\alpha^2+\omega^2} \tag{5.251}$$

5.6.5 Summary

A brief table of Laplace transforms and their properties is given in Table 5.5. Note that we assume that $t, t_0 > 0$ throughout and that initial conditions enter for $t \to 0^+$. The parameters t_0, ω, γ, and λ are real. We also assume that the required derivatives exist and integrals converge. Be careful in applying theorems for derivatives and other operations to functions with discontinuities.

Table 5.5. Brief table of Laplace transforms.

Type	$f[t]$	$\tilde{f}[s]$
definition	$f[t] = \frac{1}{2\pi i}\int_{\gamma-i\infty}^{\gamma+i\infty} ds e^{st}\tilde{f}[s]$	$\tilde{f}[s] = \int_0^\infty dt e^{-st} f[t]$
step function	$\Theta[t]$	s^{-1}
delta function	$\delta[t-t_0]$	e^{-st_0}
shifting	$f[t-t_0]\Theta[t-t_0]$	$e^{-st_0}\tilde{f}[s]$
attenuation	$e^{-\gamma t}f[t]$	$\tilde{f}[s+\gamma]$
dilatation	$f[\lambda t]$	$\lambda^{-1}\tilde{f}[s/\lambda]$
convolution	$f[t] = \int_0^t g[\tau]h[t-\tau]\,d\tau$	$\tilde{f}[s] = \tilde{g}[s]\tilde{h}[s]$
derivatives	$f'[t]$	$s\tilde{f}[s] - f[0^+]$
	$f''[t]$	$s^2\tilde{f}[s] - sf[0^+] - f'[0^+]$
integral	$\int_0^t f[t']\,dt'$	$s^{-1}\tilde{f}[s]$
powers	t^ν	$\Gamma[\nu+1]/s^{\nu+1}$
multiplication by powers	$tf[t]$	$-\tilde{f}'[s]$
	$t^{-1}f[t]$	$\int_s^\infty \tilde{f}[\sigma]\,d\sigma$
exponential	$e^{-\gamma t}\Theta[t]$	$1/(s+\gamma)$
trigonometric	$\mathrm{Sin}[\omega t]$	$\omega/(s^2+\omega^2)$
	$\mathrm{Cos}[\omega t]$	$s/(s^2+\omega^2)$
periodic	$f[t+nT] = g[t]$	$\tilde{f}[s] = (1-e^{-sT})^{-1}\int_0^T dt e^{-st} g[t]$

5.7 Green Functions via Laplace Transform

Like the Fourier transform, the Laplace transform can be very useful in analyzing the response of a linear system to external stimuli. An advantage of the Laplace transform is that initial conditions are included automatically. In this section we illustrate this technique using a few relatively simple examples.

5.7.1 Example: Series RC Circuit

The current in a simple RC circuit with a variable voltage source $V[t]$ satisfies

$$RI[t] + \frac{1}{C}\int_0^t I\left[t'\right] dt' == V[t] \tag{5.252}$$

where we assume that the circuit is quiescent for $t < 0$ (no current and discharged capacitor). Although this can be converted into a differential equation for the stored charge, it is instructive to solve the problem as an integral equation instead. The Laplace transform for this simple integral equation gives

$$R\tilde{I} + \frac{1}{C}\frac{\tilde{I}}{s} == \tilde{V} \Longrightarrow \tilde{I} = \frac{s}{1 + s\tau} C\tilde{V} \tag{5.253}$$

where $\tau = RC$ is the characteristic time constant for this circuit. Therefore, the general solution takes the form of a convolution

$$I[t] = \int_0^t G[t - t'] V[t'] \, dt' \tag{5.254}$$

where the Laplace transform of the Green function is

$$\tilde{G}[s] = C\frac{s}{1 + s\tau} \tag{5.255}$$

The inversion can be accomplished either by using the Bromwich integral or by using the basic properties of Laplace transforms, whereby

$$\begin{aligned} \mathcal{L}^{-1}\left[\frac{s}{1 + s\tau}\right] &= \tau^{-1}\frac{\partial}{\partial t}\mathcal{L}^{-1}\left[\frac{1}{s + \tau^{-1}}\right] = \tau^{-1}\frac{\partial}{\partial t} e^{-t/\tau}\Theta[t] \\ &= \tau^{-1}(-\tau^{-1} e^{-t/\tau}\Theta[t] + e^{-t/\tau}\delta[t]) \end{aligned} \tag{5.256}$$

We can drop the step function because this Green function is only used for $t > 0$ and we can evaluate the coefficient of the delta function at $t = 0$ to obtain the simpler form

$$G[t] = R^{-1}\left(\delta[t] - \frac{e^{-t/\tau}}{\tau}\right) \tag{5.257}$$

such that

$$RI[t] = V[t] - \tau^{-1}\int_0^t \operatorname{Exp}[-(t - t')/\tau] V[t'] \, dt' \tag{5.258}$$

or

$$RI[t] = V[t] - \tau^{-1} e^{-t/\tau}\int_0^t \operatorname{Exp}[t'/\tau] V[t'] \, dt' \tag{5.259}$$

For example, suppose that the voltage is constant for $t > 0$. Then

$$\begin{aligned} V[t] = V_0 \Theta[t] \Longrightarrow RI[t] &= V_0 (1 - \tau^{-1} e^{-t/\tau} \int_0^t \text{Exp}[t'/\tau]\, dt') \\ &= V_0 \left(1 - e^{-t/\tau}(e^{t/\tau} - 1)\right) \end{aligned} \tag{5.260}$$

such that

$$V[t] = V_0 \Theta[t] \Longrightarrow RI[t] = V_0 e^{-t/\tau} \tag{5.261}$$

Thus, the current is initially determined by Ohm's law for R and V_0 but decreases to zero for long times as the capacitor becomes charged. Although this derivation may be unfamiliar, the result should be familiar and agrees with our expectations. For more general $V[t]$ profiles, sometimes it may be easier to invert the Laplace transform $\tilde{I}[s]$ and at other times it is easier to perform the convolution with respect to time. These approaches are compared in the exercises.

5.7.2 Example: Damped Oscillator

Consider once again the initial-value problem for the damped harmonic oscillator

$$\left(\frac{\partial^2}{\partial t^2} + 2\gamma \frac{\partial}{\partial t} + \omega_0^2\right) x[t] == f[t], \quad x[0] = x_0, \; x'[0] = v_0 \tag{5.262}$$

where we consider only positive times, $t > 0$, that are suitable for the Laplace transform. An advantage of the Laplace transform is that the initial conditions are included automatically. Thus, the transformed equation

$$s^2 \tilde{x}[s] - s x_0 - v_0 + 2\gamma (s \tilde{x}[s] - x_0) + \omega_0^2 \tilde{x}[s] == \tilde{f}[s] \tag{5.263}$$

reduces to

$$\tilde{x}[s] = \tilde{x}_1[s] + \tilde{x}_2[s] \tag{5.264}$$

where

$$\tilde{x}_1[s] = \frac{x_0(s + 2\gamma) + v_0}{s^2 + 2\gamma s + \omega_0^2} = \frac{x_0(s+\gamma)}{(s+\gamma)^2 + (\omega_0^2 - \gamma^2)} + \frac{v_0}{(s+\gamma)^2 + (\omega_0^2 - \gamma^2)} \tag{5.265}$$

is the solution to the homogeneous equation that depends upon the initial conditions and

$$\tilde{x}_2[s] = \frac{\tilde{f}[s]}{s^2 + 2\gamma s + \omega_0^2} \tag{5.266}$$

is the particular solution to the inhomogeneous equation that depends upon the forcing term. The homogeneous solution can be inverted using

$$\mathcal{L}^{-1}\left[\frac{1}{(s+\gamma)^2+\alpha^2}\right] = e^{-\gamma t}\frac{\text{Sin}[\alpha t]}{\alpha} \tag{5.267}$$

$$\mathcal{L}^{-1}\left[\frac{s+\gamma}{(s+\gamma)^2+\alpha^2}\right] = e^{-\gamma t}\,\text{Cos}[\alpha t] \tag{5.268}$$

to obtain

$$x_1[t] = e^{-\gamma t}\left(x_0\,\text{Cos}[\alpha t] + \frac{v_0}{\alpha}\,\text{Sin}[\alpha t]\right) \tag{5.269}$$

The particular solution takes the form of a convolution

$$x_2[t] = \int_0^t G[t-t']f[t']\,dt' \tag{5.270}$$

where

$$G[t] = e^{-\gamma t}\frac{\text{Sin}[\alpha t]}{\alpha} \tag{5.271}$$

is the Green function. Notice that the lower limit of integration is $t = 0$ because for the initial-value we can imagine that the forcing term is absent for $t < 0$, implicitly using $f[t] \to f[t]\Theta[t]$ for the method of Laplace transforms. Also notice the similarity between the Laplace and Fourier representations of the Green function, which are related by $s \leftrightarrow i\omega$. Further analysis of the underdamped ($\alpha = \omega_R$), critically damped ($\alpha \to 0$), and overdamped ($\alpha = i\kappa$) situations is identical to previous results using the Fourier transform.

5.7.3 Example: Diffusion with Constant Boundary Value

Suppose that $\psi[x,t]$ satisfies a diffusion equation

$$\frac{\partial^2\psi}{\partial x^2} == \frac{1}{\kappa}\frac{\partial\psi}{\partial t} \tag{5.272}$$

for $x > 0$, subject to boundary conditions

$$x > 0 \quad \text{and} \quad t \le 0 \Longrightarrow \psi = 0 \tag{5.273}$$

$$x = 0 \quad \text{and} \quad t \ge 0 \Longrightarrow \psi = \psi_0 \tag{5.274}$$

where ψ_0 is a constant. For example, if a large block of material is brought into contact with a hotter block that is maintained at constant temperature, ψ could represent its temperature increase. Alternatively, if a block with pure chemical composition is immersed in a chemical solution, ψ could represent the concentration of a foreign chemical diffusing into the block over time. It is useful to perform a Laplace transform with respect to time

$$\tilde{\psi}[x,s] = \int_0^{\infty} dt e^{-st}\psi[x,t] \tag{5.275}$$

$$\psi[x,t] = \frac{1}{2\pi i}\int_{c-i\infty}^{c+i\infty} ds e^{st}\tilde{\psi}[x,s] \tag{5.276}$$

to obtain an ordinary differential equation of the form

$$\frac{\partial^2\tilde{\psi}[x,s]}{\partial x^2} == \frac{s}{\kappa}\tilde{\psi}[x,s], \quad \tilde{\psi}[0,s] == \frac{\psi_0}{s} \tag{5.277}$$

where the Laplace constant $\psi[x,0^+]$ vanishes for $x > 0$. The Laplace transform of the boundary condition $\psi[0,t] = \psi_0\Theta[t]$ becomes $\tilde{\psi}[0,s] = s^{-1}\psi_0$ because of the step function inherent in the Laplace method. Notice that a Laplace transform with respect to x instead would yield an inhomogeneous equation because $\psi[0^+,t] = \psi_0$ does not vanish; hence, that strategy would not reduce the complexity of the original problem.

The solution for $\tilde{\psi}[x,s]$ that satisfies the boundary condition for $\tilde{\psi}[0,s]$ can now be written by inspection as

$$\tilde{\psi}[x,s] = \frac{\psi_0}{s}\,\mathrm{Exp}\left[-x\sqrt{s/\kappa}\right] \tag{5.278}$$

where we must obviously reject the exponentially growing possibility. The problem now is to obtain $\psi[x,t]$ by inverting the Laplace transform. There is a simple pole at the origin and the square root in the exponential is made single-valued by a branch cut along the negative real axis. Using the contour sketched in Fig. 5.18, which encloses none of these features, we can write

$$\oint_C ds e^{st}\tilde{\psi}[x,s] = 0 \Longrightarrow \psi[x,t] = -\frac{\psi_0}{2\pi i}\sum_{k=2}^{6} I_k \tag{5.279}$$

where

$$I_k = \int_{C_k}\frac{ds}{s}e^{st}\,\mathrm{Exp}\left[-x\sqrt{s/\kappa}\right] \tag{5.280}$$

represents contributions from each segment of the contour. Segment C_1 corresponds to the Bromwich integral while the other segments serve to establish its value through the Cauchy integral theorem.

The contributions to I_2 and I_6 from the left half-plane vanish in the limit $R \to \infty$ because with

$$\begin{aligned} s = Re^{i\theta} &\Longrightarrow \left|\frac{e^{st}}{s}\,\mathrm{Exp}\left[-x\sqrt{s/\kappa}\right]\right| \sim R^{-1}\,\mathrm{Exp}\left[Rt\,\mathrm{Cos}[\theta]\right]\mathrm{Exp}\left[-x\sqrt{R/\kappa}\,\mathrm{Cos}\left[\tfrac{\theta}{2}\right]\right] \\ &\Longrightarrow I_2 = I_6 = 0 \end{aligned} \tag{5.281}$$

the range $\frac{\pi}{2} \le \theta \le \frac{3\pi}{2}$ ensures that the dominant exponential has a negative argument. The short horizontal segments in the right half-plane with $s = \rho \pm i\infty$, where ρ is real

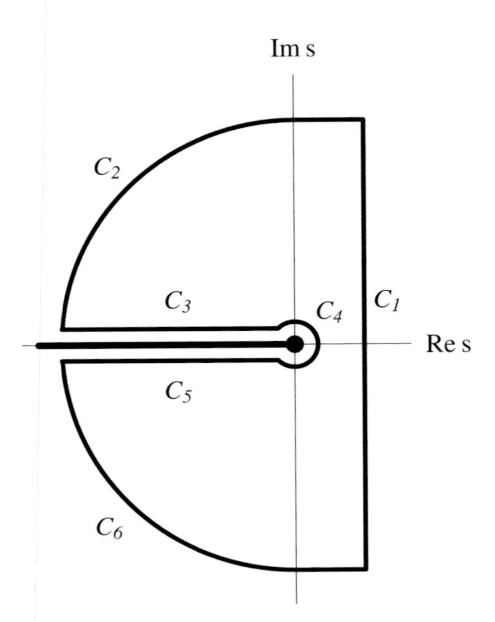

Figure 5.18. Contour for inverse Laplace transform of $\mathrm{Exp}\left[-x\sqrt{s/\kappa}\right]/s$.

and $ds = d\rho$, do not contribute either because there

$$s = \rho \pm iR \Longrightarrow \frac{e^{st}}{s}\,\mathrm{Exp}\left[-x\sqrt{s/\kappa}\right] \simeq \mp iR^{-1}\,\mathrm{Exp}\left[\rho t \pm iRt - x\sqrt{R/\kappa}e^{\pm i\pi/4}\right] \tag{5.282}$$

Separating the real and imaginary parts of the argument of the exponential

$$\frac{e^{st}}{s}\,\mathrm{Exp}\left[-x\sqrt{s/\kappa}\right] \simeq \mp iR^{-1}\,\mathrm{Exp}\left[\rho t - x\sqrt{R/2\kappa}\right]\mathrm{Exp}\left[\pm iRt + ix\sqrt{R/2\kappa}\right] \tag{5.283}$$

and retaining only the leading dependencies in R

$$s = \rho \pm iR \Longrightarrow \frac{e^{st}}{s}\,\mathrm{Exp}\left[-x\sqrt{s/\kappa}\right] \simeq \mp iR^{-1}\,\mathrm{Exp}\left[-x\sqrt{R/2\kappa}\right]\mathrm{Exp}[\pm iRt] \tag{5.284}$$

we find that the magnitudes of integrands are proportional to $R^{-1}\,\mathrm{Exp}[-\alpha R^{1/2}]$ where $\alpha > 0$. The lengths of these segments is finite and the integrands vanish quickly in the limit $R \to \infty$.

As C_3 and C_5 approach the real axis, C_4 approaches a complete circle enclosing the pole with unit residue in a negative sense, such that

$$I_4 = -2\pi i \tag{5.285}$$

The segments neighboring the real axis take the form

$$\begin{aligned} I_3 &= \int_{-\infty}^{0} \text{Exp}\left[st - xe^{i\pi/2}\sqrt{\frac{|s|}{\kappa}}\right]\frac{ds}{s} = -\int_{0}^{\infty} \text{Exp}\left[-st - ix\sqrt{\frac{|s|}{\kappa}}\right]\frac{ds}{s} \\ &= -2\int_{0}^{\infty} \text{Exp}[-\kappa t u^2 - iux]\frac{du}{u} \end{aligned} \tag{5.286}$$

$$\begin{aligned} I_5 &= \int_{0}^{-\infty} \text{Exp}\left[st - xe^{-i\pi/2}\sqrt{\frac{|s|}{\kappa}}\right]\frac{ds}{s} = \int_{0}^{\infty} \text{Exp}\left[-st + ix\sqrt{\frac{|s|}{\kappa}}\right]\frac{ds}{s} \\ &= 2\int_{0}^{\infty} \text{Exp}[-\kappa t u^2 + iux]\frac{du}{u} \end{aligned} \tag{5.287}$$

such that

$$I_3 + I_5 = 4i\int_{0}^{\infty} \frac{du}{u}\,\text{Exp}[-\kappa t u^2]\,\text{Sin}[ux] \tag{5.288}$$

To evaluate the integral

$$\xi[x] = \int_{0}^{\infty} \frac{du}{u}\,\text{Exp}[-\kappa t u^2]\,\text{Sin}[ux] \tag{5.289}$$

we first consider

$$\xi'[x] = \int_{0}^{\infty} du\,\text{Exp}[-\kappa t u^2]\,\text{Cos}[ux] = \frac{1}{2}\int_{-\infty}^{\infty} du\,\text{Exp}[-\kappa t u^2]\,\text{Exp}[iux] \tag{5.290}$$

which can be obtained by completing the square

$$\begin{aligned} \int_{-\infty}^{\infty} du\,\text{Exp}[-\kappa t u^2]\,\text{Exp}[iux] &= \text{Exp}\left[-\frac{x^2}{4\kappa t}\right]\int_{-\infty}^{\infty} du\,\text{Exp}\left[-\left(\sqrt{\kappa t}u - i\frac{x}{2\sqrt{\kappa t}}\right)^2\right] \\ &= \sqrt{\frac{\pi}{\kappa t}}\,\text{Exp}\left[-\frac{x^2}{4\kappa t}\right] \end{aligned} \tag{5.291}$$

Hence, the desired integral satisfies a differential equation

$$\xi'[x] == \frac{1}{2}\sqrt{\frac{\pi}{\kappa t}}\,\text{Exp}\left[-\frac{x^2}{4\kappa t}\right], \quad \xi[0] == 0 \tag{5.292}$$

which can be integrated to obtain

$$\xi[x] = \sqrt{\frac{\pi}{4\kappa t}}\int_{0}^{x} dx'\,\text{Exp}\left[-\frac{x'^2}{4\kappa t}\right] = \sqrt{\pi}\int_{0}^{x/2\sqrt{\kappa t}} e^{-y^2}\,dy = \frac{\pi}{2}\,\text{Erf}\left[\frac{x}{2\sqrt{\kappa t}}\right] \tag{5.293}$$

Therefore, we finally obtain

$$\psi[x,t] = \psi_0\text{Erfc}\left[\frac{x}{2\sqrt{\kappa t}}\right] \tag{5.294}$$

where Erfc is the complementary error function.

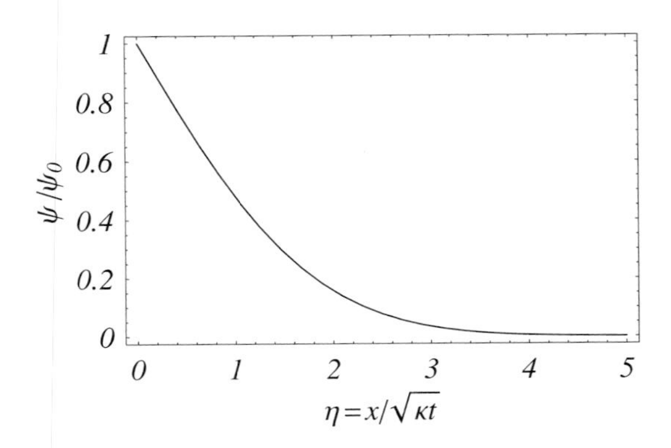

Figure 5.19. Diffusion with constant boundary value.

Notice that the natural time scale is determined by κ^{-1} while the natural distance scale is determined by $\sqrt{\kappa t}$ such that the dimensionless variable $\eta = x/\sqrt{\kappa t}$ provides a simple one-dimensional function describing both the distance and time dependence of the approach to equilibrium. Figure 5.19 illustrates the equilibration of the sample in terms of its natural dimensionless variable. This solution is characteristic of any diffusion problem (temperature, chemical concentration, etc.) for which a plane is maintained at a constant boundary value.

It is useful to examine the behavior of this function for both small and large values of its dimensionless variables using series expansions. For a fixed position, the initial time variation is determined by the solution for large η while its asymptotic approach to equilibrium is determined by the solution for small η.

```
Series[Erfc[η/2], {η, 0, 2}]
```

$$1 - \frac{\eta}{\sqrt{\pi}} + O[\eta]^3$$

```
(Series[Erfc[x], {x, ∞, 4}] /. {x → η/2}) // Normal // Simplify
```

$$\frac{2e^{-\frac{\eta^2}{4}}(-2+\eta^2)}{\sqrt{\pi}\,\eta^3}$$

Perhaps it is also useful to visualize the equilibration process using the three-dimensional Fig. 5.20. The initial temperature elevation at $x = 0$ spreads gradually to more distant areas and later approaches the final temperature asymptotically.

Integral transforms are so useful that extensive tables of Fourier and Laplace transforms have been compiled. Therefore, once one is comfortable with the basic methodology, one can often obtain the necessary transforms either from standard compilations or from *MATHEMATICA*® or another symbolic math program. For example, *MATHEMATICA*® can provide the inverse transform

```
InverseLaplaceTransform[ψ0/s Exp[-x√(s/κ)], s, t] //
  FullSimplify[#, {t > 0, x > 0, κ > 0}]&
```

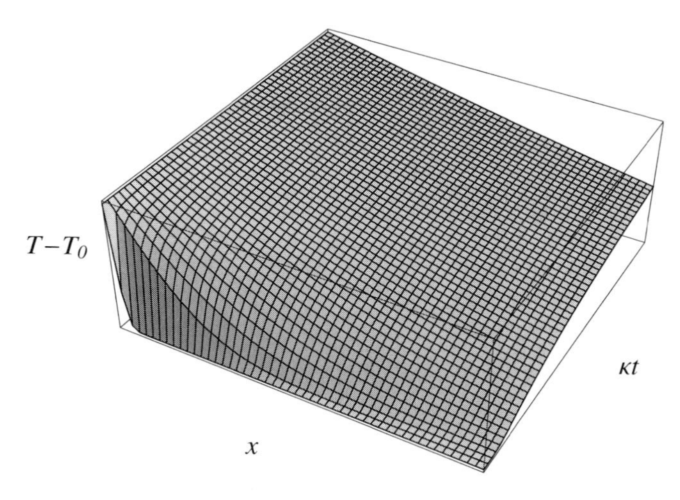

Figure 5.20. Temperature equilibration (arbitrary units).

$$\mathrm{Erfc}\left[\frac{x}{2\sqrt{t\,\kappa}}\right]\psi_0$$

needed for this problem without the pain of performing some rather tedious integrals by hand.

Problems for Chapter 5

1. Moments of form factors

The charge density $\rho_{\text{ch}}[r]$ in an atomic nucleus can be measured using electron scattering. Analysis of the angular distribution for electron scattering provides the *form factor*, which is related to the density by the Bessel transform

$$\tilde{\rho}_{\text{ch}}[q] = 4\pi \int_0^\infty dr r^2 j_0[qr]\rho_{\text{ch}}[r] \quad j_0[x] = \frac{\mathrm{Sin}[x]}{x} \tag{5.295}$$

where r is the radial coordinate, q is the momentum transfer, and j_0 is the spherical Bessel function of order zero. It is often useful to characterize distributions in terms of moments

$$M_n[\rho] = 4\pi \int_0^\infty dr r^{2+n}\rho[r] \tag{5.296}$$

and the mean-square radius

$$\langle r^2 \rangle = \frac{M_2}{M_0} \tag{5.297}$$

Finally, the nuclear charge density results from a convolution of the proton density ρ (in other words, the distribution of protons with the nucleus) and the charge density ρ_p within a proton such that

$$\rho_{\text{ch}} = \rho \otimes \rho_p \Longrightarrow \tilde{\rho}_{\text{ch}}[q] = \tilde{\rho}[q]\tilde{\rho}_p[q] \tag{5.298}$$

a) Show that a form factor can be expanded in terms of

$$\tilde{\rho}[q] = \sum_{n=0}^{\infty} a_n M_{2n} q^{2n} \tag{5.299}$$

and determine the coefficients. How can one extract M_{2n} using derivatives with respect to q^2? How does $\langle r^2 \rangle$ enter a low-q expansion?

b) Determine the general relationship between $M_{2n}[\rho_{\text{ch}}]$, $M_{2n}[\rho]$, and $M_{2n}[\rho_p]$. Find explicit formulas for $n = 0, 1$. In particular, how are the mean-square radii related?

2. Elastic form factor for ^{16}O

In the harmonic oscillator model, the radial wave functions for the nuclear $1s$ and $1p$ orbitals take the shapes

$$R_{1s}[r] \propto \text{Exp}\left[-\frac{r^2}{2b^2}\right] \quad R_{1p}[r] \propto r\,\text{Exp}\left[-\frac{r^2}{2b^2}\right] \tag{5.300}$$

where b is the oscillator constant. The radial wave functions should be normalized so that

$$\int_0^{\infty} dr r^2 |R_{nl}[r]|^2 = 1 \tag{5.301}$$

The proton density for ^{16}O, containing 2 protons in the $1s$ orbital and 6 protons in the $1p$ orbital, is then

$$\rho[r] = \frac{2R_{1s}[r]^2 + 6R_{1p}[r]^2}{4\pi} \tag{5.302}$$

The elastic form factor is given by

$$\tilde{\rho}[q] = 4\pi \int_0^{\infty} dr r^2 j_0[qr]\rho[r] \quad j_0[x] = \frac{\text{Sin}[x]}{x} \tag{5.303}$$

where q is the momentum transfer. Compute the elastic form factor and produce a semilog plot of $|\tilde{\rho}[q]|^2$ given $b = 1.8\,\text{fm}$. The momentum transfer q should be measured in units of fm^{-1} where 1 femtometer (fm) corresponds to 10^{-15} m, which is the appropriate length scale for nuclear physics. Also, determine the rms radius by examining the power series for $\tilde{\rho}[q]$ for small q.

3. Fourier transform of periodic function

Suppose that $f[t] = f[t+T]$ is strictly periodic with period T. Evaluate the Fourier transform, $\tilde{f}[\omega]$.

4. Array theorem

Suppose that

$$f[x] = \sum_{n=1}^{N} g[x - x_n] \tag{5.304}$$

consists of an array of identical distributions about a set of positions $\{x_n\}$.

a) Evaluate the Fourier transform $\tilde{f}[k]$ by expressing the summation as a convolution with a set of point sources represented by delta functions.

b) Perform the summation for a uniform array $x_n = na$, $n = 1, N$ and evaluate the intensity $|\tilde{f}[k]|^2$.

5. Fraunhofer diffraction grating

The Fraunhofer approximation to diffraction takes the form

$$S[\theta] = \left| \int_{-\infty}^{\infty} A[x]\,\mathrm{Exp}[iqx]\,dx \right|^2 \tag{5.305}$$

where $A[x]$ is an aperture function that takes unit values for open regions and vanishes at obstacles and where $q = k\,\mathrm{Sin}[\theta]$ for wave number $k = 2\pi/\lambda$. The aperture function for a grating with N identical slits takes the form

$$A[x] = \sum_{n=1}^{N} \Theta[b - 2|x - na|] \tag{5.306}$$

where a is the spacing and $b < a$ is the width. Use the convolution theorem to evaluate the diffraction pattern efficiently.

6. Convolution of a Breit–Wigner resonance with a Gaussian resolution function

Suppose that the energy dependence of the differential cross section for a nuclear reaction is described by the Breit–Wigner profile

$$S[\omega] = \frac{\Gamma}{2\pi} \frac{S_0}{(\omega - \omega_0)^2 + \frac{\Gamma^2}{4}} \tag{5.307}$$

where Γ is the full-width at half-maximum and S_0 is the integral over energy. However, the measurement of energy is smeared by a Gaussian resolution function

$$R[\omega] = \frac{1}{\sqrt{2\pi\sigma^2}}\,\mathrm{Exp}\left[-\frac{\omega^2}{2\sigma^2}\right] \tag{5.308}$$

such that the observed distribution is the convolution $Y = S \otimes R$. Use the convolution theorem to obtain an explicit formula for $Y[\omega]$; your result will probably involve the complementary error function. If you have access to suitable mathematical software, compare your symbolic result with numerical evaluation of the convolution integral. Also compare $Y[\omega]$ with $S[\omega]$. (Note that MATHEMATICA® 5.1 cannot evaluate the convolution integral symbolically.)

7. Loaded beam

The displacement of an infinitely long beam on an elastic foundation is described by

$$\left(\frac{d^4}{dx^4} + 1\right) y[x] == P[x] \tag{5.309}$$

where we assume that $y[\pm\infty] = 0$. Use the Fourier transform method to compute the Green function for this system. Then provide an integral which can be used to compute the displacement given an arbitrary $P[x]$.

8. Diffusion and radiation of heat

Heat diffuses in a medium that also radiates heat at a rate proportional to temperature. Thus, a one-dimensional spatial distribution of temperature $T[x, t]$ satisfies

$$\frac{\partial T}{\partial t} == D\frac{\partial^2 T}{\partial x^2} - \alpha T \tag{5.310}$$

where D and α are positive constants. A finite amount of heat is released at a point in an infinite bar such that

$$T[x, 0] = Q\delta[x] \tag{5.311}$$

a) Evaluate *and describe* the temperature distribution at later times.

b) Compute the total amount of heat lost to radiation between time zero and time t and interpret the result.

9. Damped oscillator subject to a step function

Use the Fourier transform to solve

$$\left(\frac{\partial^2}{\partial t^2} + 2\gamma\frac{\partial}{\partial t} + \omega_0^2\right)x[t] == f[t], \quad t < 0 \Longrightarrow x[t] = x'[t] = 0 \tag{5.312}$$

when

$$f[t] = f_0\Theta[t] \tag{5.313}$$

is a step function that "turns on" at $t = 0$. Find explicit solutions for the underdamped, overdamped, and critically damped cases. Compare these three solutions graphically and explain their behavior.

10. Damped oscillator subject to square pulse

Use the Fourier transform to solve

$$\left(\frac{\partial^2}{\partial t^2} + 2\gamma\frac{\partial}{\partial t} + \omega_0^2\right)x[t] == f[t] \tag{5.314}$$

when

$$f[t] = f_0(\Theta[t + \tau] - \Theta[t - \tau]) \tag{5.315}$$

is a square pulse. Plot the solutions for $\gamma/\omega_0 = 0.5, 1.0, 2.0$ with $\tau = 5/\omega_0$ as functions of $\omega_0 t$ and explain their general characteristics. If this were an RLC circuit, under what conditions would the output follow the input most closely?

11. Orthogonality relations for Fourier sine and cosine transforms

Derive the orthogonality relations:

$$\int_0^\infty dt\, \mathrm{Cos}[\omega t]\,\mathrm{Cos}[\omega' t] = \frac{\pi}{2}(\delta[\omega - \omega'] + \delta[\omega + \omega']) \tag{5.316}$$

$$\int_0^\infty dt\, \mathrm{Sin}[\omega t]\,\mathrm{Sin}[\omega' t] = \frac{\pi}{2}(\delta[\omega - \omega'] - \delta[\omega + \omega']) \tag{5.317}$$

12. Relationship between Fourier sine, cosine, and exponential transforms
Demonstrate the equivalence between the representations

$$f[t] = \frac{1}{2\pi}\int_{-\infty}^{\infty} d\omega e^{-i\omega t}\tilde{f}[\omega] = \frac{2}{\pi}\int_{0}^{\infty} d\omega \left(\tilde{f}_C^{(+)}[\omega]\,\mathrm{Cos}[\omega t] + \tilde{f}_S^{(-)}[\omega]\,\mathrm{Sin}[\omega t]\right) \tag{5.318}$$

where

$$f^{(\pm)}[t] = \frac{1}{2}(f[t] \pm f[-t]) \tag{5.319}$$

and

$$\tilde{f}[\omega] = \int_{-\infty}^{\infty} dt e^{i\omega t} f[t] \tag{5.320}$$

$$\tilde{f}_C^{(+)}[\omega] = \int_{0}^{\infty} dt\,\mathrm{Cos}[\omega t] f^{(+)}[t] \tag{5.321}$$

$$\tilde{f}_S^{(-)}[\omega] = \int_{0}^{\infty} dt\,\mathrm{Sin}[\omega t] f^{(-)}[t] \tag{5.322}$$

Finally, demonstrate that

$$\tilde{f}[\omega] = 2\left(\tilde{f}_C^{(+)}[\omega] + i\tilde{f}_S^{(-)}[\omega]\right) \tag{5.323}$$

13. Green functions using Fourier cosine or sine transforms
The Fourier cosine transform is defined by

$$\tilde{f}[k] = \mathcal{F}_C[f] = \int_{0}^{\infty} f[x]\,\mathrm{Cos}[kx]\,dx \tag{5.324}$$

a) Use the Fourier cosine transform to solve

$$y''[x] - \alpha^2 y[x] == 0 \quad y'[0] == b, \quad y[\infty] == 0 \tag{5.325}$$

b) Next use the Fourier cosine transform to obtain the Green function that satisfies

$$G_\alpha''[x,x'] - \alpha^2 G_\alpha[x,x'] == \delta[x-x'] \quad G_\alpha'[0,x'] == 0, \quad G_\alpha[\infty,x'] == 0 \tag{5.326}$$

and write a formal solution for the inhomogeneous equation

$$y''[x] - \alpha^2 y[x] == f[x] \quad y'[0] == b, \quad y[\infty] == 0 \tag{5.327}$$

c) Now suppose that the lower boundary conditions are replaced by $y[0] == a$ for the homogeneous problem and by $G_\alpha[0,x'] == 0$ for the Green function. How do the solutions change? Why is the Laplace transform inconvenient for this problem?

14. Born approximation for Yukawa potential

The Born approximation for elastic scattering of distinguishable particles with interaction potential V approximates the differential cross-section by $|\mathcal{M}|^2$ where the matrix element takes the form

$$\mathcal{M} = -\frac{\mu}{2\pi} \int \Phi_f^*[\vec{r}] V[\vec{r}] \Phi_i[\vec{r}] \, d^3r \tag{5.328}$$

with μ being the reduced mass and $\vec{r} = \vec{r}_1 - \vec{r}_2$ the separation vector. The wave functions are approximated by plane waves of the form

$$\Phi_i = \text{Exp}[i\vec{k}_i \cdot \vec{r}], \quad \Phi_f = \text{Exp}[i\vec{k}_f \cdot \vec{r}] \tag{5.329}$$

where the wave vectors are equal in magnitude, such that $k_f = k_i$. Evaluate $\mathcal{M}$ for the central part of the Yukawa interation between two nucleons, given by

$$V[r] = V_0 \frac{e^{-m_\pi r}}{m_\pi r} \tag{5.330}$$

where m_π is the pion mass, V_0 is a strength constant, and $r = |\vec{r}_1 - \vec{r}_2|$. Express your result both in terms of the momentum transfer $\vec{q} = \vec{k}_i - \vec{k}_f$ and in terms of the scattering angle θ. (Note: we are neglecting the spin dependence and tensor properties of the nucleon–nucleon interaction.)

15. Numerical convolution using FFT

Use the FFT function in your favorite mathematical software to reproduce the figures in the text relating to the convolution

$$x[t] = \int_{-\infty}^{\infty} G[t-\tau] F[\tau] \, d\tau \tag{5.331}$$

where

$$G[t] = \omega_R^{-1} \, \text{Exp}[-\gamma t] \, \text{Sin}[\omega_R t] \Theta[t], \quad \omega_R = \sqrt{\omega_0^2 - \gamma^2} \tag{5.332}$$

We used $\omega_0 = 0.3$ and $\gamma = 0.05$ in natural units, $N = 256$, and

$$F[t] = (t-s)^3 \, \text{Exp}\left[-\frac{(t-s)^2}{w}\right] \tag{5.333}$$

with $s = 150$ and $w = 200$. Show that a judicious choice of phase shift suppresses artifacts due to wrap-around in the working array.

16. Temporal correlation

Use the FFT function in your favorite mathematical software to reproduce the figures in the text relating to the autocorrelation function for the logistic map.

17. Van der Pol spectrum

Again assuming that mathematical software with both a differential-equation solver and FFT is available, reproduce the figures in the text that illustrate the power spectrum for a van der Pol oscillator.

18. Windowing

The spreading of the power spectrum for a discrete sinusoid is similar to that of a finite wave train using a continuous Fourier transform. In both cases one is effectively multiplying a persistent function by a time window of the form $\Theta[t]\Theta[T-t]$ that has sharp edges and reproduction of a rapid temporal transition requires Fourier components with arbitrarily high frequency. Therefore, the window function artificially enhances the power for large frequencies. More realistic estimates of the power spectrum for the input signal can be obtained by using a window function, $w[t]$, with softer features that do not distort high frequencies as badly. For the purposes of discrete Fourier transform using sampled data within a range $0 \leq t \leq T$, the window function should vanish for $t \leq 0$ and $t \geq T$, should be relatively flat over a wide central region surrounding a maximum value of unity at $T/2$, and should be symmetric about $T/2$. Many window functions with these properties can be found in the literature and their proponents often argue that their version optimizes a particularly important figure of merit, such as equivalent noise bandwidth, scallop loss, or other esoterica. The nonspecialist may as well use the simple Welch window

$$w_j = 1 - \left(\frac{2j-N-1}{N-1}\right)^2 \tag{5.334}$$

with parabolic shape. The purpose of this problem is to demonstrate numerically, with two simple examples, that use of an appropriate window function provides qualitative improvements in the power spectrum. More detailed analysis of the performance of various window functions may be found in specialized literature.

a) Show that the Welch window has the desired temporal characteristics and then display its discrete Fourier transform using $N = 256$.
b) Compare power spectra for $f[t]\Theta[t]\Theta[T-t]$ and $w[t]f[t]$ where $f[t] = \mathrm{Sin}[0.3t]$ and comment on the effect of using a softer window.
c) Apply the Welch window to improve the estimated power spectrum for the van der Pol oscillator with $\varepsilon = 10$. We suggest 4096 samples within $200 \leq t \leq 600$.

19. Laplace transform of $J_0[x]$ using an integral representation

a) Demonstrate that the integral representation

$$J_0[z] = \frac{1}{\pi}\int_0^{\pi} d\theta\, \mathrm{Cos}[z\,\mathrm{Cos}[\theta]] \tag{5.335}$$

provides a solution to Bessel's equation

$$z^2 f''[z] + z f'[z] + \left(z^2 - \nu^2\right) f[z] == 0 \tag{5.336}$$

for $\nu = 0$ and initial condition $f[0] = 1$.

b) Use this integral representation to determine the Laplace transform of $J_0[x]$.

20. Laplace transform of $J_0[x]$ using its differential equation

The Bessel function $J_0[x]$ satisfies the differential equation

$$\left(\frac{d^2}{dx^2} + \frac{1}{x}\frac{d}{dx} + 1\right) J_0[x] == 0, \quad J_0[0] = 1 \tag{5.337}$$

a) Show that the Laplace transform $\tilde{J}_0[s]$ satisfies a first-order differential equation and obtain $\tilde{J}_0[s]$. (Hint: if $f[x]$ is continuous at $x = 0$, you can use $\int_0^\infty e^{-sx} f[x]\,dx \simeq s^{-1} f[0]$ when $s \to \infty$.)

b) Use an expansion of $\tilde{J}_0[s]$ in powers of s^{-1} to develop a power series for $J_0[x]$. What is the radius of convergence?

21. Decay chain

Suppose that a radioactive nucleus decays sequentially such that the populations of the first three members of the decay chain evolve according to the rate equations

$$\frac{dN_1}{dt} = -\lambda_1 N_1 \tag{5.338}$$

$$\frac{dN_2}{dt} = \lambda_1 N_1 - \lambda_2 N_2 \tag{5.339}$$

$$\frac{dN_3}{dt} = \lambda_2 N_2 - \lambda_3 N_3 \tag{5.340}$$

Use the Laplace transform method to determine $N_i[t]$ assuming initial populations $N_i[0] = n_i$. How can we extend the procedure to longer sequences?

22. RC circuit using Laplace transform

In the text we demonstrated that the current in a simple RC circuit with a variable voltage source $V[t]$ satisfying

$$RI[t] + \frac{1}{C}\int_0^t I[t']\,dt' = V[t], \quad I[t<0] = 0 \tag{5.341}$$

can be expressed as

$$\tilde{I} = \frac{s}{1+s\tau} C\tilde{V} \tag{5.342}$$

or

$$RI[t] = V[t] - \tau^{-1} e^{-t/\tau} \int_0^t \mathrm{Exp}[t'/\tau] V[t']\,dt' \tag{5.343}$$

where $\tau = RC$ is the characteristic time constant for this circuit. For each of the following $V[t]$ profiles, demonstrate that both the inverse Laplace transform and the convolution integral give the same current $I[t]$.

a) Step function: $V[t] = V_0 \Theta[t-a]$ with $a > 0$.

b) Rectangular pulse: $V[t] = V_0(\Theta[t-a] - \Theta[t-b])$ with $b > a > 0$.

c) Sinusoidal voltage: $V[t] = V_0 \operatorname{Sin}[\omega t]\Theta[t]$. Identify the transient component and evaluate the phase shift between current and voltage oscillations after the transient has decayed away.

23. Current in RLC circuit using Laplace transform

The current in a series RLC circuit with a variable voltage source $V[t]$ satisfies

$$L\frac{dI}{dt} + RI + \frac{1}{C}\int_0^t I[t']\,dt' == V[t] \tag{5.344}$$

where we assume that the circuit is quiescent for $t < 0$ (no current and discharged capacitor). Although this equation is usually converted into a differential equation for the stored charge, it is instructive to solve the problem as an integro-differential equation instead.

a) Express the relationship between Laplace transforms $\tilde{I}$ and $\tilde{V}$ for current and voltage as a convolution and determine the Green functions for (i) underdamped, (ii) overdamped, and (iii) critically damped circuits.

b) Use the Laplace transform method to evaluate $I[t]$ when the voltage is constant during $a \le t \le b$. Sketch the behavior of underdamped and overdamped circuits.

c) Use the Laplace transform method to evaluate the current for a sinusoidal voltage switched on for $t > 0$. Identify both the transient component and the phase-shift between current and voltage oscillations.

24. Laplace transform of periodic functions

Consider a periodic function of the form $f[t + T] = f[t]$.

a) Show that the Laplace transform is given by

$$\tilde{f}[s] = \frac{1}{1 - e^{-sT}}\int_0^T f[t]e^{-st}\,dt \tag{5.345}$$

What happens in the limit $T \to \infty$?

b) Verify that the expected results are obtained for Sin[t] or Cos[t].

c) Evaluate Laplace transforms of the square wave.

d) Evaluate the Laplace transform of the sawtooth wave, $f[t] = \text{Mod}[t, T]$.

25. Rectifiers

An ideal half-wave rectifier circuit passes positive and blocks negative voltage signals while the output of an ideal full-wave rectifier is the absolute value of the input voltage. Evaluate the Laplace transform of the output for both half-wave and full-wave rectification of a sinusoidal input voltage, Sin[ωt].

26. Initial-value problems using Laplace transform

Solve the following differential equations, with specific initial conditions, using the Laplace transform.

a) $$\left(\frac{d^4}{dt^4} - 1\right)y[t] == t\,, \quad y[0] = y'[0] = y''[0] = y'''[0] = 0 \tag{5.346}$$

b) $$y'[t] + \int_0^t y[\tau]\,d\tau == e^{-t}\,, \quad y[0] = 1 \tag{5.347}$$

c) $y''[x] + xy'[x] + y[x] == 0\,,\quad y[0] = 1\,,\quad y'[0] = 0$ (5.348)

27. Difference-differential equation using Laplace transform
Suppose that $y[t]$ satisfies a difference-differential equation

$$t \geq 0 \Longrightarrow y'[t] + y[t] - y[t-1] == 0 \tag{5.349}$$

with the initial condition

$$-1 \leq t \leq 0 \Longrightarrow y[t] == y_0[t] \tag{5.350}$$

where $y_0[t]$ is a known function.

a) Show that the Laplace transform satisfies

$$\tilde{y}[s] = (1 + s - e^{-s})^{-1} \left(y_0[0] + e^{-s} \int_{-1}^{0} e^{-st} y_0[t] \, dt \right) \tag{5.351}$$

b) Compute $y[t]$ for $t \geq 0$ given $y_0[t] = 1$ and check the result.

28. Abel's integral equation
Abel's integral equation for the unknown function $y[t]$, with $t > 0$, is

$$\int_0^t y[\tau](t-\tau)^{-\alpha} \, d\tau == f[t] \tag{5.352}$$

where α is a constant in the range $0 < \alpha < 1$ and $f[t]$ is a known function.

a) Use the convolution theorem for Laplace transforms to obtain a formal solution for $y[t]$.

b) Evaluate $y[t]$ given $f[t] = 1$ and check the result.

29. Wave equation with traveling disturbance
Consider an inhomogeneous wave equation of the form

$$\left(\frac{\partial^2}{\partial x^2} - \frac{1}{c^2} \frac{\partial^2}{\partial t^2} \right) \psi[x,t] == \delta[x - vt] \tag{5.353}$$

for $t > 0$ in $0 < x < \infty$ subject to the boundary and initial conditions

$$\psi[0,t] == 0\,,\quad \psi[x,0] == \frac{\partial \psi[x,0]}{\partial t} == 0 \tag{5.354}$$

Use a Laplace transform with respect to t to solve this equation for $v < c$, $v = c$, and $v > c$. Describe and compare these scenarios.

6 Analytic Continuation and Dispersion Relations

Abstract. If two representations of analytic functions are equivalent in the overlapping regions of their domains, they represent the same function. Analytic continuation provides a systematic means for extending the definition of an analytic function into broader domains of the complex plane. Combined with physical requirements of causality, this method can be used to relate the real and imaginary parts for functions that describe the response of a physical system to an external perturbation. We use this method to analyze the relationship between refraction and absorption in optical media and dispersion relations for wave propagation. We also introduce the phenomenon of solitons in systems with nonlinear dispersion.

6.1 Analytic Continuation

6.1.1 Motivation

Often one develops a representation of an analytic function $f[z]$ in terms of a series expansion or an integral that converges in a finite domain D_1 and would like to extend that domain into a broader region of the complex plane. To illustrate the idea, we consider a very simple example. The function $f[z] = (1 - z)^{-1}$ can be expanded about any finite z_n using a Taylor series

$$f[z] = \sum_{k=0}^{\infty} \frac{(z - z_n)^k}{(1 - z_n)^{k+1}} \tag{6.1}$$

that will converge in the domain $D_n : |z - z_n| < |1 - z_n|$. Although we can easily derive the series for this simple function, in a more complicated problem we might have developed the series representation near a particular z_n without knowledge of the underlying analytic function that it represents. Thus, if we had deduced the series

$$f_1[z] = \sum_{k=0}^{\infty} z_k \tag{6.2}$$

we could then demonstrate that f_1 converges to $f[z] = (1 - z)^{-1}$ within the domain $D_1 : |z| < 1$. Similarly, by considering the region around $z_0 = i$ we might deduce a second series

$$f_2[z] = \sum_{k=0}^{\infty} \frac{(z - i)^k}{(1 - i)^{k+1}} \tag{6.3}$$

Graduate Mathematical Physics. James J. Kelly

ISBN: 3-527-40637-9

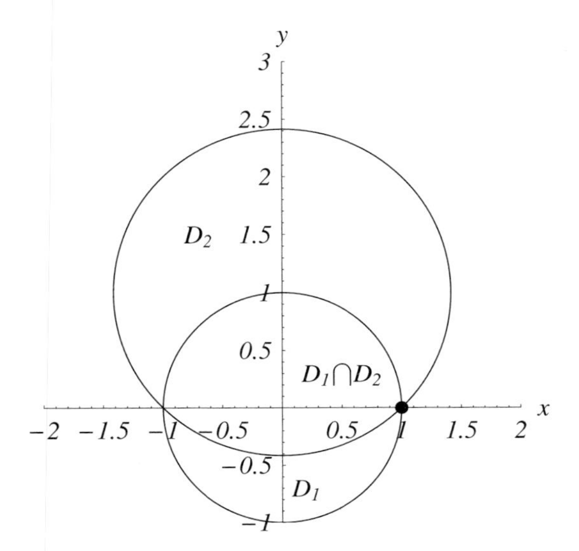

Figure 6.1. Analytic continuation using Taylor series. Here a Taylor series around $z = 0$ converges in domain D_1, while another series around $z = i$ converges in domain D_2. The heavy dot represents a singularity that limits the convergence radii for both domains. If these series are equal in the overlap $D_1 \cap D_2$, they are elements of the same analytic function in a larger domain that includes $D_1 \cup D_2$.

and then demonstrate that f_2 converges to the same $f[z] = (1 - z)^{-1}$ within the domain D_2: $|z - i| < \sqrt{2}$. Although these series look different, they represent the same function within the overlap region $D_1 \cap D_2$. We describe f_2 as an *analytic continuation* of f_1 from domain D_1 into domain D_2 and identify both f_1 and f_2 as *elements* of the same analytic function $f[z]$. Although f_1 and f_2 converge and hence are analytic only within their respective domains D_1 and D_2, the underlying function f is analytic in a domain that is at least as large as the union $D_1 \cup D_2$. Therefore, the underlying analytic function is more general than any of its representations (elements).

Generally one attempts to perform a sequence of analytic continuations which extends the domain to the largest possible region of the complex plane, but in practice this process is often limited by singularities or branch points. Taylor series often provide the simplest method, where the radius of various overlapping disks is limited only by the distance to the nearest singularity, but other methods are sometimes more efficient.

6.1.2 Uniqueness

Suppose that $f_1[z]$ and $f_2[z]$ are analytic in domains D_1 and D_2. One can show that if $D_1 \cap D_2$ contains more than one point it contains an infinite number of points and also constitutes a domain. Thus, the function $g[z] = f_1[z] - f_2[z]$ is analytic in $D_1 \cap D_2$. Now suppose that $g[z] = 0$ on an arc or within a subdomain that lies within $D_1 \cap D_2$. Recognizing that the zeros of an analytic function are isolated (recall Sec. 1.12.5) unless the function

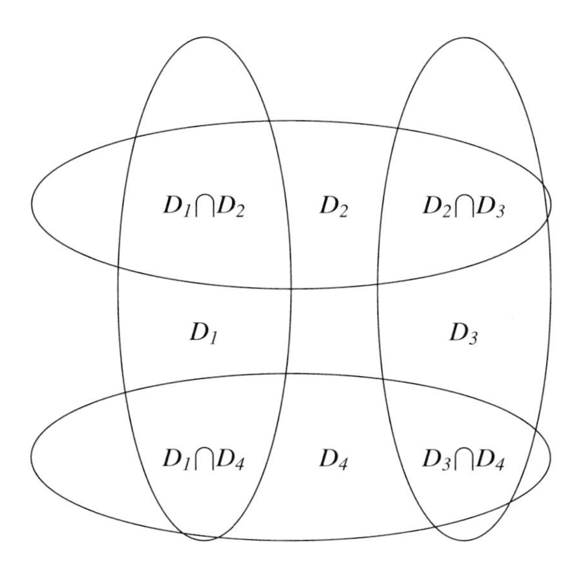

Figure 6.2. Analytic continuation using a sequence of domains $D_1 \to D_2 \to D_3 \to D_4$ that surrounds a singularity.

vanishes identically, we conclude that $g[z] = 0$ throughout $D_1 \cap D_2$, such that $f_1[z] = f_2[z]$ throughout $D_1 \cap D_2$. Hence, any pair of functions f_1 and f_2 that are equal on an arc or a subdomain within their common domain of analyticity $D_1 \cap D_2$ represents the same analytic function everywhere throughout $D_1 \cap D_2$. Therefore, a function that is analytic in domain D is uniquely determined by its values on any arc or domain contained within D.

Reconsider the function

$$f_1[z] = \sum_{k=0}^{\infty} z^k \tag{6.4}$$

which is analytic within $|z| < 1$ where it converges to $f[z] = (1 - z)^{-1}$. According to the uniqueness theorem, $f[z]$ is the only function that is analytic in the entire complex plane, except for $z = 1$, and coincides with f_1 in their common domain $|z| < 1$. Therefore, f is the unique analytic continuation of f_1 into the entire complex plane.

Now suppose that we attempt to extend a function by analytic continuation from D_1 into D_2 into D_3, etc., carefully avoiding singularities in an attempt to cover the largest possible region of the complex plane. A schematic set of overlapping domains is sketched in Fig. 6.2 representing the sequence of continuations $f_1 \to f_2 \to f_3 \to f_4$. Will $f_4[z] == f_1[z]$ in $D_1 \cap D_4$? Maybe, but maybe not – the uniqueness theorem provides no guidance on this question because there is no overlap between D_3 and D_1 or between D_2 and D_4 in the sequence sketched. Often a singularity or branch point in the nonoverlapping region will spoil the closure of such a sequence. Perhaps it will help to consider a specific example, even if it is somewhat artificial. The sequence of functions

$$f_1[z] = \sqrt{r}\, e^{i\theta_1/2} \qquad 0 \le \theta_1 < \pi \tag{6.5}$$

$$f_2[z] = \sqrt{r}\, e^{i\theta_2/2} \qquad \frac{\pi}{2} \le \theta_2 < \frac{3\pi}{2} \tag{6.6}$$

$$f_3[z] = \sqrt{r}\, e^{i\theta_3/2} \qquad \pi \le \theta_3 < 2\pi \tag{6.7}$$

$$f_4[z] = \sqrt{r}\, e^{i\theta_4/2} \qquad \frac{3\pi}{2} \le \theta_4 < \frac{5\pi}{2} \tag{6.8}$$

all represent branches of $\sqrt{z}$ continued from one half-plane into another that overlaps in one quadrant. Thus, f_2 is a continuation of f_1 into the third quadrant with $f_2 = f_1$ in the second quadrant, f_3 continues into the fourth quadrant with $f_3 = f_2$ in the third, and f_4 returns to the first quadrant with $f_4 = f_3$ in the fourth. However, even though f_1 and f_4 are both defined in the first quadrant, they are not equal:

$$f_4[z] = -f_1[z] \quad \text{in } D_4 \cap D_1 \tag{6.9}$$

Here analytic continuation fails to close because the path encloses a branch point of $\sqrt{z}$. Note that both $D_1 \cap D_3$ and $D_2 \cap D_4$ contain only one point, the branch point, which does not constitute a domain.

The theorem that an analytic function is completely determined throughout its domain of analyticity by its values on an arc appears to be extremely powerful. If physics arguments require a function to be analytic, it might appear that we could construct the entire function by measuring on such an arc, which is surely easier than measuring it everywhere in a domain. It is almost like cloning your mother from a hair follicle! However, mathematical and physical standards of knowledge are different. To apply this theorem we would have to make measurements that are perfectly accurate, which is not possible. In practice, the errors in the reconstruction process would grow quickly as the distance from the measured region increases. The larger the arc the better the convergence of the reconstruction is likely to be, but analysis of the accuracy of such a procedure requires considerable sophistication. Nevertheless, in many fields one can make much progress by measuring the function over as large an arc as possible, typically as much of the real axis as possible, and constraining the asymptotic behavior using physical principles. Thus, the theory of analytic functions finds widespread application in nuclear, particle, condensed matter, and other fields. A few basic examples will be studied later in this chapter.

6.1.3 Reflection Principle

Suppose that $f[z]$ is analytic in a domain D that includes a segment of the real axis and that $f[x]$ is real on that segment, such that $f[x] = f[x]^*$. Using the Cauchy–Riemann equations, one can show that if $f[z]$ is analytic in a domain that includes both z and z^*, then so is $f[z^*]^*$. Hence, because the two analytic functions $f[z^*]^*$ and $f[z]$ are equal on an arc within D, they are equal throughout the portion of D that is symmetric with respect to the real axis. This result is known as the *Schwarz reflection principle*.

6.1.4 Permanence of Algebraic Form

Suppose that a function $f[x]$ is defined on a portion of the real axis. If we then define $f[z]$ by replacing the real variable by a complex variable, $x \to z$, the function $f[z]$ uniquely determines the analytic function that has the same values on that portion of the real axis throughout any domain that contains it; there can be no other distinct analytic continuation from the real axis into the complex plane. For example, if $f[x]$ is represented by a power series that converges on some portion of the real axis, the same series with $x \to z$ represents an analytic function that converges in domains containing that portion of the real axis. We implicitly used that property when deriving the Euler identity by replacing x by $i\theta$ in the Taylor series for e^x (recall Sec. 1.1.3). In fact, e^z is the only entire function which reduces to e^x on the real axis. More generally, if $f[x]$ and $g[x]$ satisfy an algebraic relationship of the form $F\left[f[x], g[x]\right] == 0$, then the analytic continuations $f[z]$ and $g[z]$ satisfy an algebraic relationship $F\left[f[z], g[z]\right] == 0$ throughout their common domain of analyticity. Furthermore, if $F\left[f_1[z], g_1[z]\right] == 0$ throughout domain D_1, where F is an analytic function of its arguments, then $F\left[f_2[z], g_2[z]\right] == 0$ in domain D_2 where f_2 and g_2 are analytic continuations of f_1 and g_1 from D_1 into D_2. This property is described as *permanence of algebraic form.* Often it is simpler to prove a relationship using one representation than another, but for analytic functions we can be confident that the same relationship applies to all representations within their domains of analyticity.

6.1.5 Example: Gamma Function

The gamma function for positive real variables x is defined by the integral

$$\Gamma[x] = \int_0^\infty t^{x-1} e^{-t}\, dt \quad \text{for } x > 0 \tag{6.10}$$

and reduces to the factorial $\Gamma[n] = (n-1)!$ for positive integers. Note that we must restrict the range of x to ensure convergence of this integral representation. Continuing this function into the complex plane, we attempt to define the analytic function $\Gamma[z]$ as

$$\Gamma[z] = \int_0^\infty t^{z-1} e^{-t}\, dt \quad \text{for } \mathrm{Re}[z] > 0 \tag{6.11}$$

This definition clearly agrees with the first on the real axis, but we must prove that the extended definition is analytic in a domain that includes the positive real axis. For this we appeal to Morera's theorem (Sec. 1.10.5) which states that if $\oint_C f[z]\, dz = 0$ for every simple closed contour C in domain D, then $f[z]$ is analytic in D. First, we prove that $\Gamma[z]$ converges absolutely by evaluating

$$\left|\Gamma[z]\right| = \left|\int_0^\infty t^{z-1} e^{-t}\, dt\right| \le \int_0^\infty \left|t^{x-1} t^{iy}\right| e^{-t}\, dt \Longrightarrow \left|\Gamma[x+iy]\right| \le \Gamma[x] \tag{6.12}$$

where t is real and $z = x + iy$. Next we express contour integrals in the form

$$\oint_C \Gamma[z]\, dz = \oint_C \left(\int_0^\infty t^{z-1} e^{-t}\, dt\right) dz = \int_0^\infty \left(\oint_C t^{z-1}\, dz\right) e^{-t}\, dt = 0 \tag{6.13}$$

where absolute convergence justifies exchanging the order of integrations. Recognizing that t^{z-1} is analytic, the innermost integral vanishes and proves that $\Gamma[z]$ is analytic in the domain $\mathrm{Re}[z] > 0$.

Using integration by parts,

$$\Gamma[z+1] = \int_0^\infty t^z e^{-t}\, dt = (-t^z e^{-t})_0^\infty + z\int_0^\infty t^{z-1} e^{-t}\, dt \tag{6.14}$$

one easily obtains a recursion relation

$$\Gamma[z+1] = z\Gamma[z] \tag{6.15}$$

that can be used to continue Γ into the region with $\mathrm{Re}[z] < 0$. First,

$$\Gamma[z] = \frac{\Gamma[z+1]}{z} \qquad -1 \le \mathrm{Re}[z] \le 0, \quad z \ne 0, -1 \tag{6.16}$$

provides a definition that now extends in the domain $\mathrm{Re}[z] \ge -1$ excluding $z = 0, 1$. Then,

$$\Gamma[z] = \frac{\Gamma[z+2]}{z(z+1)} \qquad -2 \le \mathrm{Re}[z] \le -1, \quad z \ne -1, -2 \tag{6.17}$$

extends the domain a little further. By repeating this process indefinitely, we obtain an analytic continuation

$$\Gamma[z] = \frac{\Gamma[z+n+1]}{\Pi_{k=0}^{n}(z+k)} \qquad \text{with } n < -\mathrm{Re}[z] < n+1 \tag{6.18}$$

that extends the definition of $\Gamma[z]$ into the entire complex plane excluding nonpositive integers. Although this might not be the most convenient representation, the uniqueness theorem ensures that any analytic function that reproduces $\Gamma[z]$ for positive $\mathrm{Re}[z]$ will produce the same values in its range of analyticity as produced by the representation above, and hence is really the same function whatever its superficial appearance might be.

6.2 Dispersion Relations

6.2.1 Causality

Often there are physical arguments, such as causality, that justify extension of a function of a real variable into the complex plane as an analytic function of the complex variable whose real part is experimentally accessible. For example, the propagation of an electromagnetic wave through a homogeneous isotropic medium is represented by a complex refractive index of the form $\tilde{n} = n + ia$ where $n[\omega]$ and $a[\omega]$ are real functions of frequency ω that govern the phase velocity and absorption of the wave. These properties depend upon the electromagnetic properties of the constituents and the collective response of the system to an electromagnetic field. The dependence of phase velocity upon frequency is known as dispersion. Using quite general arguments based upon causality, one argues that the combined function $\tilde{n}[\omega] = n[\omega] + i a[\omega]$ is an analytic function of complex frequency whose real and imaginary components reduce to the refractive index and absorption coefficient on the physical (real) axis. The theory of analytic functions then provides some very powerful relationships between the dispersive (refractive) properties and the absorptive properties of the system, known as *dispersion relations*. Dispersion relations were first derived by Kramers and Kronig for electromagnetic theory but have since found widespread application in many areas of physics, including nuclear, particle, and condensed matter physics. Rather than develop the theory of dispersion relations for arbitrary analytic functions, we illustrate the general approach using the specific case of electromagnetic waves in a dielectric medium.

Consider a plane wave of the form $\text{Exp}\left[i(kx - \omega t)\right]$ where the wave number is given by

$$k = \tilde{n}\frac{\omega}{c} = \frac{\omega}{c}\sqrt{\mu\epsilon} = (n + i a)\frac{\omega}{c} \tag{6.19}$$

where $\tilde{n}[\omega]$ is the complex refractive index and $\epsilon[\omega]$ is the dielectric permittivity. For simplicity we will assume that the medium is nonconducting and nonmagnetic ($\mu = 1$), such that $\tilde{n}^2 = \epsilon$. Identifying the real and imaginary parts of $\tilde{n}[\omega]$ as $n[\omega]$ and $a[\omega]$, the plane wave

$$\phi \propto \text{Exp}\left[i(kx - \omega t)\right] = \text{Exp}\left[i(nx - ct)\omega/c\right]\text{Exp}[-\alpha x/2] \Longrightarrow |\phi|^2 \propto e^{-\alpha x} \tag{6.20}$$

propagates with phase velocity c/n but is absorbed with attenuation length $\alpha^{-1} = c/2\omega a$ (reduced in intensity by the factor e^{-1} in distance α^{-1}). In Gaussian units the dielectric permittivity is related to the electric susceptibility χ by

$$\epsilon = 1 + 4\pi\chi \tag{6.21}$$

where the polarization of the medium is related to the electric field by $P = \chi E$.

Next consider the electric displacement $D[\omega] = \epsilon[\omega]E[\omega]$ for a wave packet

$$D[t] = \frac{1}{2\pi}\int_{-\infty}^{\infty} d\omega e^{-i\omega t} D[\omega] = \frac{1}{2\pi}\int_{-\infty}^{\infty} d\omega e^{-i\omega t}\epsilon[\omega]E[\omega] \tag{6.22}$$

where

$$E[\omega] = \int_{-\infty}^{\infty} dt e^{i\omega t} E[t] \tag{6.23}$$

Assuming that the integrals converge and interchanging the order of integrations, we obtain

$$\begin{aligned} D[t] &= \frac{1}{2\pi}\int_{-\infty}^{\infty} d\omega e^{-i\omega t}\epsilon[\omega]\int_{-\infty}^{\infty} dt' e^{i\omega t'} E[t'] \\ &= \frac{1}{2\pi}\int_{-\infty}^{\infty} d\tau E[t-\tau]\int_{-\infty}^{\infty} d\omega e^{-i\omega\tau}\epsilon[\omega] \end{aligned} \tag{6.24}$$

where $\tau = t - t'$. Then substituting $\epsilon \to 1 + 4\pi\chi$, we find

$$D[t] = \frac{1}{2\pi}\int_{-\infty}^{\infty} d\tau E[t-\tau]\left(2\pi\delta[\tau] + 4\pi\int_{-\infty}^{\infty} d\omega e^{-i\omega\tau}\chi[\omega]\right) \tag{6.25}$$

or

$$D[t] = E[t] + \int_{-\infty}^{\infty} d\tau\, E[t-\tau]G[\tau] \tag{6.26}$$

where the first term is the incident plane wave in the absence of the medium, while the second term arises from the contribution of charges in the medium governed by a response function of the form

$$G[\tau] = \frac{1}{2\pi}\int_{-\infty}^{\infty} d\omega\, e^{-i\omega\tau}\left(\epsilon[\omega]-1\right) \tag{6.27}$$

Physically we expect that the response of the system at time t can only depend upon fields at earlier times $t' < t \Longrightarrow \tau > 0$. This notion that effects must follow their causes is known as the *causality principle*. Therefore, we expect $G[\tau]$ to vanish for $\tau < 0$ independent of any specific properties of the system and can write

$$D[t] = E[t] + \int_{0}^{\infty} d\tau E[t-\tau]G[\tau] \tag{6.28}$$

Equations of this form are typical expressions of the delayed response of a linear system to a driving force. If the damping is weak the present amplitude may be amplified by coherent contributions from many earlier vibration periods.

We also expect that $\omega \to \infty \Longrightarrow \chi \to 0 \Longrightarrow \epsilon \to 1$ for extremely high frequencies where the electromagnetic field oscillates too rapidly for any charged particles with inertia to respond. Under these conditions the integral for the response function converges for real values of ω. Suppose that we allow ω to be a complex variable and define

$$\mathcal{G}[\tau] = \frac{1}{2\pi}\oint_C e^{-i\omega\tau}\left(\epsilon[\omega]-1\right)d\omega \tag{6.29}$$

where the contour C includes the real axis. We can close the contour for $\tau < 0$ with a great semicircle around the upper half-plane, as sketched below, such that

$$\mathcal{G}[\tau] = \frac{1}{2\pi}\int_{-R}^{R} e^{-i\omega\tau}\left(\epsilon[\omega]-1\right)d\omega + iR\int_{0}^{\pi} \mathrm{Exp}\left[\tau R\,\mathrm{Sin}[\theta]\right]\mathrm{Exp}\left[-iR\tau\,\mathrm{Cos}[\theta]\right]\left(\epsilon[Re^{i\theta}]-1\right)d\theta \tag{6.30}$$

can be expressed in terms of a contribution from the real axis that approaches $G[\tau]$ in the limit $R \to \infty$ and a contribution from the great semicircle that vanishes for negative τ because ϵ is bounded and the integrand includes an exponential whose argument approaches $-\infty$. Similarly, for positive τ we close C around the lower half-plane and again find that $\mathcal{G}[\tau] \to G[\tau]$, such that

$$G[\tau] = \frac{1}{2\pi}\oint_C e^{-i\omega\tau}\left(\epsilon[\omega]-1\right)d\omega \tag{6.31}$$

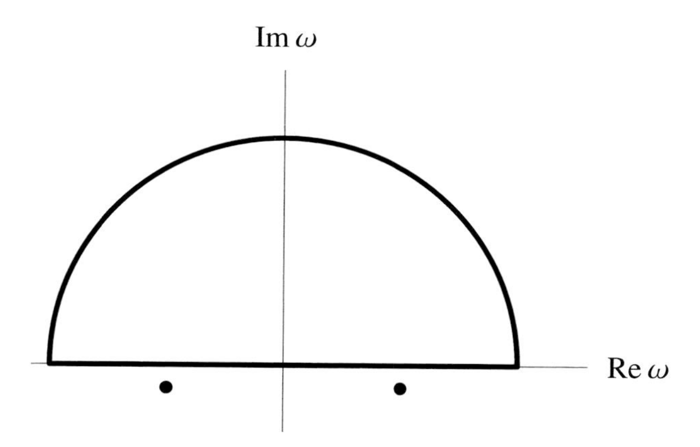

Figure 6.3. Contour for $G[\tau < 0]$. Two poles of $\epsilon[\omega]$ are indicated by heavy dots. Causality confines singularities of ϵ to the lower half-plane such that $G[\tau < 0] = 0$.

However, the causality principle

$$\tau < 0 \Longrightarrow G[\tau] = 0 \Longrightarrow \frac{1}{2\pi} \oint_C e^{-i\omega\tau} \left(\epsilon[\omega] - 1 \right) d\omega = 0 \tag{6.32}$$

requires $G[\tau]$ to vanish for $\tau < 0$. Therefore, according to the Cauchy integral theorem, we conclude that $\epsilon[\omega]$ is analytic within the upper half of the complex ω plane. Any singularities, such as poles or branch cuts, are confined to the lower half-plane and are responsible for the delayed response of the system to an external influence. The two black dots in Fig. 6.3 represent possible poles in the lower half-plane.

The convolution formula for $D[t]$ can be obtained more easily using physical reasoning than by the derivation above. Recognizing that $\epsilon[\omega]$ does not have a simple Fourier transform because it approaches unity rather than zero for infinite frequencies, we simply add and subtract that offset to write

$$D[\omega] = \epsilon[\omega] E[\omega] = E[\omega] + (\epsilon[\omega] - 1) E[\omega] \tag{6.33}$$

where the function $(\epsilon[\omega] - 1)$ is analytic in the upper half-plane and is suitable for Fourier transformation because it does approach zero for $\omega \to \infty$. Then applying the convolution theorem to the product of Fourier transforms, we can write

$$D[t] = E[t] + \int_0^\infty d\tau E[t - \tau] G[\tau] \tag{6.34}$$

where

$$G[\tau] = \frac{1}{2\pi} \int_{-\infty}^{\infty} d\omega \, e^{-i\omega\tau} (\epsilon[\omega] - 1) \tag{6.35}$$

practically by inspection.

6.2.2 Oscillator Model

Let us illustrate this argument with a toy model. Consider a particle of mass m and charge q that is attached to an atom by a spring (linear restoring force) with natural frequency ω_0 and which experiences an external electric field $E_\omega e^{-i\omega t}$. We also assume that the motion of the particle is damped, either by interactions with the surrounding medium or by its own radiation of energy. The equation of motion for a driven damped oscillator takes the form

$$\ddot{x} + 2\gamma\dot{x} + \omega_0^2 x = \frac{q}{m} E_\omega e^{-i\omega t} \tag{6.36}$$

Using

$$x[t] = x_\omega e^{-i\omega t} \Longrightarrow x_\omega = \frac{q}{m} \frac{E_\omega}{\omega_0^2 - 2i\gamma\omega - \omega^2} \tag{6.37}$$

the dipole moment becomes

$$p_\omega = \frac{q^2}{m} \frac{E_\omega}{\omega_0^2 - 2i\gamma\omega - \omega^2} \Longrightarrow \alpha = \frac{p_\omega}{E_\omega} = \frac{q^2}{m} \frac{1}{\omega_0^2 - 2i\gamma\omega - \omega^2} \tag{6.38}$$

where α is the single-particle polarizability. If we have a dilute system with N oscillators per unit volume, such that $\chi = N\alpha$, the dielectric permittivity becomes

$$\epsilon = 1 + 4\pi N\alpha = 1 + \frac{\omega_{\mathrm{p}}^2}{\omega_0^2 - 2i\gamma\omega - \omega^2}, \quad \omega_{\mathrm{p}}^2 = 4\pi N \frac{q^2}{m} \tag{6.39}$$

where ω_{p} is known as the plasma frequency. According to this oscillator model, ϵ is an analytic function except for two simple poles in the lower half-plane located at

$$\omega_\pm = -i\gamma \pm \omega_0 \sqrt{1 - \left(\frac{\gamma}{\omega_0}\right)^2} \tag{6.40}$$

The damping is usually rather weak, such that the pole positions and residues reduce to

$$\gamma \ll \omega_0 \Longrightarrow \omega_\pm \approx -i\gamma \pm \omega_0 \quad R_\pm \approx \mp \frac{\omega_{\mathrm{p}}^2}{2\omega_0} \tag{6.41}$$

Notice that damping requires $\gamma > 0$. Hence, the absence of spontaneously growing solutions confines the poles to the lower half-plane and is closely related to the property of causality in physics. The dielectric permittivity for a single mode is sketched in Fig. 6.4. If the damping is weak, absorption is confined to a narrow peak around the resonant frequency ω_0. The refractive index is slightly greater (smaller) than unity below (above) the peak. Normal dispersion with $\partial n/\partial\omega > 0$ occurs over most of the frequency range, but in the immediate vicinity of the resonant frequency there is *anomalous dispersion* with $\partial n/\partial\omega < 0$.

To evaluate $G[\tau]$ for $\tau > 0$, we must close the contour in the lower half-plane, as shown in Fig. 6.5, to ensure that the contribution of the great semicircle vanishes. Thus, the value

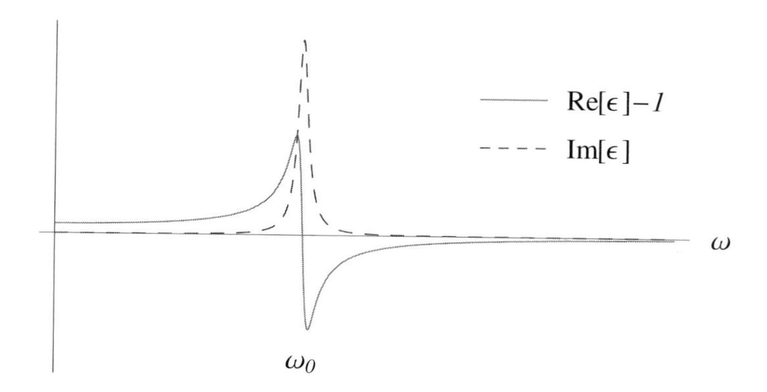

Figure 6.4. Dielectric permittivity for a single mode modeled as a weakly damped oscillator.

of the integral becomes $-2\pi i$ times the sum of the residues, where the negative sign arises because the contour is traversed in a negative (clockwise) sense.

$$G[\tau] = \frac{1}{2\pi} \oint_C e^{-i\omega\tau}(\epsilon[\omega] - 1)\, d\omega = -i(R_+ e^{-i\omega_+\tau} + R_- e^{-i\omega_-\tau}) \quad \text{for } \tau > 0 \tag{6.42}$$

Therefore, we obtain the response

$$G[\tau] = \omega_{\mathrm{p}}^2 e^{-\gamma\tau} \frac{\mathrm{Sin}[\omega_0\tau]}{\omega_0} \Theta[\tau] \tag{6.43}$$

where the Heaviside step function

$$\begin{aligned} \Theta[\tau] &= 1 \quad \text{for } \tau > 0 \\ \Theta[\tau] &= 0 \quad \text{for } \tau < 0 \end{aligned} \tag{6.44}$$

enforces causality. Although this model is simplistic, it does satisfy the physical constraints of causality and analyticity. We find that the response vanishes at $\tau \le 0$, its initial growth is governed by ω_0, and its decay is determined by γ. If the applied field is oscillatory, then the response of the system will be greatest when the driving frequency is close to the system's natural frequency and the damping is small enough that many earlier periods reinforce the response to the current period.

More generally, if the system is composed of a single type of atom with Z electrons and we assume these atoms have a spectrum of normal modes of vibration with frequencies ω_j, the dielectric permittivity can be parametrized in the form

$$\epsilon = 1 + 4\pi N \frac{e^2}{m_e} \sum_j \frac{f_j}{\omega_j^2 - 2i\gamma_j\omega - \omega^2}, \quad \sum_j f_j = Z \tag{6.45}$$

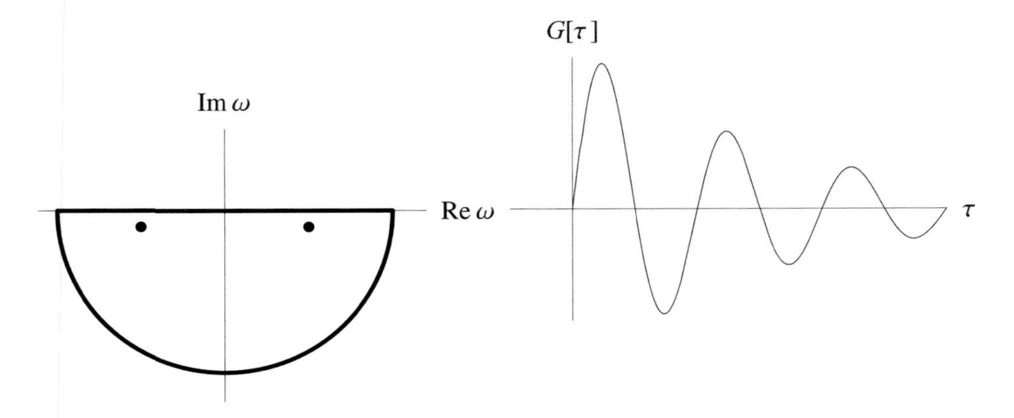

Figure 6.5. Left: $G[\tau > 0]$ contour. Right: response for $\lambda = 0.2\omega_0$

where the parameters f_j are described as *oscillator strengths*. The low-frequency response is determined by the limit $\omega \ll \omega_j$ where

$$\mathrm{Re}\,\epsilon \approx 1 + 4\pi N \frac{e^2}{m_e} \sum_j \frac{f_j}{\omega_j^2} \left(1 + \left(\frac{\omega}{\omega_j} \right)^2 \right) \tag{6.46}$$

$$\mathrm{Im}\,\epsilon \approx 4\pi N \frac{e^2}{m_e} \omega \sum_j \frac{2 f_j \gamma_j}{\omega_j^4} \tag{6.47}$$

Hence, if all normal modes have nonvanishing frequencies, the low-frequency response of dielectrics has the form

$$\mathrm{Re}\,\epsilon[\omega] \approx 1 + a + b\omega^2 \tag{6.48}$$

$$\mathrm{Im}\,\epsilon[\omega] \approx c\omega \tag{6.49}$$

where a, b, c are constants. At the other extreme, the asymptotic response to large frequencies takes the form

$$\omega \to \infty \Longrightarrow \epsilon \simeq 1 - 4\pi N \frac{e^2}{m_e \omega^2} \sum_j f_j \left(1 - \frac{2i\gamma_j}{\omega} \right) \tag{6.50}$$

such that

$$\mathrm{Re}\,\epsilon \simeq 1 - \left(\frac{\omega_{\mathrm{p}}}{\omega} \right)^2, \quad \mathrm{Im}\,\epsilon \simeq 2\bar{\gamma} \frac{\omega_{\mathrm{p}}^2}{\omega^3} \tag{6.51}$$

where

$$\omega_{\mathrm{p}}^2 = 4\pi N \frac{e^2}{m_e} \sum_j f_j = 4\pi N Z \frac{e^2}{m_e} \tag{6.52}$$

is the plasma frequency and

$$\bar{\gamma} = \frac{1}{Z} \sum_j f_j \gamma_j \tag{6.53}$$

is the mean oscillator strength. These asymptotic characteristics are expected to be more general than this semiclassical oscillator model. Therefore, the high-frequency response of real dielectrics is often characterized by the parameters

$$\omega_{\mathrm{p}}^2 = \lim_{\omega \to \infty} \left(\omega^2 (1 - \mathrm{Re}\, \epsilon[\omega]) \right) \tag{6.54}$$

$$\bar{\gamma} = \lim_{\omega \to \infty} \left(\frac{\omega^3}{2\omega_{\mathrm{p}}^2} \mathrm{Im}\, \epsilon[\omega] \right) \tag{6.55}$$

The behavior of conductors is somewhat more complicated because the frequency of the lowest mode vanishes and will be considered in the exercises.

6.2.3 Kramers–Kronig Relations

The analyticity of $\epsilon[\omega]$ in the upper half of the complex ω plane allows one to derive useful relationships between the real and imaginary, or dispersive and absorptive, components of ϵ on the physical (real) axis. Applying the Cauchy integral formula to $\epsilon - 1$, which vanishes for large ω, we write

$$\epsilon[\omega] = 1 + \frac{1}{2\pi i} \oint_C \frac{\epsilon[\omega'] - 1}{\omega' - \omega} \, d\omega' \tag{6.56}$$

where C is any contour in the upper half of the ω'-plane that encloses ω. Experimental observations are made for real frequencies, of course, but the integrand appears to be singular for real ω. Formally we evaluate the integral for real frequencies by replacing ω by $\omega + i\delta$, where δ is a small positive number, and then taking the limit $\delta \to 0^+$ from the positive side. Assuming that an infinitesimal excursion into the lower half-plane encounters no singularities, we can employ a contour that makes a very small detour around the singularity on the real axis, as sketched in Fig. 6.6. Assuming that the integrand falls sufficiently rapidly at large ω' so that the contribution of the semicircle vanishes, the contour integral can be separated into three contributions of the form

$$\oint_C \frac{\epsilon[\omega'] - 1}{\omega' - \omega} \, d\omega' = \lim_{\delta \to 0^+} \left(\int_{-\infty}^{\omega - \delta} \frac{\epsilon[\omega'] - 1}{\omega' - \omega} \, d\omega' + \int_{\omega + \delta}^{\infty} \frac{\epsilon[\omega'] - 1}{\omega' - \omega} \, d\omega' + \int_{\pi}^{2\pi} \frac{\epsilon[\omega + \delta e^{i\theta}] - 1}{\delta e^{i\theta}} i\delta e^{i\theta} \, d\theta \right) \tag{6.57}$$

where the first two terms represent the contribution of the real axis and the third is the contribution of the small semicircle parametrized by $\omega' - \omega = \delta e^{i\theta} \Longrightarrow d\omega' = i\delta e^{i\theta} \, d\theta$. Together the first two terms comprise the *Cauchy principal value*, denoted by

$$\mathcal{P} \int_{-\infty}^{\infty} \frac{\epsilon[\omega'] - 1}{\omega' - \omega} \, d\omega' = \lim_{\delta \to 0^+} \left(\int_{-\infty}^{\omega - \delta} \frac{\epsilon[\omega'] - 1}{\omega' - \omega} \, d\omega' + \int_{\omega + \delta}^{\infty} \frac{\epsilon[\omega'] - 1}{\omega' - \omega} \, d\omega' \right) \tag{6.58}$$

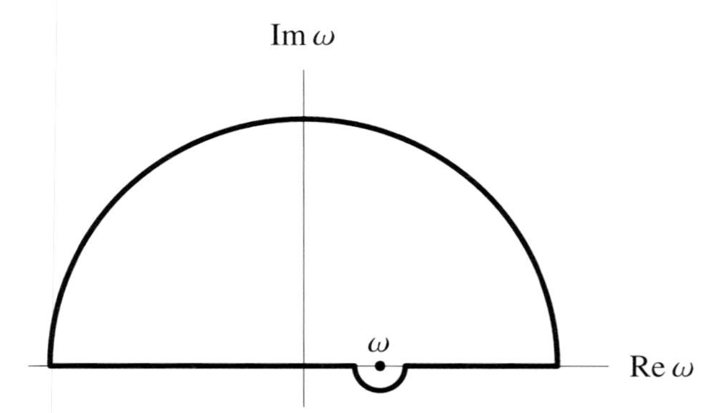

Figure 6.6. Contour in ω' used to evaluate Eq. (6.56) for ω on the positive real axis.

while the third term has the value $i\pi(\epsilon[\omega]-1)$. Therefore, we obtain the net result

$$\epsilon[\omega] = 1 + \frac{1}{i\pi}\mathcal{P}\int_{-\infty}^{\infty} \frac{\epsilon[\omega']-1}{\omega'-\omega}\,d\omega' \tag{6.59}$$

This method of evaluating contour integrals in the limit that a pole approaches the real axis, appears so frequently that a mnemonic formula

$$\lim_{\delta\to 0^+} \frac{1}{\omega'-\omega-i\delta} = \frac{\mathcal{P}}{\omega'-\omega} + i\pi\delta[\omega'-\omega] \tag{6.60}$$

is used to represent the limiting process. Here $\mathcal{P}$ is interpreted as an operator that converts an integral along the real axis to its principal value while the second term involving a Dirac delta function represents the contribution of the pole. Note that the contribution of a pole on the contour is half the contribution it would have made if it had been interior to the contour. Another way to interpret this result is that the contribution of a pole on the contour is the average of its contributions if it is displaced infinitesimally inward or infinitesimally outward with respect to the region enclosed by the contour (a fence sitter that cannot quite decide which side to fall on). Incidentally, this result does not depend upon the details of the deformation of the contour – the same result is obtained using a semicircle around the other side or using a different parametrization of the detour, provided that its length vanishes in the limit $\delta \to 0$ and that the coefficient of $(\omega'-\omega)^{-1}$ is continuous at ω.

Separating $\epsilon[\omega']$ into its real and imaginary components, we obtain the pair of equations

$$\mathrm{Re}\,\epsilon[\omega] = 1 + \frac{1}{\pi}\mathcal{P}\int_{-\infty}^{\infty} \frac{\mathrm{Im}\,\epsilon[\omega']}{\omega'-\omega}\,d\omega' \tag{6.61}$$

$$\mathrm{Im}\,\epsilon[\omega] = -\frac{1}{\pi}\mathcal{P}\int_{-\infty}^{\infty} \frac{\mathrm{Re}\,\epsilon[\omega']-1}{\omega'-\omega}\,d\omega' \tag{6.62}$$

relating the real and imaginary components of $\epsilon[\omega]$ through *dispersion integrals*. Relationships of this general form can be derived for any function that is analytic in a half-plane and falls sufficiently rapidly far from the origin. However, functions representing physical quantities often have a more limited domain. Here, for example, measurements are made only for positive frequencies. Fortunately, symmetry properties permit extension from positive ω' to the entire real axis. Recognizing that $D[t]$ and $E[t]$ are real functions, we require

$$G[\tau] = \frac{1}{2\pi} \int_{-\infty}^{\infty} d\omega e^{-i\omega\tau} (\epsilon[\omega] - 1) \tag{6.63}$$

to be real also, such that

$$G[\tau] = G^*[\tau] \Longrightarrow \epsilon[-\omega] = \epsilon^*[\omega^*] \tag{6.64}$$

On the real axis we find

$$\omega^* = \omega \Longrightarrow \operatorname{Re} \epsilon[-\omega] = \operatorname{Re} \epsilon[\omega], \quad \operatorname{Im} \epsilon[-\omega] = -\operatorname{Im} \epsilon[\omega] \tag{6.65}$$

and with some straightforward manipulations we can express the dispersion relations in terms of positive frequencies only. Therefore, we finally obtain the *Kramers–Kronig relations* in the form

$$\operatorname{Re} \epsilon[\omega] = 1 + \frac{2}{\pi} \mathcal{P} \int_0^{\infty} \frac{\omega' \operatorname{Im} \epsilon[\omega']}{\omega'^2 - \omega^2} d\omega' \tag{6.66}$$

$$\operatorname{Im} \epsilon[\omega] = -\frac{2\omega}{\pi} \mathcal{P} \int_0^{\infty} \frac{\operatorname{Re} \epsilon[\omega'] - 1}{\omega'^2 - \omega^2} d\omega' \tag{6.67}$$

These relations between the dispersive and absorptive properties of a dielectric medium are very general, depending only upon the very reliable assumptions of causality and inertial limitations of the polarization induced by extreme frequencies. Thus, if one measures the absorption spectrum the refractive index can be computed, or vice versa. Notice that refraction must be accompanied by absorption – one cannot have one without the other without violating fundamental physics. In particle physics it is often easier to measure the energy dependence of the total cross-section, which is directly related to the imaginary part of the forward scattering amplitude, than it is to measure the real part of the scattering amplitude. Hence, dispersion integrals of similar form, based upon quite general principles, provide important tools in nuclear and particle physics.

6.2.4 Sum Rules

It is useful to consider the asymptotic properties of $\epsilon[\omega]$ in the limit $\omega \to \infty$, where the dispersion relations take the form

$$\operatorname{Re} \epsilon[\omega] \simeq 1 - \frac{2}{\pi\omega^2} \int_0^{\infty} \omega' \operatorname{Im} \epsilon[\omega'] \left(1 + O\left(\left(\frac{\omega'}{\omega}\right)^2\right)\right) d\omega' \tag{6.68}$$

$$\operatorname{Im} \epsilon[\omega] \simeq \frac{2}{\pi\omega} \int_0^{\infty} (\operatorname{Re} \epsilon[\omega'] - 1) \left(1 + \left(\frac{\omega'}{\omega}\right)^2 + O\left(\left(\frac{\omega'}{\omega}\right)^4\right)\right) d\omega' \tag{6.69}$$

Assuming that the integral converges, the asymptotic behavior of the real part becomes

$$\mathrm{Re}\,\epsilon \simeq 1 - \left(\frac{\omega_{\mathrm{p}}}{\omega}\right)^2 \tag{6.70}$$

where the definition of the plasma frequency

$$\omega_{\mathrm{p}}^2 = \lim_{\omega\to\infty} \left(\omega^2(1 - \mathrm{Re}\,\epsilon[\omega])\right) \tag{6.71}$$

is consistent with the earlier definition motivated by the oscillator model. Therefore, we obtain a *sum rule* of the form

$$\omega_{\mathrm{p}}^2 = \frac{2}{\pi}\int_0^\infty \omega\,\mathrm{Im}\,\epsilon[\omega]\,d\omega \tag{6.72}$$

The interpretation of the asymptotic behavior of the imaginary part is trickier. The present result suggests that its limiting form is proportional to ω^{-1}, but convergence of the sum rule for ω_{p} requires $\mathrm{Im}\,\epsilon[\omega]$ to decrease asymptotically more rapidly than ω^{-1} which is consistent with the ω^{-3} asymptotic behavior found above for the oscillator model. In fact, using causality one can show more generally (see exercises) that for large frequencies

$$\mathrm{Im}\,\epsilon \simeq -\frac{G''[0]}{\omega^3} \tag{6.73}$$

and thus we expect to find

$$\int_0^\infty (\mathrm{Re}\,\epsilon[\omega] - 1)\,d\omega = 0 \tag{6.74}$$

and must carry one more term in the expansion of $\mathrm{Im}\,\epsilon$. Therefore, the asymptotic behavior takes the form

$$\mathrm{Im}\,\epsilon \simeq 2\bar{\gamma}\frac{\omega_{\mathrm{p}}^2}{\omega^3} \tag{6.75}$$

where

$$\bar{\gamma} = \frac{1}{\pi\omega_{\mathrm{p}}^2}\int_0^\infty (\mathrm{Re}\,\epsilon[\omega] - 1)\omega^2\,d\omega \tag{6.76}$$

provides a second sum rule.

If we assume that the asymptotic behavior of $\mathrm{Re}\,\epsilon$ applies when $\omega > \omega_{\mathrm{c}}$, such that

$$\omega > \omega_{\mathrm{c}} \Longrightarrow \mathrm{Re}\,\epsilon = 1 - \left(\frac{\omega_{\mathrm{p}}}{\omega}\right)^2 + O\left(\left(\frac{\omega_p}{\omega_{\mathrm{c}}}\right)^4\right) \tag{6.77}$$

we find

$$\int_0^\infty (\mathrm{Re}\,\epsilon[\omega] - 1)\,d\omega = \int_0^{\omega_{\mathrm{c}}} \mathrm{Re}\,\epsilon[\omega]\,d\omega - \omega_{\mathrm{c}} + \frac{\omega_{\mathrm{p}}^2}{\omega_{\mathrm{c}}} + O\left(\frac{\omega_{\mathrm{p}}^4}{\omega_{\mathrm{c}}^3}\right) \tag{6.78}$$

Therefore,

$$\int_0^\infty (\operatorname{Re}\epsilon[\omega]-1)\,d\omega = 0 \Longrightarrow \frac{1}{\omega_c}\int_0^{\omega_c}\operatorname{Re}\epsilon[\omega]\,d\omega = 1+\left(\frac{\omega_p}{\omega_c}\right)^2+O\left(\left(\frac{\omega_p}{\omega_c}\right)^4\right) \tag{6.79}$$

is known as a *superconvergence relation.*

Sum rules play an important role in nuclear and particle physics. For some theories, such as QCD, calculations are often so intractable that it becomes difficult to test the theory by comparing energy-dependent predictions with experimental data. Sum rules, on the other hand, depend only upon very general assumptions, such as causality and symmetry properties. Therefore, provided that one can perform measurements over a large enough range of energy to ensure convergence of the experimental integral, sum rules test the underlying assumptions of a theory without need for detailed calculations of scattering processes.

6.3 Hilbert Transform

The Kramers–Kronig formula is an example of a Hilbert transform. Suppose that $f[z]$ is analytic on the real axis and in the upper half-plane and that $|f[z]|$ converges uniformly to zero on the surrounding great semicircle. The Cauchy integral formula can then be used to write

$$f[z] = \frac{1}{2\pi i}\oint_C \frac{f[s]}{s-z}\,ds = \frac{1}{2\pi i}\int_{-\infty}^{\infty}\frac{f[s]}{s-x}\,ds + \lim_{R\to\infty}\frac{R}{2\pi}\int_0^{\pi}\frac{f[Re^{i\theta}]}{Re^{i\theta}-x}\,d\theta \tag{6.80}$$

where C consists of the real axis plus the great semicircle and z is in the interior of C. The contribution of the great semicircle vanishes because $|f[z]| \to 0$. Thus, separating $f[z] = \operatorname{Re} f[z] + i \operatorname{Im} f[z]$ into real and imaginary components, we obtain the pair of equations

$$\operatorname{Re} f[z] = \frac{1}{2\pi}\int_{-\infty}^{\infty}\frac{\operatorname{Im} f[s]}{s-x}\,ds \tag{6.81}$$

$$\operatorname{Im} f[z] = -\frac{1}{2\pi}\int_{-\infty}^{\infty}\frac{\operatorname{Re} f[s]}{s-x}\,ds \tag{6.82}$$

that relate the real component in the upper half-plane to an integral of the imaginary component on the real axis, or vice versa.

Similarly, for a point on the real axis we deform C using an infinitesimal semicircular detour to write

$$\begin{aligned} f[x] &= \frac{1}{2\pi i}\oint_C \frac{f[s]}{s-x}\,ds \\ &= \frac{\mathcal{P}}{2\pi i}\int_{-\infty}^{\infty}\frac{f[s]}{s-x}\,ds + \lim_{\epsilon\to 0}\frac{1}{2\pi}\int_{-\pi}^{0} f[x+\epsilon e^{i\theta}]\,d\theta \\ &\quad + \lim_{R\to\infty}\frac{R}{2\pi}\int_0^{\pi}\frac{f[Re^{i\theta}]}{Re^{i\theta}-x}e^{i\theta}\,d\theta \\ &= \frac{\mathcal{P}}{2\pi i}\int_{-\infty}^{\infty}\frac{f[s]}{s-x}\,ds + \tfrac{1}{2}f[x] \end{aligned} \tag{6.83}$$

and again discard the great semicircle to obtain

$$f[x] = \frac{\mathcal{P}}{\pi i} \int_{-\infty}^{\infty} \frac{f[s]}{s - x} ds \tag{6.84}$$

Note that the analysis is practically unchanged for functions that are analytic and decay as $z \to \infty$ in the lower half-plane instead. Separating $f[x]$ into real and imaginary components, we find

$$\text{Re}\, f[x] = \frac{\mathcal{P}}{\pi} \int_{-\infty}^{\infty} \frac{\text{Im}\, f[s]}{s - x} ds \tag{6.85}$$

$$\text{Im}\, f[x] = -\frac{\mathcal{P}}{\pi} \int_{-\infty}^{\infty} \frac{\text{Re}\, f[s]}{s - x} ds \tag{6.86}$$

Notice that these formulas differ by a factor of 2 from those above because the contribution for a singularity on the contour is the average of contributions for nearby singularities on either side (half in and half out). If we now consider the real and imaginary components on the real axis to be two real functions of a real variable, we obtain the *Hilbert transform pair*

$$\psi[x] = \frac{\mathcal{P}}{\pi} \int_{-\infty}^{\infty} \frac{\tilde{\psi}[s]}{s - x} ds \Longleftrightarrow \tilde{\psi}[s] = \frac{\mathcal{P}}{\pi} \int_{-\infty}^{\infty} \frac{\psi[x]}{x - s} dx \tag{6.87}$$

One can assemble a table of Hilbert transforms by decomposing analytic functions that decay sufficiently rapidly as $z \to \infty$ into real and imaginary components on the real axis. For example, Cos$[x]$ and Sin$[x]$ constitute a Hilbert transform pair because Exp$[iz]$ is analytic and satisfies the convergence requirements in the upper half-plane. However, the real strength of this technique is to the analysis of experimental data where physical arguments, such as causality, guarantee the analyticity and convergence properties such that parametrization of measurements for one component can be used to reconstruct the other.

6.4 Spreading of a Wave Packet

In quantum mechanics or optics one often constructs wave packets as a superposition of plane waves such that

$$\psi[x, t] = \int_{-\infty}^{\infty} \frac{dk}{2\pi} \tilde{\psi}[k] \,\text{Exp}[i(kx - \omega t)] \tag{6.88}$$

where $\tilde{\psi}[k]$ is the amplitude of a plane wave with wave number k and where the frequency $\omega = \omega[k]$ is a function of k. The limitation to one spatial dimension is made here for simplicity only. If

$$\tilde{\psi}[k] = \tilde{f}[k - k_0] \tag{6.89}$$

exhibits a relatively narrow peak at k_0, it becomes useful to express the Fourier integral in the form

$$\psi[x, t] = \text{Exp}[i(k_0 x - \omega_0 t)] \int_{-\infty}^{\infty} \frac{dk}{2\pi} \tilde{f}[k - k_0] \,\text{Exp}\big[i\big((k - k_0)x - (\omega - \omega_0)t\big)\big] \tag{6.90}$$

where $\omega_0 = \omega[k_0]$. The contributions from k appreciably different from k_0 will tend to interfere destructively because the phase of the complex exponential varies rapidly. Expanding the temporal frequency as

$$\omega[k] \approx \omega_0 + \omega_1(k - k_0) + \omega_2 \frac{(k - k_0)^2}{2} + \cdots \tag{6.91}$$

we can write

$$\psi[x, t] \approx \text{Exp}[i(k_0 x - \omega_0 t)] \int_{-\infty}^{\infty} \frac{d\kappa}{2\pi} \tilde{f}[\kappa] \, \text{Exp}\left[i\left(\kappa(x - \omega_1 t) - \frac{\omega_2 \kappa^2}{2} t \right) \right] \tag{6.92}$$

where $\kappa = k - k_0$. Notice that if $\omega_2 = 0$, one obtains

$$\omega_2 = 0 \Longrightarrow \psi[x, t] = f[x - v_g t] \, \text{Exp}[i(k_0 x - \omega_0 t)] \tag{6.93}$$

where

$$f[x] = \int_{-\infty}^{\infty} \frac{d\kappa}{2\pi} \tilde{f}[\kappa] \, \text{Exp}[i\kappa x], \quad \tilde{f}[\kappa] = \int_{-\infty}^{\infty} dx e^{-i\kappa x} f[x] \tag{6.94}$$

is the initial shape of the wave packet at time $t = 0$ and where $v_g = \omega_1 = \omega'[k_0]$ is the group velocity at the peak of the momentum distribution. Under these conditions, the intensity

$$|\psi[x, t]|^2 = \left| f[x - v_g t] \right|^2 \tag{6.95}$$

propagates as a wave with velocity v_g without changing shape. Therefore, if $\omega[k]$ is linear, the propagation is described as nondispersive because the shape of the wave packet is preserved. However, if $\omega_2 \neq 0$, the wave packet will usually spread with time.

The phase and group velocities are related by

$$v_p = \frac{\omega}{k}, \quad v_g = \frac{d\omega}{dk} \Longrightarrow v_g = v_p + k \frac{dv_p}{dk} \Longrightarrow \frac{v_g}{v_p} = 1 + \frac{k}{v_p} \frac{dv_p}{dk} \tag{6.96}$$

which for a narrow spectral distribution should be evaluated near the peak, k_0. From optics we describe situations where $dv_p / dk < 0 \Longrightarrow v_g < v_p$ as *normal dispersion* and $dv_p / dk > 0 \Longrightarrow v_g > v_p$ as *anomalous dispersion*.

Consider a Gaussian wave packet of the form

$$\tilde{f}[\kappa] = \left(4\pi\sigma^2\right)^{1/4} \text{Exp}\left[-\frac{\sigma^2 \kappa^2}{2} \right] \tag{6.97}$$

$$f[x] = \left(\pi\sigma^2\right)^{-1/4} \text{Exp}\left[-\frac{x^2}{2\sigma^2} \right] \tag{6.98}$$

normalized according to

$$\int_{-\infty}^{\infty} dx |f[x]|^2 = \int_{-\infty}^{\infty} \frac{d\kappa}{2\pi} |\tilde{f}[\kappa]|^2 = 1 \tag{6.99}$$

such that

$$\psi[x,t] = \left(4\pi\sigma^2\right)^{1/4} \text{Exp}\left[i\left(k_0 x - \omega_0 t\right)\right] \cdot \int_{-\infty}^{\infty} \frac{d\kappa}{2\pi} \text{Exp}\left[-\frac{\left(\sigma^2 + i\omega_2 t\right)\kappa^2}{2} + i\kappa(x - \omega_1 t)\right] \tag{6.100}$$

Completing the square, we obtain a wave packet of similar form

$$\psi[x,t] = \left(\frac{\sigma^2}{\pi}\right)^{1/4} \frac{\text{Exp}[i(k_0 x - \omega_0 t)]}{(\sigma^2 + i\omega_2 t)^{1/2}} \text{Exp}\left[-\frac{(x-\omega_1 t)^2}{2(\sigma^2 + i\omega_2 t)}\right] \tag{6.101}$$

except that the width

$$\sigma^2 \longrightarrow \sigma^2 + i\omega_2 t \tag{6.102}$$

becomes time dependent and complex. Using

$$\frac{1}{\sigma^2 + i\omega_2 t} = \frac{1 - i\tau}{\sigma^2(1+\tau^2)} \quad \text{with } \tau = \frac{\omega_2 t}{\sigma^2} \tag{6.103}$$

the intensity profile takes the form

$$|\psi[x,t]|^2 = \left(\pi\sigma^2(1+\tau^2)\right)^{-1/2} \text{Exp}\left[-\frac{(x-\omega_1 t)^2}{\sigma^2(1+\tau^2)}\right] \tag{6.104}$$

or

$$|\psi[x,t]|^2 = \left(\pi\sigma^2\left(1 + \left(\frac{\omega_2 t}{\sigma^2}\right)^2\right)\right)^{-1/2} \text{Exp}\left[-\frac{(x-\omega_1 t)^2}{\sigma^2\left(1 + \left(\frac{\omega_2 t}{\sigma^2}\right)^2\right)}\right] \tag{6.105}$$

Therefore, we find that a Gaussian wave packet remains Gaussian in shape but that its width increases with time according to

$$\sigma \longrightarrow \sigma\left(1 + \left(\frac{\omega_2 t}{\sigma^2}\right)^2\right)^{1/2} \tag{6.106}$$

The spreading of the intensity profile for a typical wave packet is shown in Fig. 6.7.

It is important to recognize that the width increases much more rapidly for a narrow than for a broad wave packet. A brief pulse contains a broad spectrum of frequencies, but high frequencies propagate more rapidly than low frequencies in a medium with normal dispersion ($\omega_2 > 0$). Thus, the high-frequency components race ahead of the lower frequencies and broaden the pulse. This effect can be seen by examining the real and imaginary components separately, as shown in Figs. 6.8 and 6.9. Notice that the two figures have different spatial scales – the pulse in Fig. 6.8 is initially twice as wide as the pulse in Fig. 6.9, but after some time the latter is much wider than the former. It should also be obvious that the higher frequencies dominate the leading side while the lower frequencies

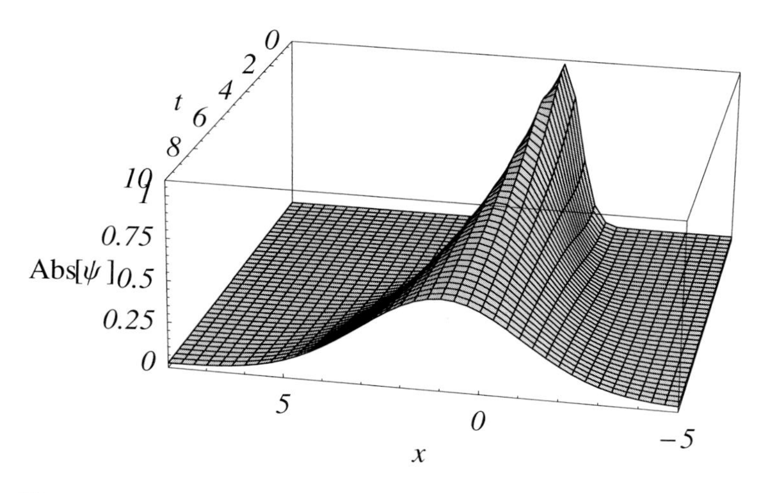

Figure 6.7. Spreading of a Gaussian wave packet.

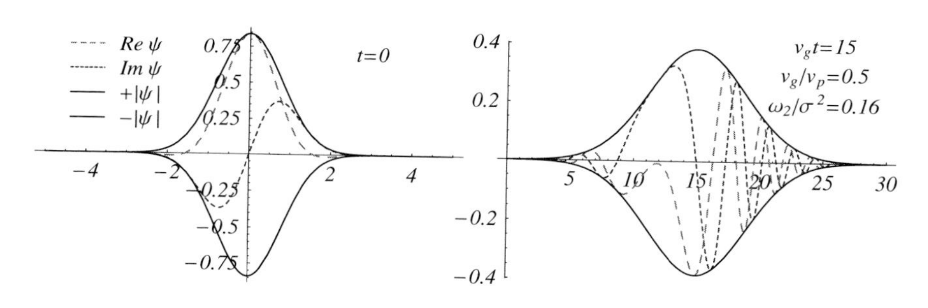

Figure 6.8. Spreading of the real and imaginary contributions to the intensity of a relatively broad wave packet.

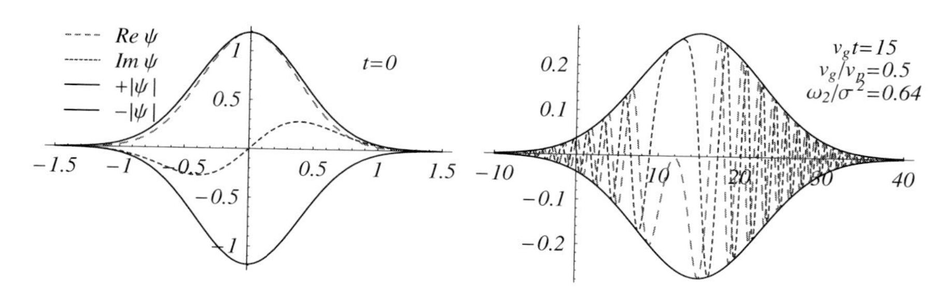

Figure 6.9. Spreading of the real and imaginary contributions to the intensity of a relatively narrow wave packet.

dominate the trailing side of the pulse. We chose a ratio $v_g/v_p = 0.5$ between group and phase velocities that is typical of light in a plastic scintillator or voltage in a coaxial cable. Our choice of dispersive coefficient, ω_2, was somewhat arbitrary, but it is clear that narrow signals can be distorted quite rapidly by dispersive media.

The relationship $\omega = \omega[k]$ between frequency and wave number is often called a dispersion relation because it governs the spreading of wave packets. A linear relation is nondispersive, but nonlinear components of the dispersion relation generally produce dispersion of the wave packet because Fourier components with different frequencies propagate with different phase velocities. However, the shapes of nongaussian wave packets may change with time in more complicated ways. Furthermore, in nonlinear media there are sometimes special profiles, called *solitons*, for which the distribution of Fourier amplitudes $\tilde{f}[\kappa]$ is matched to $\omega[\kappa]$ such that the wave packet actually retains its initial shape as it propagates.

6.5 Solitons

Consider a classical field $\phi = \phi[x,t]$ that satisfies a one-dimensional wave equation of the form

$$\sigma\frac{\partial^2\phi}{\partial t^2} - \tau\frac{\partial^2\phi}{\partial x^2} = -\frac{\partial V}{\partial\phi} \tag{6.107}$$

where σ and τ are inertial and stiffness constants and where

$$V[\phi] = -\tfrac{1}{2}A\phi^2 + \tfrac{1}{4}B\phi^4 \tag{6.108}$$

with positive constants A and B is a potential that exhibits a spontaneously broken symmetry. Almost by inspection we recognize that $V[\phi]$ has a local maximum at $\phi = 0$ and symmetric minima at $\phi = \pm\sqrt{A/B}$ with depth $-A^2/4B$. Thus, it is useful to express $V[\phi]$ in the form

$$V_m = \frac{A^2}{4B},\ \phi_m = \sqrt{\frac{A}{B}} \Longrightarrow V[\phi] = -V_m\left(2\left(\frac{\phi}{\phi_m}\right)^2 - \left(\frac{\phi}{\phi_m}\right)^4\right) \tag{6.109}$$

which is plotted in Fig. 6.10. Although the potential is symmetric with respect to ϕ, the equilibrium state with $\phi = 0$ is unstable and is not the ground state. There are two stable ground states at $\phi = \pm\phi_m$ that do not share the symmetry of the potential – at low temperature the system will be found near either the positive or the negative state and this choice violates the symmetry of the equation of motion. The situation in which the ground state does not share the symmetry of the dynamical equations is described as a *spontaneously broken symmetry*. For example, the dynamics of a ferromagnet are rotationally symmetric, but in the ground state the constituent magnetic moments are aligned in some arbitrary direction that violates rotational invariance.

In this section we demonstrate that, in addition to familiar wave solutions for small-amplitude oscillations about either of the minima, this highly dispersive nonlinear differential equation also possesses propagating soliton solutions that induce transitions between

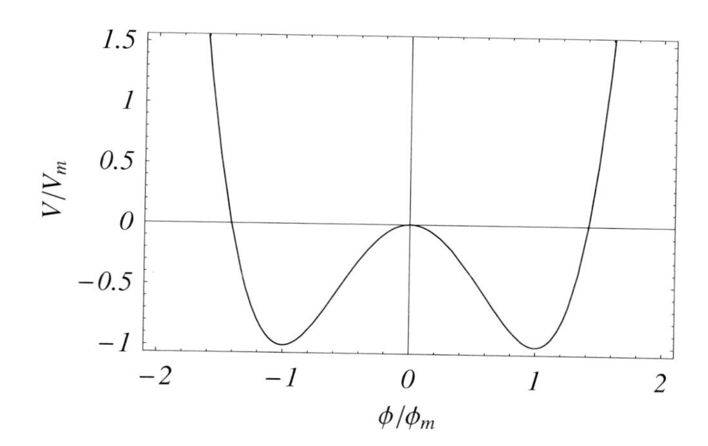

Figure 6.10. A potential that exhibits spontaneous symmetry breaking.

the stable states. We do not attempt to develop general methods for finding such solutions, which are beyond the scope of this course, but will be content to discuss the general properties of stipulated solutions in order to gain some appreciation for the phenomena. Models of this type find widespread application in condensed matter and particle physics.

It is useful to define the intrinsic wave speed in the absence of V as $c = \sqrt{\tau/\sigma}$ and to divide out ϕ_m to obtain a wave equation for $\psi = \phi/\phi_m$ that takes the form

$$\frac{1}{c^2}\frac{\partial^2\psi}{\partial t^2} - \frac{\partial^2\psi}{\partial x^2} == \beta(\psi - \psi^3) \tag{6.110}$$

where $\beta = 4V_m/\tau\phi_m^2 = A/\tau$. First we demonstrate that there exist approximate solutions of the form

$$\psi[x,t] \approx \pm 1 + \alpha\,\mathrm{Sin}[kx - \omega t + \delta] \tag{6.111}$$

with $\alpha \ll 1$ that represent small-amplitude oscillations of ψ around one of the potential minima. Substituting the trial solution and expanding to lowest order in α, we find

$$-\left(\frac{\omega^2}{c^2} - k^2\right)\alpha\,\mathrm{Sin}[kx - \omega t + \delta] == \beta(-2\alpha\,\mathrm{Sin}[kx - \omega t + \delta] + \cdots) \tag{6.112}$$

Therefore, we obtain a highly nonlinear dispersion relation of the form

$$\omega^2 = c^2(k^2 + 2\beta) \tag{6.113}$$

with phase velocity

$$v_{\mathrm{p}} = \frac{\omega}{k} = c\sqrt{1 + \frac{2\beta}{k^2}} = c\sqrt{1 + \frac{2A}{k^2\tau}} \tag{6.114}$$

that is independent of the oscillation amplitude. Note, however, that because the equation is nonlinear the accuracy of the solution does depend upon amplitude – the smaller the

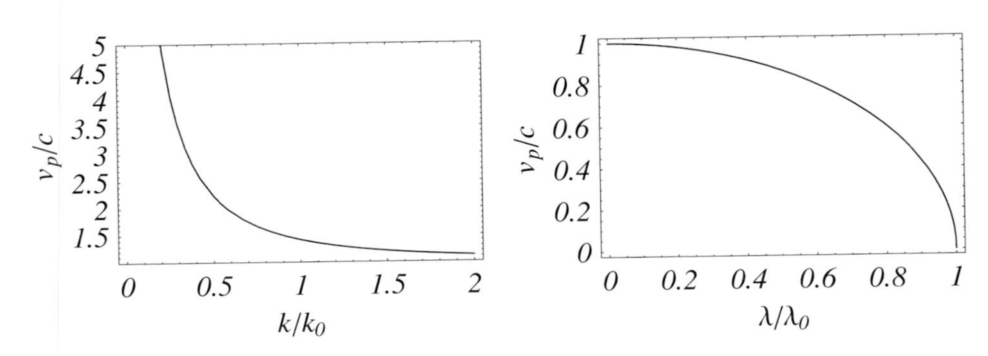

Figure 6.11. Dispersion relations for Eq. (6.110). Left: small-amplitude waves. Right: soliton solutions.

amplitude the more accurate the solution. Also notice that the phase velocity for long wavelengths is increased dramatically by the harmonic part of the potential. The left side of Fig. 6.11 shows the phase velocity where $k_0 = \sqrt{2A/\tau}$.

Next, we demonstrate that there exist exact solutions of the form

$$\psi[x,t] = \psi_0 \, \mathrm{Tanh}[\xi(x \pm vt)] \tag{6.115}$$

for suitable choices of ψ_0, ξ, and v. Using

$$\frac{d^2}{d^2 z} \mathrm{Tanh}[z] = -2 \frac{\mathrm{Tanh}[z]}{\mathrm{Cosh}[z]^2} \tag{6.116}$$

direct substitution gives

$$2\psi_0 \xi^2 \left(1 - \frac{v^2}{c^2}\right) \frac{\mathrm{Tanh}[\xi(x \pm vt)]}{\mathrm{Cosh}[\xi(x \pm vt)]^2} == \beta(\psi_0 \, \mathrm{Tanh}[\xi(x \pm vt)] - \psi_0^3 \, \mathrm{Tanh}[\xi(x \pm vt)]^3) \tag{6.117}$$

or

$$\frac{v^2}{c^2} == 1 - \frac{\beta}{2\xi^2} (\mathrm{Cosh}[\xi(x \pm vt)]^2 - \psi_0^2 \, \mathrm{Sinh}[\xi(x \pm vt)]^2) \tag{6.118}$$

The space and time dependencies of the right-hand side can be eliminated by choosing $\psi_0 = \pm 1$, such that

$$v = c\sqrt{1 - \frac{\beta}{2\xi^2}} \tag{6.119}$$

where we choose the positive root because the trial solution already includes the propagation direction. A soliton of this form propagates without changing shape, but the amplitude is not arbitrary and the speed depends upon the slope parameter ξ. Unlike small-amplitude

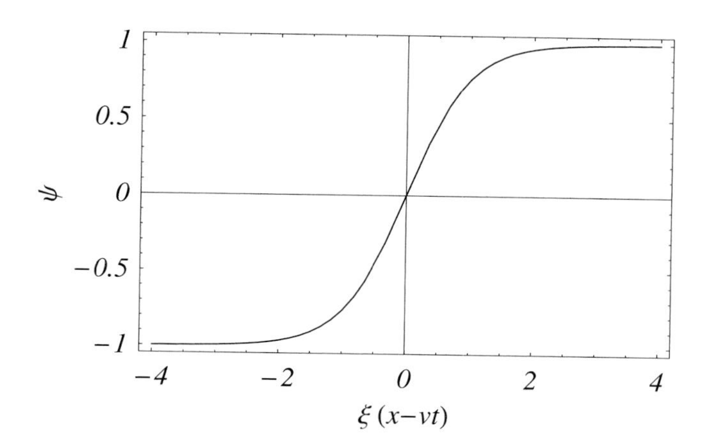

Figure 6.12. A soliton solution to Eq. (6.110).

waves which propagate with $v > c$, large-amplitude solitons propagate with velocity $v < c$. The dependence of the propagation velocity upon the slope parameter is a form of dispersion, but unlike a familiar wave the amplitude of the soliton is not arbitrary because the equation is nonlinear.

A snapshot of the soliton that propagates in the positive x-direction is sketched in Fig. 6.12. The soliton is a propagating disturbance that switches the local field from one stable state to the other as it passes and is often described as a *kink*. The distance over which the transition occurs is characterized by $\lambda = \xi^{-1}$, the inverse of the slope of the kink at the origin. Thus, we can express the phase velocity in the form

$$v = c\sqrt{1 - \left(\frac{\lambda}{\lambda_0}\right)^2} \tag{6.120}$$

where $\lambda_0 = \sqrt{2B\tau/A^2}$ is a characteristic distance. The smaller the distance over which the transition occurs the stronger the kink and the greater its speed. Very strong kinks with $\lambda \to 0$ approach the asymptotic speed c while weak kinks with $\lambda \to \lambda_0$ propagate very slowly. In fact, the existence of this type of solution with real v requires

$$\lambda < \lambda_0 \Longrightarrow \beta < 2\xi^2 \Longrightarrow \xi > \sqrt{A/2\tau} \tag{6.121}$$

The soliton velocity is also shown in Fig. 6.11.

Solitons are commonly found in systems described by nonlinear equations; in fact, linear equations are often just an approximation for small amplitudes. However, if the compensation between dispersion and nonlinearity needed for solitons to propagate without changing shape were too delicate, it could be destroyed by external disturbances or by imperfections or impurities in real systems. In fact, stability is crucial to the observation of solitons, which otherwise would be little more than mathematical curiosities. Although it is beyond the scope of this course, one can often demonstrate explicitly that solitons are stable with respect to perturbations and propagate over long distances with little distortion even under nonideal circumstances. When analytical techniques fail, numerical simulations can be used to show that suitable initial profiles evolve into one or more solitons which then propagate stably.

Problems for Chapter 6

1. Analytic continuation of simple functions

The following functions are defined by definite integrals which converge in domain D. For each, specify the original domain D, construct an analytic continuation into the largest possible region of the complex plane, and specify the analytic domain for the parent functions.

mathindent=0pt

$$\text{a)}\quad f[z] = \int_0^\infty e^{-zt}\,\mathrm{Sin}[t]\,dt \tag{6.122}$$

$$\text{b)}\quad f_n[z] = \int_0^\infty e^{-zt} t^n\,dt \quad \text{for integer } n \geq 0\,. \tag{6.123}$$

2. Euler transformation

Suppose that $f[z]$ is represented as an alternating series of the form

$$f[z] = \sum_{n=0}^{\infty} (-)^n a_n z^n \tag{6.124}$$

that converges for $|z| < R$. In this problem we explore a procedure for analytic continuation that uses differencing of coefficients to rearrange a power series in order to obtain a representation that converges more rapidly and in a larger domain. Additional variations of this method are omitted for the sake of brevity.

a) Show that

$$(1+z)f[z] = a_0 + z\sum_{n=0}^{\infty} (-)^n \delta a_n z^n \tag{6.125}$$

where $\delta a_n = a_n - a_{n+1}$. Continuing this process, show that

$$(1+z)f[z] = a_0 + \zeta\delta a_0 + \zeta^2 \sum_{n=0}^{\infty} (-)^n \delta^2 a_n z^n \tag{6.126}$$

where

$$\zeta = \frac{z}{1+z}, \quad \delta^m a_n = \delta(\delta^{m-1} a_n) \tag{6.127}$$

and obtain thereby a series expansion for $g[\zeta] = (1+z)f[z]$. Find an explicit expression for $\delta^m a_n$.

b) To demonstrate the usefulness of the Euler transformation, demonstrate that the Euler series for $g[\zeta]$ converges much more rapidly near $\zeta \approx \frac{1}{2}$ than does the Taylor series for $f[z] = \text{Log}[1+z]$ near $z \approx 1$. In fact, show that $g[\zeta]$ converges for $\text{Re}[z] > -\frac{1}{2}$ independent of $\text{Im}[z]$, which is a much larger region than for the original Taylor series.

3. High-frequency response

The temporal response of a dielectric medium to an applied electric field was expressed in the form

$$D[t] = E[t] + \int_{-\infty}^{\infty} d\tau E[t-\tau] G[\tau] \tag{6.128}$$

$$G[\tau] = \frac{1}{2\pi} \int_{-\infty}^{\infty} d\omega e^{-i\omega\tau} (\epsilon[\omega] - 1) \tag{6.129}$$

Express $\epsilon[\omega]$ in terms of a Fourier transform and perform a Taylor series of $G[\tau]$ applicable for $\tau \to 0^+$. Note that this is a somewhat unusual series because $G[\tau] = 0$ for $\tau \leq 0$, so that derivatives at $\tau = 0$ are understood as limits for $\tau \to 0^+$. How is the high-frequency behavior of $\epsilon[\omega]$ related to the time dependence of G? Compare your results with those of the semiclassical oscillator model.

4. Absorption band

Suppose that a material absorbs electromagnetic radiation only in the band $\omega_1 < \omega < \omega_2$. For simplicity assume that $\text{Im}\,\epsilon = \eta$ is constant within, and vanishes outside, the absorption band. Use the Kramers–Kronig model to evaluate $\text{Re}\,\epsilon$. Sketch this function and compare with the single-mode model.

5. Subtracted dispersion relations

If $f[z]$ does not vanish as $z \to \infty$, or does not diminish rapidly enough to ensure good convergence of the dispersion integrals, one can apply the technique to the function $z^{-1} f[z]$ instead. More generally, dispersion relations for a function of the form $g[z] = \left(f[z] - f[x_0]\right)/(z - x_0)$ will converge more rapidly than for $f[z]$ itself, especially if an optimum choice of x_0 is made. Derive dispersion relations for $g[x]$, assuming that $g[z]$ is analytic in the upper half-plane and vanishes on a great semicircle, using a contour with detours on the real axis around both x and x_0. Then deduce the dispersion relations for $f[x]$.

6. Hilbert transforms

Given

$$v[x] = -\frac{\mathcal{P}}{\pi} \int_{-\infty}^{\infty} \frac{u[s]}{s-x} ds = \frac{1}{1+x^2} \tag{6.130}$$

determine $u[x]$ and construct the corresponding analytic function $f[z] = u[x,y] + iv[x,y]$. Then verify that $f[z]$ satisfies the necessary requirements for $\{u, v\}$ to constitute a Hilbert transform pair. Note that $u[x] \equiv u[x,0]$ and $v[x] \equiv v[x,0]$.

7. Dispersion of a Gaussian wave function

Suppose that at time $t = 0$ an electron is represented by a wave packet of the form

$$\tilde{\psi}[k] = \left(4\pi\sigma_0^2\right)^{1/4} \text{Exp}\left[-\frac{\sigma_0^2 \left(k - k_0\right)^2}{2}\right] \tag{6.131}$$

where k_0 and σ_0 are constants. Assume that the velocity is nonrelativistic, such that the kinetic energy is given by $\varepsilon = \hbar\omega = (\hbar k)^2/2m$.

a) Evaluate the wave function in space and show that it is normalized properly.

b) Find expressions for the phase and group velocities.

c) Evaluate the time dependencies for the mean position, $\bar{x}$, and root-mean-square (rms) width, σ, where

$$\bar{x}[t] = \langle\psi[x,t]|x|\psi[x,t]\rangle = \int_{-\infty}^{\infty} \psi^*[x,t]\psi[x,t]x\, dx \tag{6.132}$$

$$\sigma^2[t] = \langle\psi[x,t]|(x-\bar{x})^2|\psi[x,t]\rangle = \int_{-\infty}^{\infty} \psi^*[x,t]\psi[x,t](x-\bar{x})^2\, dx \tag{6.133}$$

d) Suppose that an electron with 10 keV kinetic energy is initially confined to $\sigma_0 = 1$ Å, about the size of an atom. How much time is required for appreciable spreading of the wave packet? How far does it travel in that time? How much time does it take for the wave function to become broader than a typical laboratory?

8. Plane electromagnetic waves in conducting media

One can show that electromagnetic waves satisfy a wave equation of the form

$$\left(\nabla^2 - \frac{\epsilon\mu}{c^2}\frac{\partial^2}{\partial t^2} - \frac{4\pi\sigma\mu}{c^2}\frac{\partial}{\partial t}\right)\vec{E} == 0 \tag{6.134}$$

when the permittivity ϵ, permeability μ, and conductivity σ are constant. The electric field for a plane wave with frequency ω is the real part of

$$\vec{E}\left[\vec{r},t\right] = \vec{E}_0\, \text{Exp}\left[i\left(\vec{k}\cdot\vec{r} - \omega t\right)\right] \tag{6.135}$$

a) Separate the wave number $k = (n + i\gamma)\omega/c$ into real and imaginary parts. Discuss the dependence of n and γ upon frequency for both low and high conductivity.

b) The electric and magnetic fields are related by

$$\vec{\nabla}\times\vec{E} == -\frac{1}{c}\frac{\partial\vec{B}}{\partial t} \tag{6.136}$$

Evaluate and discuss the dependencies of the relative magnitude and phase of the magnetic and electric fields for plane waves upon frequency.

9. Signal transmission in coaxial cable

A coaxial transmission line consists of two concentric cylindrical conductors separated by a dielectric. The voltage and current changes across a length Δz of the inner conductor are described by

$$\frac{\Delta V}{\Delta z} == -RI - L\frac{\partial I}{\partial t} \tag{6.137}$$

$$\frac{\Delta I}{\Delta z} == -GV - C\frac{\partial V}{\partial t} \tag{6.138}$$

where R, L, and C are the resistance, inductance, and capacitance per unit length and G is the conductance per unit length for leakage current between the conductors. R, L, G usually vary relatively slowly with frequency.

a) Show that both the voltage and the current satisfy decoupled equations of the form

$$\left(\frac{1}{v_0^2}\frac{\partial^2}{\partial t^2} - \frac{\partial^2}{\partial z^2} + \alpha\frac{\partial}{\partial t} + \beta\right)\begin{pmatrix} V \\ I \end{pmatrix} == 0 \tag{6.139}$$

Determine the phase velocity for an ideal lossless cable with $R \to 0$, $G \to 0$. Under what conditions do signals propagate without dispersion?

b) Using plane-wave solutions of the form $V[z,t] \propto \mathrm{Exp}[i(kz - \omega t)]$, evaluate the dispersion relation $k[\omega] = \kappa + i\gamma$ and determine the frequency dependencies of the wave number, κ, and attenuation coefficient, γ. Also determine the phase velocity. Discuss the limiting cases of

(a) an ideal lossless cable,

(b) high frequencies, and

(c) $G \to 0$.

c) Express the characteristic impedance $Z = V/I$ for sinusoidal solutions in terms of the parameters (R, L, C, G) and the frequency ω. Determine the phase difference between the voltage and the current for high frequencies.

d) A logic pulse

$$V[0,t] = V_0\Theta[t] - V_0\Theta[t - T] \tag{6.140}$$

is injected at $z = 0$ and propagates on the cable for $z > 0$. Produce a Fourier representation for the wave form at later times; however, do not waste much time attempting to evaluate the integral, which is very difficult.

10. Landau damping

One can show that longitudinal density waves in plasma described by $n[x,t] = n_0\,\mathrm{Exp}\left[i(kx - \omega t)\right]$ satisfy a dispersion relation of the form

$$\left(\frac{k}{\omega_p}\right)^2 == \int_{-\infty}^{\infty} \frac{\partial f/\partial v}{v - \omega/k}\,dv \tag{6.141}$$

where $\omega_p = n_0 e^2/\varepsilon_0 m$ is the plasma frequency, n_0 is the average electron density, m is the electron mass, and

$$f[v] = (2\pi v_T^2)^{-1/2} \operatorname{Exp}\left[-\frac{1}{2}\left(\frac{v}{v_T}\right)^2\right] \tag{6.142}$$

is the Maxwell velocity distribution expressed in terms of the characteristic thermal velocity $v_T = \sqrt{k_B T/m}$. Treat the wave number k as a real variable and $\omega[k]$ as a complex-valued function of k to be obtained as a solution to this integral equation. Assume that $\omega_p \gg k v_T$.

a) Notice that the integrand would have a pole on the integration path if ω were real, but ω may actually be complex. Alternatively, we can assume that the pole is practically upon the real axis and use a semicircular detour to avoid it. Show that the dispersion relation takes the form

$$\left(\frac{k}{\omega_p}\right)^2 == \mathcal{P}\int_{-\infty}^{\infty} \frac{f'[v]}{v - \omega/k}\, dv \pm i\pi f'\left[\frac{\omega}{k}\right] \tag{6.143}$$

depending upon the choice of detour and provide a physical criterion for selection of the proper sign.

b) Approximate the principal-value integral using the assumptions $\omega \approx \omega_p$ and $\omega_p \gg k v_T$.

c) Find an approximate solution for $\omega[k]$. Which contour deformation is required?

11. Faster than light?

Suppose that a one-dimensional electromagnetic wave is described by

$$\psi[x,t] = \int_{-\infty}^{\infty} \frac{d\omega}{2\pi} \tilde{\psi}[\omega] \operatorname{Exp}\left[i\frac{\omega}{c}(n[\omega]x - ct)\right] \tag{6.144}$$

where $n[\omega]$ is the complex refractive index and that

$$\psi[x > 0, t < 0] = 0 \tag{6.145}$$

Thus, the spectral amplitude is given by

$$\tilde{\psi}[\omega] = \int_0^{\infty} dt \psi[0,t] \operatorname{Exp}[i\omega t] \tag{6.146}$$

where the frequency ω is real. It is useful to extend $\tilde{\psi}[\omega] \to \tilde{\psi}[z]$ into the complex plane. On physical grounds we expect $n[\omega] \to 1$ for very large frequencies and we assume that $n[z]$ is analytic in the upper half-plane with $n[z] \to 1$ when $|z| \to \infty$. Evaluate the signal $\psi[x,t]$ for $x > ct$. Many systems have frequency bands where the phase velocity exceeds light speed because $n[\omega] < 1$; how does that situation affect the result?

12. KdV equation

Water waves on the surface of a shallow channel satisfy the Korteweg–deVries equation

$$\frac{\partial \psi}{\partial t} + \psi \frac{\partial \psi}{\partial x} + \alpha \frac{\partial^3 \psi}{\partial x^3} == 0 \tag{6.147}$$

where the variables have been scaled conveniently and where certain approximations that are needed for the derivation remain implicit. This equation nicely illustrates how a suitable balance between dispersion and nonlinearity can produce a soliton.

a) Suppose that the amplitude is sufficiently small to omit the nonlinear term, such that

$$\frac{\partial \phi}{\partial t} + \alpha \frac{\partial^3 \phi}{\partial x^3} == 0 \tag{6.148}$$

Show that there exist solutions of the form

$$\phi[x,t] = \int \tilde{\phi}[k]\,\mathrm{Exp}[i(kx - \omega t)]\, dk \tag{6.149}$$

and deduce $\omega[k]$ for this linear approximation.

b) Next, suppose that α is small, such that

$$\frac{\partial \eta}{\partial t} + \eta \frac{\partial \eta}{\partial x} == 0 \tag{6.150}$$

Show that there exist traveling waves of the form $\eta[x,t] = \eta[x - vt]$ where v depends upon the amplitude η. Describe *qualitatively* the time development of η given that $\eta[x,0]$ initially has a bell shape. You will not be able to construct a familiar single-valued function, but this method does provide an algorithm suitable for numerical evaluation.

c) Next, show that if we assume a traveling wave solution of the form $\psi[x,t] = \psi[x - vt]$, the KdV equation can be integrated once. Evaluate the constant of integration when $\psi[x,t] = 0$ for $t \to \pm\infty$.

d) Finally, show that there exist solitons of the form

$$\psi[x,t] = A\,\mathrm{Sech}[B(x - vt)]^2 \tag{6.151}$$

for suitable choices of A and B. Provide a physical description of this solution.

13. Sine–Gordon equation

The sine–Gordon equation

$$\frac{\partial^2 \psi[x,t]}{\partial t^2} - \frac{\partial^2 \psi[x,t]}{\partial^2 x^2} == -\,\mathrm{Sin}\big[\psi[x,t]\big] \tag{6.152}$$

arises in models of systems with periodic potentials. In this form all variables are dimensionless.

a) Deduce the potential $V[\psi]$ for which the inhomogeneous term is given by $-\partial V/\partial\psi$. Identify the stable and unstable equilibrium states of the field. Show that there exist position-independent solutions that describe small-amplitude oscillations about a stable equilibrium state of the field.

b) Construct approximate solutions that describe traveling waves of small-amplitude oscillations about the stable states and deduce the corresponding dispersion relation.

c) Show that there exist soliton solutions of the form

$$\psi_{\pm} = 2n\pi + 4\text{ArcTan}\left[\text{Exp}\left[\mp\frac{x - vt}{\sqrt{1 - v^2}}\right]\right] \tag{6.153}$$

Sketch and provide a physical interpretation of these solitons. (It is sufficient to verify the solutions by substitution and to use *MATHEMATICA*® or equivalent software to perform the algebra.)

7 Sturm–Liouville Theory

Abstract. The Sturm–Liouville operator with suitable boundary conditions provides a self-adjoint system with real eigenvalues and a complete set of orthogonal eigenfunctions. We explore the general properties of such systems and their role in constructing Green functions for physical systems. Perturbative and variational methods are also introduced.

7.1 Introduction: The General String Equation

The displacement $\Psi[x,t]$ of a finite, but not necessarily uniform, string can be described by a Lagrangian of the form

$$L = \int_a^b \mathcal{L}\, dx \tag{7.1}$$

where the Lagrangian density $\mathcal{L}$ takes the form

$$\mathcal{L} = \mathcal{T} - \mathcal{V} \tag{7.2}$$

$$\mathcal{T} = \tfrac{1}{2}\sigma[x]\left(\frac{\partial\Psi}{\partial t}\right)^2 = \tfrac{1}{2}\sigma[x]\Psi_t^2 \tag{7.3}$$

$$\mathcal{V} = \tfrac{1}{2}\tau[x]\left(\frac{\partial\Psi}{\partial x}\right)^2 + \tfrac{1}{2}\kappa[x]\Psi^2 = \tfrac{1}{2}\tau[x]\Psi_x^2 + \tfrac{1}{2}\kappa[x]\Psi^2 \tag{7.4}$$

Here σ is the linear mass density, τ is the tension, and κ is the stiffness parameter for an additional linear restoring force that might be produced by the coupling of the string to another system. It is useful to define generalized velocities

$$\Psi_t = \frac{\partial\Psi}{\partial t}, \quad \Psi_x = \frac{\partial\Psi}{\partial x} \tag{7.5}$$

based upon the generalized coordinate Ψ. The equation of motion is determined by minimizing the action, whereby

$$\delta\int_{t_1}^{t_2} L\, dt == 0 \Longrightarrow \int_{t_1}^{t_2} dt \int_a^b dx\left(\frac{\partial\mathcal{L}}{\partial\Psi}\delta\Psi + \frac{\partial\mathcal{L}}{\partial\Psi_t}\delta\Psi_t + \frac{\partial\mathcal{L}}{\partial\Psi_x}\delta\Psi_x\right) == 0 \tag{7.6}$$

where the endpoints in both space and time are held constant, as are coupling functions σ, τ, κ. Substituting

$$\delta\Psi_t = \frac{\partial\delta\Psi}{\partial t}, \quad \delta\Psi_x = \frac{\partial\delta\Psi}{\partial x} \tag{7.7}$$

Graduate Mathematical Physics. James J. Kelly

ISBN: 3-527-40637-9

into

$$\int_{t_1}^{t_2} dt \int_a^b dx \left(\frac{\partial \mathcal{L}}{\partial \Psi} \delta\Psi + \frac{\partial \mathcal{L}}{\partial \Psi_t} \delta\Psi_t + \frac{\partial \mathcal{L}}{\partial \Psi_x} \delta\Psi_x \right)$$
$$= \int_{t_1}^{t_2} dt \int_a^b dx \left(\frac{\partial \mathcal{L}}{\partial \Psi} \delta\Psi + \frac{\partial \mathcal{L}}{\partial \Psi_t} \frac{\partial \delta\Psi}{\partial t} + \frac{\partial \mathcal{L}}{\partial \Psi_x} \frac{\partial \delta\Psi}{\partial x} \right) \quad (7.8)$$

and using the boundary conditions

$$x = a, b \quad \text{or} \quad t = t_1, t_2 \Longrightarrow \delta\Psi = 0 \quad (7.9)$$

to integrate by parts, we find

$$\delta \int_{t_1}^{t_2} L\, dt == 0 \Longrightarrow \int_{t_1}^{t_2} dt \int_a^b dx \left(\frac{\partial \mathcal{L}}{\partial \Psi} - \frac{\partial}{\partial t} \frac{\partial \mathcal{L}}{\partial \Psi_t} - \frac{\partial}{\partial x} \frac{\partial \mathcal{L}}{\partial \Psi_x} \right) \delta\Psi == 0 \quad (7.10)$$

This equation must be satisfied for independent variations $\delta\Psi$, which then requires that the coefficient of $\delta\Psi$ vanishes anywhere in the spacetime interior. Thus, the continuum Euler–Lagrange equations take the form

$$\frac{\partial}{\partial t} \frac{\partial \mathcal{L}}{\partial \Psi_t} + \frac{\partial}{\partial x} \frac{\partial \mathcal{L}}{\partial \Psi_x} - \frac{\partial \mathcal{L}}{\partial \Psi} == 0 \quad (7.11)$$

Therefore, using

$$\frac{\partial \mathcal{L}}{\partial \Psi} = -\kappa[x]\Psi \quad (7.12)$$

$$\frac{\partial \mathcal{L}}{\partial \Psi_t} = \sigma[x]\Psi_t \quad (7.13)$$

$$\frac{\partial \mathcal{L}}{\partial \Psi_x} = -\tau[x]\Psi_x \quad (7.14)$$

we finally obtain the *general string equation*

$$\frac{\partial}{\partial x} \left(\tau[x] \frac{\partial \Psi}{\partial x} \right) - \sigma[x] \frac{\partial^2 \Psi}{\partial t^2} - \kappa[x]\Psi == 0 \quad (7.15)$$

The normal modes of vibration can be analyzed by hypothesizing the sinusoidal time dependence

$$\Psi[x, t] = \psi[x]\, \text{Exp}[-i\omega t] \quad (7.16)$$

such that

$$\frac{d}{dx} \left(\tau[x] \frac{d\psi[x]}{dx} \right) + \omega^2 \sigma[x]\psi[x] - \kappa[x]\psi[x] == 0 \quad (7.17)$$

represents the spatial dependence. In addition, one must impose boundary conditions on the normal modes. For example, if the string is clamped at both ends one would require

$$\text{clamped} \Longrightarrow \psi[a] == \psi[b] == 0 \tag{7.18}$$

while if the string were free at either end we would require the derivative to vanish there instead. In general, nontrivial solutions that are consistent simultaneously with two separate boundary conditions are possible only for particular characteristic values of the frequency. These characteristic frequencies, ω_n, are called *eigenvalues* and the corresponding spatial functions, ψ_n, are called *eigenfunctions* or *normal modes*. Thus, the nonuniform string is described by an eigenvalue problem of the form

$$\frac{d}{dx}\left(\tau[x]\frac{d\psi_n[x]}{dx}\right) - \kappa[x]\psi_n[x] + \omega_n^2\sigma[x]\psi_n[x] == 0 \tag{7.19}$$

This equation reduces to the familiar case of a vibrating string in the special case of constant τ, constant σ, and vanishing κ, such that

$$\tau[x] \to \tau, \quad \sigma[x] \to \sigma, \quad \kappa[x] \to 0 \Longrightarrow \frac{d^2\psi_n}{dx^2} == -k_n^2\psi_n \tag{7.20}$$

where

$$\omega_n \to k_n c, \quad c^2 = \frac{\tau}{\sigma} \tag{7.21}$$

Finally, if a string of length l is clamped at both ends, the eigenvalues are simply $k_n = n\pi/l$ where n is an integer. The solutions

$$\psi[0] == \psi[l] == 0 \Longrightarrow \psi_n[x] = \sqrt{\frac{2}{l}}\,\text{Sin}[\frac{nx}{l}] \tag{7.22}$$

form a complete orthonormal set with

$$\int_0^l \psi_n[x]\psi_m[x]\,dx = \delta_{n,m} \tag{7.23}$$

and can be used to represent the spatial dependence of any piecewise continuous function

$$f[x] = \sum_{n=1}^{\infty} a_n\psi_n[x] \Longrightarrow a_n = \int_0^l \psi_n[x]f[x]\,dx \tag{7.24}$$

as a Fourier sine series. These familiar and useful properties can be generalized to the nonuniform string, with suitable modifications to accommodate variations of tension or density. (Note: the term *orthonormal* means orthogonal and normalized.)

The general string equation is one example of a *Sturm–Liouville* eigenvalue problem. Let

$$\mathcal{D} = \frac{d}{dx}p[x]\frac{d}{dx} - q[x] \Longleftrightarrow \mathcal{D}\psi = \frac{d}{dx}\left(p[x]\frac{d\psi[x]}{dx}\right) - q[x]\psi[x] \tag{7.25}$$

represent a linear differential operator based upon the coefficient functions $p[x]$ and $q[x]$. Based upon the mechanical problem discussed above, we could describe $p[x]$ as a force density and $q[x]$ as an inertial density. An eigenvalue equation for a system with *weight* or *metric function* $w[x]$ that is analogous to the mass density then takes the form

$$(\mathcal{D} + \lambda w)\psi == 0 \Longleftrightarrow \frac{d}{dx}\left(p[x]\frac{d\psi[x]}{dx}\right) - q[x]\psi[x] + \lambda w[x]\psi[x] == 0 \tag{7.26}$$

where one expects nontrivial solutions, ψ_n, for discrete eigenvalues, $\lambda \to \lambda_n$, when suitable boundary conditions are applied for a finite interval. Many physics problems can be represented in this manner. Therefore, in this chapter we shall study the general properties of one-dimensional Sturm–Liouville systems and the application of eigenvalue expansions to the solution of inhomogeneous equations of the general form

$$(\mathcal{D} + \lambda w)\psi[x] == f[x] \tag{7.27}$$

The chapter on *boundary-value problems* will later apply these techniques to problems in two or three spatial dimensions using separable coordinate systems. After separation of variables one generally obtains Sturm–Liouville systems for each spatial dependence and these systems are connected by means of separation constants that depend upon the eigenvalues for the other equations.

7.2 Hilbert Spaces

In the broadest sense a *Hilbert space* is any infinite-dimensional linear vector space with an inner product. An abstract linear vector space $\mathcal{S}$ is a set of elements (vectors) with addition and multiplication operations that satisfy the following properties.

1. closure under addition: $\forall f, g \in \mathcal{S}, h = f + g \in \mathcal{S}$
2. commutativity of addition: $f + g = g + f$
3. associativity of addition: $f + (g + h) = (f + g) + h$
4. existence of additive identity: $\exists 0 \in \mathcal{S} \ni f + 0 = f$
5. closure under multiplication by scalar: $\forall f \in \mathcal{S}, \alpha f \in \mathcal{S}$
6. commutativity of multiplication by scalar: $\alpha(\beta f) = (\alpha\beta) f$
7. distributive law: $\alpha(f + g) = \alpha f + \alpha g$
8. associative law: $(\alpha + \beta) f = \alpha f + \beta f$

9. existence of multiplicative identity: $\exists 1 \in \mathcal{S} \ni 1f = f$.

10. existence of additive inverse: $\forall f \in \mathcal{S}\ \exists g \in \mathcal{S} \ni f + g = 0$.

The inner product $\langle f \mid g \rangle$ for a real Hilbert space must satisfy:

1. $\langle f \mid g \rangle = \langle g \mid f \rangle$
2. $\langle f \mid g + h \rangle = \langle f \mid g \rangle + \langle f \mid h \rangle$
3. $\langle f \mid f \rangle \geq 0$ with equality iff $f = 0$,

while for a complex Hilbert space we require:

1. $\langle f \mid g \rangle = \langle g \mid f \rangle^*$
2. $\langle f \mid g + h \rangle = \langle f \mid g \rangle + \langle f \mid h \rangle$
3. $\langle f \mid f \rangle \geq 0$ with equality iff $f = 0$.

A linear *Hermitian* operator $\mathcal{L}$ acting upon the elements $f, g \in \mathcal{S}$ is defined by the requirements

$$g = \mathcal{L}f \Longrightarrow g \in \mathcal{S} \tag{7.28}$$

$$\mathcal{L}(\alpha f + \beta g) = \alpha \mathcal{L}f + \beta \mathcal{L}g \tag{7.29}$$

$$\langle f \mid \mathcal{L}g \rangle = \langle g \mid \mathcal{L}f \rangle^* = \langle \mathcal{L}f \mid g \rangle \tag{7.30}$$

where α and β are arbitrary constants. The complex conjugation properties are often combined in the statement

$$\mathcal{L}^\dagger = \mathcal{L} \tag{7.31}$$

where $\mathcal{L}^\dagger$ is the Hermitian adjoint of $\mathcal{L}$. For an ordinary vector space, $\mathcal{L}$ is a square matrix and $\mathcal{L}^\dagger$ is the complex conjugate of its transpose; hence, a Hermitian matrix is identical to its Hermitian adjoint and is described as *self-adjoint*. The notation

$$\mathcal{L}^\dagger = \mathcal{L} \Longrightarrow \langle \mathcal{L}f \mid g \rangle = \langle f \mid \mathcal{L}g \rangle = \langle f \mid \mathcal{L} \mid g \rangle \tag{7.32}$$

expresses the fact that a Hermitian $\mathcal{L}$ can act to the right or its adjoint $\mathcal{L}^\dagger$ can act to the left without changing the value of the matrix element.

Two elements of $\mathcal{S}$ are described as *orthogonal* when their inner product

$$f \perp g \Longleftrightarrow \langle f \mid g \rangle = 0 \tag{7.33}$$

vanishes. It is often useful to generalize the notion of orthogonality to include a metric function w, where in a conventional vector space w is a positive-definite diagonal matrix, meaning a diagonal matrix whose elements are real, positive, and nonzero. A matrix of this type is clearly self-adjoint, such that

$$\langle f \mid w \mid g \rangle^* = \langle g \mid w \mid f \rangle \Longleftrightarrow w = w^\dagger \tag{7.34}$$

satisfies the complex conjugate property required for the inner product of a Hilbert space. Thus, two elements are described as *orthogonal with respect to* w when

$$f \perp g \Longleftrightarrow \langle f \mid w \mid g \rangle = 0 \tag{7.35}$$

This notion of orthogonality with respect to a metric function will prove useful in more general Hilbert spaces also.

For the purposes of this chapter we define a particular type of Hilbert space consisting of functions of a single real variable that are smooth and continuous on a finite interval $a \le x \le b$ and satisfy boundary conditions of the form

$$\mathcal{B}_a f == \mathcal{B}_b f \tag{7.36}$$

where

$$\mathcal{B}_a = \sum_{j=1}^{n} \alpha_j \left(\frac{\partial}{\partial x} \right)^j, \quad \mathcal{B}_b = \sum_{j=1}^{n} \beta_j \left(\frac{\partial}{\partial x} \right)^j \tag{7.37}$$

are linear differential operators evaluated at the endpoints of the domain. Hence, we require the elements $f[x] \in \mathcal{S}$ to possess at least n continuous derivatives in the interior of the domain. We will refer to real Hilbert spaces with boundary conditions of this type as *Sturm–Liouville spaces* and complex Hilbert spaces of this type as *Hermitian spaces*. The simplest boundary conditions can be further classified as

$$\text{Dirichlet:} \quad f[a] == f[b] == 0 \tag{7.38}$$

$$\text{Neumann:} \quad f'[a] == f'[b] == 0 \tag{7.39}$$

$$\text{periodic:} \quad f[a] == f[b], \, f'[a] == f'[b] \tag{7.40}$$

More generally, boundary conditions with $\mathcal{B}f == 0$ are described as homogeneous.

Similarly, we define the inner product for a real function space as

$$\langle f \mid g \rangle = \int_a^b f[x] g[x] \, dx = \langle g \mid f \rangle \tag{7.41}$$

while for a complex function space we define

$$\langle f \mid g \rangle = \int_a^b f^*[x] g[x] \, dx = \langle g \mid f \rangle^* \tag{7.42}$$

Notice that these definitions implicitly impose another constraint on $\mathcal{S}$ – the existence of an inner product requires the elements of $\mathcal{S}$ to be square integrable. Two elements of $\mathcal{S}$ are described as orthogonal with respect to the *metric function* or *weight* $w[x]$ when

$$f \perp g \Longleftrightarrow \langle f \mid w \mid g \rangle = 0 \tag{7.43}$$

where $w[x] \ge 0$ is real and nonnegative within the domain of $\mathcal{S}$. In fact, we generally assume that $w[x]$ is free of interior zeros. Usually the requirement of square integrability is less demanding with than without the metric function. When using a nontrivial metric function, we often refer to the matrix element

$$\langle f \mid w \mid g\rangle = \int_a^b f[x]g[x]w[x]\,dx = \langle g \mid w \mid f\rangle \tag{7.44}$$

for real or

$$\langle f \mid w \mid g\rangle = \int_a^b f^*[x]g[x]w[x]\,dx = \langle g \mid w \mid f\rangle^* \tag{7.45}$$

for complex functions as the inner product, leaving the phrase "with respect to w" understood. For the general string equation where $w[x]$ is the mass density $\sigma[x]$, we interpret the combination $\sigma[x]\,dx = dm$ as the mass of a differential element of string. The product of two functions $f[x]$ and $g[x]$ is naturally weighted by the mass dm carried by an infinitesimal length dx.

A linear differential operator $\mathcal{D}$ of order n is defined by

$$\mathcal{D} = \sum_{j=1}^{n} d_j[x]\left(\frac{\partial}{\partial x}\right)^j \tag{7.46}$$

where each $d_j[x]$ is a fixed function of x that is independent of the element of $\mathcal{S}$ to which it is applied. We assume that the functions $d_j[x]$ are finite and continuous within the domain of $\mathcal{S}$. A differential operator is *self-adjoint* in a Sturm–Liouville space when

$$\int_a^b f[x](\mathcal{D}g[x])\,dx = \int_a^b g[x](\mathcal{D}f[x])\,dx \Longleftrightarrow \langle f \mid \mathcal{D} \mid g\rangle = \langle g \mid \mathcal{D} \mid f\rangle \tag{7.47}$$

or in a Hermitian space when

$$\int_a^b f^*[x](\mathcal{D}g[x])\,dx = \left(\int_a^b g^*[x](\mathcal{D}f[x])\,dx\right)^* \Longleftrightarrow \langle f \mid \mathcal{D} \mid g\rangle = \langle g \mid \mathcal{D} \mid f\rangle^* \tag{7.48}$$

such that $\mathcal{D}$ can be shifted from one side to the other without changing the value of the matrix element.

7.2.1 Schwartz Inequality

The Schwartz inequality

$$|\langle f \mid w \mid g\rangle|^2 \leq \langle f \mid w \mid f\rangle\langle g \mid w \mid g\rangle \tag{7.49}$$

where equality pertains iff $f = g$ should be familiar from linear algebra and applies to any Hilbert space, but we review the derivation to be sure. The theorem is obviously true for the trivial case in which either vector is null, so we assume that neither is. Let $h = af + bg$ where a, b are arbitrary scalars, such that

$$h = af + bg \Longrightarrow \langle h \mid w \mid h \rangle = a^* a \langle f \mid w \mid f \rangle + ab^* \langle f \mid w \mid g \rangle + a^* b \langle g \mid w \mid f \rangle + b^* b \langle g \mid w \mid g \rangle \geq 0 \quad (7.50)$$

If we choose

$$a = \langle g \mid w \mid g \rangle \Longrightarrow \langle g \mid w \mid g \rangle \langle f \mid w \mid f \rangle + b^* \langle f \mid w \mid g \rangle + b \langle g \mid w \mid f \rangle + b^* b \geq 0 \quad (7.51)$$

and then choose

$$b = -\langle f \mid w \mid g \rangle \Longrightarrow \langle g \mid w \mid g \rangle \langle f \mid w \mid f \rangle - \langle g \mid w \mid f \rangle \langle f \mid w \mid g \rangle \geq 0 \quad (7.52)$$

we obtain the Schwartz inequality.

7.2.2 Gram–Schmidt Orthogonalization

Suppose that we possess a set of m linearly independent but possibly nonorthogonal vectors $\{u_j, j = 1, m\}$ and wish to construct orthogonal linear combinations

$$\phi_i = \sum_{j=1}^{m} a_{i,j} u_j \quad (7.53)$$

such that

$$i \neq j \Longrightarrow \langle \phi_i \mid w \mid \phi_j \rangle = 0 \quad (7.54)$$

A simple procedure constructs the set $\{\phi_i\}$ one at a time, subtracting off the projections upon previous vectors. Choose

$$\phi_1 = u_1 \quad (7.55)$$

and then assign

$$\phi_2 = u_2 - \frac{\langle \phi_1 \mid w \mid u_2 \rangle}{\langle \phi_1 \mid w \mid \phi_1 \rangle} \phi_1 \quad (7.56)$$

such that

$$\langle \phi_1 \mid w \mid \phi_2 \rangle = \langle u_1 \mid w \mid u_2 \rangle - \frac{\langle u_1 \mid w \mid u_2 \rangle}{\langle u_1 \mid w \mid u_1 \rangle} \langle u_1 \mid w \mid u_1 \rangle = 0 \quad (7.57)$$

Continuing in this manner

$$\phi_3 = u_3 - \frac{\langle \phi_1 \mid w \mid u_3 \rangle}{\langle \phi_1 \mid w \mid \phi_1 \rangle} \phi_1 - \frac{\langle \phi_2 \mid w \mid u_3 \rangle}{\langle \phi_2 \mid w \mid \phi_2 \rangle} \phi_2 \quad (7.58)$$

ensures

$$\langle \phi_1 \mid w \mid \phi_3 \rangle = \langle u_1 \mid w \mid u_3 \rangle - \frac{\langle u_1 \mid w \mid u_3 \rangle}{\langle \phi_1 \mid w \mid \phi_1 \rangle}\langle \phi_1 \mid w \mid \phi_1 \rangle - \frac{\langle \phi_2 \mid w \mid u_3 \rangle}{\langle \phi_2 \mid w \mid \phi_2 \rangle}\langle \phi_1 \mid w \mid \phi_2 \rangle = 0 \tag{7.59}$$

$$\langle \phi_2 \mid w \mid \phi_3 \rangle = \langle \phi_2 \mid w \mid u_3 \rangle - \frac{\langle u_1 \mid w \mid u_3 \rangle}{\langle \phi_1 \mid w \mid \phi_1 \rangle}\langle \phi_2 \mid w \mid \phi_1 \rangle - \frac{\langle \phi_2 \mid w \mid u_3 \rangle}{\langle \phi_2 \mid w \mid \phi_2 \rangle}\langle \phi_2 \mid w \mid \phi_2 \rangle = 0 \tag{7.60}$$

This procedure can now be repeated enough times to produce m independent orthogonal vectors of the form

$$\phi_1 = u_1, \quad \phi_{i+1} = u_{i+1} - \sum_{j=1}^{i} \frac{\langle \phi_j \mid w \mid u_{i+1} \rangle}{\langle \phi_j \mid w \mid \phi_j \rangle}\phi_j \tag{7.61}$$

Usually we prefer an orthonormal basis, where the vectors are both mutually orthogonal and normalized, such that

$$\varphi_i = e^{i\delta_i}\langle \phi_i \mid w \mid \phi_i \rangle^{-1}\phi_i \tag{7.62}$$

where δ_i are real phases which may be chosen at our convenience.

7.2.2.1 Example: Legendre Polynomials

Suppose that we wish to construct a set of polynomials $\phi_n[x]$ on the interval $(-1, 1)$ that are orthogonal with respect to the inner product

$$\langle f \mid g \rangle = \int_{-1}^{1} f[x]\, g[x]\, dx \tag{7.63}$$

We can start with a nonorthogonal basis $u_n = x^n$ with $n \geq 0$ and apply the Gram–Schmidt orthogonalization procedure

$$\phi_0 = 1, \quad \langle \phi_0 \mid \phi_0 \rangle = 2 \tag{7.64}$$

$$\phi_1 = x - \frac{\langle \phi_0 \mid u_1 \rangle}{\langle \phi_0 \mid \phi_0 \rangle}\phi_0 = x - \frac{1}{2}\int_{-1}^{1} t\, dt = x, \quad \langle \phi_1 \mid \phi_1 \rangle = \int_{-1}^{1} t^2\, dt = \frac{2}{3} \tag{7.65}$$

$$\phi_2 = x^2 - \frac{\langle \phi_0 \mid u_2 \rangle}{\langle \phi_0 \mid \phi_0 \rangle}\phi_0 - \frac{\langle \phi_1 \mid u_2 \rangle}{\langle \phi_1 \mid \phi_1 \rangle}\phi_1 = x^2 - \frac{1}{2}\int_{-1}^{1} t^2\, dt - x\int_{-1}^{1} t^3\, dt = x^2 - \frac{1}{3} \tag{7.66}$$

to generate as many terms as our patience permits. Obviously, this is a job for a machine.

```
Table[LegendreP[n, x], {n, 0, 5}]
```

$$\left\{1, x, -\frac{1}{2} + \frac{3x^2}{2}, -\frac{3x}{2} + \frac{5x^3}{2}, \frac{3}{8} - \frac{15x^2}{4} + \frac{35x^4}{8}, \frac{15x}{8} - \frac{35x^3}{4} + \frac{63x^5}{8}\right\}$$

The Legendre polynomials are normally defined using the conventional normalization $P_n[1] = 1$.

7.3 Properties of Sturm–Liouville Systems

7.3.1 Self-Adjointness

An important example of a self-adjoint operator is the *Sturm–Liouville operator*

$$\mathcal{L} = \frac{d}{dx}\left(p[x]\frac{d}{dx}\right) - q[x] \Longleftrightarrow \mathcal{L}f[x] = p[x]f''[x] + p'[x]f'[x] - q[x]f[x] \tag{7.67}$$

where $p[x]$ and $q[x]$ are real functions and where we can assume, without loss of generality, that $p[x]$ is nonnegative. Suppose that $f[x]$ and $g[x]$ are two piecewise smooth functions that both satisfy specific boundary conditions at $x = a, b$ but are otherwise arbitrary. Such an operator is self-adjoint whenever these boundary conditions ensure that

$$\int_a^b f[x]^* \mathcal{L}g[x]\, dx = \left(\int_a^b g[x]^* \mathcal{L}f[x]\, dx\right)^* \Longleftrightarrow \left(f^* p\frac{dg}{dx}\right)_a^b = \left(gp\frac{df^*}{dx}\right)_a^b \tag{7.68}$$

(This result is obtained by integrating both sides by parts.) Thus, we require boundary conditions which will ensure

$$p[b](f[b]^* g'[b] - g[b]f'[b]^*) = p[a](f[a]^* g'[a] - g[a]f'[a]^*) \tag{7.69}$$

for any $f, g \in \mathcal{S}$. Boundary conditions which ensure that a differential operator is self-adjoint are sometimes described as self-adjoint boundary conditions.

First, consider a *regular Sturm–Liouville system* for which $p[x]$ is strictly positive throughout the interval, having no interior zeros and being nonzero at both endpoints. Perhaps the simplest boundary conditions with this property are the *Dirichlet conditions* $f[a] == f[b] == 0$. Also simple are the *Neumann conditions* $f'[a] == f'[b] == 0$. More generally, one can show that any linear conditions of the form

$$\mathcal{B}_a f = \alpha_0 f[a] + \alpha_1 f'[a] == 0 \tag{7.70}$$
$$\mathcal{B}_b f = \beta_0 f[b] + \beta_1 f'[b] == 0 \tag{7.71}$$

where α_j and β_j are fixed real constants (independent of f), will suffice to ensure that $\mathcal{L}$ is self-adjoint. Such conditions are described as *unmixed* because each equation only involves values at one endpoint and as *homogeneous* because $\mathcal{B}_a f == \mathcal{B}_b f == 0$. A boundary condition that involves both a value and a derivative at the same point is described as *intermediate* (between Dirichlet and Neumann). When $p[a] = p[b]$, the *periodic* boundary conditions

$$f[a] == f[b], \quad f'[a] == f'[b] \tag{7.72}$$

will also produce a self-adjoint system. Periodic boundary conditions compare values at the two endpoints and may be described as *mixed*. Sometimes it is possible to employ more complicated mixed boundary conditions, but we will not.

Next, suppose that $p[x]$ vanishes at one or both of the endpoints. At first glance it might appear that one need not impose boundary conditions upon the elements of $\mathcal{S}$, but since

$\mathcal{L}$ usually arises in connection with a differential equation of the form

$$\mathcal{L}f[x] + \lambda w[x]f[x] == 0 \tag{7.73}$$

we must recognize that a zero of $p[x]$ is a singular point of the differential equation and that conditions, such as finiteness, will be needed to handle the singularities. If $p[x]$ were to display interior roots, we would have to have to apply Sturm–Liouville methods to the intervening regions separately and hope that some method could be devised to bridge the singular points. However, space does not permit a more general exploration of *irregular Sturm–Liouville problems*.

Similar methods can often be used to demonstrate that an operator is self-adjoint on an infinite domain, either $0 \leq x < \infty$ or $-\infty < x < \infty$. Under these conditions one usually needs $p \to 0$ as $x \to \pm\infty$ and the appropriate boundary conditions require the solution to remain finite in order to ensure that the integrated terms vanish and that the operator is self-adjoint. Similarly, one can extend many of the properties developed here to operators that satisfy

$$\int_a^b f^*[\vec{r}](\mathcal{D}g[\vec{r}])\,dV = \left(\int_a^b g^*[\vec{r}](\mathcal{D}f[\vec{r}])\,dV\right)^* \tag{7.74}$$

in two or more dimensions. The integrated terms encountered in the demonstration that the inner product is self-adjoint with respect to suitable boundary conditions are then described as *surface terms*. If the surface terms must vanish for any pair of functions (f, g) that satisfy the boundary conditions, then $\mathcal{D}$ is self-adjoint with respect to those boundary conditions.

The Sturm–Liouville operator is just one example, albeit an important one, of a self-adjoint operator. A *Sturm–Liouville system* is defined by its operator, $\mathcal{L}$, or equivalently its coefficient functions $p[x]$ and $q[x]$, its boundary conditions $\mathcal{B}_a$ and $\mathcal{B}_b$, and its weight function $w[x]$. Many of the important special functions in physics originate in Sturm–Liouville systems.

7.3.2 Reality of Eigenvalues and Orthogonality of Eigenfunctions

An eigenvalue problem is defined by

$$\mathcal{D}u_i[x] + \lambda_i w[x]u_i[x] == 0 \tag{7.75}$$

where $\mathcal{D}$ is a self-adjoint operator and $u_i \in \mathcal{S}$ satisfies self-adjoint boundary conditions at the endpoints of (a, b). Generally one can satisfy both boundary conditions simultaneously only for very particular values of λ_i, called characteristic values or *eigenvalues*. The corresponding solution u_i is then called a characteristic function or a normal mode or an *eigenfunction*. Consider two eigenfunctions satisfying

$$\mathcal{D}u_i[x] + \lambda_i w[x]u_i[x] == 0 \Longrightarrow u_j[x]^*\mathcal{D}u_i[x] == -\lambda_i w[x]u_j[x]^* u_i[x] \tag{7.76}$$

$$\mathcal{D}u_j[x] + \lambda_j w[x]u_j[x] == 0 \Longrightarrow u_i[x]^*\mathcal{D}u_j[x] == -\lambda_j w[x]u_i[x]^* u_j[x] \tag{7.77}$$

such that

$$\begin{aligned}\int_a^b u_j[x]^* \mathcal{D} u_i[x]\, dx - \left(\int_a^b u_i[x]^* \mathcal{D} u_j[x]\, dx\right)^* \\ = -\lambda_i \int_a^b u_j[x]^* u_i[x] w[x]\, dx + \left(\lambda_j \int_a^b u_i[x]^* u_j[x] w[x]\, dx\right)^* \\ = (\lambda_j^* - \lambda_i) \int_a^b u_j[x]^* u_i[x] w[x]\, dx\end{aligned} \tag{7.78}$$

where the left-hand side vanishes for any self-adjoint $\mathcal{D}$ and where we use the fact that the weight w is real. Thus, recognizing that the normalization integral

$$\langle \phi_i \mid w \mid \phi_i \rangle = \int_a^b u_i[x]^* u_i[x] w[x]\, dx \geq 0 \tag{7.79}$$

is positive-definite for any nontrivial (i.e., nonzero) eigenfunction, we find that the eigenvalues must be real, such that

$$i = j \Longrightarrow \lambda_i = \lambda_i^* \Longrightarrow \lambda_i \in \mathbb{R} \tag{7.80}$$

Furthermore, if $\lambda_i \neq \lambda_j$ the integral on the right-hand side must vanish, such that

$$\lambda_i \neq \lambda_j \Longrightarrow \langle u_j \mid w \mid u_i \rangle = 0 = \int_a^b u_j[x]^* u_i[x] w[x]\, dx \tag{7.81}$$

Therefore, eigenfunctions with different eigenvalues are *orthogonal with respect to the weight function*. On the other hand, if there are m linearly independent eigenfunctions with the same eigenvalue, which is described as *m-fold degenerate*, it is possible to form m independent linear combinations of the form

$$\phi_i[x] = \sum_{j=1}^{m} a_{i,j} u_j[x] \tag{7.82}$$

that are mutually orthogonal (with respect to w), such that

$$i \neq j \Longrightarrow \langle \phi_i \mid w \mid \phi_j \rangle = 0 \tag{7.83}$$

using the Gram–Schmidt orthogonalization procedure. The eigenvalues can then be obtained from

$$\lambda_i = -\frac{\langle \phi_i \mid \mathcal{D} \mid \phi_i \rangle}{\langle \phi_i \mid w \mid \phi_i \rangle} \tag{7.84}$$

where it is not necessary to normalize the eigenfunctions. On the other hand, it is always possible and usually more convenient to construct an orthonormal basis satisfying

$$\mathcal{D}\varphi_i[x] + \lambda_i w[x] \varphi_i[x] == 0, \quad \langle \varphi_i \mid w \mid \varphi_j \rangle = \delta_{i,j} \tag{7.85}$$

Note that it is sometimes convenient to distinguish between the orthonormal set $\{\varphi_i\}$ and more primitive sets $\{\phi_i\}$ before normalization and $\{u_i\}$ before orthogonalization within degenerate subspaces.

7.3.3 Discreteness of Eigenvalues

One can also show that the set of eigenvalues $\{\lambda_i\}$ is infinite, bounded from below, and has at most one accumulation point at ∞. Thus, one can arrange the eigenvalues in an increasing sequence $\lambda_0 < \lambda_1 < \lambda_2 \cdots$ where we omit duplicated values. Often some of the eigenvalues will be duplicated several times; a value that is repeated m times is described as *m-fold degenerate* and then corresponds to m linearly independent eigenfunctions. We normally assume that the Gram–Schmidt orthogonalization procedure has been used to construct m linearly independent, mutually orthogonal eigenvectors for each degenerate eigenvalue. Degeneracies often result from a symmetry, such as reflection or rotational symmetry.

7.3.4 Completeness of Eigenfunctions

Suppose that $\psi[x]$ is an arbitrary piecewise smooth function in (a, b) with a finite number of discontinuities and that it satisfies the boundary conditions imposed upon a self-adjoint operator $\mathcal{D}$. Let $\{\varphi_n\}$ represent the orthonormal eigenfunctions of $\mathcal{L}$, such that

$$\mathcal{D}\varphi_n + \lambda_n w \varphi_n == 0 \tag{7.86}$$

$$\langle \varphi_n \mid w \mid \varphi_m \rangle = \int_a^b \varphi_n[x]^* \varphi_m[x] w[x] \, dx = \delta_{n,m} \tag{7.87}$$

where $w[x]$ is the weight function for this system. We can then expand

$$\psi[x] = \sum_n a_n \varphi_n[x] \tag{7.88}$$

$$a_n = \langle \varphi_n \mid w \mid \psi \rangle = \int_a^b \varphi_n[x]^* \psi[x] w[x] \, dx \tag{7.89}$$

to any desired degree of accuracy. More rigorously, we define the mean-square error in a truncated expansion with N terms as

$$\Delta_N \psi = \int_a^b \left| \psi[x] - \sum_{n=1}^{N} a_n \varphi_n[x] \right|^2 w[x] \, dx \tag{7.90}$$

One can show that the mean-square error approaches zero as the number of terms increases, such that

$$\lim_{N \to \infty} \Delta_N \psi = 0 \tag{7.91}$$

This condition is described as *convergence in the mean* and is less restrictive than uniform convergence, but it suffices for practically any application in physics. Indeed, we cannot expect uniform convergence if ψ has a discontinuity; the error in the eigenfunction expansion at a discontinuity is sometimes called the *Gibb's phenomenon*. Note that uniform convergence can be established for analytic functions ψ by comparison with a Laurent series and expansion of φ_n in power series.

Next consider the eigenfunction expansion of a delta function

$$\delta[x-\xi] = \sum_n a_n \varphi_n[x] \tag{7.92}$$

$$a_n = \int_a^b \varphi_n[x]^* \delta[x-\xi] w[x]\, dx = w[\xi]\varphi_n[\xi]^* \tag{7.93}$$

whereby

$$\delta[x-\xi] = \sum_n \varphi_n[x] w[\xi] \varphi_n[\xi]^* \tag{7.94}$$

Given that the weight function $w[x]$ is positive-definite within the relevant interval, this is sometimes written in the more symmetric form

$$\delta[x-\xi] = \sum_n \varphi_n[x] \sqrt{w[x]w[\xi]}\varphi_n[\xi]^* \tag{7.95}$$

This somewhat peculiar expansion represents the *completeness* of the eigenfunction expansion. If we take a sufficiently large number of terms, the oscillations of $\varphi_n[x]$ and $\varphi_n[\xi]$ delicately conspire, with the aid of the density function $w[x]$, to assemble a good approximation to a delta function that vanishes everywhere except at the single point $x = \xi$ but still has unit area. Thus, any function

$$f[x] = \int_a^b \delta[x-\xi] f[\xi]\, d\xi \tag{7.96}$$

represented as a convolution over a delta function can be expanded as

$$f[x] = \int_a^b \sum_n \varphi_n[x] w[\xi] \varphi_n[\xi]^* f[\xi]\, d\xi = \sum_n a_n \varphi_n[x] \tag{7.97}$$

where

$$a_n = \langle \varphi_n \mid w \mid f \rangle = \int_a^b \varphi_n[\xi]^* f[\xi] w[\xi]\, d\xi \tag{7.98}$$

Notice that these properties apply to any self-adjoint operator – nowhere did we use the specific Sturm–Liouville form. Therefore, any self-adjoint operator $\mathcal{D}$ has real eigenvalues and a complete orthonormal set of eigenfunctions.

7.3.4.1 Example: Fourier Series

The Fourier series is represented by

$$\mathcal{L} = \frac{d^2}{dx^2}, \quad w[x] = 1, \quad \varphi[0] == \varphi[L] \tag{7.99}$$

such that

$$(\mathcal{L} + \lambda_n w)\varphi_n == 0 \Longrightarrow \varphi_n[x] = L^{-1/2}\,\mathrm{Exp}[ik_n x], \quad \lambda_n = k_n^2, \quad k_n = \frac{2n\pi}{L} \tag{7.100}$$

where n is an integer. The orthonormality and completeness relations take the form

$$\langle\varphi_n \mid \varphi_m\rangle = \frac{1}{L}\int_0^L \mathrm{Exp}[i(k_m - k_n)x]\,dx = \delta_{n,m} \tag{7.101}$$

$$\delta[x-\xi] = \frac{1}{L}\sum_{n=-\infty}^{\infty} \mathrm{Exp}[ik_n(x-\xi)] \tag{7.102}$$

Thus, an arbitrary piecewise smooth function $f[x]$ satisfying periodic boundary conditions, $f[0] == f[L]$, can be represented by the discrete Fourier series

$$f[x] = \sqrt{\frac{1}{L}}\sum_n a_n\,\mathrm{Exp}[ik_n x] \tag{7.103}$$

$$a_n = \sqrt{\frac{1}{L}}\int_a^b \mathrm{Exp}[-ik_n x]f[x]\,dx \tag{7.104}$$

7.3.5 Parseval's Theorem

Suppose that

$$\psi[x] = \sum_n a_n\varphi_n[x] \tag{7.105}$$

$$a_n = \langle\varphi_n \mid w \mid \psi\rangle = \int_a^b \varphi_n[x]^*\psi[x]w[x]\,dx \tag{7.106}$$

is expanded in a complete orthonormal set of eigenfunctions. The normalization of ψ is given by

$$\langle\psi \mid w \mid \psi\rangle = \int_a^b \psi[x]^*\psi[x]w[x]\,dx = \int_a^b \sum_{m,n} a_m^* a_n \varphi_m[x]^*\varphi_n[x]w[x]\,dx \tag{7.107}$$

such that

$$\langle\psi \mid w \mid \psi\rangle = \sum_n |a_n|^2 = \sum_n \langle\psi \mid w \mid \varphi_n\rangle\langle\varphi_n \mid w \mid \psi\rangle \tag{7.108}$$

represents a generalization of Parseval's theorem for Fourier series to general Hermitian systems.

7.3.6 Reality of Eigenfunctions

Although the coefficient functions for the Sturm–Liouville operator are real, one should not assume that the eigenfunctions are necessarily real. For example, the complex functions

$$\varphi_n[x] = L^{-1/2}\,\mathrm{Exp}[i k_n x], \quad k_n = \frac{2n\pi}{L} \tag{7.109}$$

are eigenfunctions of

$$\left(\frac{d^2}{dx^2} + k_n^2\right)\varphi_n[x] == 0, \quad \varphi[0] == \varphi[L] \tag{7.110}$$

even though the eigenvalue equation is real. On the other hand, the conjugate functions $\varphi_n[x]^*$ are also solutions to the same equation

$$\left(\frac{d^2}{dx^2} + k_n^2\right)\varphi_n[x]^* == 0, \quad \varphi[0]^* == \varphi[L]^* \tag{7.111}$$

with the same eigenvalues. The linearity of the equation permits one to form the real combinations

$$\begin{aligned} u_n[x] &= \frac{\varphi_n[x] + \varphi_n[x]^*}{2} = L^{-1/2}\,\mathrm{Cos}[k_n x], \\ v_n[x] &= \frac{\varphi_n[x] - \varphi_n[x]^*}{2i} = L^{-1/2}\,\mathrm{Sin}[k_n x] \end{aligned} \tag{7.112}$$

that also satisfy the same equations

$$\left(\frac{d^2}{dx^2} + k_n^2\right)u_n[x] == 0, \quad u_n[0] == u_n[L] \tag{7.113}$$

$$\left(\frac{d^2}{dx^2} + k_n^2\right)v_n[x] == 0, \quad v_n[0] == v_n[L] \tag{7.114}$$

with the same eigenvalues. Therefore, we may construct real orthogonal eigenfunctions for any self-adjoint differential operator with real coefficients. This property is often helpful in analyzing Sturm–Liouville systems.

7.3.7 Interleaving of Zeros

Consider two real eigenfunctions satisfying

$$\begin{aligned} &\mathcal{L}\psi_1[x] + \lambda_1 w[x]\psi_1[x] == 0 \implies \\ &\qquad \psi_2[x]\left(\frac{d}{dx}\left(p[x]\psi_1'[x]\right) - q[x]\psi_1[x]\right) == -\lambda_1 w[x]\psi_2[x]\psi_1[x] \end{aligned} \tag{7.115}$$

$$\begin{aligned} &\mathcal{L}\psi_2[x] + \lambda_2 w[x]\psi_2[x] == 0 \implies \\ &\qquad \psi_1[x]\left(\frac{d}{dx}\left(p[x]\psi_2'[x]\right) - q[x]\psi_2[x]\right) == -\lambda_2 w[x]\psi_1[x]\psi_2[x] \end{aligned} \tag{7.116}$$

where $\lambda_1 \neq \lambda_2$. Subtract these equations and integrate from a to x, such that

$$\left(\psi_2[t]p[t]\frac{d\psi_1[t]}{dt} - \psi_1[t]p[t]\frac{d\psi_2[t]}{dt}\right)_a^x = (\lambda_2 - \lambda_1)\int_a^x \psi_2[t]\psi_1[t]w[t]\,dt \tag{7.117}$$

Self-adjoint boundary conditions ensure that the lower limit of integration does not contribute to the left-hand side, such that

$$p[x]\left(\psi_2[x]\frac{d\psi_1[x]}{dx} - \psi_1[x]\frac{d\psi_2[x]}{dx}\right) = (\lambda_2 - \lambda_1)\int_a^x \psi_2[t]\psi_1[t]w[t]\,dt \tag{7.118}$$

Suppose that we choose ξ to be a zero of ψ_1, such that

$$p[\xi]\psi_2[\xi]\psi_1'[\xi] = (\lambda_2 - \lambda_1)\int_a^\xi \psi_2[t]\psi_1[t]w[t]\,dt \tag{7.119}$$

Both p and w are positive in (a, b). Further suppose that ξ is the root of ψ_1 which is nearest to the lower endpoint a and that we choose the sign of ψ_1 to be positive for $a < x < \xi$, such that $\psi_1'[\xi] < 0$. The sign of the left-hand side is then opposite to the sign of $\psi_2[\xi]$. If $\lambda_2 > \lambda_1$ there must be a sign change in $\psi_2[x]$ for $a < x < \xi$ in order to obtain a negative value for the integral on the right-hand side. Therefore, larger eigenvalues produce stronger oscillations and a smaller spacing between roots of the corresponding eigenfunctions.

This argument can be extended by integrating in the subinterval (x_1, x_2) between two consecutive roots of ψ_1, such that

$$\left(\psi_2[t]p[t]\frac{d\psi_1[t]}{dt} - \psi_1[t]p[t]\frac{d\psi_2[t]}{dt}\right)_{x_1}^{x_2} = \left(\lambda_2 - \lambda_1\right)\int_{x_1}^{x_2} \psi_2[t]\psi_1[t]w[t]\,dt \tag{7.120}$$

Integrating the original eigenvalue equation over a subinterval gives

$$\mathcal{L}\psi[x] + \lambda w[x]\psi[x] == 0 \Longrightarrow \left(p[t]\frac{d\psi[t]}{dt}\right)_{x_1}^{x_2} = \int_{x_1}^{x_2} (\lambda w[t] - q[t])\psi[t]\,dt \tag{7.121}$$

It is useful to define

$$g_\lambda[x] = \lambda w[x] - q[x] \tag{7.122}$$

such that

$$\left(p[t]\frac{d\psi[t]}{dt}\right)_{x_1}^{x_2} = \int_{x_1}^{x_2} g_\lambda[t]\psi[t]\,dt \tag{7.123}$$

Suppose that x_1 and x_2 are two consecutive roots of ψ. If $g_\lambda[x] > 0$ in (x_1, x_2), the slope increases in the interval such that $\psi[x]$ has an approximately exponential behavior. On the other hand, if $g_\lambda[x] < 0$ in (x_1, x_2), the slope decreases, and $\psi[x]$ has an oscillatory behavior.

$$\lambda w[x] > q[x] \Longrightarrow \left\langle \frac{1}{\psi}\frac{d^2\psi}{dx^2} \right\rangle > 0 \tag{7.124}$$

$$\lambda w[x] < q[x] \Longrightarrow \left\langle \frac{1}{\psi}\frac{d^2\psi}{dx^2} \right\rangle < 0 \tag{7.125}$$

7.3.7.1 Example: Fourier Series

The differential operator

$$\mathcal{L} = \frac{d^2}{dx^2}, \quad f[0] == f[\pi] == 0 \tag{7.126}$$

is self-adjoint and the eigenvalue equation has solutions

$$(\mathcal{L} + \lambda_n) u_n[x] == 0 \Longrightarrow u_n = \mathrm{Sin}[nx], \quad \lambda_n = n^2 \tag{7.127}$$

which satisfy the orthonormality condition

$$\langle u_n \mid u_m \rangle = \int_0^\pi \mathrm{Sin}[nx]\,\mathrm{Sin}[mx]\,dx = \frac{\pi}{2}\delta_{n,m} \tag{7.128}$$

The roots are obviously interleaved.

7.3.8 Comparison Theorems

Suppose that $u[x]$ and $v[x]$ satisfy

$$\left(\frac{d}{dx}p[x]\frac{d}{dx} + g_1[x]\right)u[x] == 0 \tag{7.129}$$

$$\left(\frac{d}{dx}p[x]\frac{d}{dx} + g_2[x]\right)v[x] == 0 \tag{7.130}$$

where $p[x] > 0$ and $g_1[x] \le g_2[x]$ in the interval $a \le x \le b$. One can then show that there is at least one root of $v[x]$ between any two roots of $u[x]$ in this interval, such that $v[x]$ oscillates more rapidly than $u[x]$. To prove this result, known as *Sturm's first comparison theorem*, we argue by contradiction. First observe that

$$\frac{d}{dx}\left(p(u'v - uv')\right) = uv(g_2 - g_1) \tag{7.131}$$

Next, suppose that $x_{1,2}$ are two successive zeros of $u[x]$ such that $a \le x_1 < x_2 \le b$. Integration of the preceding equation between the roots of u gives

$$\left(p(u'v - uv')\right)_{x_1}^{x_2} = \int_{x_1}^{x_2} uv(g_2 - g_1)\,dx \tag{7.132}$$

or

$$p[x_2]u'[x_2]v[x_2] - p[x_1]u'[x_1]v[x_1] = \int_{x_1}^{x_2} uv(g_2 - g_1)\,dx \tag{7.133}$$

If v does not have a root within this interval, we may assume without loss of generality that it is positive throughout the interval. Similarly, we may also assume that u is positive between its roots because the differential equations are linear and we are free to multiply solutions by a constant that will achieve the desired sign. Thus, the right-hand side is nonnegative because $g_2 \geq g_1$ in this interval and vanishes if and only if $g_2 = g_1$ over the entire interval. However, the assumption that u is positive between roots at $x_{1,2}$ requires that $u'[x_2] < 0$ and $u'[x_1] > 0$, such that the left-hand side is negative. Therefore, the assumption that $v[x]$ does not change sign in this interval leads to a contradiction and we can conclude that $v[x]$ must have at least one root in the interval $x_1 < x < x_2$. More generally, *Picone's modification* states that if

$$\left(\frac{d}{dx}p_1[x]\frac{d}{dx} + g_1[x]\right)u[x] == 0 \tag{7.134}$$

$$\left(\frac{d}{dx}p_2[x]\frac{d}{dx} + g_2[x]\right)v[x] == 0 \tag{7.135}$$

where $0 \leq p_2 \leq p_1$ and $g_1 \leq g_2$ in $a \leq x \leq b$, then Sturm's first comparison theorem still applies.

Recognizing that the Sturm–Liouville eigenvalue equation can be expressed in the form

$$\left(\frac{d}{dx}p[x]\frac{d}{dx} + g[x]\right)y[x] == 0 \tag{7.136}$$

where $g[x] = \lambda w[x] - q[x]$ with $w[x] > 0$, we immediately see that increasing λ makes the solution $y[x]$ more oscillatory. Suppose that $y[x]$ satisfies

$$\left(\frac{d}{dx}p[x]\frac{d}{dx} + \lambda w[x] - q[x]\right)y[x] == 0, \quad y[a] == 0 \tag{7.137}$$

and that $u[x]$ satisfies

$$\left(\frac{d}{dx}p_{\max}\frac{d}{dx} + \lambda w_{\min} - q_{\max}\right)u[x] == 0, \quad u[a] == 0 \tag{7.138}$$

where the weight functions are replaced by their maximum or minimum values within the interval $a \leq x \leq b$, as indicated. The solutions for u are simply sine functions with period

$$T_{\max} = 2\pi\sqrt{\frac{p_{\max}}{\lambda w_{\min} - q_{\max}}} \tag{7.139}$$

and, according to Picone's modification, there is at least one root of y between successive roots of u. However, because the period $T_{\max}$ decreases as the eigenvalue λ increases, it is clear that the spacing between roots of y decreases as $\lambda \to \infty$. The maximum separation between zeros of y is $T_{\max}/2$. By applying a similar analysis to

$$\left(\frac{d}{dx}p_{\min}\frac{d}{dx} + \lambda w_{\max} - q_{\min}\right)v[x] == 0, \quad v[a] == 0 \tag{7.140}$$

we find that the minimum separation between zeros of y is given by $T_{\min}/2$ where

$$T_{\min} = 2\pi\sqrt{\frac{p_{\min}}{\lambda w_{\max} - q_{\min}}} \tag{7.141}$$

Therefore, if x_n is the n-th zero of y it must be found in the interval

$$n\pi \sqrt{\frac{p_{\min}}{\lambda w_{\max} - q_{\min}}} < x_n - a < n\pi \sqrt{\frac{p_{\max}}{\lambda w_{\min} - q_{\max}}} \tag{7.142}$$

The interval we found for x_n can be used to deduce bounds upon the corresponding eigenvalue. Suppose that the boundary conditions require $y[a] == y[b] == 0$ and let $x_n \to b$ and $\lambda \to \lambda_n > 0$. Simple manipulations then provide the bracketing conditions

$$\frac{1}{w_{\max}}\left(\left(\frac{n\pi}{b-a}\right)^2 p_{\min} + q_{\min}\right) < \lambda_n < \frac{1}{w_{\min}}\left(\left(\frac{n\pi}{b-a}\right)^2 p_{\max} + q_{\max}\right) \tag{7.143}$$

Although the numerical factors may be different, similar brackets can be deduced for more general boundary conditions and often provide useful bounds when exact eigenvalues are difficult to obtain or when very precise results are not needed. Such bounds can also provide limits and starting conditions for numerical methods of computing eigenvalues. Furthermore, for large n one finds

$$n \to \infty \Longrightarrow \frac{p_{\min}}{w_{\max}}\left(\frac{n\pi}{b-a}\right)^2 < \lambda_n < \frac{p_{\max}}{w_{\min}}\left(\frac{n\pi}{b-a}\right)^2 \tag{7.144}$$

and concludes that λ_n scales with n^2 for large n. This scaling is a general property of Sturm–Liouville systems that is independent of the specific boundary conditions. Notice that the spacing between eigenvalues cannot be made infinitesimal unless the interval is infinite. Thus, the spectrum of eigenvalues might be continuous for an infinite interval, but is discrete for a finite interval. Nevertheless, the eigenvalues increase without limit, eventually scaling with n^2. The lowest eigenvalue has the smallest number of nodes within $a \le x \le b$ that is consistent with the boundary conditions and increasing n increases the number of nodes. Therefore, one can index the eigenfunctions according to the number of nodes. When several eigenvalues are degenerate, a second index might be required to distinguish between orthogonal eigenfunctions with the same number of nodes.

7.4 Green Functions

We have already seen that Green functions provide a powerful method for solving linear inhomogeneous equations in which the output of a system can be expressed as a convolution of the input with its response to a point source. Many physical systems can be represented in terms of self-adjoint operators. Therefore, it will often be useful to develop representations of the Green function in terms of the eigenfunctions for self-adjoint systems. In this section we develop a couple of these methods.

7.4.1 Interface Matching

Suppose that

$$\mathcal{L} = \frac{d}{dx}\left(p[x]\frac{d}{dx}\right) - q[x] \tag{7.145}$$

with suitable boundary conditions is a Sturm–Liouville operator that is self-adjoint with respect to the density function $w[x]$ on the interval $a \leq x \leq b$. We seek a Green function $G_\lambda[x, \xi]$ that satisfies

$$(\mathcal{L} + \lambda w[x])G_\lambda[x, \xi] == \delta[x - \xi] \Longleftrightarrow$$
$$\left(\frac{d}{dx}\left(p[x]\frac{d}{dx}\right) - q[x] + \lambda w[x]\right)G_\lambda[x, \xi] == \delta[x - \xi] \tag{7.146}$$

with the same boundary conditions. The subscript acknowledges the parametric dependence of the Green function upon λ. We expect $G_\lambda[x, \xi]$ to be continuous in value, but the delta function produces a discontinuity in slope. The magnitude of this discontinuity is determined by integrating the differential equation across the interface

$$\int_a^b \left(\frac{d}{dx}\left(p[x]\frac{d}{dx}\right) - q[x] + \lambda w[x]\right)G_\lambda[x, \xi]\, dx == 1 \tag{7.147}$$

such that

$$\lim_{\varepsilon \to 0} p[\xi](G'_\lambda[\xi + \varepsilon, \xi] - G'_\lambda[\xi - \varepsilon, \xi]) == 1 \tag{7.148}$$

where p, q, and w are continuous. These conditions can be expressed more compactly as

$$\Delta G_\lambda[x, \xi] == 0, \quad \Delta G'_\lambda[x, \xi] == \frac{1}{p[\xi]} \tag{7.149}$$

where Δ represents the change across the interface.

Further, suppose that we possess two independent solutions to the simpler homogeneous problems

$$(\mathcal{L} + \lambda w[x])y_a[x] == 0, \quad \mathcal{B}_a y_a[x] == 0 \tag{7.150}$$
$$(\mathcal{L} + \lambda w[x])y_b[x] == 0, \quad \mathcal{B}_b y_b[x] == 0 \tag{7.151}$$

with a common parameter λ such that the *Wronskian*

$$W_\lambda[x] = y_a[x]y'_b[x] - y_b[x]y'_a[x] \tag{7.152}$$

is nonzero throughout the region $a \leq x \leq b$. Notice that these solutions do not need to satisfy both boundary conditions simultaneously; y_a satisfies the lower and y_b the upper boundary condition. The equation for the Green function is also homogeneous in the two separate regions $x < \xi$ and $x > \xi$. Thus, we seek a Green function with piecewise-continuous representation

$$x < \xi \Longrightarrow G_\lambda[x, \xi] = A[\xi]y_a[x] \tag{7.153}$$
$$x > \xi \Longrightarrow G_\lambda[x, \xi] = B[\xi]y_b[x] \tag{7.154}$$

that automatically satisfies both boundary conditions. The coefficients are determined by applying the matching conditions

$$\Delta G_\lambda[x, \xi] == 0 \Longrightarrow B[\xi]y_b[\xi] - A[\xi]y_a[\xi] == 0 \tag{7.155}$$

$$\Delta G'_\lambda[x, \xi] == \frac{1}{p[\xi]} \Longrightarrow B[\xi]y'_b[\xi] - A[\xi]y'_a[\xi] == \frac{1}{p[\xi]} \tag{7.156}$$

at the interface, such that

$$A[\xi] = \frac{y_b[\xi]}{p[\xi]W_\lambda[\xi]}, \quad B[\xi] = \frac{y_a[\xi]}{p[\xi]W_\lambda[\xi]} \tag{7.157}$$

Therefore, we finally obtain the Green function in the form

$$G_\lambda[x, \xi] = \frac{y_a[x_<]y_b[x_>]}{p[\xi]W_\lambda[\xi]} \tag{7.158}$$

where $x_<$ is the smaller and $x_>$ is the larger of x and ξ. One can show that the product $p[\xi]W_\lambda[\xi]$ is actually independent of ξ for Sturm–Liouville systems; that exercise is left to the reader. It is then clear that a Sturm–Liouville Green function satisfies the *reciprocity condition*

$$G_\lambda[x, \xi] = G_\lambda[\xi, x] \tag{7.159}$$

showing that the effect of a point source at ξ on the response at x is the same as the effect of a point source at x on the response at ξ.

The solution to a more general inhomogeneous problem of the form

$$(\mathcal{L} + \lambda w[x])\psi[x] == w[x]f[x] \tag{7.160}$$

with the same boundary conditions can now be expressed in the convolution form

$$\psi[x] = \int_a^b G_\lambda[x, \xi]f[\xi]w[\xi]\,d\xi \tag{7.161}$$

where it is not necessary to include a solution to the homogeneous equation because the Green function incorporates the boundary conditions automatically in its very construction. If we divide the convolution into lower and upper regions, the convolution integral takes the form

$$\psi[x] = (p[x]W_\lambda[x])^{-1}\left(y_b[x]\int_a^x y_a[\xi]f[\xi]w[\xi]\,d\xi + y_a[x]\int_x^b y_b[\xi]f[\xi]w[\xi]\,d\xi\right) \tag{7.162}$$

where pW_λ can be extracted because it is constant.

This method fails if $W_\lambda[\xi] = 0$. A Wronskian vanishes when two solutions are not linearly independent. Suppose that y_1 and y_2 satisfy the boundary conditions at both endpoints simultaneously. Under those circumstances, $\lambda = \lambda_n$ actually represents one the eigenvalues of the Sturm–Liouville problem represented by

$$(\mathcal{L} + \lambda_n w[x])\varphi_n[x] == 0\,, \quad \mathcal{B}\varphi_n[x] == 0 \tag{7.163}$$

and the Green function G_{λ_n} does not exist. We will return to this problem again later. Similarly, the method also fails if $p[x] = 0$ within the interval. The roots of $p[x]$ are singular points for the underlying differential equation. Hence, Sturm–Liouville methods are generally limited to the intervals between singular points and it can be difficult to connect adjacent intervals. Although special care may also be needed if $p[x]$ vanishes at one or both of the endpoints, we will assume that $p[x]$ is nonzero in the interior.

7.4.1.1 Example: Vibrating String

The Green function for a vibrating string clamped at both ends satisfies

$$\left(\frac{d^2}{dx^2} + k^2\right)G_k[x, \xi] == \delta[x - \xi]\,, \quad G[0, \xi] == G[l, \xi] == 0 \tag{7.164}$$

Rather than apply the general formula, let us apply the method of interface matching directly. On the two sides of the interface we write

$$x < \xi \Longrightarrow G_k[x, \xi] = a\,\mathrm{Sin}[kx] \tag{7.165}$$

$$x > \xi \Longrightarrow G_k[x, \xi] = b\,\mathrm{Sin}[k(x - l)] \tag{7.166}$$

and identify the discontinuity in first derivative as

$$\lim_{\varepsilon \to 0}(G'_k[\xi + \varepsilon, \xi] - G'_k[\xi - \varepsilon, \xi]) == 1 \tag{7.167}$$

to form the equations

$$0 == b\,\mathrm{Sin}[k(\xi - l)] - a\,\mathrm{Sin}[k\xi] \tag{7.168}$$

$$1 == k(b\,\mathrm{Cos}[k(\xi - l)] - a\,\mathrm{Cos}[k\xi]) \tag{7.169}$$

Thus, we obtain

$$x < \xi \Longrightarrow G_k[x, \xi] = \frac{\mathrm{Sin}[kx]\,\mathrm{Sin}[k(\xi - l)]}{k\,\mathrm{Sin}[kl]} \tag{7.170}$$

$$x > \xi \Longrightarrow G_k[x, \xi] = \frac{\mathrm{Sin}[k\xi]\,\mathrm{Sin}[k(x - l)]}{k\,\mathrm{Sin}[kl]} \tag{7.171}$$

These expressions can be combined into the more compact form

$$G_k[x, \xi] = \frac{\mathrm{Sin}[kx_<]\,\mathrm{Sin}[k(x_> - l)]}{k\,\mathrm{Sin}[kl]} \tag{7.172}$$

where $x_<$ is the smaller and $x_>$ is the larger of x and ξ. The solution to the more general inhomogeneous problem

$$\left(\frac{d^2}{dx^2} + k^2\right)\psi[x] == f[x], \quad \psi[0] == \psi[l] == 0 \tag{7.173}$$

now takes the form

$$\psi[x] = \frac{1}{k\,\mathrm{Sin}[kl]}\left(\mathrm{Sin}[k(x-l)]\int_0^x \mathrm{Sin}[k\xi]f[\xi]\,d\xi + \mathrm{Sin}[kx]\int_x^l \mathrm{Sin}[k(\xi-l)]f[\xi]\,d\xi\right) \tag{7.174}$$

Notice that the Green function does not exist when

$$k \to \frac{n\pi}{l} \Longrightarrow \mathrm{Sin}[kl] = 0 \tag{7.175}$$

for integer n. Thus, the Green function does not exist when $k = k_n$ is one of the eigenvalues of the homogeneous problem.

The present solution does, of course, agree with the general formula. Identifying

$$p = 1, \quad q = 0, \quad w = 0, \quad \lambda = k^2 \tag{7.176}$$

and

$$\begin{aligned} y_a &= \mathrm{Sin}[kx] \\ y_b &= \mathrm{Sin}[k(x-l)] \Longrightarrow W_k[x] &= k(\mathrm{Sin}[kx]\,\mathrm{Cos}[k(x-l)] - \mathrm{Sin}[k(x-l)]\,\mathrm{Cos}[kx]) \\ & &= k\,\mathrm{Sin}[kl] \end{aligned} \tag{7.177}$$

we would have written

$$G_k[x,\xi] = \frac{y_a[x_<]y_b[x_>]}{p[\xi]W_\lambda[\xi]} = \frac{\mathrm{Sin}[kx_<]\,\mathrm{Sin}[k(x_> - l)]}{k\,\mathrm{Sin}[kl]} \tag{7.178}$$

without being the wiser for it. Once the method is familiar, one can usually apply it rather quickly and in so doing may notice important special features of a particular problem – it is usually better to apply the method to each problem rather than simply plug into the formula.

It is instructive to examine the evolution of G_k as k increases. Figures 7.1–7.4 display $G_k[x,\xi]$ for several values of k midway between adjacent eigenvalues. The cusp along the $x == \xi$ diagonal is responsible for the characteristic $\mathcal{L}G == \delta[x-\xi]$ behavior that defines a Green's function and is clearly evident for small k. In fact, in that limit we obtain

$$k \to 0 \Longrightarrow G_k[x,\xi] \to \frac{x_<(x_> - l)}{l} \tag{7.179}$$

However, as k increases, oscillations off the main diagonal become important also.

7.4.2 Eigenfunction Expansion of Green Function

Suppose that

$$\mathcal{L} = \frac{d}{dx}\left(p[x]\frac{d}{dx}\right) - q[x] \tag{7.180}$$

with suitable boundary conditions is self-adjoint with respect to the weight function $w[x]$ on the interval $a \le x \le b$ and let ξ represent a source point within that interval. The

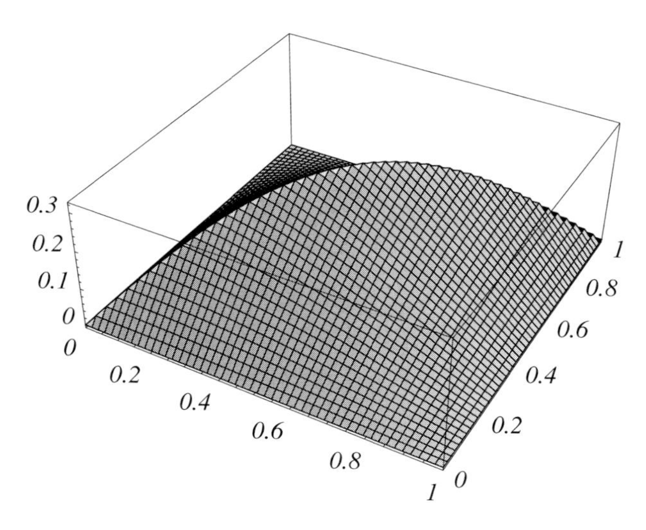

Figure 7.1. $G_k[x, \xi]$ for the vibrating string with $k = \pi/2$ and $l = 1$ (arbitrary units).

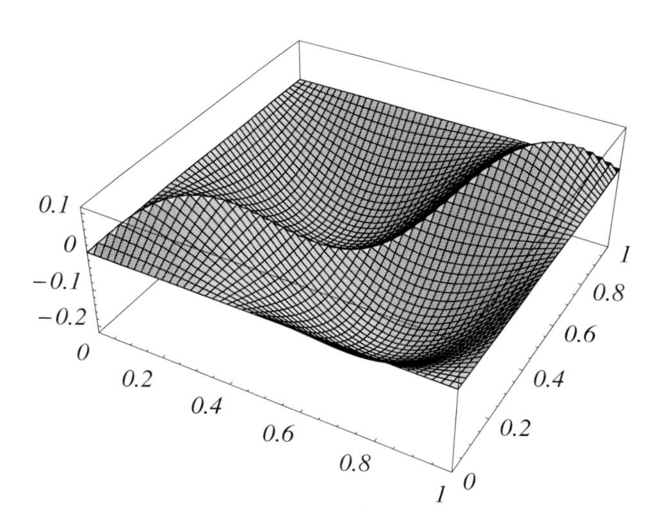

Figure 7.2. $G_k[x, \xi]$ for the vibrating string with $k = 3\pi/2$ and $l = 1$ (arbitrary units).

response of a Sturm–Liouville system to a point source is described by a Green function $G_\lambda[x, \xi]$ that satisfies the differential equation

$$(\mathcal{L} + \lambda w[x])G_\lambda[x, \xi] == \delta[x - \xi] \Longleftrightarrow \tag{7.181}$$

$$\frac{d}{dx}(p[x]G'_\lambda[x, \xi]) + (\lambda w[x] - q[x])G_\lambda[x, \xi] == \delta[x - \xi] \tag{7.182}$$

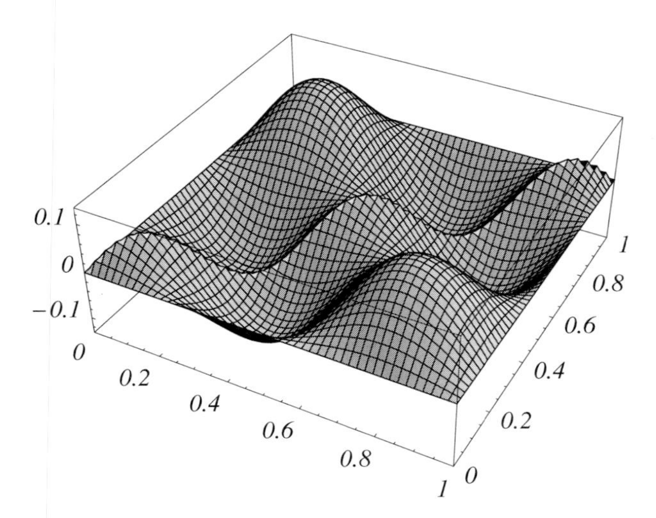

Figure 7.3. $G_k[x, \xi]$ for the vibrating string with $k = 5\pi/2$ and $l = 1$ (arbitrary units).

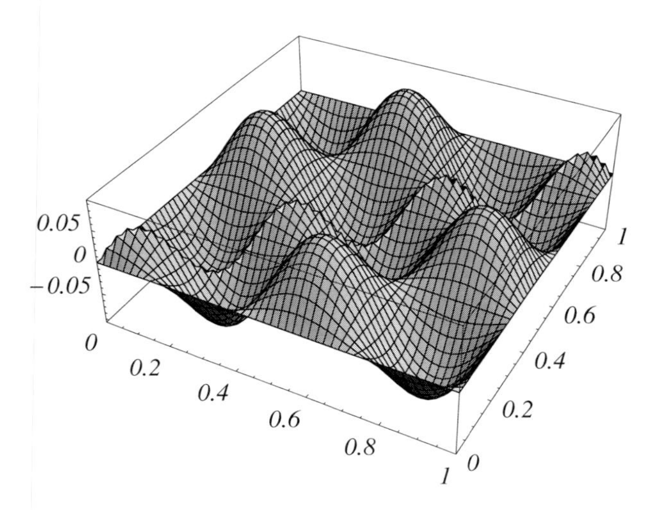

Figure 7.4. $G_k[x, \xi]$ for the vibrating string with $k = 7\pi/2$ and $l = 1$ (arbitrary units).

with the same self-adjoint boundary conditions. Here λ is treated as a fixed parameter. Let $\{\varphi_n[x]\}$ represent a complete orthonormal set of eigenfunctions that satisfy

$$(\mathcal{L} + \lambda_n w[x])\varphi_n[x] == 0 \tag{7.183}$$

$$\langle \varphi_n \mid w \mid \varphi_m \rangle = \delta_{n,m} \Longleftrightarrow \int_a^b \varphi_n[x]^* \varphi_m[x] w[x] \, dx = \delta_{n,m} \tag{7.184}$$

subject to appropriate boundary conditions. If we substitute an eigenfunction expansion of the Green function

$$G_\lambda[x, \xi] = \sum_n c_n \varphi_n[x] \tag{7.185}$$

into its defining differential equation, we obtain

$$\sum_n c_n (\mathcal{L} + \lambda w[x]) \varphi_n[x] == \delta[x - \xi] \tag{7.186}$$

This can be simplified considerably by using the Sturm–Liouville equation for $\varphi_n[x]$ to write

$$\sum_n c_n (\lambda - \lambda_n) w[x] \varphi_n[x] == \delta[x - \xi] \tag{7.187}$$

Provided that λ does not coincide with any of the eigenvalues $\{\lambda_n\}$, we can isolate the coefficient c_m by multiplying both sides by $\varphi_m[x]^*$ and integrating over the interval (a, b) to obtain

$$\int_a^b \sum_n c_n (\lambda - \lambda_n) \varphi_m[x]^* \varphi_n[x] w[x] \, dx == \varphi_m[\xi]^* \tag{7.188}$$

such that

$$\lambda \neq \lambda_m \Longrightarrow c_m = \frac{\varphi_m[\xi]^*}{\lambda - \lambda_m} \tag{7.189}$$

Therefore, the Green function takes the form

$$\lambda \neq \lambda_n \Longrightarrow G_\lambda[x, \xi] = \sum_n \frac{\varphi_n[x] \varphi_n[\xi]^*}{\lambda - \lambda_n} \tag{7.190}$$

where eigenfunctions φ_n with λ_n near λ tend to contribute most strongly. As a sanity check, we apply the operator $\mathcal{L} + \lambda w$ to both sides

$$\begin{aligned}
(\mathcal{L} + \lambda w[x]) G_\lambda[x, \xi] &= \sum_n \frac{\varphi_n[\xi]^*}{\lambda - \lambda_n} (\mathcal{L} + \lambda w[x]) \varphi_n[x] \\
&= \sum_n \frac{\varphi_n[\xi]^*}{\lambda - \lambda_n} (\lambda - \lambda_n) w[x] \varphi_n[x] \\
&= \sum_n w[x] \varphi_n[x] \varphi_n[\xi]^* \\
&= \delta[x - \xi]
\end{aligned} \tag{7.191}$$

and recover the completeness relation expressed as an eigenfunction expansion of the delta function.

Notice that the reciprocity condition, $G_\lambda[x, \xi] = G_\lambda[\xi, x]$, is apparent in this representation also. It is sometimes convenient to include a symmetric function of (x, ξ) related to the weight function on the right-hand side of the equation for the Green function. For example, in spherical coordinates one might employ $\delta[r-\xi]/r\xi$ on the right-hand side. Other times it is useful to include normalization factors, such as -4π for electrostatics. As always, it is better to learn the method and to apply it to each particular problem rather than to employ the general formulas derived here.

Now consider the more general situation in which $\psi[x]$ satisfies an inhomogeneous equation of the form

$$(\mathcal{L}+\lambda w[x])\psi[x] == w[x]f[x] \tag{7.192}$$

where the boundary conditions for $\psi[x]$ make $\mathcal{L}$ self-adjoint with respect to $w[x]$ on the interval (a,b). It is useful to expand both f and ψ in terms of the eigenfunctions $\{\varphi_n\}$ of $\mathcal{L}$ according to

$$f[x]=\sum_n f_n\varphi_n[x] \Longrightarrow f_n=\langle\varphi_n \mid w \mid f\rangle=\int_a^b \varphi_n[x]^* f[x]w[x]\,dx \tag{7.193}$$

$$\psi[x]=\sum_n \psi_n\varphi_n[x] \Longrightarrow \psi_n=\langle\varphi_n \mid w \mid \psi\rangle=\int_a^b \varphi_n[x]^* \psi[x]w[x]\,dx \tag{7.194}$$

such that

$$\sum_n \psi_n(\mathcal{L}+\lambda w[x])\varphi_n[x] == w[x]\sum_n f_n\varphi_n[x] \tag{7.195}$$

Use of the eigenfunction equation then gives

$$\mathcal{L}\varphi_n[x]=-\lambda_n w[x]\varphi_n[x] \Longrightarrow \sum_n \psi_n(\lambda-\lambda_n)w[x]\varphi_n[x] == w[x]\sum_n f_n\varphi_n[x] \tag{7.196}$$

and we project the expansion coefficient ψ_n by multiplying both sides by $\varphi_m[x]^*$ and integrating over (a,b) to obtain

$$\lambda \neq \lambda_n \Longrightarrow \psi_n=\frac{1}{\lambda-\lambda_n}\int_a^b \varphi_n[x]^* f[x]w[x]\,dx=\frac{\langle\varphi_n \mid w \mid f\rangle}{\lambda-\lambda_n}=\frac{f_n}{\lambda-\lambda_n} \tag{7.197}$$

where the coefficient of φ_n should be recognized as the overlap (inner product) of the driving term $f[x]$ with the normal mode φ_n. This result can be represented in the form of a convolution integral

$$\lambda \neq \lambda_n \Longrightarrow \psi[x]=\int_a^b G_\lambda[x,\xi]f[\xi]w[\xi]\,d\xi=\sum_n \frac{\varphi_n[x]}{\lambda-\lambda_n}\int_a^b \varphi_n[\xi]^* f[\xi]w[\xi]\,d\xi \tag{7.198}$$

where $G_\lambda[x,\xi]$ describes the response at x produced by a source at ξ. Notice that it is not necessary to add a solution to the homogeneous equation because the Green function already incorporates the boundary conditions. To verify this solution, we simply apply the operator $(\mathcal{L} + \lambda w)$ to both sides

$$\begin{aligned}(\mathcal{L} + \lambda w[x])\psi[x] &= \int_a^b (\mathcal{L} + \lambda w[x])G_\lambda[x,\xi]f[\xi]w[\xi]\,d\xi \\ &= \int_a^b \delta[x-\xi]f[\xi]w[\xi]\,d\xi = f[x]w[x]\end{aligned} \tag{7.199}$$

and recover the original equation. Therefore, the Green function $G_\lambda[x,\xi]$ offers a simple and insightful solution to practically any inhomogeneous equation of this kind provided that λ is not one of the eigenvalues of $\mathcal{L}$.

More care is needed when λ does match one of the eigenvalues because the Green function does not exist under those circumstances. If λ is close to one of the eigenvalues, such that $\lambda \approx \lambda_m$, the eigenfunction expansion can be approximated by a single dominant term

$$\lambda \approx \lambda_m \Longrightarrow \psi[x] \approx \frac{f_m}{\lambda - \lambda_m}\varphi_m[x] + B[x] \tag{7.200}$$

that diverges as $\lambda \to \lambda_m$ and a smooth background contribution. In fact, for fixed x it is useful to think of ψ as a function of λ. The eigenvalues of $\mathcal{L}$ then correspond to the poles of ψ with residues $f_m\varphi_m[x]$. Often these poles represent resonances or normal modes of the dynamical system and the coefficients f_n represent the coupling of the driving term to those resonances; in other words, f_n measures the strength with which the driving term can excite mode n. The nonresonant background

$$\lambda \approx \lambda_m \Longrightarrow B[x] = \sum_{n \neq m} \frac{f_n}{\lambda_m - \lambda_n}\varphi_n[x] \tag{7.201}$$

represents the contributions from all other more distant resonances and is usually much weaker. The response of the system is very strong near its resonances unless the driving term is orthogonal to the normal mode, such that $f_m = 0$. Thus, if $\lambda = \lambda_m$ and $f_m = 0$, we can write a formal solution

$$\lambda = \lambda_m,\ \langle \varphi_m \mid f\rangle = 0 \Longrightarrow \psi[x] = c_m\varphi_m[x] + \sum_{n \neq m} \frac{f_n}{\lambda_m - \lambda_n}\varphi_n[x] \tag{7.202}$$

where c_m is an arbitrary constant. However, solutions of this type are inherently unstable and probably not very useful because we cannot determine c_m or keep it constant – any perturbation of the physical system or any numerical error in the mathematical computation, no matter how small, would permit an uncontrollably large contamination by $\varphi_m[x]$. A small perturbation δf yields a small but finite overlap ϵ

$$f \longrightarrow f + \delta f, \quad \lambda \approx \lambda_m \Longrightarrow \psi \longrightarrow \psi + \frac{\epsilon_m}{\lambda - \lambda_m}\varphi_m[x] \tag{7.203}$$

where

$$\epsilon_m = \int_a^b \varphi_m[x]^* \delta f[x] w[x]\,dx \tag{7.204}$$

The values of ϵ_m or λ_m often exhibit small but random fluctuations in time, leading to large unstable fluctuations of the output when λ is nearly resonant. Alternatively, numerical methods may amplify round-off errors when λ is near an eigenvalue. Therefore, although the eigenfunction expansion of the Green function provides a valuable formal solution, considerable care must be exercised near any of the eigenvalues.

7.4.3 Example: Vibrating String

The Green function for a vibrating string clamped at both ends satisfies

$$\left(\frac{d^2}{dx^2} + \lambda\right)G[x, \xi] == \delta[x - \xi], \quad G[0, \xi] == G[l, \xi] == 0 \tag{7.205}$$

The eigenfunctions for the corresponding homogeneous equation

$$\left(\frac{d^2}{dx^2} + \lambda_n\right)\varphi_n[x] == 0, \quad \varphi_n[0] == \varphi_n[l] == 0 \tag{7.206}$$

are

$$\varphi_n[x] = \sqrt{\frac{2}{l}}\,\mathrm{Sin}[k_n x], \quad k_n = \frac{n\pi}{l} \tag{7.207}$$

with eigenvalues $\lambda_n = k_n^2$. Thus, the eigenfunction expansion of the Green function takes the form

$$G_k[x, \xi] = \sum_{n=0}^{\infty} \frac{\varphi_n[x]\varphi_n[\xi]}{\lambda - \lambda_n} = \frac{2}{l}\sum_{n=0}^{\infty} \frac{\mathrm{Sin}[k_n x]\,\mathrm{Sin}[k_n \xi]}{k^2 - k_n^2} \tag{7.208}$$

It is not immediately obvious that this result is consistent with our earlier findings using interface matching, Eq. (7.172). To demonstrate equivalence, we perform an eigenfunction expansion of the previous result. It is convenient to factor out the normalization factors, such that

$$G_k[x, \xi] = \frac{2}{l}\sum_{n=1}^{\infty} c_n\,\mathrm{Sin}[k_n x] \quad \text{with} \quad c_n = \int_0^l \mathrm{Sin}[k_n x]G_k[x, \xi]\,dx \tag{7.209}$$

The integrals can be evaluated by hand using various trigonometric identities in a tedious but straightforward manner, with a result

```
Sin[k (ξ - l)]/(k Sin[k l]) Integrate[Sin[n π/l x] Sin[k x], {x, 0, ξ}] +
  Sin[k ξ]/(k Sin[k l]) Integrate[Sin[n π/l x] Sin[k (x - l)], {x, ξ, l}] //
  Simplify[#, n ∈ Integers]&
```

$$\frac{l^2\,\mathrm{Sin}\left[\frac{n\pi\xi}{l}\right]}{-n^2\pi^2 + k^2 l^2}$$

that agrees with the present result. (Don't try this at home!)

7.5 Perturbation Theory

Often one wishes to investigate the effect of a small change in the dynamics or one can divide a differential operator into a solvable part and a small correction or *perturbation*. Suppose that

$$\mathcal{L} = \mathcal{L}_0 + \epsilon \mathcal{L}_1 \tag{7.210}$$

where $\mathcal{L}_0$ represents a dominant and solvable part while $\mathcal{L}_1$ represents a correction scaled by the small parameter ϵ. Let $\{\varphi_n^{(0)}\}$ represent a complete set of eigenfunctions for $\mathcal{L}_0$ satisfying

$$(\mathcal{L}_0 + \lambda_n^{(0)} w[x])\varphi_n^{(0)} == 0 \tag{7.211}$$

$$\langle \varphi_m^{(0)} \mid w \mid \varphi_n^{(0)} \rangle = \delta_{m,n} \tag{7.212}$$

and let $\{\varphi_n\}$ represent the eigenfunctions for the complete system, such that

$$(\mathcal{L} + \lambda_n w[x])\varphi_n == 0 \Longrightarrow (\mathcal{L}_0 + \lambda_n w[x])\varphi_n == -\epsilon \mathcal{L}_1 \varphi_n \tag{7.213}$$

Treating $\epsilon\, \mathcal{L}_1 \varphi_n$ as an inhomogeneous term, this equation can be solved, at least formally, by using the Green function for $\mathcal{L}_0$, such that

$$\varphi_n[x] = -\epsilon \int_a^b G_{\lambda_n}^{(0)}[x, \xi] \mathcal{L}_1[\xi] \varphi_n[\xi]\, d\xi \tag{7.214}$$

where

$$(\mathcal{L}_0 + \lambda_n w[x]) G_{\lambda_n}^{(0)}[x, \xi] == \delta[x - \xi] \Longrightarrow G_{\lambda_n}^{(0)}[x, \xi] = \sum_m \frac{\varphi_m^{(0)}[x] \varphi_m^{(0)}[\xi]^*}{\lambda_n - \lambda_m^{(0)}} \tag{7.215}$$

expresses the unperturbed Green function $G_{\lambda_n}^{(0)}$ in terms of unperturbed eigenfunctions $\varphi_n^{(0)}$ and eigenvalues $\lambda_n^{(0)}$. Note that we assume implicitly that $\mathcal{L}_1$ shifts the eigenvalues slightly so that $\lambda_n - \lambda_n^{(0)} \propto \epsilon$ and does not vanish. We can now write this formal solution as

$$\varphi_n[x] = -\epsilon \sum_m \varphi_m^{(0)}[x] \frac{\langle \varphi_m^{(0)} \mid \mathcal{L}_1 \mid \varphi_n \rangle}{\lambda_n - \lambda_m^{(0)}} \tag{7.216}$$

where

$$\langle \varphi_m^{(0)} \mid \mathcal{L}_1 \mid \varphi_n \rangle = \int_a^b \varphi_m^{(0)}[x]^* \mathcal{L}_1 \varphi_n[x]\, dx \tag{7.217}$$

represents a matrix element of $\mathcal{L}_1$ that couples φ_n to $\varphi_m^{(0)}$. However, this formal result is not especially useful because φ_n appears in the right-hand side as part of an infinite series of integrals that we cannot evaluate without already having the solution we seek. Nor do we know λ_n yet.

The fundamental assumption of *perturbation theory* is that changes in the eigenvalues and eigenfunctions can be developed as power series with respect to the small parameter ϵ, such that

$$\varphi_n = \sum_{m=0}^{\infty} \epsilon^m \varphi_n^{(m)}, \quad \lambda_n = \sum_{m=0}^{\infty} \epsilon^m \lambda_n^{(m)} \tag{7.218}$$

Thus, the first-order eigenfunction can be expanded in terms of unperturbed eigenfunctions as

$$\varphi_n \approx \varphi_n^{(0)} + \epsilon \sum_{m\neq n} c_{n,m}^{(1)} \varphi_m^{(0)} \tag{7.219}$$

such that

$$\epsilon \to 0 \Longrightarrow \varphi_n \to \varphi_n^{(0)} \tag{7.220}$$

provided that the expansion coefficients $c_{n,m}$ remain finite, as expected. Notice that it is convenient to employ the normalization

$$\langle \varphi_n^{(0)} \mid w \mid \varphi_n \rangle = 1 \tag{7.221}$$

and to exclude $\varphi_n^{(0)}$ from the correction term; attempting to maintain unit normalization leads to unnecessary complications. Substituting this approximation into the eigenvalue equation gives

$$(\mathcal{L}_0 + \epsilon\mathcal{L}_1 + \lambda_n w[x])\varphi_n == 0 \Longrightarrow \tag{7.222}$$

$$(\epsilon\mathcal{L}_1 + (\lambda_n - \lambda_n^{(0)})w[x])\varphi_n^{(0)} \approx -\epsilon \sum_{m\neq n} c_{n,m}^{(1)} (\epsilon\mathcal{L}_1 + (\lambda_n - \lambda_m^{(0)})w[x])\varphi_m^{(0)} \tag{7.223}$$

or

$$(\mathcal{L}_1 + \lambda_n^{(1)} w[x])\varphi_n^{(0)} = -\sum_{m\neq n} c_{n,m}^{(1)} (\lambda_n - \lambda_m^{(0)}) w[x] \varphi_m^{(0)} \tag{7.224}$$

where terms of order ϵ^2 have been dropped. Multiplying both sides by $\varphi_n^{(0)}$ and integrating, we obtain the first-order correction to the eigenvalue

$$\lambda_n^{(1)} = -\langle \varphi_n^{(0)} \mid \mathcal{L}_1 \mid \varphi_n^{(0)} \rangle \tag{7.225}$$

in terms of a diagonal matrix element of $\mathcal{L}_1$ that uses the unperturbed eigenfunctions. This can be evaluated because the $\{\varphi_n^{(0)}\}$ are presumed known. Similarly, we substitute the first approximations for the eigenfunctions and eigenvalues into the formal solution to obtain

$$\sum_{m\neq n} c_{n,m}^{(1)} \varphi_m^{(0)} \approx -\sum_{k\neq n} \frac{\varphi_k^{(0)}[x]}{\lambda_n^{(0)} - \lambda_k^{(0)}} \left(\langle \varphi_k^{(0)} \mid \mathcal{L}_1 \mid \varphi_n^{(0)} \rangle + \epsilon \sum_{m\neq n} c_{n,m}^{(1)} \langle \varphi_k^{(0)} \mid \mathcal{L}_1 \mid \varphi_m^{(0)} \rangle \right) \tag{7.226}$$

Once again we drop higher-order terms and use orthogonality to isolate the coefficient

$$m \neq n \Longrightarrow c_{n,m}^{(1)} = \frac{\langle \varphi_m^{(0)} \mid \mathcal{L}_1 \mid \varphi_n^{(0)} \rangle}{\lambda_m^{(0)} - \lambda_n^{(0)}} \Longrightarrow \varphi_n \approx \varphi_n^{(0)} + \epsilon \sum_{m\neq n} \varphi_m^{(0)} \frac{\langle \varphi_m^{(0)} \mid \mathcal{L}_1 \mid \varphi_n^{(0)} \rangle}{\lambda_m^{(0)} - \lambda_n^{(0)}} \tag{7.227}$$

in a form that is also calculable directly.

It should be clear that this procedure can be iterated indefinitely, substituting the current approximation back into the formal solutions to obtain corrections to the eigenfunctions and eigenvalues to one higher power of ϵ. However, the procedure becomes more complicated if some of the unperturbed eigenvalues are degenerate because the denominators involving differences between eigenvalues will vanish. The problem can be handled by forming, in degenerate subspaces, linear combinations that are diagonal with respect to $\mathcal{L}_1$, but we will not pursue *degenerate perturbation theory* here.

7.5.1 Example: Bead at Center of a String

Suppose that a string of length L under tension τ with mass density σ carries a small bead of mass m at its center. The normal-mode equation takes the form

$$\left(\frac{d^2}{dx^2} + k_n^2\left(1 + \mu\delta\left[x - \frac{L}{2}\right]\right)\right)\varphi_n[x] == 0, \quad \varphi_n[0] == \varphi_n[L] == 0 \tag{7.228}$$

where $\mu = m/\sigma$ is a small parameter with dimensions of length; hence, we apply perturbation theory when $\mu \ll L$. Obviously, we identify the unperturbed solutions as

$$\left(\frac{d^2}{dx^2} + k_n^{(0)2}\right)\varphi_n^{(0)}[x] == 0, \quad \varphi_n[0] == \varphi_n[L] == 0 \Longrightarrow \varphi_n^{(0)}[x] = \sqrt{\frac{2}{L}}\,\mathrm{Sin}[k_n^{(0)}x] \tag{7.229}$$

where

$$k_n^{(0)} = \omega_n^{(0)}\sqrt{\frac{\sigma}{\tau}} = \frac{n\pi}{L} \tag{7.230}$$

The first-order corrections to the eigenvalues are given by

$$k_n^{(1)2} = -\left\langle\varphi_n^{(0)}\middle|\mu\delta\left[x - \frac{L}{2}\right]\middle|\varphi_n^{(0)}\right\rangle = -\frac{2\mu}{L}\,\mathrm{Sin}\left[\frac{n\pi}{2}\right]^2 = -\frac{2\mu}{L}\,\mathrm{Mod}[n, 2] \tag{7.231}$$

where Mod[n, 2] is the remainder for n divided by 2 and has the values of 1 for odd n and 0 for even n. Notice that the bead does not affect the frequencies for even n because those eigenfunctions have a node at $L/2$ anyway – if the bead is stationary, it might as well be absent. Combining with the zeroth-order eigenvalue and expanding with respect to μ, we find

$$k_n \approx \frac{n\pi}{L} - \frac{\mu}{L}\,\mathrm{Mod}[n, 2] \tag{7.232}$$

Similarly, the perturbed eigenfunctions take the form

$$\varphi_n \approx \varphi_n^{(0)} + \sum_{m \neq n}\varphi_m^{(0)}\frac{\left\langle\varphi_m^{(0)}\middle|\mu\delta\left[x - \frac{L}{2}\right]\middle|\varphi_n^{(0)}\right\rangle}{\lambda_m^{(0)} - \lambda_n^{(0)}} = \varphi_n^{(0)} + \frac{2\mu}{\pi}\sum_{m \neq n}\varphi_m^{(0)}\frac{\mathrm{Sin}\left[\frac{m\pi}{2}\right]\mathrm{Sin}\left[\frac{n\pi}{2}\right]}{m - n} \tag{7.233}$$

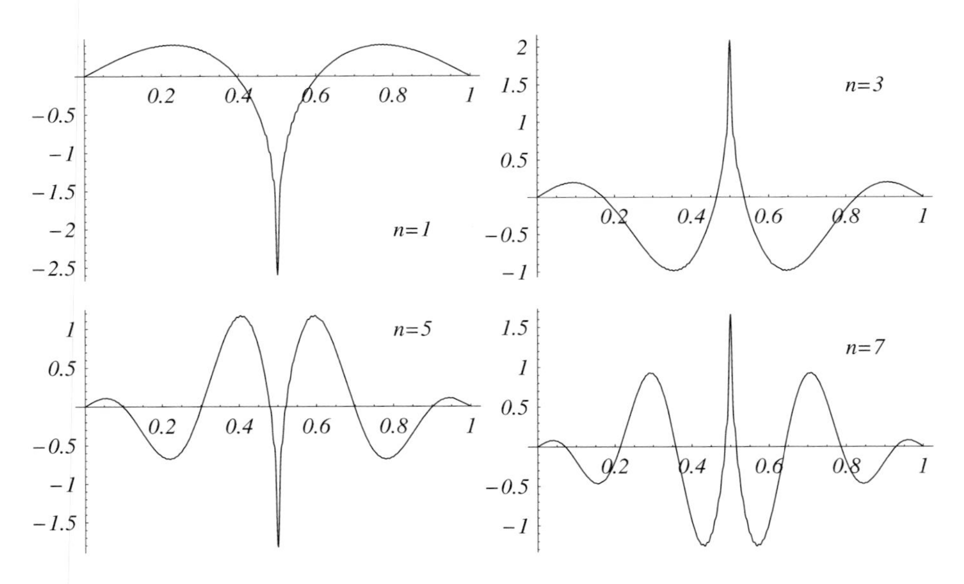

Figure 7.5. $\delta\varphi_n[x]$ for bead at $x = L/2$ in units where $L = 1$.

or

$$\varphi_n \approx \varphi_n^{(0)} + \frac{2\mu}{\pi} \text{Mod}[n, 2] \sum_{m \neq n} \varphi_m^{(0)} \frac{(-)^{(m+n)/2} \text{Mod}[m, 2]}{m - n} \tag{7.234}$$

and are unchanged for even n. Notice that the admixture of terms with $m \neq n$ is proportional to $(m - n)^{-1}$, so that nearby terms contribute most; nevertheless, this expansion converges relatively slowly.

To illustrate the effects of this perturbation, it is useful to express our result in the form

$$\varphi_n[x] \approx \varphi_n^{(0)}[x] + \frac{2\mu}{\pi} \delta\varphi_n[x] \tag{7.235}$$

Figure 7.5 displays the lowest several $\delta\varphi_n[x]$ in units where $L = 1$. These functions are qualitatively similar to Green functions with central cusps produced by the small bead, but exist for $k = k_n^{(0)}$.

7.6 Variational Methods

At the beginning of this chapter we derived the Sturm–Liouville equation using a variational argument and here we would like to extend this approach to a variational analysis of its eigenvalues. Consider the functional

$$\Lambda[\psi] = \frac{\int_a^b \left(p[x] \left(\frac{d\psi[x]}{dx} \right)^2 + q[x]\psi[x]^2 \right) dx}{\int_a^b \psi[x]^2 w[x] \, dx} \tag{7.236}$$

where ψ is any continuous function that complies with appropriate self-adjoint boundary conditions and where $p[x]$ and $w[x]$ are nonnegative in $a \leq x \leq b$. This functional is known as the *Rayleigh quotient*. Since we have shown that both the real and imaginary components satisfy the same Sturm–Liouville equation, we may assume without loss of generality that $\psi[x]$ is real and not use complex conjugates in the definition of Λ; this will simplify the algebra. First, we demonstrate that functions ψ that render $\Lambda[\psi]$ stationary with respect to small variations $\psi[x] \to \psi[x] + \delta\psi[x]$ consistent with the boundary conditions are eigenfunctions of the Sturm–Liouville operator. Recognizing that

$$\Lambda = \frac{A}{B} \Longrightarrow \delta\Lambda = \frac{B\delta A - A\delta B}{B^2} = \frac{\delta A - \Lambda\delta B}{B} \tag{7.237}$$

and that B is positive-definite, we require

$$\delta\Lambda == 0 \Longrightarrow \delta A - \Lambda\delta B == 0 \tag{7.238}$$

The variation of the numerator

$$\delta A = 2\int_a^b \left(p[x]\left(\frac{d\psi[x]}{dx}\right)\frac{d\delta\psi[x]}{dx} + q[x]\psi[x]\delta\psi[x]\right)dx \tag{7.239}$$

can be simplified using partial integration

$$\tfrac{1}{2}\delta A = \left(p[x]\left(\frac{d\psi[x]}{dx}\right)\delta\psi[x]\right)_a^b - \int_a^b \left(\frac{d}{dx}\left(p[x]\frac{d\psi[x]}{dx}\right) - q[x]\psi[x]\right)\delta\psi[x]\,dx \tag{7.240}$$

and the self-adjoint boundary conditions

$$\left(p[x]\frac{d\psi[x]}{dx}\delta\psi[x]\right)_a^b == 0 \tag{7.241}$$

to write

$$\delta A = -2\int_a^b \left(\frac{d}{dx}\left(p[x]\frac{d\psi[x]}{dx}\right) - q[x]\psi[x]\right)\delta\psi[x]\,dx \tag{7.242}$$

The variation of the denominator

$$\delta B = 2\int_a^b \psi[x]\delta\psi[x]w[x]\,dx \tag{7.243}$$

is already simple. Combining these terms now gives

$$\delta A - \Lambda\delta B == 0 \Longrightarrow \int_a^b \left(\frac{d}{dx}\left(p[x]\frac{d\psi[x]}{dx}\right) - q[x]\psi[x] + \Lambda w[x]\psi[x]\right)\delta\psi[x]\,dx == 0 \tag{7.244}$$

If this integral is to vanish for arbitrary $\delta\psi[x]$, the coefficient of $\delta\psi[x]$ must vanish identically, such that

$$\frac{d}{dx}\left(p[x]\frac{d\psi[x]}{dx}\right) - q[x]\psi[x] + \Lambda w[x]\psi[x] == 0 \tag{7.245}$$

Thus, if we identify $\Lambda[\psi]$ with the eigenvalue λ we recover the Sturm–Liouville eigenvalue equation

$$\frac{d}{dx}\left(p[x]\frac{d\psi_n[x]}{dx}\right) - q[x]\psi_n[x] + \lambda_n w[x]\psi_n[x] == 0 \tag{7.246}$$

Therefore, if $\Lambda[\psi_n]$ is stationary with respect to small variations $\psi_n \to \psi_n + \delta\psi$ then $\psi_n[x]$ is an eigenfunction of the Sturm–Liouville operator with eigenvalue $\lambda_n = \Lambda[\psi_n]$.

Suppose that

$$\psi[x] = \sum_{n=1}^{\infty} c_n \varphi_n[x] \tag{7.247}$$

is expanded in a complete orthonormal set of eigenfunctions $\{\varphi_n\}$ satisfying

$$\langle \varphi_m \mid w \mid \varphi_n \rangle = \int_a^b \varphi_m[x]\varphi_n[x]w[x]\,dx = \delta_{m,n} \tag{7.248}$$

The numerator of the Rayleigh quotient then takes the form

$$\int_a^b (p[x](\frac{d\psi[x]}{dx})^2 + q[x]\psi[x]^2)\,dx$$
$$= \sum_{m,n} c_m c_n \int_a^b (p[x]\varphi'_m[x]\varphi'_n[x] + q[x]\varphi_m[x]\varphi_n[x])\,dx \tag{7.249}$$

and can be simplified using partial integration of the first term and the boundary conditions to obtain

$$\Lambda \sum_n c_n^2 = -\sum_{m,n} c_m c_n \int_a^b \varphi_m[x]\left(\frac{d}{dx}(p[x]\varphi'_n[x]) - q[x]\varphi_n[x]\right)dx \tag{7.250}$$

The term in parentheses can be simplified further using the Sturm–Liouville eigenvalue equation, leaving

$$\Lambda \sum_n c_n^2 = -\sum_{m,n} c_m c_n \int_a^b \varphi_m[x](-\lambda_n w[x]\varphi_n[x])\,dx \tag{7.251}$$

Thus, we find that

$$\Lambda[\psi] = \frac{\sum_n c_n^2 \lambda_n}{\sum_n c_n^2} = -\frac{\langle \psi \mid \mathcal{L} \mid \psi \rangle}{\langle \psi \mid w \mid \psi \rangle} \tag{7.252}$$

where

$$\mathcal{L} = \frac{d}{dx}p[x]\frac{d}{dx} - q[x] \tag{7.253}$$

is the Sturm–Liouville operator. Therefore, the Rayleigh quotient can be interpreted as the average of the eigenvalues weighted by the intensity of the corresponding eigenfunction in ψ. Here intensity is interpreted as the square of the expansion coefficient and represents the contribution of that term to the normalization integral for ψ, as given by Parseval's theorem. This result should be familiar from Fourier analysis but applies to any self-adjoint $\mathcal{L}$.

Our analysis of the interleaving of roots demonstrated that there exists a smallest eigenvalue; hence, the smallest eigenvalue must represent an absolute minimum for $\Lambda[\psi]$. Suppose that $\psi[x]$ is a trial function that is intended to approximate $\varphi_1[x]$. The trial function can be formally expanded in terms of eigenfunctions as

$$\psi[x] = \varphi_1[x] + \sum_{n=2}^{\infty} a_n \varphi_n[x] \tag{7.254}$$

where each a_n is small if ψ is a decent approximation to φ_1. Notice that we are free to use

$$\langle \varphi_1 \mid w \mid \psi \rangle = 1 \tag{7.255}$$

because $\Lambda[\psi]$ is independent of normalization. Then

$$\Lambda[\psi] = \frac{\lambda_1 + \sum_{n=2}^{\infty} a_n^2 \lambda_n}{1 + \sum_{n=2}^{\infty} a_n^2} \approx \lambda_1 + \sum_{n=2}^{\infty} a_n^2 (\lambda_n - \lambda_1) \tag{7.256}$$

demonstrates that the error in the approximation to λ_1 due to the error $(\psi - \varphi_1)$ is second-order with respect to a_n. Thus, even if the trial function is not especially good, the estimate $\lambda_1 \approx \Lambda[\psi]$ may still be pretty accurate. Furthermore, the true value is smaller than that obtained with any trial function (other than φ_1 itself). Therefore, if we construct a trial function that resembles φ_1 based upon intuition and which includes a parameter, minimization of $\Lambda[\psi]$ with respect to that parameter should produce a good approximation to λ_1. Naturally, a better trial function provides a better estimate of λ_1. This procedure is known as the *Rayleigh–Ritz method*.

7.6.1 Example: Vibrating String

The almost universal first example of this technique is the vibrating string. Although we already know the eigenvalues and eigenfunctions for

$$\left(\frac{d^2}{dx^2} + k_n^2 \right) \varphi_n == 0, \quad \varphi_n[0] == \varphi_n[L] == 0 \tag{7.257}$$

with Rayleigh quotient

$$\Lambda[\psi] = \frac{\int_0^L \psi'[x]^2 \, dx}{\int_0^L \psi[x]^2 \, dx} \tag{7.258}$$

suppose that we feign ignorance and employ a zero-parameter trial solution of the form

$$\psi[x] = x(L - x) \tag{7.259}$$

which satisfies the boundary conditions and is symmetric with respect to $L/2$, as expected for the lowest eigenfunction. Thus,

$$\psi'[x] = L - 2x \Longrightarrow \Lambda[\psi] = \frac{\int_0^L (L-2x)^2\,dx}{\int_0^L x^2(L-x)^2\,dx} = \frac{10}{L^2} \tag{7.260}$$

is only 1.3 % larger than the exact answer of π^2/L^2. A better approximation can be obtained by including a parameter. Consider, for example, a trial function of the form

$$\psi[x\_] = x(L-x)\left(1+\alpha\left(x-\frac{L}{2}\right)^2\right);$$

Although the evaluation of Λ is straightforward, we will use *MATHEMATICA*® to perform that unenlightening task

$$\Lambda = \frac{\text{Integrate}\left[\psi'[x]^2, \{x, 0, L\}\right]}{\text{Integrate}\left[\psi[x]^2, \{x, 0, L\}\right]} //\text{Simplify}$$

$$\frac{6\left(560 + 56L^2\alpha + 11L^4\alpha^2\right)}{L^2\left(336 + 24L^2\alpha + L^4\alpha^2\right)}$$

and then minimize Λ with respect to the parameter α, to obtain

$$\text{sol} = \text{Solve}\left[\partial_\alpha \Lambda == 0, \alpha\right]$$

$$\left\{\left\{\alpha \to \frac{4\left(-49 + 4\sqrt{133}\right)}{13L^2}\right\}, \left\{\alpha \to -\frac{4\left(49 + 4\sqrt{133}\right)}{13L^2}\right\}\right\}$$

$$\frac{\Lambda}{\pi^2} /.\text{sol} //N$$

$$\left\{\frac{1.00001}{L^2}, \frac{10.348}{L^2}\right\}$$

Obviously one of the solutions corresponds to a maximum and should be rejected, while the other brings us within one part in 10^5 of the exact answer. Of course, a poor choice of trial function, such as the addition of a term that is odd with respect to the midpoint, would bring little or no improvement in the variational estimate. By respecting the symmetry of the problem, this judicious choice of trial function offers a very accurate estimate.

Problems for Chapter 7

1. Laguerre polynomials

Suppose that $L_n[x]$ is a polynomial of order n and that polynomials of different order are orthogonal with respect to the exponential weight function $w[x] = e^{-x}$ on the interval $0 \le x < \infty$, such that

$$\int_0^\infty L_n[x]L_m[x]e^{-x}\,dx = \delta_{n,m} \tag{7.261}$$

a) Construct the lowest three Laguerre polynomials using orthonormality conditions.

b) Alternatively, show that these functions satisfy a Sturm–Liouville equation that can be expressed in the form

$$xy''[x] + g[x]y'[x] + \lambda_n y[x] == 0 \tag{7.262}$$

and demonstrate that the eigenvalue must be a positive integer for the corresponding eigenfunction to be bounded as $x \to \infty$.

2. Minimization of mean-square error

Suppose that a piecewise smooth function $f[x]$ is approximated by the truncated expansion

$$f_N[x] = \sum_{m=0}^{N} c_m \varphi_m[x] \tag{7.263}$$

where the eigenfunctions $\{\varphi_m\}$ are orthonormal with respect to the weight function $w[x]$ on the interval $a \leq x \leq b$, such that

$$\int_a^b \varphi_m[x]^* \varphi_n[x] w[x]\, dx = \delta_{m,n} \tag{7.264}$$

Show that the mean-square error

$$\delta_N^2 = \int_a^b |f[x] - f_N[x]|^2 w[x]\, dx \tag{7.265}$$

is minimized when

$$c_m = \int_a^b \varphi_m[x]^* f[x] w[x]\, dx \tag{7.266}$$

independent of N. Note that this independence of the optimum coefficients c_m from the number of terms N is a special property of eigenfunction expansions that would not be satisfied if the coefficients of arbitrary functions, such as x^m, were determined by least-squares fitting. (Hint: consider independent variations of c_m and c_m^*.)

3. Intermediate boundary conditions

Show that a regular Sturm–Liouville operator (with $p[a] \neq 0$ and $p[b] \neq 0$) is self-adjoint in $a \leq x \leq b$ for any nontrivial intermediate boundary conditions of the form

$$\alpha_0 f[a] + \alpha_1 f'[a] == 0 \tag{7.267}$$

$$\beta_0 f[b] + \beta_1 f'[b] == 0 \tag{7.268}$$

where the coefficients α_i and β_i for $i = 0, 1$ are real.

4. Eigenfunctions for string with intermediate boundary conditions
The eigenfunctions for a vibrating string with intermediate boundary conditions satisfy

$$\left(\frac{d^2}{dx^2} + k_n^2\right)\varphi_n[x] == 0 \tag{7.269}$$

$$\alpha_0\varphi_n[0] + \alpha_1\varphi_n'[0] == 0 \tag{7.270}$$

$$\beta_0\varphi_n[L] + \beta_1\varphi_n'[L] == 0 \tag{7.271}$$

Derive a transcendental equation for the eigenvalues k_n and discuss a graphical procedure for determining numerical values.

5. Sturm–Liouville Wronskian
Suppose that y_1 and y_2 are two solutions to the Sturm–Liouville equations

$$(\mathcal{L} + \lambda w[x])y_1[x] == 0 \tag{7.272}$$

$$(\mathcal{L} + \lambda w[x])y_2[x] == 0 \tag{7.273}$$

where

$$\mathcal{L} = \frac{d}{dx}\left(p[x]\frac{d}{dx}\right) - q[x] \tag{7.274}$$

Evaluate their Wronskian

$$W[x] = y_1[x]y_2'[x] - y_2[x]y_1'[x] \tag{7.275}$$

and show that $p[x]W[x]$ is constant. (Hint: evaluate $W'[x]$.)

6. Bounds on eigenvalues
Consider the eigenvalue problem

$$(1 + x^2)y''[x] + 2xy'[x] + \lambda(1 + x^2)y[x] == 0 \tag{7.276}$$

$$y'[0] == 0, \quad y[1] == 0 \tag{7.277}$$

in the interval $0 \le x \le 1$. Show that the lowest eigenvalue is in the range $\frac{\pi^2}{8} \le \lambda_1 \le \frac{\pi^2}{2}$.

7. Green function for vibrating string
The Green function for a vibrating string clamped at both ends satisfies

$$\left(\frac{d^2}{dx^2} + k^2\right)G_k[x, x'] == \delta[x - x'], \quad G_k[0, x'] == G_k[l, x'] == 0 \tag{7.278}$$

a) Write down the formal eigenfunction expansion of $G_k[x, x']$. Under what conditions does the Green function exist?

b) Demonstrate that the Green function can also be expressed in closed form as

$$G_k[x, x'] = \frac{\mathrm{Sin}[kx_<]\,\mathrm{Sin}[k(x_> - l)]}{k\,\mathrm{Sin}[kl]} \tag{7.279}$$

where $x_<$ is the smaller and $x_>$ is the larger of x and x'.

c) Prove that these representations are equivalent.

8. Green function for $y'' - k^2 y == f[x]$ with $y[0] = y[L] = 0$
Consider the differential equation

$$y''[x] - k^2 y[x] == f[x], \quad y[0] == y[L] == 0 \tag{7.280}$$

where $k > 0$.

a) Construct a closed form for the Green function and write a general form for the solution to the inhomogeneous equation.
b) Construct the Green function as a Fourier sine series and again write a general form for the solution to the inhomogeneous equation.
c) Demonstrate that these two representations are equivalent.

9. Green function for stopped pipe
Acoustical vibrations in a stopped pipe satisfy

$$y''[x] + k^2 y[x] == f[x], \quad y'[0] == 0, \quad y[L] == 0 \tag{7.281}$$

a) Construct a closed form for the Green function and write a general form for the solution of the inhomogeneous equation.
b) Construct the Green function as an eigenfunction series and again write a general form for the solution of the inhomogeneous equation.
c) Demonstrate that these representations are equivalent.

10. Green function for open pipe
Acoustical vibrations in an open pipe satisfy

$$y''[x] + k^2 y[x] == f[x], \quad y'[0] == y'[L] == 0 \tag{7.282}$$

a) Construct a closed form for the Green function and write a general form for the solution of the inhomogeneous equation.
b) Construct the Green function as an eigenfunction series and again write a general form for the solution of the inhomogeneous equation.
c) Demonstrate that these representations are equivalent.

11. Green function for periodic orbits
Suppose that small-amplitude oscillations around a stable orbit in a synchrotron satisfy a differential equation of the form

$$y''[x] + \omega^2 y[x] == f[x], \quad 0 \le x \le 2\pi \tag{7.283}$$

subject to periodic boundary conditions

$$y[0] == y[2\pi], \quad y'[0] == y'[2\pi] \tag{7.284}$$

where $f[x]$ is a localized disturbance.

a) Obtain the eigenvalues ω_m and eigenfunctions $\varphi_m[x]$ for the homogeneous equation and write a formal expansion for its Green function.

b) Obtain an explicit closed-form solution for the Green function.

c) Write a formal expression for the solution to the inhomogeneous equation.

d) Discuss the behavior of the system near a resonance with $\omega \approx \omega_m$.

12. Fredholm alternative theorem

Suppose that you have in your possession the eigenvalues and eigenvectors $\{\lambda_m, \varphi_m[x]\}$ that satisfy a homogeneous Sturm–Liouville problem of the form

$$(\mathcal{L} + \lambda_m w[x])\varphi_m[x] == 0\,, \quad \mathcal{L} = \frac{d}{dx}(p[x]\frac{d}{dx}) - q[x] \tag{7.285}$$

and wish to solve the inhomogeneous equation

$$(\mathcal{L} + \lambda_m w[x])\psi[x] == f[x] \tag{7.286}$$

by expanding ψ in terms of φ_m. The boundary conditions are homogeneous but for our present purposes need not be specified further. What condition must f satisfy for this strategy to succeed? Write a formal solution assuming that f satisfies your condition. What happens if f fails to satisfy this condition? The general solution to this problem is called the Fredholm alternative theorem.

13. Green function for a 4th order equation

Consider a linear fourth-order differential equation of the form

$$\left(\frac{d^4}{dx^4} + \lambda\right)y[x] == 0\,, \quad y[0] == y''[0] == y[\pi] == y''[\pi] == 0 \tag{7.287}$$

a) Show that this equation is self-adjoint for the specified boundary conditions.

b) Obtain the normalized eigenfunctions $\phi_n[x]$ for $0 \le x \le \pi$ and corresponding eigenvalues λ_n.

c) Construct an eigenfunction expansion for the Green function that satisfies

$$\left(\frac{d^4}{dx^4} + \lambda\right)G_\lambda[x, x'] == \delta[x - x'] \tag{7.288}$$

with the same boundary conditions.

d) Use the Green function to produce a series solution to the inhomogeneous equation

$$\left(\frac{d^4}{dx^4} + \lambda\right)y[x] == x\,, \quad y[0] == y''[0] == y[\pi] == y''[\pi] == 0 \tag{7.289}$$

14. Shifting the eigenvalue spectrum
Suppose that

$$\mathcal{L}_1 = \frac{d}{dx} p_1[x] \frac{d}{dx} - q_1[x] \Longrightarrow (\mathcal{L}_1 + \alpha_n w_1[x]) \psi_n[x] == 0 \tag{7.290}$$

has an eigenvalue spectrum $\{\alpha_n, n = 0, \infty\}$ with $\alpha_0 \neq 0$. Construct a related Sturm–Liouville operator

$$\mathcal{L}_2 = \frac{d}{dx} p_2[x] \frac{d}{dx} - q_2[x] \Longrightarrow (\mathcal{L}_2 + \beta_n w_2[x]) \psi_n[x] == 0 \tag{7.291}$$

that has the same eigenfunctions and a spectrum of eigenvalues $\{\beta_n\}$ that is shifted with respect to $\{\alpha_n\}$ such that $\beta_0 = 0$. How are $\{p_2, q_2, w_2\}$ related to $\{p_1, q_1, w_1\}$?

15. An eigenvalue problem with mixed boundary conditions
Show that the eigenvalue problem

$$y'' + \lambda y == 0, \quad y[0] = 0, \quad y'[0] == y[1] \tag{7.292}$$

with mixed boundary conditions has only one real eigenvalue and determine the corresponding eigenfunction. Why does this problem not have an denumerably infinite spectrum of solutions?

16. Normal modes for stretched string with one end attached to spring
Suppose that a string of length L under tension T has one end fixed while the other is attached to a ring that slides without friction on a fixed rod and that the ring is attached to a spring with spring constant α. Thus, the transverse displacement $y[x,t]$ satisfies

$$\rho \frac{\partial^2 y[x,t]}{\partial t^2} - T \frac{\partial^2 y[x,t]}{\partial x^2} == 0, \quad y[0,t] == 0, \quad \left(\frac{\partial y[x,t]}{\partial x} \right)_{x=L} == -\frac{\alpha}{T} y[L,t] \tag{7.293}$$

where ρ is the mass density.

a) Show that there exist normal modes of the form

$$y[x,t] = e^{-i\omega_n t} f[k_n x] \tag{7.294}$$

and determine the relationship between ω_n and the parameters of the problem (L, T, α, ρ). Discuss the special cases of very weak and very strong springs.

b) Demonstrate that the normal modes are orthogonal.

17. Hanging string
A string of length L with uniform mass density σ hangs under its own weight. Here we investigate small-amplitude transverse vibrations in a plane, assuming that the tension $\tau[x]$ is not affected by this motion.

a) Evaluate the equilibrium tension $\tau[x]$ and construct the normal-mode equation. What are the appropriate boundary conditions?

b) Use the substitution $s^2 = 1 - x/L$ to show that the normal modes satisfy the Bessel equation and find an expression for the eigenfrequencies.

c) Construct the general solution to the initial-value problem.

18. Twirling string; simplified version
A string of length L with uniform mass density σ is attached at one end to a thin rod that rotates with constant angular velocity Ω in the horizontal plane. Here we investigate small-amplitude vertical vibrations neglecting gravity and assuming that the tension $\tau[x]$ is not affected by transverse vibrations.

a) Evaluate the equilibrium tension $\tau[x]$ and construct the normal-mode equation. What are the appropriate boundary conditions?

b) Show that the normal modes satisfy the Legendre equation and find an expression for the eigenfrequencies. Determine the lowest three eigenfrequencies and sketch the corresponding eigenfunctions.

c) Construct the general solution to the initial-value problem.

19. Small-amplitude oscillations of a twirling string; general version
A string of length L with uniform mass density σ is attached at one end to a thin rod that rotates with constant angular velocity Ω in the horizontal plane. Here we investigate small-amplitude transverse oscillations with respect to the equilibrium state of uniform rotation. Neglect gravity and assume that the tension is not affected by transverse vibrations.

a) Construct the equations of motion for vertical and transverse horizontal oscillations in a coordinate system that rotates with angular velocity $\vec{\Omega} = \Omega\hat{z}$. What are the appropriate boundary conditions? (Hints: it is useful to parametrize the coordinates in terms of the distance $0 \leq s \leq 1$ along the string relative to L. The coordinates of a mass element $dm = \sigma L\, ds$ are then functions of s and t. One can apply Newton's second law to the two ends of mass element requiring the tension to be tangential. The positions can then be expressed relative to the rotating coordinate system. Assume that the tension depends upon s but is independent of time.)

b) Show that the normal modes satisfy the Legendre equation and find expressions for the eigenfrequencies for both horizontal and vertical modes relative to Ω. Why are these frequencies different? Determine the lowest three eigenfrequencies and sketch the corresponding eigenfunctions. Does the lowest horizontal mode represent an actual oscillation?

c) Construct the general solution to the initial-value problem.

20. Clamped string with central bead
Suppose that a string of length L under uniform tension τ is clamped at both ends and that the mass density σ is uniform except for a small bead of mass m affixed to its center. It is convenient to define the range as $-a \leq x \leq a$ where $a = L/2$ and to define $\mu = m/\sigma$. Neglect gravity.

a) Construct the unnormalized eigenfunctions and deduce the equations satisfied by the eigenfrequencies. Compare with the ordinary string ($m = 0$) and explain why symmetric and antisymmetric modes behave differently.

b) Describe a graphical procedure for determining the eigenfrequencies for symmetric modes. Use this construction to compare eigenfrequencies for symmetric and antisymmetric modes for large μ or for large k. Derive a simple approximation for the splitting between symmetric and antisymmetric modes under these conditions.

c) Sketch the first few symmetric eigenfunctions and evaluate the displacement of the central bead.

d) Verify explicitly that the eigenfunctions are orthogonal with respect to

$$w[x] = 1 + \mu\delta[x] \tag{7.295}$$

where $\mu = m/\sigma$ has dimensions of length.

21. An upper bound on the first root of $J_0[x]$
Evaluate the Rayleigh quotient $\Lambda[\phi]$ for the eigenvalue problem

$$y'' + \frac{y'}{x} + \lambda y == 0, \quad |y[0]| < \infty, \quad y[1] == 0 \tag{7.296}$$

using the test function $\phi[x] = \alpha(1 - x)$ and deduce an upper limit for the smallest root of $J_0[x]$.

8 Legendre and Bessel Functions

Abstract. The properties of Legendre and Bessel functions are developed in considerable detail, including: generating functions, recursion relations, orthonormality, series and integral representations. These functions are important for a wide variety of physics problems based upon equations featuring the Laplacian operator and will play a central role in our subsequent study of boundary-value problems.

8.1 Introduction

Many of the special functions we employ in physics arise by applying in appropriate coordinate systems the technique of separation of variables to reduce partial differential equations, such as the Laplace, diffusion, or Schrödinger equations, to systems of ordinary second-order differential equations. With appropriate boundary conditions these differential equations are often self-adjoint and their solutions exhibit the general properties of Sturm–Liouville systems. Other properties can be developed using integral representations, which permit analytic continuation, or by using *generating functions* of the form

$$F[t,x] = \sum_{n=0}^{\infty} f_n[x] t^n \tag{8.1}$$

where the coefficient in a power-series expansion with respect to one variable is a function of the second variable.

There is a vast literature on the hundreds of special functions that have been studied during the last 300 years or so. Obviously, an exhaustive study is impossible within the confines of one chapter of a single-semester course; careers have been devoted to this topic. This author is not expert in this subject and is not capable of memorizing all of the relationships he employs frequently, much less those he does not, and expects that few readers would be willing or able to do so either; it is an important but dry subject. Instead, he relies on compendia like Abramowitz and Stegun or classic texts for details that are not readily at hand. Nevertheless, it is important to be familiar with several of the most important methods for studying such functions and their most important generic properties. For this purpose we will concentrate on those functions that are most useful for problems in electrodynamics, diffusion, and quantum mechanics at the first-year graduate level. Specifically, we will study Legendre and Bessel functions of several types. These functions will be used in the chapter on boundary-value problems to solve a variety of interesting and important physics problems. Our derivations here will not always provide all of the gory details, but some of the exercises will request the missing steps.

Graduate Mathematical Physics. James J. Kelly

ISBN: 3-527-40637-9

8.2 Legendre Functions

8.2.1 Generating Function for Legendre Polynomials

The Legendre polynomials $P_n[x]$ where $x = \text{Cos}[\theta]$ arise often in the representation of the dependence of a physical quantity on a polar angle θ. Perhaps the most direct method for obtaining these polynomials is through the Green function for the Poisson equation

$$\nabla^2 \frac{1}{|\vec{r} - \vec{r}'|} = -4\pi\delta[\vec{r} - \vec{r}'] \tag{8.2}$$

If $r > r'$, we may expand

$$\frac{1}{|\vec{r} - \vec{r}'|} = r^{-1}\left(1 - 2\frac{r'}{r}x + \left(\frac{r'}{r}\right)^2\right)^{-1/2} \tag{8.3}$$

as a power series in $t = r'/r$ whose coefficients depend upon $x = \hat{r}\cdot\hat{r}'$. Obviously, when $r < r'$ we employ a similar expansion with the roles of r and r' exchanged. These expansions can be unified in the form

$$\frac{1}{|\vec{r} - \vec{r}'|} = r_>^{-1}(1 - 2xt + t^2)^{-1/2} \tag{8.4}$$

where $r_< = \min[r, r']$ is the smaller and $r_> = \max[r, r']$ is the larger of r and r' and where $t = r_</r_>$ is in the range $0 \le t \le 1$. Thus, the *generating function*

$$g[t, x] = \frac{1}{\sqrt{1 - 2xt + t^2}} = \sum_{n=0}^{\infty} P_n[x]t^n \tag{8.5}$$

exhibits a uniformly convergent power series with respect to t whose coefficients $P_n[x]$ are identified as Legendre polynomials of degree n with respect to x. Once these functions have been constructed, the Green function for the Poisson equation becomes

$$\frac{1}{|\vec{r} - \vec{r}'|} = \sum_{n=0}^{\infty} \frac{r_<^n}{r_>^{n+1}} P_n[x] \tag{8.6}$$

where $r_<$ is the smaller and $r_>$ the larger of r and r'.

Before proceeding with explicit construction of the Legendre polynomials, we obtain some of their basic properties directly from the generating function. Recognizing that

$$g[t, \pm 1] = \frac{1}{1 \mp t} = \sum_{n=0}^{\infty} (\pm t)^n \tag{8.7}$$

we find

$$P_n[\pm 1] = (\pm 1)^n \tag{8.8}$$

Similarly, using

$$g[t,-x] = g[-t,x] \Longrightarrow P_n[-x] = (-)^n P_n[x] \tag{8.9}$$

we conclude that the *parity* of P_n is $(-)^n$. Finally, using

$$g[0,x] = 1\,, \quad \left(\frac{\partial g[t,x]}{\partial t}\right)_{t=0} = x \tag{8.10}$$

we find results for the two lowest Legendre polynomials

$$P_0 = 1\,, \quad P_1 = x \tag{8.11}$$

that are consistent with the properties developed so far and which form the seeds for construction of a series by recursion.

We can obtain *recursion relations* by differentiating the generating function with respect to either of its variables. Differentiation with respect to t gives

$$\begin{aligned} \frac{\partial g[t,x]}{\partial t} &= \frac{x-t}{(1-2xt+t^2)^{3/2}} = \sum_{n=0}^{\infty} P_n[x] n t^{n-1} \\ &\Longrightarrow \frac{x-t}{1-2xt+t^2} \sum_{n=0}^{\infty} P_n[x] t^n = \sum_{n=0}^{\infty} P_n[x] n t^{n-1} \end{aligned} \tag{8.12}$$

where the last step uses the expansion of the generating function to eliminate fractional powers. Collecting like powers of t gives

$$\sum_{n=0}^{\infty} t^n \big((n+1)P_{n+1} - (2n+1)xP_n + nP_{n-1} \big) = 0 \tag{8.13}$$

Notice that the coefficient of P_{-1} vanishes, so that polynomials of negative order do not actually occur. If this relationship is to be satisfied for any value of t, the coefficients for each power t^n must vanish separately. Therefore, we obtain a three-term recursion relation of the form

$$(n+1)P_{n+1} - (2n+1)xP_n + nP_{n-1} = 0 \tag{8.14}$$

that can be used to construct the entire sequence P_n given $P_0 = 1$, $P_1 = x$. This is sometimes described as a *pure recursion relation* because it does not involve derivatives. Direct calculation provides Table 8.1 and Fig. 8.1. Notice that each P_n is a polynomial of order n containing only even or odd powers according to its parity.

A recursion relation for derivatives P_n' can be obtained by differentiating the generating function with respect to x, such that

$$\frac{\partial g[t,x]}{\partial x} = \frac{t}{(1-2xt+t^2)^{3/2}} = \sum_{n=0}^{\infty} P_n'[x] t^n \Longrightarrow \frac{t}{1-2xt+t^2} \sum_{n=0}^{\infty} P_n[x] t^n = \sum_{n=0}^{\infty} P_n'[x] t^n \tag{8.15}$$

Table 8.1. Lowest few Legendre polynomials.

n	$P_n[x]$
0	1
1	x
2	$\frac{1}{2}(-1+3x^2)$
3	$\frac{1}{2}x(-3+5x^2)$
4	$\frac{1}{8}(3-30x^2+35x^4)$
5	$\frac{1}{8}x(15-70x^2+63x^4)$
6	$\frac{1}{16}(-5+105x^2-315x^4+231x^6)$
7	$\frac{1}{16}x(-35+315x^2-693x^4+429x^6)$
8	$\frac{1}{128}(35-1260x^2+6930x^4-12\,012x^6+6435x^8)$

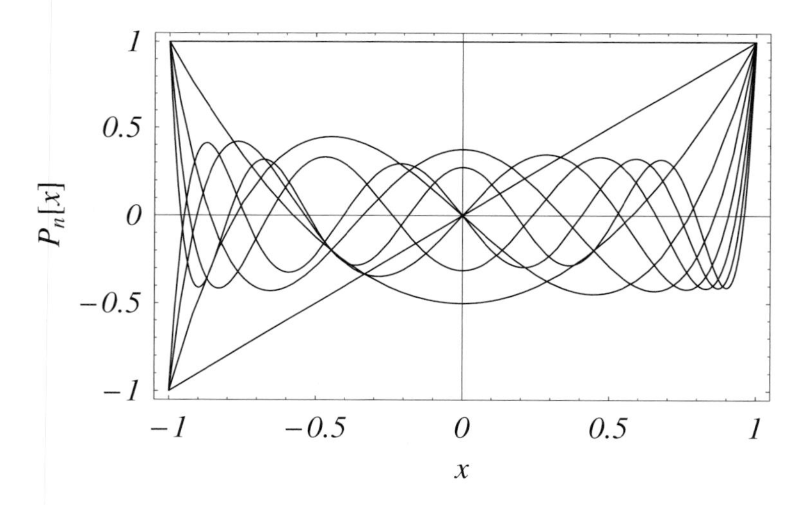

Figure 8.1. Legendre polynomials.

Once again collecting like powers of t^n we find

$$P'_{n+1} - 2xP'_n + P'_{n-1} = P_n \tag{8.16}$$

Using the previous recursion relation

$$\begin{aligned} (n+1)P_{n+1} - (2n+1)xP_n + nP_{n-1} &= 0 \\ \Longrightarrow (n+1)P'_{n+1} - (2n+1)(xP'_n + P_n) + nP'_{n-1} &= 0 \end{aligned} \tag{8.17}$$

we can eliminate P'_n to obtain

$$P'_{n+1} - P'_{n-1} = (2n+1)P_n \tag{8.18}$$

Many similar relationships can now be obtained by manipulation of these basic recursion relations. For example, by solving the simultaneous equations

$$P'_{n+1} - P'_{n-1} = (2n+1)P_n \tag{8.19}$$

$$(n+1)P'_{n+1} + nP'_{n-1} = (2n+1)(xP'_n + P_n) \tag{8.20}$$

one finds

$$P'_{n+1} = xP'_n + (n+1)P_n \tag{8.21}$$

$$P'_{n-1} = xP'_n - nP_n \tag{8.22}$$

Multiplying by x and manipulating the index using $n \to n \pm 1$

$$xP'_n = x^2 P'_{n-1} + nxP_{n-1} \tag{8.23}$$

$$xP'_n = x^2 P'_{n+1} - (n+1)xP_{n+1} \tag{8.24}$$

one obtains the so-called *ladder relations*

$$(n+1)P_{n+1} = (n+1)xP_n - (1-x^2)P'_n \tag{8.25}$$

$$nP_{n-1} = nxP_n + (1-x^2)P'_n \tag{8.26}$$

that can be used to develop Legendre functions using either upward or downward recursion.

Finally, the recursion relations can be used to derive a second-order differential equation for $P_n[x]$. We begin by expressing one of the ladder relations in the form

$$(1-x^2)P'_n = nP_{n-1} - nxP_n \tag{8.27}$$

Differentiating this expression

$$(1-x^2)P''_n - 2xP'_n = nP'_{n-1} - nP_n - nxP'_n \tag{8.28}$$

and eliminating P'_{n-1}, we finally obtain a second-order differential equation

$$(1-x^2)P''_n[x] - 2xP'_n[x] + n(n+1)P_n[x] == 0 \tag{8.29}$$

that is a special case of Legendre's differential equation

$$(1-z^2)y''[z] - 2zy'[z] + \nu(\nu+1)y[z] == 0 \tag{8.30}$$

that applies when $\nu \to n$ is a nonnegative integer. Therefore, the regular solutions to Legendre's equation for integer $\nu \to n$ and real $z \to x$ on the interval $[-1, 1]$ are known as Legendre polynomials. More general solutions will be considered later.

When discussing Legendre functions, it is convenient to use n or l to represent nonnegative integers, $x = \text{Cos}[\theta]$ to represent real numbers in the range $|x| \le 1$, and to permit ν or z to represent extensions of the degree or argument to complex numbers.

8.2.2 Series Representation and Rodrigues' Formula

Using the Taylor series

$$(1-z)^{-1/2} = \sum_{n=0}^{\infty} \frac{(2n)!}{2^{2n}(n!)^2} z^n \tag{8.31}$$

the Legendre generating function becomes

$$g[t,x] = \frac{1}{\sqrt{1-2xt+t^2}} = \sum_{n=0}^{\infty} \frac{(2n)!}{2^{2n}(n!)^2}(2xt-t^2)^n \tag{8.32}$$

Then using the binomial expansion, we obtain

$$\begin{aligned} g[t,x] &= \sum_{n=0}^{\infty} \frac{(2n)!}{2^{2n}(n!)^2} \sum_{k=0}^{n} \frac{n!}{k!(n-k)!}(2xt)^{n-k}(-t^2)^k \\ &= \sum_{n=0}^{\infty}\sum_{k=0}^{n}(-)^k \frac{(2n)!}{2^{2n}n!k!(n-k)!}(2x)^{n-k}t^{n+k} \end{aligned} \tag{8.33}$$

An absolutely convergent double sum of this form can be rearranged using a theorem proven in the exercises that states

$$\sum_{p=0}^{\infty}\sum_{q=0}^{p} a_{q,p-q} = \sum_{r=0}^{\infty}\sum_{s=0}^{\lfloor r/2 \rfloor} a_{s,r-2s} \tag{8.34}$$

where

$$r \text{ even} \implies \lfloor r/2 \rfloor = \frac{r}{2} \tag{8.35}$$

$$r \text{ odd} \implies \lfloor r/2 \rfloor = \frac{r-1}{2} \tag{8.36}$$

is the floor function. After a little algebra we obtain

$$g[t,x] = \sum_{m=0}^{\infty}\sum_{k=0}^{\lfloor m/2 \rfloor}(-)^k \frac{(2m-2k)!}{2^{2m-2k}(m-k)!k!(m-2k)!}(2x)^{m-2k}t^m \tag{8.37}$$

Therefore, identifying the Legendre polynomial P_m with the coefficient of t^m, we obtain a series representation

$$P_n[x] = \sum_{k=0}^{\lfloor n/2 \rfloor}(-)^k \frac{(2n-2k)!}{2^n(n-k)!k!(n-2k)!}x^{n-2k} \tag{8.38}$$

of explicit, albeit cumbersome, form.

This series can be expressed in the form

$$P_n[x] = \sum_{k=0}^{\lfloor n/2 \rfloor} (-)^k \frac{1}{2^n (n-k)!k!} \left(\frac{d}{dx}\right)^n x^{2n-2k} \tag{8.39}$$

Using $k > \frac{n}{2} \implies 2n - 2k < n$, we recognize that n derivatives would eliminate any term of the form x^{2n-2k} with $k > \lfloor n/2 \rfloor$. Therefore, we are free to extend the upper limit of the summation to n to obtain

$$P_n[x] = \left(\frac{d}{dx}\right)^n \sum_{k=0}^{n} (-)^k \frac{x^{2n-2k}}{2^n (n-k)!k!} = \frac{1}{2^n n!} \left(\frac{d}{dx}\right)^n \sum_{k=0}^{n} (-)^k \frac{n!}{(n-k)!k!} (x^2)^{n-k} \tag{8.40}$$

where the final sum is simply the binomial expansion of $\left(x^2 - 1\right)^n$. Therefore, we obtain *Rodrigues' formula*

$$P_n[x] = \frac{1}{2^n n!} \left(\frac{d}{dx}\right)^n (x^2 - 1)^n \tag{8.41}$$

for the Legendre polynomials. Notice that the parity $P_n[-x] = (-)^n P_n[x]$ is obvious using Rodrigues' formula.

8.2.3 Schläfli's Integral Representation

Extending Rodrigues' formula into the complex plane

$$P_n[z] = \frac{1}{2^n n!} \left(\frac{d}{dz}\right)^n (z^2 - 1)^n \tag{8.42}$$

and applying the Cauchy integral formula to

$$(z^2 - 1)^n = \frac{1}{2\pi i} \oint \frac{(t^2 - 1)^n}{t - z} \, dt \tag{8.43}$$

we obtain the *Schläfli integral representation*

$$P_n[z] = \frac{2^{-n}}{2\pi i} \oint \frac{(t^2 - 1)^n}{(t - z)^{n+1}} \, dt \tag{8.44}$$

where the contour encloses z. An advantage of this integral representation is that it provides an analytic function for positive integer n that reduces to the Legendre polynomials for $z \to x$ in the range $-1 \le x \le 1$.

8.2.4 Legendre Expansion

The Legendre equation can be expressed in a form

$$\left(\frac{d}{dx}\left((1 - x^2)\frac{d}{dx}\right) + n(n+1)\right) P_n[x] == 0 \tag{8.45}$$

that is manifestly self-adjoint. Orthogonality between Legendre polynomials is immediately evident, but the normalization must still be evaluated. This is simplest using the generating function to evaluate

$$\frac{1}{1-2xt+t^2} = \sum_{n,m} P_n[x]P_m[x]t^{n+m} \tag{8.46}$$

such that

$$\int_{-1}^{1} \frac{dx}{1-2xt+t^2} = \sum_{n,m} t^{n+m} \int_{-1}^{1} P_n[x]P_m[x]\,dx \tag{8.47}$$

The integral is obtained by an elementary change of variables

$$\int_{-1}^{1} \frac{dx}{1-2xt+t^2} = \frac{1}{2t}\int_{(1-t)^2}^{(1+t)^2} \frac{dy}{y} = \frac{1}{t}\,\mathrm{Log}\left[\frac{1+t}{1-t}\right] = 2\sum_{n=0}^{\infty} \frac{t^{2n}}{2n+1} \tag{8.48}$$

where in the last step a power series permits identification of the normalization integral as

$$\int_{-1}^{1} P_n[x]P_m[x]\,dx = \frac{2}{2n+1}\delta_{n,m} \tag{8.49}$$

Therefore, the Sturm–Liouville theorem tell us that a piecewise continuous $f[x]$ can be expanded on the interval $|x| < 1$ in a Legendre series

$$f[x] = \sum_{n=0}^{\infty} a_n P_n[x] \tag{8.50}$$

$$a_n = \frac{2n+1}{2}\int_{-1}^{1} dx\, P_n[x]f[x] \tag{8.51}$$

that converges in the mean. Expansions of this type will prove quite useful in solving boundary-value problems based upon second-order partial differential equations.

Of course, one of the most important Legendre expansions

$$\delta[x-x'] = \sum_{n=0}^{\infty} a_n P_n[x] \tag{8.52}$$

provides a representation of the delta function. Using the orthogonality relation, the coefficients are simply

$$a_n = \frac{2n+1}{2}\int_{-1}^{1} dx\, P_n[x]\delta[x-x'] = \frac{2n+1}{2}P_n[x'] \tag{8.53}$$

Therefore, we obtain the completeness relation

$$\delta[x-x'] = \sum_{n=0}^{\infty} \frac{2n+1}{2}P_n[x]P_n[x'] \tag{8.54}$$

8.2.5 Associated Legendre Functions

Separating the Laplacian operator in spherical polar coordinates, one obtains an equation of the form

$$\left(\frac{1}{\mathrm{Sin}[\theta]}\frac{d}{d\theta}\left(\mathrm{Sin}[\theta]\frac{d}{d\theta}\right)+l(l+1)-\frac{m^2}{\mathrm{Sin}[\theta]^2}\right)y\left[\mathrm{Cos}[\theta]\right]==0 \tag{8.55}$$

in terms of the polar angle θ. Thus, defining $x = \mathrm{Cos}[\theta]$ we obtain an *associated Legendre equation*

$$\left((1-x^2)\frac{d^2}{dx^2}-2x\frac{d}{dx}+\left(l(l+1)-\frac{m^2}{1-x^2}\right)\right)y[x]==0 \tag{8.56}$$

that is similar to the ordinary Legendre equation except for the additional contribution proportional to m^2. Rather than attempting a frontal assault, it is sufficient to demonstrate that given a solution $y_l[x]$ to the ordinary Legendre equation, one can generate a solution to the associated Legendre equation using a transformation

$$y_{lm}[x]=(1-x^2)^{m/2}\left(\frac{d}{dx}\right)^m y_l[x] \tag{8.57}$$

that is motivated by Rodrigues' formula. If we write the Legendre equation for a specific l as

$$(1-x^2)y^{(2)}-2xy^{(1)}+l(l+1)y^{(0)}==0 \tag{8.58}$$

where $y^{(n)}$ denotes the nth derivative, and differentiate the equation once, we obtain

$$(1-x^2)y^{(3)}-4xy^{(2)}+\left(l(l+1)-2\right)y^{(1)}==0 \tag{8.59}$$

A second differentiation then gives

$$(1-x^2)y^{(4)}-6xy^{(3)}+\left(l(l+1)-6\right)y^{(2)}==0 \tag{8.60}$$

It is useful to define $u_m = y^{(m)}$ such that

$$(1-x^2)u_m''-2(m+1)xu_m'+\left(l(l+1)-m(m+1)\right)u_m==0 \tag{8.61}$$

is valid for $0 \le m \le 2$. Assuming that the equation is valid for a particular m and differentiating again, we obtain

$$(1-x^2)u_m^{(3)}-2(m+2)xu_m^{(2)}+\left(l(l+1)-(m+1)(m+2)\right)u_m^{(1)}==0 \tag{8.62}$$

and use $u_m^{(k)} = u_{m+1}^{(k-1)}$ to rewrite this equation as

$$(1-x^2)u_n''-2nxu_n'+\left(l(l+1)-n(n+1)\right)u_n==0 \tag{8.63}$$

where $n = m+1$. Therefore, by induction, the second-order differential equation applies to any u_m with $0 \le m \le l$. However, this equation is not yet in the desired form and is not self-adjoint. The final step is to define

$$v_m = (1 - x^2)^{m/2} u_m \Longrightarrow u_m = (1 - x^2)^{-m/2} v_m \tag{8.64}$$

and to show that v_m satisfies a self-adjoint equation. Allowing *MATHEMATICA*® to evaluate the derivatives

```
u[x_] = (1 - x^2)^(-m/2) v[x];
u'[x]
m x(1 - x^2)^(-1-m/2) v[x] + (1 - x^2)^(-m/2) v'[x]
u''[x] // Simplify
(1 - x^2)^(-2-m/2) (m (1 + (1 + m)x^2) v[x] + (-1 + x^2) (-2m x v'[x] + (-1 + x^2) v''[x]))
```

and substituting into the differential equation for u_m

```
A = (1 - x^2)u''[x] - 2x(m + 1)u'[x] + (l (l + 1) - m(m + 1))u[x] // Simplify
(1 - x^2)^(-1-m/2) (- (m^2 + (-1 + x^2) l (1 + l)) v[x] + (-1 + x^2) (2x v'[x] + (-1 + x^2) v''[x]))
A(1 - x^2)^(m/2) // Simplify
((m^2 + (-1 + x^2) l (1 + l)) v[x] - (-1 + x^2) (2x v'[x] + (-1 + x^2) v''[x])) / (-1 + x^2)
```

we conclude that y_{lm} does indeed satisfy the associated Legendre equation for $0 \leq m \leq l$ if y_l satisfies the ordinary Legendre equation.

The associated Legendre polynomials can now be expressed in terms of Rodrigues' formula as

$$P_l[x] = \frac{1}{2^l l!} \left(\frac{d}{dx}\right)^l (x^2 - 1)^l \Longrightarrow P_{l,m}[x] = (-)^m \frac{(1 - x^2)^{m/2}}{2^l l!} \left(\frac{d}{dx}\right)^{l+m} (x^2 - 1)^l \tag{8.65}$$

or

$$P_{l,m}[x] = (-)^m (1 - x^2)^{m/2} \left(\frac{d}{dx}\right)^m P_l[x] \tag{8.66}$$

where the *Condon–Shortley phase* $(-)^m$ proves convenient for quantum mechanics. Note that $P_{l,0}[x] = P_l[x]$. Here l is denoted the *degree* and m the *order* of $P_{l,m}$. Using Rodrigues' formula we immediately find

$$P_{l,m}[-x] = (-)^{l+m} P_{l,m}[x] \tag{8.67}$$

and

$$m \neq 0 \Longrightarrow P_{l,m}[\pm 1] = 0 \tag{8.68}$$

Although our derivation assumed that m is nonnegative, one can show that the generalized Rodrigues' formula offers a valid solution any integer in the range $-l \leq m \leq l$. However, the solutions for negative m are related to those for positive m according to

$$P_{l,-m}[x] = (-)^m \frac{(l - m)!}{(l + m)!} P_{l,m}[x] \tag{8.69}$$

and thus are not independent and we do not obtain a normal completeness relation. Nevertheless, the solutions for negative m will be useful for spherical harmonics in which the ϕ dependence distinguishes the sign of m.

Some attention must be given to the phase convention for associated Legendre polynomials. Many authors of textbooks on mathematical physics, such as Arfken or Butkov, omit the factor of $(-)^m$ in the relationship between P_l and $P_{l,m}$ on the grounds that it is an unnecessary complication at this stage. However, this phase is convenient for applications of spherical harmonics in quantum mechanics, which are based upon associated Legendre functions. Therefore, other authors of recent textbooks, like Lea, do include this factor in $P_{l,m}$. Hence, it is incumbent upon the reader to be cognizant of the conventions chosen by the author. (*Caveat emptor!*) We have chosen to include the $(-)^m$ phase for three practical reasons: (1) it is used by the most useful compilation of mathematical formulas, that by Abramowitz and Stegun; (2) my favorite mathematical software, MATHEMATICA®, also uses this phase in its definition of associated Legendre functions; and (3) it provides a more natural extension to complex arguments.

The orthogonality properties for associated Legendre functions can also be developed by using Rodrigues' formula to write

$$\int_{-1}^{1} P_{l,m}[x]P_{l',m}[x]\,dx = \frac{(-)^{l+l'}}{2^{l+l'}\,l!l'!}\int_{-1}^{1}(1-x^2)^m\left(\left(\frac{d}{dx}\right)^{l+m}(1-x^2)^l\right) \times\left(\left(\frac{d}{dx}\right)^{l'+m}(1-x^2)^{l'}\right)dx \tag{8.70}$$

where it is important to note that m is the same for both Legendre functions. We assume, without loss of generality, that $l \le l'$ and integrate by parts $l'+m$ times to obtain

$$\int_{-1}^{1} P_{l,m}[x]P_{l',m}[x]\,dx = \frac{(-)^{l+l'}(-)^{l'+m}}{2^{l+l'}\,l!l'!}\int_{-1}^{1}(1-x^2)^{l'}\left(\frac{d}{dx}\right)^{l'+m} \times\left((1-x^2)^m\left(\left(\frac{d}{dx}\right)^{l+m}(1-x^2)^l\right)\right)dx \tag{8.71}$$

where the integrated terms vanish at both limits. The integrand can be simplified using Leibnitz's expansion of multiple derivatives of a product

$$\left(\frac{d}{dx}\right)^n(f[x]g[x]) = \sum_{m=0}^{n}\binom{n}{m} f^{(n-m)}[x]g^{(m)}[x] \tag{8.72}$$

in terms of binomial coefficients and products of derivatives. Thus, we obtain

$$\left(\frac{d}{dx}\right)^{l'+m}\left((1-x^2)^m\left(\left(\frac{d}{dx}\right)^{l+m}(1-x^2)^l\right)\right) = \sum_{k=0}^{l'+m}\binom{l'+m}{k}\left(\left(\frac{d}{dx}\right)^{l'+m-k}(1-x^2)^m\right) \times\left(\left(\frac{d}{dx}\right)^{l+m+k}(1-x^2)^l\right) \tag{8.73}$$

Notice that because $(1-x^2)^\mu$ has only 2μ nonvanishing derivatives, nonvanishing terms require

$$\left.\begin{aligned} l'+m-k &\le 2m \\ l+m+k &\le 2l \end{aligned}\right\} \Longrightarrow l' \le l, \quad l'-m \le k \le l-m \tag{8.74}$$

Yet we assumed $l \leq l'$. Therefore, this integral vanishes for $l' \neq l$, demonstrating the orthogonality of associated Legendre functions with the same m but different l values. Furthermore, the normalization for $l = l'$ can be obtained from the single term with $k = l - m$, such that

$$\begin{aligned}\int_{-1}^{1} \left(P_{l,m}[x]\right)^2 dx &= \frac{(-)^{l+m}}{\left(2^l l!\right)^2} \binom{l+m}{l-m} \int_{-1}^{1} (1-x^2)^l \left(\left(\frac{d}{dx}\right)^{2m} (1-x^2)^m\right) \\ &\quad \times \left(\left(\frac{d}{dx}\right)^{2l} (1-x^2)^l\right) dx \\ &= \frac{1}{\left(2^l l!\right)^2} \binom{l+m}{l-m} (2m)!(2l)! \int_{-1}^{1} (1-x^2)^l\, dx\end{aligned} \tag{8.75}$$

Although the remaining integral can be evaluated most simply as a finite sum by expanding the integrand, more clever techniques are needed to obtain the simple closed-form result

$$\int_{-1}^{1} (1-x^2)^l\, dx = \frac{2^{2l+1}(l!)^2}{(2l+1)!} \tag{8.76}$$

and are relegated to the problems at the end of the chapter. Collecting these results, we finally obtained the orthogonality integral

$$\int_{-1}^{1} P_{l,m}[x] P_{l',m}[x]\, dx = \frac{2}{(2l+1)} \frac{(l+m)!}{(l-m)!} \delta_{l,l'} \tag{8.77}$$

Having two indices, the associated Legendre functions exhibit a wider variety of recursion relations than the ordinary Legendre functions. Some recursion relations vary the degree, others the order, and many others vary both. Among these are

$$P_{l,m+2}[x] + (m+1)\frac{2x}{1-x^2} P_{l,m+1}[x] + \left(l(l+1) - m(m+1)\right) P_{l,m}[x] == 0 \tag{8.78}$$

$$x P_{l,m} + (l-m+1)(1-x^2)^{1/2} P_{l,m-1} - P_{l-1,m} == 0 \tag{8.79}$$

$$P_{l-1,m+1} - (2l+1)(1-x^2)^{1/2} P_{l,m} - P_{l+1,m+1} == 0 \tag{8.80}$$

$$(l-m+1) P_{l+1,m} - (2l+1) x P_{l,m} + (l+m) P_{l-1,m} == 0 \tag{8.81}$$

but we leave the derivations as exercises. Note that the choice of phase affects some of the signs in these recursion relations. A simple method for deriving recursion relations for $m \geq 0$ is to begin either with the differential equation or with one of the recursion relations for Legendre polynomials, differentiate m times, and multiply by a power of $(1-x^2)^{1/2}$ to produce an analogous relation for associated Legendre functions. With further analysis one can show that the same relations apply under more general conditions, complex $x \to z$ and nonintegral or even complex $l, m \to \nu, \mu$.

Rather than develop further properties of associated Legendre functions, I prefer to proceed directly to the spherical harmonics in which the ϕ dependence distinguishes the sign of m.

8.2.6 Spherical Harmonics

For problems in three dimensions it is convenient to define angular basis functions with a simple orthonormality with respect to solid angle, such that

$$\int Y_{l,m}[\hat{r}]Y^*_{l',m'}[\hat{r}]\,d\Omega = \int_0^{2\pi}\int_{-1}^{1} Y_{l,m}[\theta,\phi]Y^*_{l',m'}[\theta,\phi]\,d\cos\theta\,d\phi = \delta_{l,l'}\delta_{m,m'} \tag{8.82}$$

The dependence of the spherical harmonics $Y_{l,m}[\theta,\phi]$ upon the polar angle is obviously related to the associated Legendre functions while the dependence upon the azimuthal angle is proportional to Fourier functions $e^{im\phi}$. Thus, we define

$$Y_{l,m}[\theta,\phi] = \left(\frac{2l+1}{4\pi}\frac{(l-m)!}{(l+m)!}\right)^{1/2} P_{l,m}[\cos\theta]e^{im\phi} \tag{8.83}$$

where the ugly factor provides the desired normalization. Using the standard definition for $P_{l,m}$ with $m < 0$, we obtain the conjugation property

$$Y^*_{l,m}[\theta,\phi] = (-)^m Y_{l,-m}[\theta,\phi] \tag{8.84}$$

Similarly, under inversion of the coordinate system

$$\hat{r} \to -\hat{r} \Longrightarrow \theta \to \pi - \theta,\ \ \phi \to \pi + \phi \tag{8.85}$$

and using

$$P_{l,m}[-x] = (-)^{l+m}P_{l,m}[x],\qquad \mathrm{Exp}[im(\phi+\pi)] = (-)^m\,\mathrm{Exp}[im\phi] \tag{8.86}$$

to write

$$Y_{l,m}[\pi-\theta,\pi+\phi] = (-)^l Y_{l,m}[\theta,\phi] \tag{8.87}$$

we identify the *parity*

$$Y_{l,m}[-\hat{r}] = (-)^l Y_{l,m}[\hat{r}] \tag{8.88}$$

as $(-)^l$. It is also useful to know that

$$Y_{l,m}[\theta,0] = \left(\frac{2l+1}{4\pi}\right)^{1/2} P_l[\cos\theta]\,\delta_{m,0} \tag{8.89}$$

$$Y_{l,m}[\hat{z}] = Y_{l,m}[0,0] = \left(\frac{2l+1}{4\pi}\right)^{1/2}\delta_{m,0} \tag{8.90}$$

The degree l is usually related to orbital angular momentum while the order m is the projection of the total angular momentum on the polar axis.

8.2.7 Multipole Expansion

Recognizing that the Legendre functions are complete for the polar angle while the trigonometric functions are complete for the azimuthal angle, we expect the spherical harmonics to provide a complete orthonormal basis for functions defined on the unit sphere, such that

$$f[\hat{r}] = \sum_{l=0}^{\infty} \sum_{m=-l}^{l} f_{l,m} Y_{l,m}[\hat{r}] \tag{8.91}$$

$$\begin{aligned} f_{l,m} &= \int f[\hat{r}] Y_{l,m}[\hat{r}]^* \, d\Omega \\ &= \int_0^{2\pi} \int_{-1}^{1} f[\theta, \phi] Y_{l,m}[\theta, \phi]^* \, d\cos\theta \, d\phi \\ &= \int_0^{2\pi} \int_0^{\pi} f[\theta, \phi] Y_{l,m}[\theta, \phi]^* \, \mathrm{Sin}[\theta] \, d\theta \, d\phi \end{aligned} \tag{8.92}$$

where the coefficients $f_{l,m}$ are constants. Suppose that the angular delta function

$$\delta[\hat{r} - \hat{r}'] = \delta[\cos\theta - \cos\theta'] \delta[\phi - \phi'] = \sum_{l=0}^{\infty} \sum_{m=-l}^{l} A_{l,m}[\hat{r}'] Y_{l,m}[\hat{r}] \tag{8.93}$$

is expanded with respect to unprimed variables using expansion coefficients which are functions of the primed variables, here treated as parameters. Using the orthogonality properties, the expansion coefficients become

$$A_{l,m}[\hat{r}'] = \int \delta[\hat{r} - \hat{r}'] Y_{l,m}[\hat{r}]^* \, d\Omega = Y_{l,m}[\hat{r}']^* \tag{8.94}$$

Therefore, the completeness relation for spherical harmonics takes the form

$$\delta[\hat{r} - \hat{r}'] = \sum_{l=0}^{\infty} \sum_{m=-l}^{l} Y_{l,m}[\hat{r}] Y_{l,m}[\hat{r}']^* = \sum_{l,m} Y_{l,m}[\hat{r}] Y_{l,m}[\hat{r}']^* \tag{8.95}$$

where the final sum uses a short-hand notation that indicates summation with respect to both variables over their natural ranges, namely $0 \le l < \infty$, $-l \le m \le l$.

Notice that it does not matter which of the spherical harmonics in the completeness relation is conjugated because this combination is real. The summation over m resembles the summation over components used to evaluate the scalar product of two vectors. In fact, $Y_{l,m}$ represents a component of a spherical tensor of rank l and this summation represents a scalar product between two spherical tensors of the same rank. Thus, it is convenient to define a scalar product of the form

$$Y_l[\hat{r}] \cdot Y_l[\hat{r}'] \equiv \sum_{m=-l}^{l} Y_{l,m}[\hat{r}] Y_{l,m}[\hat{r}']^* \tag{8.96}$$

and to express the completeness relation in the more compact form

$$\delta[\hat{r} - \hat{r}'] = \sum_{l} Y_l[\hat{r}] \cdot Y_l[\hat{r}'] \tag{8.97}$$

An interesting current application of multipole expansions is to angular variations of the cosmic microwave background. The relic microwave background is approximately uniform, with an average temperature of 2.72 K, but small fluctuations carry information about the expansion of the early universe and can be used to test theories of inflation. Thus, one writes

$$T[\theta,\phi] = \sum_l T_{l,m} Y_l[\theta,\phi], \quad A_l = \sum_{m=-l}^{l} |T_{l,m}|^2 \tag{8.98}$$

where $4\pi A_0$ represents the average temperature, A_1 is related to the motion of the sun relative to the preferred reference frame, and higher multipole strengths carry the desired cosmological information.

Next consider a scalar function of $\vec{r}$ expanded in the form

$$f[\vec{r}] = \sum_{l=0}^{\infty} \sum_{m=-l}^{l} f_{l,m}[r] Y_{l,m}[\hat{r}] \tag{8.99}$$

$$f_{l,m}[r] = \int f[\vec{r}] Y_{l,m}[\hat{r}]^* \, d\Omega \tag{8.100}$$

where the coefficients are radial functions. For example, if $f[\vec{r}]$ represents a density, the expansion coefficients would be described as *multipole densities*. If f is real, such that

$$f^* = f \Longrightarrow f_{l,m}^* = (-)^m f_{l,-m} \tag{8.101}$$

we find that the multipole densities share the conjugation properties of spherical harmonics.

8.2.8 Addition Theorem

Suppose that two vectors, $\vec{r}$ and $\vec{r}'$, are described by polar and azimuthal angles (θ,ϕ) and (θ',ϕ') and that γ represents the angle between them, such that $\cos\gamma = \hat{r}\cdot\hat{r}'$. Using the completeness of spherical harmonics, we can expand $P_l[\cos\gamma]$ in terms $Y_{l,m}[\theta,\phi]$ according to

$$P_l[\cos\gamma] = \sum_{l'=0}^{\infty} \sum_{m=-l'}^{l'} A_{l',m}[\theta',\phi'] Y_{l',m}[\theta,\phi] \tag{8.102}$$

where the expansion coefficients

$$A_{l'm}[\theta',\phi'] = \int Y_{l',m}[\theta,\phi]^* P_l[\cos\gamma] \, d\Omega \tag{8.103}$$

depend upon the other pair of angles. Recognizing that $P_l[\cos\gamma]$ is real and using the conjugation property of spherical harmonics

$$
\begin{aligned}
P_l[\cos\gamma] = P_l[\cos\gamma]^* &\Longrightarrow \sum_{l'=0}^{\infty}\sum_{m=-l'}^{l'} A_{l',m}[\theta',\phi']Y_{l',m}[\theta,\phi] \\
&= \sum_{l'=0}^{\infty}\sum_{m=-l'}^{l'} (-)^m A_{l',m}[\theta',\phi']^* Y_{l',m}[\theta,\phi] \\
&= \sum_{l'=0}^{\infty}\sum_{m=-l'}^{l'} (-)^m A_{l',-m}[\theta',\phi']^* Y_{l',m}[\theta,\phi]
\end{aligned}
\tag{8.104}
$$

we observe that the coefficients

$$
A_{l',m}\left[\theta',\phi'\right] = (-)^m A_{l',-m}\left[\theta',\phi'\right]^* \tag{8.105}
$$

share the same conjugation property. (Orthogonality of spherical harmonics ensures that this relation applies term by term.) Furthermore, $\hat{r}\cdot\hat{r}'$ is symmetric with respect to the exchange of primed and unprimed coordinates. Therefore, the expansion should take the form

$$
P_l[\hat{r}\cdot\hat{r}'] = \sum_{l'=0}^{\infty}\sum_{m=-l'}^{l'} B_{l',m}Y_{l',m}[\theta,\phi]Y_{l',m}[\theta',\phi']^* \tag{8.106}
$$

where the coefficients are independent of angles and this combination of spherical harmonics is real. Finally, because $P_l[\hat{r}\cdot\hat{r}']$ is a scalar function its value must be independent of the orientation of the coordinate system. Suppose that we choose a coordinate system in which $\hat{r}' = \hat{z}$ such that

$$
\begin{aligned}
P_l\left[\hat{r}\cdot\hat{r}'\right] = P_l[\cos\theta] &= \sum_{l'=0}^{\infty}\sum_{m=-l'}^{l'} B_{l',m}Y_{l',m}[\theta,\phi]\left(\frac{2l'+1}{4\pi}\right)^{1/2}\delta_{m,0} \\
&= \sum_{l'=0}^{\infty} B_{l',m}\left(\frac{2l'+1}{4\pi}\right)P_{l'}[\cos\theta]
\end{aligned}
\tag{8.107}
$$

The orthogonality of Legendre functions limits the rhs to the single term with $l' = l$. Therefore, we obtain the *addition theorem*

$$
P_l[\hat{r}\cdot\hat{r}'] = \frac{4\pi}{2l+1}\sum_{m=-l}^{l} Y_{l,m}[\hat{r}]Y_{l,m}[\hat{r}']^* \tag{8.108}
$$

or, using the scalar product,

$$
P_l[\hat{r}\cdot\hat{r}'] = \frac{4\pi}{2l+1}Y_l[\hat{r}]\cdot Y_l[\hat{r}'] \tag{8.109}
$$

The addition theorem proves invaluable in a wide variety of problems. Some of the most important examples are based upon the multipole expansion of

$$\frac{1}{|\vec{r}-\vec{r}'|} = \sum_{l=0}^{\infty} \frac{r_<^l}{r_>^{l+1}} P_l[\hat{r}\cdot\hat{r}'] \tag{8.110}$$

which we can now express in the form

$$\frac{1}{|\vec{r}-\vec{r}'|} = \sum_{l=0}^{\infty} \frac{4\pi}{2l+1} \frac{r_<^l}{r_>^{l+1}} Y_l[\hat{r}]\cdot Y_l[\hat{r}'] \tag{8.111}$$

8.2.8.1 Example: Far-Field Solution to Poisson's Equation

The electrostatic potential $\psi[\vec{r}]$ around a charge distribution $\rho[\vec{r}]$ satisfies Poisson's equation

$$\nabla^2\psi[\vec{r}] = -4\pi\rho[\vec{r}] \Longrightarrow \psi[\vec{r}] = \int G[\vec{r},\vec{r}']\rho[\vec{r}']\,d^3r' \tag{8.112}$$

where the Green function for a point charge is simply

$$\nabla^2 G[\vec{r},\vec{r}'] = -4\pi\delta[\vec{r}-\vec{r}'] \Longrightarrow G[\vec{r},\vec{r}'] = \frac{1}{|\vec{r}-\vec{r}'|} = \sum_{l=0}^{\infty} \frac{4\pi}{2l+1} \frac{r_<^l}{r_>^{l+1}} Y_l[\hat{r}]\cdot Y_l[\hat{r}'] \tag{8.113}$$

such that

$$\psi[\vec{r}] = \int \frac{\rho[\vec{r}']}{|\vec{r}-\vec{r}'|}\,d^3r' \tag{8.114}$$

The multipole expansion of the charge density takes the form

$$\rho[\vec{r}] = \sum_{l,m} \rho_{l,m}[r] Y_{l,m}[\hat{r}] \tag{8.115}$$

$$\rho_{l,m}[r] = \int \rho[\vec{r}] Y_{l,m}[\hat{r}]\,d\Omega \tag{8.116}$$

Suppose that r is much larger than any r' with appreciable charge density. A far-field solution for the electrostatic potential of a localized charge distribution can then be expanded as

$$r \gg r' \Longrightarrow \psi[\vec{r}] \simeq \sum_{l,m} \frac{4\pi}{2l+1} \frac{q_{l,m}}{r^{l+1}} Y_{l,m}[\hat{r}] \tag{8.117}$$

where the spherical multipole moments $q_{l,m}$ are given by

$$q_{l,m} = \int r^l Y_{l,m}[\hat{r}]\rho[\vec{r}]\,d^3r = \int_0^{\infty} \rho_{l,m}[r] r^{l+2}\,dr \tag{8.118}$$

Recognizing that the contributions of higher multipole moments are suppressed by increasing powers of r, one expects that it is sufficient for very large r to employ only the lowest few multipoles. Therefore, the multipole expansion provides a very efficient solution for asymptotically large distances.

8.2.9 Legendre Functions of the Second Kind

Generally a second-order differential equation must possess two linearly independent solutions, but our anticipated application to the polar angle dependence of physics problems with $x = \mathrm{Cos}[\theta]$ in the range $-1 \leq x \leq 1$ selects $P_n[x]$ as the most relevant of the two solutions. The (sometimes) implicit requirement of analyticity at $x = \pm 1$ is actually a boundary condition that forces the eigenvalue in the differential equation

$$(1 - x^2)y''[x] - 2xy'[x] + \lambda y[x] == 0 \tag{8.119}$$

to assume the values $\lambda = n(n+1)$ where n is an integer and selects the corresponding eigenfunctions $P_n[x]$ that are regular at the singular points of the differential equation (where the coefficient of y'' vanishes). However, if the polar angle is limited to a more restricted range $|x| \leq a < 1$, the second solution is required. Linear homogeneous boundary conditions at the smaller angular limits still yield an eigenvalue problem, but the eigenvalues are more general and the eigenfunctions are linear combinations of the two solutions.

We apply the *Frobenius method* to construct solutions from power series about the origin. Let

$$y[x] = x^r \sum_{n=0}^{\infty} a_n x^n \tag{8.120}$$

where the appropriate value of r has yet to be determined. Direct substitution gives

$$\sum_{n=0}^{\infty} a_n \left((n + r)(n + r - 1)(1 - x^2)x^{n+r-2} - 2(n + r)x^{n+r} + \lambda x^{n+r} \right) == 0 \tag{8.121}$$

or

$$\sum_{n=-2}^{\infty} a_{n+2}(n + r + 2)(n + r + 1)x^{n+r} == \sum_{n=0}^{\infty} a_n \left((n + r)(n + r + 1) - \lambda \right) x^{n+r} \tag{8.122}$$

The left-hand side has two powers that are absent from the right, namely $n \in \{-2, -1\}$, but the coefficients of both vanish if we choose $r = 0$. In order that this equation be satisfied for all x within the radius of convergence, we equate the coefficients for like powers of x to obtain a recursion relation

$$\frac{a_{n+2}}{a_n} = \frac{n(n + 1) - \lambda}{(n + 2)(n + 1)} \tag{8.123}$$

that determines two solutions, one for even and another for odd n. Therefore, the general solution takes the form

$$\begin{aligned} y[x] &= a_0 \left(1 - \lambda \frac{x^2}{2!} - (6 - \lambda)\lambda \frac{x^4}{4!} - (6 - \lambda)(20 - \lambda)\lambda \frac{x^6}{6!} - \cdots \right) \\ &+ a_1 \left(x + (2 - \lambda)\frac{x^3}{3!} + (2 - \lambda)(12 - \lambda)\frac{x^5}{5!} + (2 - \lambda)(12 - \lambda)(30 - \lambda)\frac{x^7}{7!} + \cdots \right) \end{aligned} \tag{8.124}$$

where a_0 and a_1 are arbitrary.

Convergence is a key issue. The ratio test

$$\lim_{n\to\infty}\left|\frac{a_{n+2}}{a_n}R^2\right| < 1 \Longrightarrow R < 1 \tag{8.125}$$

informs us that solutions for arbitrary λ diverge on the unit circle. (Although the ratio test is technically inconclusive at $|x| = 1$, the integral test is conclusively divergent.) However, divergence is avoided when $\lambda = l(l+1)$ for nonnegative integer l causes the power series to terminate in a polynomial of order l that is identified as the familiar Legendre polynomials $P_l[x]$. If l is even, the function associated with a_0 is proportional to P_l while the function associated with a_1 diverges at $x = \pm 1$, whereas if l is odd a_1 produces P_l while a_0 produces a divergent function. The Legendre functions of the second kind, $Q_l[x]$, are identified with the divergent series except for a conventional normalization yet to be specified.

The lowest few Q_l are illustrated in Fig. 8.2. Obviously, these functions become more oscillatory as l increases and have parity $Q_l[-x] = (-)^{l+1}Q_l[x]$ opposite that of the corresponding P_l. For integer order one can factor the logarithmic divergence at the end points from polynomial portions to obtain

$$Q_l[z] = \frac{1}{2}P_l[z]\,\mathrm{Log}\left[\frac{1+z}{1-z}\right] - \sum_{m=0}^{\lfloor l/2 \rfloor}\frac{2l-4m-1}{(2m+1)!!(l-m)}P_{l-1-2m}[z] \tag{8.126}$$

Closer analysis shows that $P_\nu[+1] = 1$ for all ν, but $P_\nu[x]$ diverges logarithmically at $x \to -1$ for $\nu \neq l$. However, $Q_\nu[x]$ exhibits the opposite behavior: it is divergent at both singular points of the differential equation, $x = \pm 1$, for $\nu = l$ but is finite at $x = -1$ for $\nu = (2l+1)/2$.

8.2.10 Relationship to Hypergeometric Functions

With suitable variable transformations, the *hypergeometric equation*

$$\begin{aligned} &t(1-t)u''[t] + \big(\gamma - (\alpha+\beta+1)t\big)u'[t] - \alpha\beta u[t] == 0 \\ &\qquad \Longrightarrow y[t] = AF[\alpha,\beta,\gamma,t] + Bt^{1-\gamma}F[1-\gamma+\alpha, 1-\gamma+\beta, 2-\gamma, t] \end{aligned} \tag{8.127}$$

contains as special cases many of the second-order differential equations of interest to mathematical physics. Using standard methods, one can show that the power series

$$F[\alpha,\beta,\gamma,t] = \sum_{n=0}^{\infty}\frac{(\alpha)_n(\beta)_n}{(\gamma)_n}\frac{t^n}{n!} \tag{8.128}$$

where

$$(\alpha)_0 = 1 \tag{8.129}$$

$$(\alpha)_n = \alpha(\alpha+1)\cdots(\alpha+n-1) \tag{8.130}$$

converges for $|t| < 1$. Furthermore, when α or β is a nonpositive integer, the series terminates and the hypergeometric function F reduces to a polynomial. Although neither time

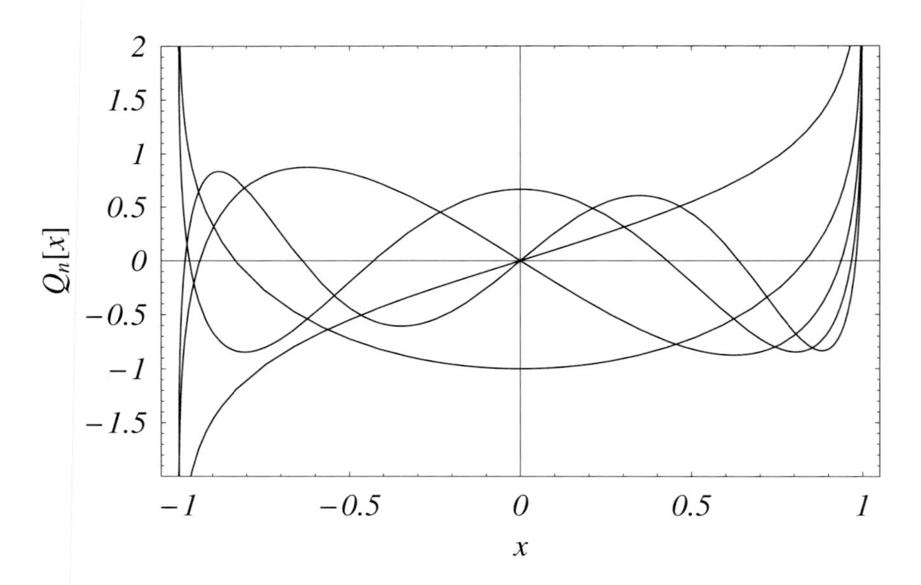

Figure 8.2. Legendre functions of second kind.

nor space permits full development here of the theory of hypergeometric functions, which can be found in more advanced texts, it is still of interest to express the Legendre functions in terms of hypergeometric functions.

Consider the variable transformation

$$t = \tfrac{1}{2}(1 - z) \Longrightarrow u'[t] = -2y'[z], \quad u''[t] = 4y''[z], \quad t(1 - t) = \tfrac{1}{4}(1 - z^2) \tag{8.131}$$

such that

$$(1 - z^2)y''[z] + \big(-2\gamma + (\alpha + \beta + 1)(1 - z)\big)y'[z] - \alpha\beta y[z] == 0 \tag{8.132}$$

By choosing

$$\alpha + \beta + 1 - 2\gamma == 0 \tag{8.133}$$

$$\alpha + \beta + 1 == 2 \qquad \Longrightarrow \gamma = 1, \quad \alpha = -\nu, \quad \beta = \nu + 1 \tag{8.134}$$

$$\alpha\beta == -\nu(\nu + 1) \tag{8.135}$$

we obtain Legendre's equation

$$(1 - z^2)y''[z] - 2zy'[z] + \nu(\nu + 1)y[z] == 0 \tag{8.136}$$

and identify

$$|z - 1| < 2 \Longrightarrow P_\nu[z] = F\left[-\nu, \nu + 1, 1, \frac{1 - z}{2}\right] \tag{8.137}$$

$$P_{\nu,\mu}[z] = \frac{1}{\Gamma[1 - \mu]}\left(\frac{z + 1}{z - 1}\right)^{\mu/2} F\left[-\nu, \nu + 1, 1 - \mu, \frac{1 - z}{2}\right] \tag{8.138}$$

as the Legendre function of the first kind in term of a Taylor series around $z = 1$ with the conventional normalization $P_\nu[1] = 1$. Alternatively, the substitutions

$$t = z^{-2}, \quad \alpha = \frac{\nu+2}{2}, \quad \beta = \frac{\nu+1}{2}, \quad \gamma = \frac{\nu+3}{2} \tag{8.139}$$

also transform the hypergeometric equation into Legendre's equation, permitting the Legendre functions of the second kind to be identified as

$$|z| > 1 \Longrightarrow Q_\nu[z] = \frac{\sqrt{\pi}}{(2z)^{\nu+1}} \frac{\Gamma[\nu+1]}{\Gamma\left[\nu+\frac{3}{2}\right]} F\left[\frac{\nu+2}{2}, \frac{\nu+1}{2}, \frac{\nu+3}{2}, z^{-2}\right] \tag{8.140}$$

$$Q_{\nu,\mu}[z] = \frac{\sqrt{\pi}}{2^{\nu+1}} \frac{(z^2-1)^{\mu/2}}{z^{\nu+\mu+1}} \frac{\Gamma[\nu+\mu+1]}{\Gamma\left[\nu+\frac{3}{2}\right]} e^{i\mu\pi} F\left[\frac{\nu+\mu+2}{2}, \frac{\nu+\mu+1}{2}, \frac{\nu+3}{2}, z^{-2}\right] \tag{8.141}$$

when ν is not a negative integer. The first factor is made single-valued by a cut on $(-\infty, +1)$ and the normalization is conventional. Definitions for larger regions can be obtained by analytic continuation, but we refer the reader to standard references for further details.

8.2.11 Analytic Structure of Legendre Functions

The most general form of the associated Legendre equation is

$$(1-z^2)y''[z] - 2zy'[z] + \left(\nu(\nu+1) - \frac{\mu^2}{1-z^2}\right)y[z] == 0$$
$$\Longrightarrow y[z] = AP_{\nu,\mu}[z] + BQ_{\nu,\mu}[z] \tag{8.142}$$

where $\{z, \nu, \mu\}$ are permitted to be complex. Most of our results for Legendre functions of real arguments in the range $-1 < x < 1$ can be extended to the unit disk $|z| < 1$ simply by replacing $x \to z$ in series representations, recursion relations, and other formulas even if the degree ν or order μ is permitted to be complex. However, for arbitrary z one must pay close attention to the choice of branch cuts. The most common, although not unique, choices are listed in Table 8.2. Each is then an entire function of z in the cut plane, and customary values on the real axis for $-1 < x < 1$ can be defined using suitable averages of values above and below the cut where $\epsilon \to 0^+$.

Table 8.2. Branch cuts for Legendre functions.

Function	Cut	Value for $-1 < x < 1$
$P_\nu[z]$	$(-\infty, -1)$	$P_\nu[x]$
$Q_\nu[z]$	$(-\infty, +1)$	$Q_\nu[x] = (Q_\nu[x+i\epsilon] + Q_\nu[x-i\epsilon])/2$
$P_{\nu,\mu}[z]$	$(-\infty, +1)$	$P_{\nu,\mu}[x] = \text{Exp}[\pm i\pi\mu/2]P_{\nu,\mu}[x \pm i\epsilon]$
$Q_{\nu,\mu}[z]$	$(-\infty, +1)$	$Q_{\nu,\mu}[x] = \text{Exp}[-i\pi\mu](\text{Exp}[-i\pi\mu/2]Q_{\nu,\mu}[x+i\epsilon] + \text{Exp}[i\pi\mu/2]Q_{\nu,\mu}[x-i\epsilon])/2$

If we define

$$f_{\nu,\mu}[z] = AP_{\nu,\mu}[z] + BQ_{\nu,\mu}[z] \tag{8.143}$$

where A, B are constant coefficients independent of $\{\nu, \mu, z\}$, we find that the recursion relations

$$f_{\nu,\mu+2} + (\mu + 1)\frac{2z}{\left(z^2 - 1\right)^{1/2}}f_{\nu,\mu+1} - \left(\nu(\nu + 1) - \mu(\mu + 1)\right)f_{\nu,\mu} == 0 \tag{8.144}$$

$$zf_{\nu,\mu} + (\nu - \mu + 1)\left(z^2 - 1\right)^{1/2} f_{\nu,\mu-1} - f_{\nu-1,\mu} == 0 \tag{8.145}$$

$$f_{\nu+1,\mu+1} - (2\nu + 1)\left(z^2 - 1\right)^{1/2} f_{\nu,\mu} - f_{\nu-1,\mu+1} == 0 \tag{8.146}$$

$$(\nu - \mu + 1)f_{\nu+1,\mu} - (2\nu + 1)zf_{\nu,\mu} + (\nu + \mu)f_{\nu-1,\mu} == 0 \tag{8.147}$$

appear slightly different but that the previous results for $z \to x$ with $-1 < x < 1$ will be recovered if $(z^2 - 1)^{1/2}$ is also handled with a cut on $(-\infty, +1)$ in a consistent manner. Similarly, the Rodrigues formulas now take the form

$$P_{\nu,m}[z] = (z^2 - 1)^{m/2}\left(\frac{d}{dz}\right)^m P_\nu[z] \Longrightarrow P_{\nu,m}[x] = (-)^m(1 - x^2)^{m/2}\left(\frac{d}{dx}\right)^m P_\nu[x] \tag{8.148}$$

$$Q_{\nu,m}[z] = (z^2 - 1)^{m/2}\left(\frac{d}{dz}\right)^m Q_\nu[z] \Longrightarrow Q_{\nu,m}[x] = (-)^m(1 - x^2)^{m/2}\left(\frac{d}{dx}\right)^m Q_\nu[x] \tag{8.149}$$

for nonnegative integer m. These results offer another justification for retention of the $(-)^m$ phase when the argument $x = \mathrm{Cos}[\theta]$ is in the range $(-1, 1)$.

Generalizing the Schläfli integral representation of $P_n[z]$, one can show that

$$P_\nu[z] = \frac{2^{-\nu}}{2\pi i}\oint_{C_1}\frac{(t^2 - 1)^\nu}{(t - z)^{\nu+1}}\,dt \tag{8.150}$$

$$Q_\nu[z] = \frac{2^{-\nu}}{4i\,\mathrm{Sin}[\nu\pi]}\oint_{C_2}\frac{(t^2 - 1)^\nu}{(z - t)^{\nu+1}}\,dt \tag{8.151}$$

provide solutions to Legendre's differential equation for arbitrary $n \to \nu$, but the contours for noninteger ν must be chosen carefully because the integrands have branch points at $t = -1, z, +1$. Suitable contours are sketched in Fig. 8.3. Legendre functions of the first kind are obtained using contour C_1 that encloses both $t = z$ and $t = 1$ with one cut below the real axis for $t \leq -1$ and a second between $t = 1$ and $t = z$. After integration with respect to t, the cut between $t = +1$ and z becomes irrelevant but $P_\nu[z]$ is left with a branch point at $z = -1$ for noninteger ν and a cut below the negative real axis from there to $-\infty$. With these choices we find that $P_\nu[1] = 1$ for any ν. Alternatively, the bow tie contour C_2 provides a related integral representation for $Q_\nu[z]$. For C_2, z must be outside both loops so that the phase changes, incurred by circumnavigating the branch points, cancel. A more detailed analysis shows that a single-valued definition of Q_ν requires a cut from $z = +1$ to $z = -\infty$. Similar integral representations can also be constructed for associated Legendre functions, but we will be content to stop here.

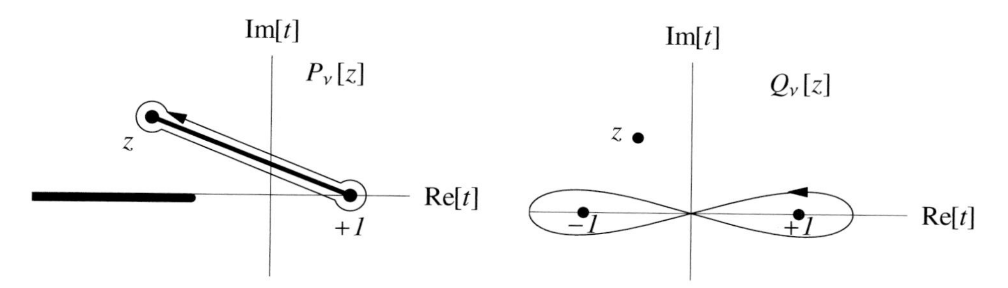

Figure 8.3. Contours for integral representations of Legendre functions.

Finally, one can also show that, for fixed z, P_ν is an entire function of ν while Q_ν is a meromorphic function of ν with simple poles at negative integers. This property of Legendre functions is useful in quantum scattering theory where significant insights can be obtained by treating the orbital angular momentum as a complex variable. More detailed development of the properties of Legendre functions of the second kind can be obtained, when needed, from standard references. These developments are omitted here because we will not find much use for these functions in the remainder of the course. Suffice it to say that most properties can be derived from the Schläfli integral representation.

8.3 Bessel Functions

8.3.1 Cylindrical

Variants of Bessel's differential equation arise frequently when the method of separation of variables is applied in cylindrical or spherical coordinates to partial differential equations based upon the Laplacian operator. However, it is instructive to begin our study of these functions using the generating function instead of the differential equation, similar to the approach used for Legendre functions. The generating function for cylindrical Bessel functions of the first kind, $J_n[x]$, takes the form of a Laurent expansion with respect to t

$$g[t,x] = \text{Exp}\left[\frac{x}{2}(t - t^{-1})\right] = \sum_{n=-\infty}^{\infty} J_n[x]t^n \tag{8.152}$$

whose coefficients are the desired functions of x. Bessel functions with negative n are related to those with positive n using the symmetry

$$g[-t^{-1}, x] = g[t,x] \Longrightarrow J_{-n}[x] = (-)^n J_n[x] \tag{8.153}$$

Similarly, we obtain the reflection symmetry

$$g[-t,-x] = g[t,x] \Longrightarrow J_n[-x] = (-)^n J_n[x] \tag{8.154}$$

such that

$$J_{-n}[-x] = J_n[x] \tag{8.155}$$

for integer n.

Some of the basic recursion relations can be obtained, as before, by differentiating $g[t, x]$ with respect to either variable. First consider

$$\frac{\partial g[t,x]}{\partial t} = \frac{x}{2t^2}(t^2+1)\,\text{Exp}\left[\frac{x}{2}(t-t^{-1})\right] = \sum_{n=-\infty}^{\infty} nJ_n[x]t^{n-1} \tag{8.156}$$

and substitute the generating function to write

$$\frac{x}{2t^2}(t^2+1)\sum_{n=-\infty}^{\infty} J_n[x]t^n = \sum_{n=-\infty}^{\infty} nJ_n[x]t^{n-1}$$

$$\Longrightarrow \sum_{n=-\infty}^{\infty}\left(\frac{x}{2}J_n + \frac{x}{2}J_{n+2} - (n+1)J_{n+1}\right)t^n = 0 \tag{8.157}$$

If this series is to vanish for arbitrary t, the coefficients must vanish separately for each term. Adjusting the indexing, we obtain the three-term recursion relation

$$\frac{2n}{x}J_n[x] = J_{n-1}[x] + J_{n+1}[x] \tag{8.158}$$

Similarly, differentiation with respect to x

$$\frac{\partial g[t,x]}{\partial x} = \tfrac{1}{2}(t-t^{-1})\sum_{n=-\infty}^{\infty} J_n[x]t^n = \sum_{n=-\infty}^{\infty} J_n'[x]t^n \tag{8.159}$$

provides a recursion relation

$$2J_n'[x] = J_{n-1}[x] - J_{n+1}[x] \tag{8.160}$$

for the derivative, almost by inspection. Manipulation of these recursion relations provides several additional relationships:

$$J_{n\pm 1}[x] = \frac{n}{x}J_n[x] \mp J_n'[x] \tag{8.161}$$

$$\frac{d}{dx}\left(x^{\pm n}J_n[x]\right) = \pm x^{\pm n}J_{n\mp 1}[x] \tag{8.162}$$

A power series for J_n is obtained by factoring the generating function and expanding each factor according to

$$\begin{aligned} g[t,x] &= \text{Exp}\left[\frac{xt}{2}\right]\text{Exp}\left[-\frac{x}{2t}\right] = \left(\sum_{n=0}^{\infty}\frac{1}{n!}\left(\frac{xt}{2}\right)^n\right)\left(\sum_{m=0}^{\infty}\frac{1}{m!}\left(-\frac{x}{2t}\right)^m\right) \\ &= \sum_{n,m=0}^{\infty}\frac{(-)^m}{n!m!}\left(\frac{x}{2}\right)^{n+m} t^{n-m} \end{aligned} \tag{8.163}$$

The coefficients for each nonnegative power of t are identified as

$$n \geq 0 \Longrightarrow J_n[x] = \sum_{m=0}^{\infty}\frac{(-)^m}{(n+m)!m!}\left(\frac{x}{2}\right)^{n+2m} \tag{8.164}$$

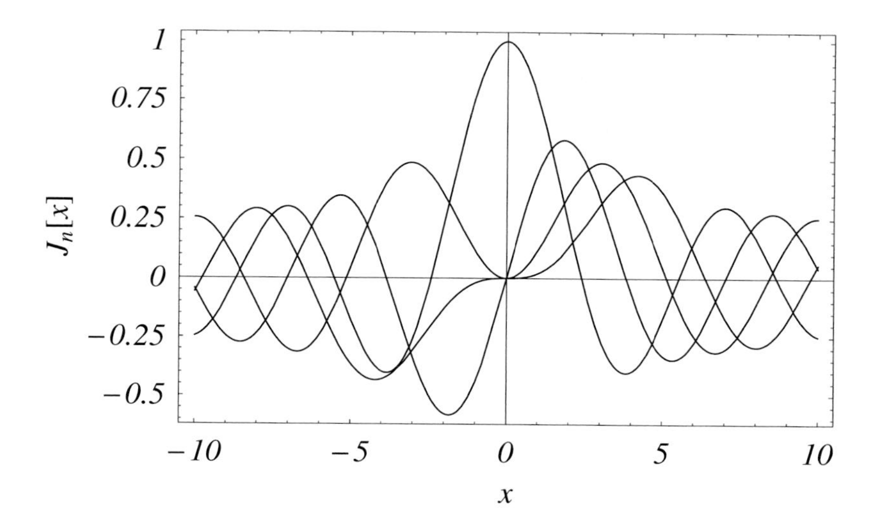

Figure 8.4. Bessel functions of first kind.

or

$$n \geq 0 \Longrightarrow J_n[x] = \left(\frac{x}{2}\right)^n \sum_{m=0}^{\infty} \frac{1}{(n+m)!m!} \left(-\frac{x^2}{4}\right)^m \tag{8.165}$$

where the leading power has been factored out and the series written as an even function of x. Similarly, for negative powers we obtain

$$J_{-n}[x] = \sum_{m=0}^{\infty} \frac{(-)^m}{(m-n)!m!} \left(\frac{x}{2}\right)^{2m-n} \tag{8.166}$$

However, because the coefficients vanish unless $m \geq n$, we use $m \to n + k$ to write

$$J_{-n}[x] = \sum_{k=0}^{\infty} \frac{(-)^{n+k}}{(n+k)!m!} \left(\frac{x}{2}\right)^{n+2k} \Longrightarrow J_{-n}[x] = (-)^n J_n[x] \tag{8.167}$$

and again conclude that changing the sign of the index does not yield an independent function. One can show that the power series is convergent for all x, even if it is inefficient for large x.

Figure 8.4 displays the first few Bessel functions of the first kind. The functions oscillate with slowly decreasing amplitudes. Their roots are interleaved, as expected for solutions of a Sturm–Liouville system.

Notice that $g[t, z]$ is an analytic function of either variable, except for the essential singularity at $t = 0$. Thus,

$$J_n[z] = \sum_{m=0}^{\infty} \frac{(-)^m}{(n+m)!m!} \left(\frac{z}{2}\right)^{n+2m} \tag{8.168}$$

is an analytic function whose radius of convergence is infinite. We can generalize this series

$$J_\nu[z] = \left(\frac{z}{2}\right)^\nu \sum_{m=0}^{\infty} \frac{1}{\Gamma[\nu+m+1]\Gamma[m+1]} \left(-\frac{z^2}{4}\right)^m \tag{8.169}$$

to arbitrary ν using a branch point at the origin for noninteger ν and intepreting the factorials as gamma functions. The cut is normally taken below the negative real axis. However, the symmetry between positive and negative integer order n does not apply for noninteger ν. Near the origin one finds

$$J_\nu[z] \approx \Gamma[\nu+1]^{-1} \left(\frac{z}{2}\right)^\nu \tag{8.170}$$

provided that ν is not a negative integer. Recognizing that the first $n-1$ coefficients of the power series vanish when $\nu = -n$, we find

$$J_{-n}[z] = (-)^n J_n[z] \approx \Gamma[n+1]^{-1} \left(-\frac{z}{2}\right)^n \tag{8.171}$$

as expected.

It is now a simple matter to verify that the basic recursion relations

$$\frac{2\nu}{z} J_\nu[z] = J_{\nu-1}[z] + J_{\nu+1}[z] \tag{8.172}$$

$$2J'_\nu[z] = J_{\nu-1}[z] - J_{\nu+1}[z] \tag{8.173}$$

continue to apply for noninteger ν and complex z. The differential equation satisfied by $J_\nu[z]$ can be deduced from the recursion relations (or from the series). Suppose that $f_\nu[z]$ is a family of functions that satisfy the basic recursion relations

$$2\nu f_\nu[z] = z(f_{\nu-1}[z] + f_{\nu+1}[z]) \tag{8.174}$$

$$2f'_\nu[z] = f_{\nu-1}[z] - f_{\nu+1}[z] \tag{8.175}$$

but are not necessarily limited to cylindrical Bessel functions of the first kind. From these one deduces

$$f_{\nu+1}[z] = \frac{\nu}{z} f_\nu[z] - f'_\nu[z] \tag{8.176}$$

$$f_{\nu-1}[z] = \frac{\nu}{z} f_\nu[z] + f'_\nu[z] \tag{8.177}$$

such that

$$z f'_\nu[z] = z f_{\nu-1}[z] - \nu f_\nu[z] \Longrightarrow z f''_\nu[z] - z f'_{\nu-1}[z] - f_{\nu-1}[z] + (\nu+1) f'_\nu[z] = 0 \tag{8.178}$$

or

$$z^2 f''_\nu[z] - z^2 f'_{\nu-1}[z] - z f_{\nu-1}[z] + (\nu+1) z f'_\nu[z] = 0 \tag{8.179}$$

Substituting

$$z f'_{\nu-1}[z] = (\nu - 1) f_{\nu-1}[z] - z f_\nu[z] \tag{8.180}$$

$$z f_{\nu-1}[z] = \nu f_\nu[z] + z f'_\nu[z] \tag{8.181}$$

such that

$$z^2 f'_{\nu-1}[z] = \left(\nu(\nu - 1) - z^2\right) f_\nu[z] + (\nu - 1) z f'_\nu[z] \tag{8.182}$$

one obtains

$$z^2 f''_\nu[z] - \left(\left(\nu(\nu - 1) - z^2\right) f_\nu[z] + (\nu - 1) z f'_\nu[z]\right) - (\nu f_\nu[z] + z f'_\nu[z]) + (\nu + 1) z f'_\nu[z] = 0 \tag{8.183}$$

Therefore, we finally obtain Bessel's differential equation

$$z^2 f''_\nu[z] + z f'_\nu[z] + (z^2 - \nu^2) f_\nu[z] == 0 \tag{8.184}$$

for which $J_\nu[z]$ are solutions that are regular at the origin for integer ν. Since this is a second-order differential equation, there must exist a second linearly independent set of solutions that also satisfy the same recursion relations.

It is important to recognize that the same recursion relations apply to any solution of Bessel's equation or to any linear combination of solutions with the same ν, provided that the coefficients are independent of both ν and z. Thus, the recursion relations

$$f_{\nu-1}[z] + f_{\nu+1}[z] = \frac{2\nu}{z} f_\nu[z] \tag{8.185}$$

$$f_{\nu-1}[z] - f_{\nu+1}[z] = 2 f'_\nu[z] \tag{8.186}$$

$$\frac{d}{dz}\left(z^{\pm\nu} f_\nu[z]\right) = \pm z^{\pm\nu} f_{\nu\mp1}[z] \tag{8.187}$$

are general properties of cylindrical Bessel functions. (We leave it as an exercise to the reader to derive the third formula from the first two.)

If we divide by z, Bessel's equation appears to be self-adjoint with respect to the weight function z provided that we use appropriate boundary conditions. If we assume that $\nu \geq 0$ (the usual situation), the requirement that the solution be finite at the origin selects J_ν while requiring either a node or a vanishing derivative at a fixed radius R provides an eigenvalue equation for either Dirichlet or Neumann boundary conditions. Thus, it is useful to change variables such that

$$\frac{1}{\xi}\frac{d}{d\xi}(\xi f'[\xi]) + \left(k^2 - \frac{\nu^2}{\xi^2}\right) f[\xi] == 0, \quad f[0] \text{ finite} \implies f[\xi] \propto J_\nu[k\xi] \tag{8.188}$$

where the boundary conditions

$$\text{Dirichlet: } f[R] == 0 \implies J_\nu[k_{\nu,n}\xi] == 0 \tag{8.189}$$

$$\text{Neumann: } f'[R] == 0 \implies J'_\nu[k_{\nu,n}\xi] == 0 \tag{8.190}$$

yield sets of eigenfunctions indexed by the positive integer n. It is now a simple matter, following the procedures of the Sturm–Liouville chapter, to demonstrate orthogonality between two solutions with the same ν but different eigenvalues. Therefore, we obtain eigenfunction expansions of the form

$$F[\xi] = \sum_n A_n J_\nu \left[k_{\nu,n}\xi\right] \tag{8.191}$$

with the aid of the orthonormality relations

$$J_\nu[k_{\nu,n}R] = 0 \Longrightarrow \int_0^R J_\nu[k_{\nu,m}\xi]J_\nu[k_{\nu,n}\xi]\xi\, d\xi = \frac{R^2}{2}J_{\nu+1}[k_{\nu,n}R]^2\delta_{m,n} \tag{8.192}$$

$$J'_\nu[k_{\nu,n}R] = 0 \Longrightarrow \int_0^R J_\nu[k_{\nu,m}\xi]J_\nu[k_{\nu,n}\xi]\xi\, d\xi = \frac{R^2}{2}\left(1 - \left(\frac{\nu}{k_{\nu,n}R}\right)^2\right)J_\nu[k_{\nu,n}R]^2\delta_{m,n} \tag{8.193}$$

that are developed in the exercises. The appropriate choice of basis depends upon the boundary conditions. Note that summation is performed over n for fixed $\nu \geq 0$. The appropriate value of ν is determined by other aspects of the problem, such as angular momentum. An important special case is the limit $R \to \infty$ for which the spacing between eigenvalues becomes infinitesimal. With careful attention to the limit, one can prove that

$$\int_0^\infty J_\nu[k\xi]J_\nu[k'\xi]\xi\, d\xi = \frac{\delta[k-k']}{\sqrt{kk'}} \tag{8.194}$$

$$\int_0^\infty J_\nu[k\xi]J_\nu[k\xi']k\, dk = \frac{\delta[\xi-\xi']}{\sqrt{\xi\xi'}} \tag{8.195}$$

for $\nu \geq 0$. The right-hand side is the appropriate delta function for cylindrical coordinates with a linear weight factor; here the denominator is written in a symmetric form and balances the weight factor such that

$$\int_0^\infty d\xi\, \xi f[\xi]\frac{\delta[\xi-\xi']}{\sqrt{\xi\xi'}} = f[\xi'] \tag{8.196}$$

The details are left to the exercises.

The generating function provides a Schläfli integral representation

$$g[t,z] = \mathrm{Exp}\left[\frac{z}{2}\left(t - t^{-1}\right)\right] = \sum_{n=-\infty}^{\infty} J_n[z]t^n \Longrightarrow J_n[z] = \frac{1}{2\pi i}\oint \mathrm{Exp}\left[\frac{z}{2}\left(t - t^{-1}\right)\right]\frac{dt}{t^{n+1}} \tag{8.197}$$

where the contour encloses the origin in a positive sense. For a circular contour, one obtains *Bessel's integral*

$$t = e^{i\theta} \Longrightarrow J_n[z] = \frac{1}{2\pi}\int_0^{2\pi} \mathrm{Exp}[i(z\,\mathrm{Sin}[\theta] - n\theta)]\, d\theta \tag{8.198}$$

or, more simply,

$$J_n[z] = \frac{1}{\pi}\int_0^{\pi} \mathrm{Cos}\left[z\,\mathrm{Sin}[\theta] - n\theta\right] d\theta \tag{8.199}$$

where z may be complex but n is integral.

8.3.2 Hankel Functions

Hankel functions are obtained by generalizing Bessel's integral representation to noninteger order ν. Consider

$$f_\nu[z] = \frac{1}{2\pi i} \int \text{Exp}\left[\frac{z}{2}(t - t^{-1})\right] \frac{dt}{t^{\nu+1}} \tag{8.200}$$

where, except for avoiding the origin, we leave the integration path arbitrary for the moment. If we apply Bessel's differential operator to both sides

$$\begin{aligned} z^2 f_\nu''[z] + z f_\nu'[z] + (z^2 - \nu^2) f_\nu[z] \\ = \frac{1}{2\pi i} \int \left(\frac{z^2}{4}(t - t^{-1})^2 + \frac{z}{2}(t - t^{-1}) + z^2 - \nu^2\right) \text{Exp}\left[\frac{z}{2}(t - t^{-1})\right] \frac{dt}{t^{\nu+1}} \end{aligned} \tag{8.201}$$

the resulting integrand is a perfect differential, such that

$$z^2 f_\nu''[z] + z f_\nu'[z] + \left(z^2 - \nu^2\right) f_\nu[z] = \frac{1}{2\pi i} \Delta F_\nu[t, z] \tag{8.202}$$

where

$$F_\nu[t, z] = t^{-\nu} \left(\frac{z}{2}(t + t^{-1}) + \nu\right) \text{Exp}\left[\frac{z}{2}(t - t^{-1})\right] \tag{8.203}$$

and where Δ indicates the difference between values at the endpoints of integration. When $\nu \to n$ is an integer, $F_n[t, z]$ is single-valued and the right-hand side vanishes for any closed path; hence, f_ν is a solution to Bessel's equation under those conditions. (Paths which do not enclose the origin, though, degenerate to the trivial solution $f_\nu \to 0$.) When ν is not integral, we require a branch cut to interpret the integrated term. It is customary to cut the t-plane below the negative real axis. We must also choose a contour for which $F_\nu[t, z]$ has the same value at both ends for any z within a usefully large domain. The simplest method is to choose semi-infinite contours that exploit the limiting properties

$$t \to 0^+ \Longrightarrow F_\nu[t, z] \sim t^{-\nu} \left(\frac{z}{2t}\right) \text{Exp}\left[-\frac{z}{2t}\right] \longrightarrow 0 \quad \text{for } \text{Re}[z] > 0 \tag{8.204}$$

$$t \to -\infty \Longrightarrow F_\nu[t, z] \sim t^{-\nu} \frac{zt}{2} \text{Exp}\left[\frac{zt}{2}\right] \longrightarrow 0 \quad \text{for } \text{Re}[z] > 0 \tag{8.205}$$

Thus, the two contours shown in Fig. 8.5 provide integral representations of the *Hankel functions*

$$\text{Re}[z] > 0 \Longrightarrow H_\nu[z] = \frac{1}{\pi i} \int_C \text{Exp}\left[\frac{z}{2}(t - t^{-1})\right] \frac{dt}{t^{\nu+1}} \tag{8.206}$$

where the normalization is conventional. Similar representations for $\text{Im}[z] < 0$ can be constructed by choosing a different cut for t and rotating the contours accordingly, as detailed in the exercises. The Hankel functions satisfy the same recursion relations as the

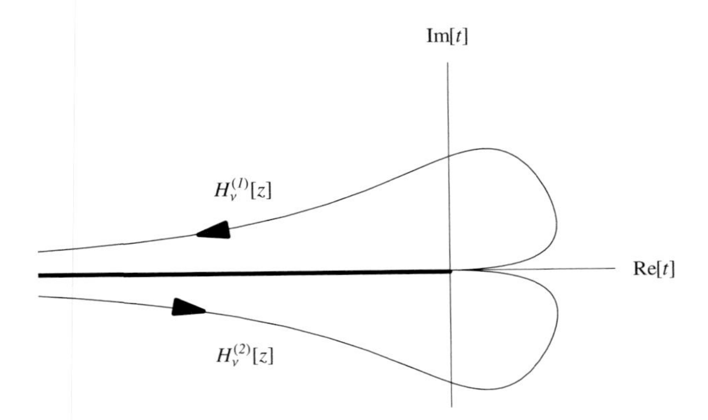

Figure 8.5. Contours for integral representation of Hankel functions.

Bessel functions because they are solutions to the same differential equation derived from those recursion relations.

Note that the same integrand provides two independent solutions to Bessel's equation, $H_\nu^{(1)}[z]$ and $H_\nu^{(2)}[z]$, depending upon whether the negative real axis is approached from above or from below. These solutions are distinct even for integral ν, despite the fact that the cut is not needed, because neither contour is closed. Thus, the ordinary Bessel function, $J_\nu[z]$, must be a linear combination of these Hankel functions. For simplicity, suppose that $\nu \to n$ actually is an integer. Recognizing that the portions of the two contours that lie along, or infinitesimally close to, the positive real axis are traversed in opposite directions and cancel, the two contours can be joined to form a closed contour for $H_n^{(1)}[z] + H_n^{(2)}[z]$ that can be deformed to that used for Bessel's integral representation of $J_n[z]$, namely a closed contour enclosing the origin in a counterclockwise sense. Thus, we obtain

$$J_n[z] = \tfrac{1}{2}(H_n^{(1)}[z] + H_n^{(2)}[z]) \tag{8.207}$$

and expect this same relationship to apply continuously to nonintegral ν. After all, Bessel functions for fixed z can be interpreted as continuous functions of ν even if that argument is customarily represented as a subscript instead of as a second variable. Alternatively, one can develop a power series by expanding the factor of $\text{Exp}[-z/2t]$ and using the integral representation of the Γ function to obtain the coefficients for powers of z, and obtaining thereby the known series for $J_\nu[z]$. Therefore,

$$J_\nu[z] = \tfrac{1}{2}(H_\nu^{(1)}[z] + H_\nu^{(2)}[z]) \tag{8.208}$$

provides a definition for arbitrary ν and, more importantly, an integral representation

$$\text{Re}[z] > 0 \Longrightarrow J_\nu[z] = \frac{1}{2\pi i}\int_C \text{Exp}\left[\frac{z}{2}(t - t^{-1})\right]\frac{dt}{t^{\nu+1}} \tag{8.209}$$

based upon the contour in Fig. 8.6.

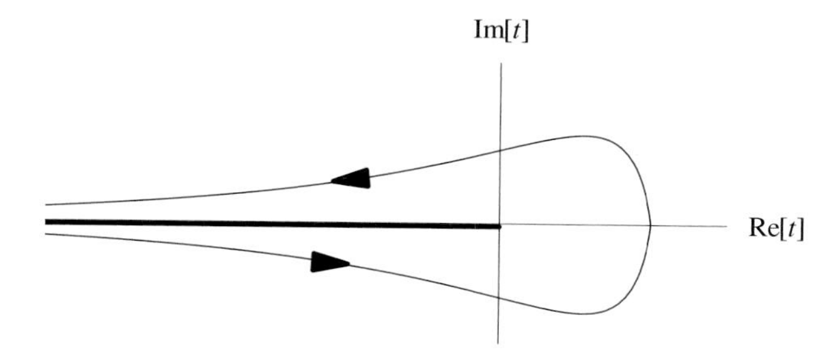

Figure 8.6. Contour for $J_\nu[z]$.

The Hankel functions and their spherical analogs are especially useful for analyzing scattering solutions in the far field. Using the method of steepest descent, one finds

$$H_\nu^{(1)}[z] \simeq \sqrt{\frac{2}{\pi}} z^{-1/2} \operatorname{Exp}\left[i\left(z - (2\nu + 1)\frac{\pi}{4}\right)\right] \tag{8.210}$$

$$H_\nu^{(2)}[z] \simeq \sqrt{\frac{2}{\pi}} z^{-1/2} \operatorname{Exp}\left[-i\left(z - (2\nu + 1)\frac{\pi}{4}\right)\right] \tag{8.211}$$

for $\operatorname{Re}[z] > 0$. (This exercise is left to the student as an opportunity to refresh his/her skills with the method of steepest descent.) Thus, we identify

$$H_n^{(1)}[k\xi] \operatorname{Exp}[-i\omega t] \simeq \sqrt{\frac{2}{\pi k\xi}} \operatorname{Exp}[i(k\xi - \omega t)] \operatorname{Exp}\left[-i(2n + 1)\frac{\pi}{4}\right] \tag{8.212}$$

as an outgoing cylindrical wave associated with the angular function $P_n\left[\operatorname{Cos}[\theta]\right]$ and

$$H_n^{(2)}[k\xi] \operatorname{Exp}[-i\omega t] \simeq \sqrt{\frac{2}{\pi k\xi}} \operatorname{Exp}[-i(k\xi + \omega t)] \operatorname{Exp}\left[i(2n + 1)\frac{\pi}{4}\right] \tag{8.213}$$

as the corresponding incoming wave for large radius ξ. Here k is the wave number and the asymptotic condition requires $k\xi \gg 1$ and modest n; for large n a somewhat more sophisticated analysis may be needed to obtain the required accuracy.

8.3.3 Neumann Functions

Having obtained two independent solutions, the Hankel functions, and identifying Bessel functions of the first kind as a linear combination of Hankel functions, we can now define Bessel functions of the second kind as the orthogonal linear combination

$$N_\nu[z] = \frac{1}{2i}\left(H_\nu^{(1)}[z] - H_\nu^{(2)}[z]\right) \tag{8.214}$$

where this choice of phase yields *Neumann functions*. (These functions are often labeled Y_ν in the literature and sometimes have different phase conventions.) The Neumann functions obey the same recursion relations as Bessel and Hankel functions. From the asymptotic behaviors

$$J_\nu[z] \simeq \sqrt{\frac{2}{\pi}} z^{-1/2} \operatorname{Cos}\left[z - (2\nu + 1)\frac{\pi}{4}\right] \tag{8.215}$$

$$N_\nu[z] \simeq \sqrt{\frac{2}{\pi}} z^{-1/2} \operatorname{Sin}\left[z - (2\nu + 1)\frac{\pi}{4}\right] \tag{8.216}$$

we recognize the Bessel functions as the cylindrical analogs of trigonometric functions.

Alternatively, we can define Neumann functions as linear combinations of J_ν and $J_{-\nu}$, which are independent solutions for nonintegral ν. To determine the appropriate coefficients, we use

$$J_{-\nu}[z] = \tfrac{1}{2}(H^{(1)}_{-\nu}[z] + H^{(2)}_{-\nu}[z]) \tag{8.217}$$

and analyze

$$H_{-\nu}[z] = \frac{1}{\pi i} \int_C \operatorname{Exp}\left[\frac{z}{2}(t - t^{-1})\right] \frac{dt}{t^{-\nu+1}} \tag{8.218}$$

The substitution

$$t = \frac{e^{\pm i\pi}}{s}, \quad dt = -\frac{e^{\pm i\pi}}{s^2}\, ds \Longrightarrow H_{-\nu}[z] = \frac{\operatorname{Exp}[\pm i\pi\nu]}{\pi i} \int_{C'} \operatorname{Exp}\left[\frac{z}{2}(s - s^{-1})\right] \frac{ds}{s^{\nu+1}} \tag{8.219}$$

preserves the integrand up to a phase. In order to preserve the contours also, we must choose the upper sign for $H^{(1)}$ and the lower sign for $H^{(2)}$. Thus, we obtain the symmetries

$$H^{(1)}_{-\nu}[z] = e^{i\pi\nu} H^{(1)}_\nu[z] \tag{8.220}$$

$$H^{(2)}_{-\nu}[z] = e^{-i\pi\nu} H^{(2)}_\nu[z] \tag{8.221}$$

such that

$$J_\nu[z] = \tfrac{1}{2}(H^{(1)}_\nu[z] + H^{(2)}_\nu[z]) \tag{8.222}$$

$$J_{-\nu}[z] = \tfrac{1}{2}(e^{i\pi\nu} H^{(1)}_\nu[z] + e^{-i\pi\nu} H^{(2)}_\nu[z]) \tag{8.223}$$

requires

$$H^{(1)}_\nu[z] = i \operatorname{Csc}[\nu\pi](e^{i\pi\nu} J_\nu[z] - J_{-\nu}[z]) \tag{8.224}$$

$$H^{(2)}_\nu[z] = -i \operatorname{Csc}[\nu\pi](e^{-i\pi\nu} J_\nu[z] - J_{-\nu}[z]) \tag{8.225}$$

for nonintegral ν. Finally, substituting these results into the definition for N_ν, we obtain

$$N_\nu[z] = \operatorname{Cot}[\nu\pi] J_\nu[z] - \operatorname{Csc}[\nu\pi] J_{-\nu}[z] \tag{8.226}$$

after a little algebra. This is the most common definition for Neumann functions, but appears to be singular for integer ν. However, under those circumstances we can apply l'Hôpital's rule to the limit $\nu \to n$ to obtain a result

$$N_n[z] = \frac{1}{\pi} \lim_{\nu \to n} \left(\frac{\partial J_\nu[z]}{\partial \nu} - (-)^n \frac{\partial J_{-\nu}[z]}{\partial \nu} \right) \tag{8.227}$$

that is continuous with respect to ν.

We now have the means to analyze the behavior of $N_\nu[z]$ near the origin, where

$$J_\nu[z] = \left(\frac{z}{2}\right)^\nu \sum_{m=0}^{\infty} \frac{1}{\Gamma[\nu + m + 1]\Gamma[m + 1]} \left(-\frac{z^2}{4}\right)^m \tag{8.228}$$

$$J_{-\nu}[z] = \left(\frac{z}{2}\right)^{-\nu} \sum_{m=0}^{\infty} \frac{1}{\Gamma[-\nu + m + 1]\Gamma[m + 1]} \left(-\frac{z^2}{4}\right)^m \tag{8.229}$$

provide the necessary expansions. Thus,

$$\begin{aligned} N_\nu[z] = {} & \text{Cot}[\nu\pi] \left(\frac{z}{2}\right)^\nu \sum_{m=0}^{\infty} \frac{1}{\Gamma[\nu + m + 1]\Gamma[m + 1]} \left(-\frac{z^2}{4}\right)^m \\ & - \text{Csc}[\nu\pi] \left(\frac{z}{2}\right)^{-\nu} \sum_{m=0}^{\infty} \frac{1}{\Gamma[-\nu + m + 1]\Gamma[m + 1]} \left(-\frac{z^2}{4}\right)^m \end{aligned} \tag{8.230}$$

for nonintegral ν. Although combining these expansions does not seem to offer an attractive formula for the coefficients, it is immediately clear that Neumann functions are singular at the origin. Hence, it is often sufficient to display only the leading terms. For simplicity suppose that $\text{Re}[\nu] > 0$ and that ν is nonintegral, such that

$$\text{Re}[\nu] > 0 \Longrightarrow N_\nu[z] \approx -\frac{\text{Csc}[\nu\pi]}{\Gamma[1 - \nu]} \left(\frac{z}{2}\right)^{-\nu} \tag{8.231}$$

Using the identity

$$\Gamma[z]\Gamma[1 - z] = \frac{\pi}{\text{Sin}[\nu\pi]} \tag{8.232}$$

we obtain

$$\text{Re}[\nu] > 0 \Longrightarrow N_\nu[z] \approx -\frac{\Gamma[\nu]}{\pi} \left(\frac{z}{2}\right)^{-\nu} \tag{8.233}$$

Thus, N_ν has a branch point at the origin with power-law divergence for $\nu \neq 0$. For the special case $\nu = 0$, we use

$$\frac{\partial z^\nu}{\partial \nu} = z^\nu \,\text{Log}[z] \tag{8.234}$$

and the definition of the *digamma function*

$$\psi[z] = \frac{\partial \,\text{Log}\big[\Gamma[z]\big]}{\partial z} = \frac{\Gamma'[z]}{\Gamma[z]} \tag{8.235}$$

to write

$$N_n[z] = \frac{1}{\pi} \lim_{\nu \to n} \left(\frac{\partial J_\nu[z]}{\partial \nu} - (-)^n \frac{\partial J_{-\nu}[z]}{\partial \nu} \right) \tag{8.236}$$

$$\frac{\partial J_\nu[z]}{\partial \nu} = J_\nu[z] \operatorname{Log}\left[\frac{z}{2}\right] - \left(\frac{z}{2}\right)^\nu \sum_{m=0}^{\infty} \frac{\psi[\nu + m + 1]}{\Gamma[\nu + m + 1]\Gamma[m + 1]} \left(-\frac{z^2}{4}\right)^m \tag{8.237}$$

$$\frac{\partial J_{-\nu}[z]}{\partial \nu} = -J_{-\nu}[z] \operatorname{Log}\left[\frac{z}{2}\right] + \left(\frac{z}{2}\right)^{-\nu} \sum_{m=0}^{\infty} \frac{(-)^m \psi[-\nu + m + 1]}{\Gamma[-\nu + m + 1]\Gamma[m + 1]} \left(-\frac{z^2}{4}\right)^m \tag{8.238}$$

and to obtain

$$N_n[z] = \frac{1}{\pi} \lim_{\nu \to n} \left(2J_n[z] \operatorname{Log}\left[\frac{z}{2}\right] - \left(\frac{z}{2}\right)^n \sum_{m=0}^{\infty} \frac{\psi[n + m + 1]}{\Gamma[n + m + 1]\Gamma[m + 1]} \left(-\frac{z^2}{4}\right)^m \right.$$
$$\left. - \left(-\frac{z}{2}\right)^{-n} \sum_{m=0}^{\infty} \frac{\psi[-\nu + m + 1]}{\Gamma[-\nu + m + 1]\Gamma[m + 1]} \left(-\frac{z^2}{4}\right)^m \right) \tag{8.239}$$

where the final summation requires further attention because both numerator and denominator appear to diverge for integer n. We leave the final steps as an exercise for the student and quote a final result

$$N_n[z] = \frac{1}{\pi} \left(2J_n[z] \operatorname{Log}\left[\frac{z}{2}\right] - \left(\frac{z}{2}\right)^{-n} \sum_{m=0}^{n-1} \frac{\Gamma[n - m]}{\Gamma[m + 1]} \left(\frac{z^2}{4}\right)^m \right.$$
$$\left. - \left(\frac{z}{2}\right)^n \sum_{m=0}^{\infty} \frac{\psi[n + m + 1] + \psi[m + 1]}{\Gamma[n + m + 1]\Gamma[m + 1]} \left(-\frac{z^2}{4}\right)^m \right) \tag{8.240}$$

in which the coefficients are manifestly finite, even if rather cumbersome. For the present purposes it is sufficient to recognize that the divergence of N_0 near the origin

$$N_0[z] \approx \frac{2}{\pi} \operatorname{Log}\left[\frac{z}{2}\right] \tag{8.241}$$

is logarithmic.

Figure 8.7 illustrates the Neumann functions. Aside from the divergence near the origin, these functions also oscillate with slowly decreasing amplitude for large x. Note that we cannot easily display the values for $x < 0$ because of the logarithmic contribution. The divergence near the origin is logarithmic or stronger: logarithmic for N_0 or power law (z^{-n}) for N_n with $n > 0$.

So why do we need two kinds of Bessel functions? A general solution to a second-order differential equation, such as Bessel's equation, is a linear combination of both kinds. Unless the boundary conditions eliminate the Neumann functions by requiring finite values at the origin, any expansion based upon eigenfunctions for Bessel's equation will require both regular and irregular solutions.

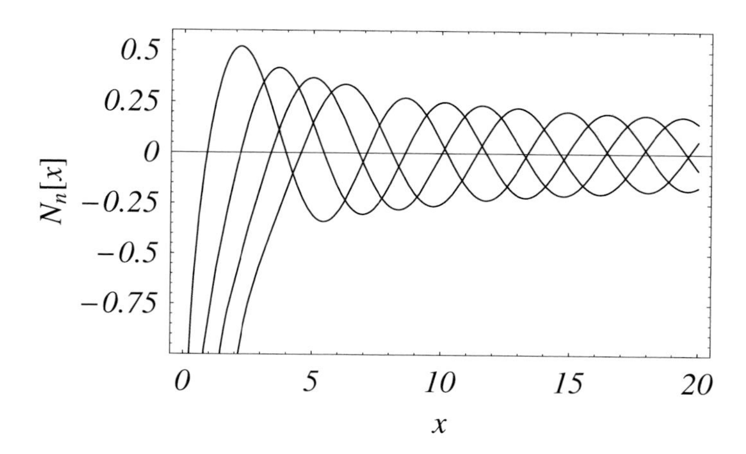

Figure 8.7. Bessel functions of second kind, for $n = 0 - 3$.

8.3.4 Modified Bessel Functions

Modified Bessel functions are solutions to a differential equation

$$\frac{1}{\xi}\frac{d}{d\xi}\left(\xi f'[\xi]\right) - \left(k^2 + \frac{\nu^2}{\xi^2}\right) f[\xi] == 0 \Longrightarrow f[\xi] = AI_\nu[k\xi] + BK_\nu[k\xi] \tag{8.242}$$

which is related to the ordinary Bessel equation by the substitution $k \to ik$ which transforms a wave equation into a diffusion equation. Consequently, we anticipate solutions of the form $J_\nu[ik\xi]$ and $N_\nu[ik\xi]$. It is customary to define modified Bessel functions with the phase and normalization conventions

$$I_\nu[z] = i^{-\nu} J[iz] \tag{8.243}$$

$$K_\nu[z] = \frac{\pi}{2} i^{\nu+1} H_\nu^{(1)}[iz] = \frac{\pi}{2} i^{\nu+1} \left(J_\nu[z] + iN_\nu[iz]\right) \tag{8.244}$$

where the first kind, I_ν, is regular while the second kind, K_ν, is irregular at the origin. Figure 8.8 illustrates the first few modified Bessel functions for integer order and positive arguments. The phase conventions ensure that both functions are real for positive x. These functions are sometimes described as *hyperbolic Bessel functions* or as Bessel functions of imaginary arguments, but the latter is not really an appropriate description because their extension to complex arguments provides many of their most important properties (as for practically all functions discussed in this course!). These functions do not have nonzero real roots and are not mutually orthogonal.

One can develop all of the important properties of modified Bessel functions using straightforward, but often tedious, analysis based upon cylindrical Bessel functions, but in a belated attempt for brevity we forgo the details and merely quote some of the results.

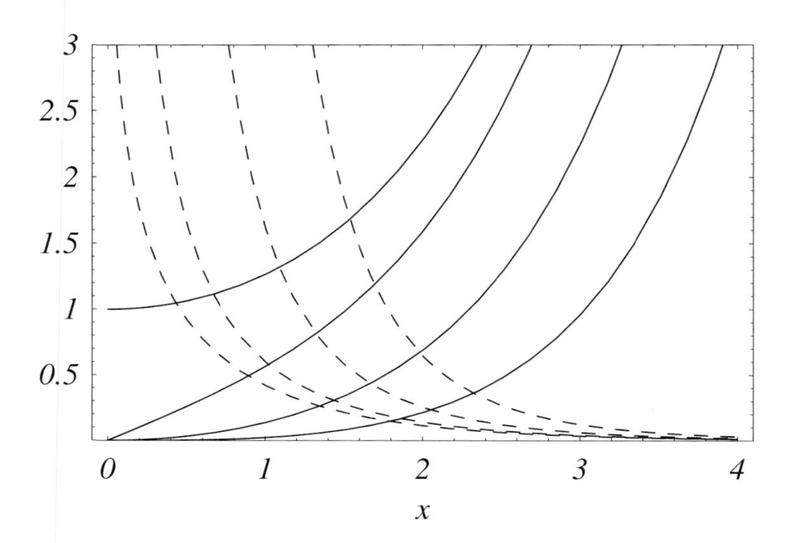

Figure 8.8. I_n solid, K_n dashed.

The basic recursion relations are

$$I_{\nu-1}[z] - I_{\nu+1}[z] = \frac{2\nu}{z} I_\nu[z] \tag{8.245}$$

$$K_{\nu-1}[z] - K_{\nu+1}[z] = -\frac{2\nu}{z} K_\nu[z] \tag{8.246}$$

$$I_{\nu\pm1}[z] \pm \frac{\nu}{z} I_\nu[z] = \frac{dI_\nu[z]}{dz} \tag{8.247}$$

$$K_{\nu\pm1}[z] \mp \frac{\nu}{z} K_\nu[z] = -\frac{dK_\nu[z]}{dz} \tag{8.248}$$

$$\frac{d}{dz}\left(z^{\pm\nu} I_\nu[z]\right) = z^{\pm\nu} I_{\nu\mp1}[z] \tag{8.249}$$

$$\frac{d}{dz}\left(z^{\pm\nu} K_\nu[z]\right) = -z^{\pm\nu} K_{\nu\mp1}[z] \tag{8.250}$$

where the sign differences between relationships for I_ν and K_ν arise from the ν dependence of the conventional phases. The leading asymptotic forms are

$$\mathrm{Re}[z] \geq 0 \Longrightarrow I_\nu[z] \simeq \frac{z^{-1/2} e^z}{\sqrt{2\pi}}, \quad K_\nu[z] \simeq \sqrt{\frac{\pi}{2}} z^{-1/2} e^{-z} \tag{8.251}$$

while the power series for I_ν takes the form

$$I_\nu[z] = \left(\frac{z}{2}\right)^\nu \sum_{m=0}^{\infty} \frac{1}{\Gamma[\nu+m+1]\Gamma[m+1]} \left(\frac{z^2}{4}\right)^m \tag{8.252}$$

Like the Neumann functions, one can show

$$K_\nu[z] = \frac{\pi}{2}\,\mathrm{Csc}[\nu\pi]\left(I_{-\nu}[z] - I_\nu[z]\right) \tag{8.253}$$

for nonintegral ν or

$$K_n[z] = \frac{(-)^n}{2}\lim_{\nu\to n}\left(\frac{\partial I_{-\nu}[z]}{\partial\nu} - \frac{\partial I_\nu[z]}{\partial\nu}\right) \tag{8.254}$$

for integer order. The power series for K_ν is cumbersome, but we note that the divergence near the origin is given by

$$\mathrm{Re}[\nu] > 0 \Longrightarrow K_\nu[z] \approx \frac{\Gamma[\nu]}{2}\left(\frac{z}{2}\right)^{-\nu} \tag{8.255}$$

$$K_0[z] \approx -\gamma - \mathrm{Log}\left[\frac{z}{2}\right] \tag{8.256}$$

where Euler's gamma constant is given by

$$\gamma = \lim_{m\to\infty}\left(\sum_{k=1}^{m}\frac{1}{k} - \mathrm{Log}[m]\right) \approx 0.577\,216\ldots \tag{8.257}$$

8.3.5 Spherical Bessel Functions

Next we consider spherical Bessel functions. In spherical coordinates the radial part of the wave equation takes the form

$$u''[r] + \left(k^2 - \frac{l(l+1)}{r^2}\right)u[r] == 0 \Longrightarrow \frac{u[r]}{kr} = A j_l[kr] + B n_l[kr] \tag{8.258}$$

where

$$j_l[x] = \sqrt{\frac{\pi}{2}}x^{-1/2}J_{l+\frac{1}{2}}[x] \tag{8.259}$$

$$n_l[x] = \sqrt{\frac{\pi}{2}}x^{-1/2}N_{l+\frac{1}{2}}[x] \tag{8.260}$$

are *spherical Bessel functions* of order l, with j_l regular and n_l irregular at the origin. Normally l is a nonnegative integer and the radial variable is nonnegative for physics applications, but it can still be helpful to generalize these definitions to include complex arguments and arbitrary order. Thus, the *spherical Bessel equation* then takes the form

$$f''[z] + \frac{2}{z}f'[z] + \left(1 - \frac{\nu(\nu+1)}{z^2}\right)f[z] == 0 \Longrightarrow f[z] = A j_\nu[z] + B n_\nu[z] \tag{8.261}$$

with solutions

$$j_\nu[z] = \sqrt{\frac{\pi}{2}}z^{-1/2}J_{\nu+\frac{1}{2}}[z] \tag{8.262}$$

$$n_\nu[z] = \sqrt{\frac{\pi}{2}}z^{-1/2}N_{\nu+\frac{1}{2}}[z] \tag{8.263}$$

whose major properties can be derived easily from those for cylindrical Bessel functions.

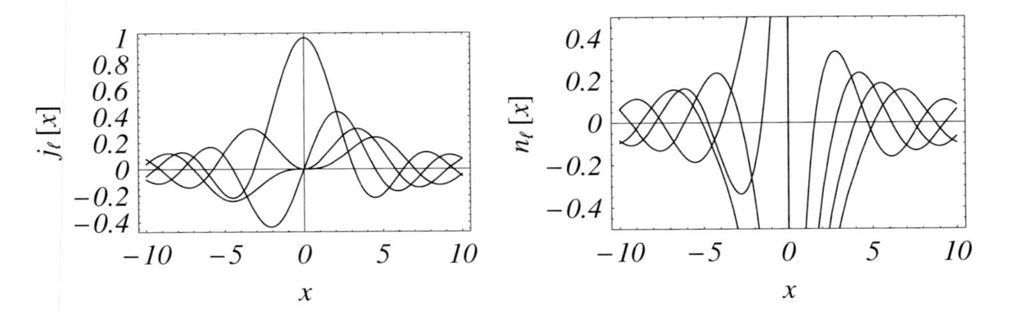

Figure 8.9. Spherical Bessel functions.

Spherical Bessel functions, like their cylindrical cousins, are entire functions of the complex variable ν even though most physics applications only require nonnegative integers $\nu \to l \geq 0$. For integer l, $j_l[z]$ is an entire function while $n_l[z]$ has a pole of order $l + 1$ at the origin. For arbitrary ν, both functions have a branch point at the origin and a cut normally taken below the negative real axis, but are analytic everywhere else. Although most authors express the proportionality factor between j_ν and $J_{\nu+1/2}$ as $\sqrt{\pi/2z}$, that notation is ambiguous on the negative real axis. Therefore, we prefer to use $z^{-1/2}$ instead of $\sqrt{1/z}$ to ensure the proper phases on the negative real axis. With our convention the parity of $j_l[x]$ is $(-)^l$, just as it is for $J_l[x]$, whereas the more common $\sqrt{1/z}$ notation results in an additional negative sign for $x < 0$. Figure 8.9 shows that spherical Bessel functions for integer l and real arguments are oscillatory with interleaved roots.

The asymptotic behavior of spherical Bessel functions are trigonometric functions

$$z j_\nu[z] \simeq \mathrm{Sin}\left[z - \frac{\nu\pi}{2}\right] \tag{8.264}$$

$$z n_\nu[z] \simeq -\mathrm{Cos}\left[z - \frac{\nu\pi}{2}\right] \tag{8.265}$$

similar to those for cylindrical Bessel functions. In most physics applications we encounter spherical Bessel functions in the form $j_l[kr]$ where k is a positive wave number and r a positive radius, such that their oscillation amplitudes decrease asymptotically according to $(kr)^{-1}$. However, for more general arguments the imaginary parts of either z or ν result in exponentially growing contributions that must be handled carefully and may limit the usefulness of these asymptotic formulas.

The basic recursion relations take the form

$$f_{\nu-1}[z] + f_{\nu+1}[z] = \frac{2\nu+1}{z} f_\nu[z] \tag{8.266}$$

$$\nu f_{\nu-1}[z] - (\nu+1) f_{\nu+1}[z] = (2\nu+1) f'_\nu[z] \tag{8.267}$$

$$\frac{d}{dz}(z^{\nu+1} f_\nu[z]) = z^{\nu+1} f_{\nu-1}[z] \tag{8.268}$$

$$\frac{d}{dz}(z^{-\nu} f_\nu[z]) = -z^{-\nu} f_{\nu+1}[z] \tag{8.269}$$

where $f \in \{j, n, h^{(1)}, h^{(2)}\}$ is any family of solutions to the spherical Bessel equation or any linear combination with coefficients independent of ν or z. These can be obtained from the corresponding results for cylindrical Bessel functions or from the differential equation. Notice that

$$j_0[z] = \frac{\mathrm{Sin}[z]}{z} \tag{8.270}$$

$$n_0[z] = -\frac{\mathrm{Cos}[z]}{z} \tag{8.271}$$

provide solutions to the spherical Bessel equation for $\nu = 0$. By inductively applying the recursion relations, we then obtain *Rayleigh's formulas* for nonnegative integers

$$j_l[z] = (-z)^l \left(\frac{1}{z}\frac{d}{dz}\right)^l \frac{\mathrm{Sin}[z]}{z} \tag{8.272}$$

$$n_l[z] = -(-z)^l \left(\frac{1}{z}\frac{d}{dz}\right)^l \frac{\mathrm{Cos}[z]}{z} \tag{8.273}$$

$$h_l^{(1)}[z] = -i(-z)^l \left(\frac{1}{z}\frac{d}{dz}\right)^l \frac{e^{iz}}{z} \tag{8.274}$$

$$h_l^{(2)}[z] = i(-z)^l \left(\frac{1}{z}\frac{d}{dz}\right)^l \frac{e^{-iz}}{z} \tag{8.275}$$

that resemble Rodrigues' formula. Explicit formulas can now be developed for positive integers, which is the most important situation.

Using

$$J_\nu[z] = \left(\frac{z}{2}\right)^\nu \sum_{m=0}^{\infty} \frac{1}{\Gamma[\nu+m+1]\Gamma[m+1]}\left(-\frac{z^2}{4}\right)^m \tag{8.276}$$

one finds

$$j_\nu[z] = \frac{\sqrt{\pi}}{2}\left(\frac{z}{2}\right)^\nu \sum_{m=0}^{\infty} \frac{1}{\Gamma\left[\nu+m+\frac{3}{2}\right]\Gamma[m+1]}\left(-\frac{z^2}{4}\right)^m \tag{8.277}$$

immediately. For integer $\nu \to l$ we use

$$\Gamma\left[l+\frac{1}{2}\right] = \sqrt{\pi}\frac{(2l-1)!!}{2^n} \tag{8.278}$$

to obtain

$$j_l[z] = z^l \sum_{m=0}^{\infty} \frac{1}{(2l+2m+1)!!m!}\left(-\frac{z^2}{2}\right)^m \tag{8.279}$$

Similarly, for integer order we use

$$N_{l+\frac{1}{2}}[z] = \mathrm{Cot}\left[\left(l+\frac{1}{2}\right)\pi\right] J_{l+\frac{1}{2}}[z] - \mathrm{Csc}\left[\left(l+\frac{1}{2}\right)\pi\right] J_{-l-\frac{1}{2}}[z] = (-)^{l+1} J_{-l-\frac{1}{2}}[z] \tag{8.280}$$

such that

$$n_l[z] = (-)^{l+1}\sqrt{\frac{\pi}{2}}z^{-1/2}\left(\frac{z}{2}\right)^{-l-\frac{1}{2}}\sum_{m=0}^{\infty}\frac{1}{\Gamma\left[m-l+\frac{1}{2}\right]m!}\left(-\frac{z^2}{4}\right)^m \tag{8.281}$$

and

$$\Gamma\left[\frac{1}{2}-l\right] = \sqrt{\pi}\frac{(-2)^l}{(2l-1)!!} \tag{8.282}$$

to obtain

$$n_l[z] = -z^{-l-1}\sum_{m=0}^{\infty}\frac{(2l-2m-1)!!}{m!}\left(\frac{z^2}{2}\right)^m \tag{8.283}$$

Thus, the limiting behavior near the origin is described by

$$|z| \ll 1 \Longrightarrow j_l[z] \approx \frac{z^l}{(2l+1)!!}, \quad n_l[z] \approx -\frac{(2l-1)!!}{z^{l+1}} \tag{8.284}$$

where

$$n!! = n\,(n-2)!! \tag{8.285}$$

$$(-1)!! = 0!! = 1!! = 1 \tag{8.286}$$

is the double factorial function for nonnegative integers.

8.4 Fourier–Bessel Transform

The plane wave e^{ikz}, where $z = r\,\mathrm{Cos}[\theta]$ is the distance along the polar axis, can be expanded in a Legendre series according to

$$\mathrm{Exp}\left[ikr\,\mathrm{Cos}[\theta]\right] = \sum_{l=0}^{\infty} f_l[kr]P_l\left[\mathrm{Cos}[\theta]\right] \tag{8.287}$$

where the coefficients f_l depend upon kr, treated as a parameter for the purpose of the Legendre expansion. Using the orthogonality properties of Legendre polynomials, we find

$$f_l[kr] = \frac{2l+1}{2}\int_{-1}^{1}\mathrm{Exp}[ikrx]P_l[x]\,dx \tag{8.288}$$

The most direct method to evaluate these coefficients is to expand the exponential

$$f_l[kr] = \frac{2l+1}{2}\sum_{m=0}^{\infty}\frac{(ikr)^m}{m!}\int_{-1}^{1}x^m P_l[x]\,dx \tag{8.289}$$

and to use a result

$$m + n \text{ even} \implies \int_{-1}^{1} x^m P_l[x]\, dx = 2 \frac{m!}{(m-l)!} \frac{(m-l-1)!!}{(m+l+1)!!} \tag{8.290}$$

that is proven in the exercises using Rodrigues' formula and partial integration. Thus, a leading power $(kr)^l$ can be factored out and the remaining series contains only even powers, $(kr)^{2m}$ for $m \geq 0$, such that

$$\begin{aligned} m \to l + 2j \implies f_l[kr] &= (2l+1)(ikr)^l \sum_{j=0}^{\infty} (-)^j (kr)^{2j} \frac{(2j-1)!!}{(2j)!(2j+2l+1)!!} \\ &= (2l+1)(ikr)^l \sum_{j=0}^{\infty} \frac{1}{j!(2j+2l+1)!!} \left(-\frac{k^2 r^2}{2} \right)^j \\ &= (2l+1) i^l j_l[kr] \end{aligned} \tag{8.291}$$

is proportional to the power series for $j_l[kr]$. Therefore, we obtain *Rayleigh's expansion*

$$\text{Exp}\,[ikr\, \text{Cos}[\theta]] = \sum_{l=0}^{\infty} (2l+1) i^l j_l[kr] P_l\, [\text{Cos}[\theta]] \tag{8.292}$$

for a plane wave. (This is sometimes named Bauer's formula instead.) Finally, the Legendre addition theorem

$$P_l[\hat{r} \cdot \hat{r}'] = \frac{4\pi}{2l+1} Y_l[\hat{r}] \cdot Y_l[\hat{r}'] \tag{8.293}$$

provides a more symmetrical form

$$\text{Exp}[i\vec{k} \cdot \vec{r}] = 4\pi \sum_{l=0}^{\infty} i^l j_l[kr] Y_l[\hat{k}] \cdot Y_l[\hat{r}] \tag{8.294}$$

which depends upon the angle between the $\hat{k}$ and $\hat{r}$ directions and is independent of the orientation of the coordinate axes. This very important formula finds myriad applications, including radiation and scattering theory, and should be committed to memory.

Consider a three-dimensional Fourier transform

$$\tilde{f}[\vec{k}] = \int f[\vec{r}]\, \text{Exp}[i\vec{k} \cdot \vec{r}]\, d^3 r \tag{8.295}$$

where the integral extends over all space. It is useful to perform a multipole expansion

$$f[\vec{r}] = \sum_{l=0}^{\infty} \sum_{m=-l}^{l} f_{l,m}[r] Y_{l,m}[\hat{r}] \tag{8.296}$$

$$f_{l,m}[r] = \int f[\vec{r}] Y_{l,m}[\hat{r}]^*\, d\Omega \tag{8.297}$$

for the angular dependence of $f[\vec{r}]$. Applying the Rayleigh expansion and assuming that the order of integration and summation can be interchanged, such that

$$\tilde{f}[\vec{k}] = 4\pi \sum_{l=0}^{\infty} i^l \int f[\vec{r}] j_l[kr] Y_l[\hat{k}] \cdot Y_l[\hat{r}] \, d^3r \tag{8.298}$$

and using the orthogonality of spherical harmonics, we obtain

$$\tilde{f}[\vec{k}] = 4\pi \sum_{l=0}^{\infty} \sum_{m=-l}^{l} i^l \tilde{f}_{l,m}[k] Y_{l,m}[\hat{k}] \tag{8.299}$$

where

$$\tilde{f}_{l,m}[k] = \int_0^\infty f_{l,m}[r] j_l[kr] r^2 \, dr \tag{8.300}$$

is the Fourier–Bessel transform of the radial dependence of the l, m multipole.

Alternatively, we could apply the Rayleigh expansion to the inverse Fourier transform

$$f[\vec{r}] = \int \tilde{f}[\vec{k}] \, \mathrm{Exp}[-i\vec{k} \cdot \vec{r}] \frac{d^3k}{(2\pi)^3} \tag{8.301}$$

to obtain

$$f_{l,m}[r] = \frac{2}{\pi} \int_0^\infty \tilde{f}_{l,m}[k] j_l[kr] k^2 \, dk \tag{8.302}$$

where the numerical coefficient is simply $(4\pi)^2/(2\pi)^3$. Substituting the definition for $\tilde{f}_{l,m}[k]$ into

$$f_{l,m}[r] = \frac{2}{\pi} \int_0^\infty dk k^2 \int_0^\infty dr' r'^2 f_{l,m}[r'] j_l[kr'] j_l[kr] \tag{8.303}$$

we require

$$\frac{2}{\pi} \int_0^\infty dk k^2 j_l[kr] j_l[kr'] = \frac{\delta[r - r']}{rr'} \tag{8.304}$$

to produce the necessary radial delta function. Although this is an admittedly heuristic derivation, the same orthogonality condition can be obtained more rigorously by evaluating the limit of the orthogonality integral for Bessel functions over a finite range as that range becomes infinite. Thus, the completeness of the Fourier–Bessel transform is expressed in spherical coordinates as

$$\delta[\vec{r} - \vec{r}'] = \frac{\delta[r - r']}{rr'} \delta[\hat{r} - \hat{r}'] \tag{8.305}$$

$$\frac{\delta[r - r']}{rr'} = \frac{2}{\pi} \int_0^\infty dk k^2 j_l[kr] j_l[kr'] \tag{8.306}$$

$$\delta[\hat{r} - \hat{r}'] = \sum_l Y_l[\hat{r}] \cdot Y_l[\hat{r}'] \tag{8.307}$$

8.4.1 Example: Fourier–Bessel Expansion of Nuclear Charge Density

The differential cross-section for scattering of a high-energy electron from the charge density of an atomic nucleus takes the schematic form

$$\frac{d\sigma}{d\Omega} = \sigma_{\text{pt}} F_l^2[q] \tag{8.308}$$

where σ_{pt} is the cross-section for a point charge, l is the angular momentum transfer, q is the momentum transfer, and $F_l[q]$ is the *form factor*. The form factor

$$F_l[q] = \int_0^\infty \rho_l[r] j_l[qr] r^2\, dr \tag{8.309}$$

is the Fourier–Bessel transform

$$\rho_l[r] = \left\langle f \left\| \sum_{k=1}^{Z} \frac{\delta[r - r_k]}{r r_k} Y_l[\hat{r}] \cdot Y_l[\hat{r}_k] \right\| i \right\rangle \tag{8.310}$$

of the transition charge density, which is a reduced matrix element of the charge density operator between initial and final states i and f. We neglect convection current, magnetization, and other complications in this introductory discussion. In its simplest form the charge density operator sums over all Z protons in the nucleus where the radial delta function specifies their positions. Given measurements of the form factor for a set of momentum transfers in the range $q_{\min} \le q \le q_{\max}$, how does one reconstruct the radial charge density $\rho_l[r]$? If the measurements produced a continuous function over an infinite range, one would simply use the orthonormality relation for spherical Bessel functions to invert the Fourier transform according to

$$\rho_l[q] = \frac{2}{\pi} \int_0^\infty F_l[q] j_l[qr] q^2\, dq \tag{8.311}$$

but experiments are limited in range and provide only discrete points with finite precision. A more practical method is to expand the radial density in a complete set of basis functions

$$\rho_l[r] = \sum_{n=1}^{\infty} a_n \rho_{l,n}[r] \Longrightarrow F_l[q] = \sum_{n=1}^{\infty} a_n \tilde{\rho}_{l,n}[q] \tag{8.312}$$

where

$$\tilde{\rho}_{l,n}[q] = \int_0^\infty \rho_{l,n}[r] j_l[qr] r^2\, dr \tag{8.313}$$

and to use the method of least-squares to fit the coefficients a_n to the data.

Recognizing that the charge density occupies a finite volume, a common choice of radial basis functions is the Fourier–Bessel expansion (FBE)

$$\rho_{l,n}[r] = j_l[q_{l,n} r]\Theta[r - R], \quad j_l[q_{l,n} R] = 0 \tag{8.314}$$

with Dirichlet boundary conditions. Here R should be large enough to comfortably enclose practically all of the charge but not so large that there are too many $q_{l,n} < q_{\max}$ to determine

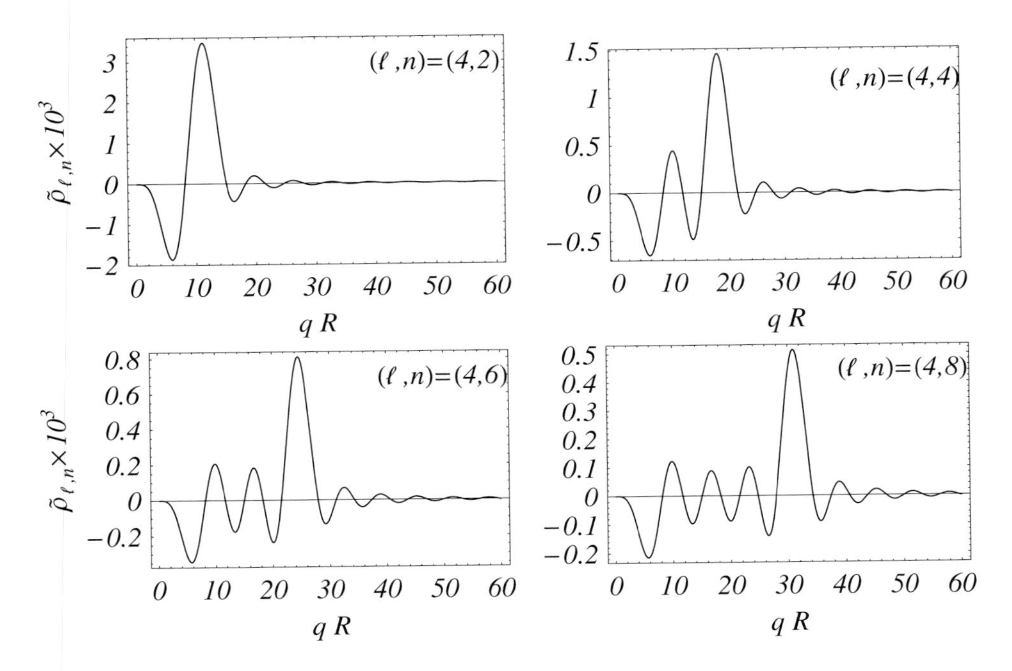

Figure 8.10. FBE form factor basis.

the corresponding expansion coefficients. An important feature of this expansion is that the basis functions for the form factor

$$\tilde{\rho}_{l,n}[q] = \int_0^R j_l[q_{l,n}r]j_l[qr]r^2\,dr = R^2 \frac{q_{l,n}j_l[qR]j_l'[q_{l,n}R] - qj_l[q_{l,n}R]j_l'[qR]}{q^2 - q_{l,n}^2} \tag{8.315}$$

exhibit a strong peak around $q_{l,n}$ and small oscillations elsewhere. Therefore, reasonable approximations to the expansion coefficients can be obtained by inspection of the data for $q \approx q_{l,n}$. Representative basis functions are displayed in Fig. 8.10 for $l = 4$. The dominant peak moves out as n increases. As R increases it moves inward and becomes taller and narrower, approaching a delta function in the continuum limit $R \to \infty$. Clearly experimental data cannot determine the amplitude for oscillations with $q \gg q_{\max}$; therefore, physics arguments are needed to estimate the uncertainty in the reconstructed radial density due to the unmeasured form factor for large momentum transfers, $q > q_{\max}$, known as *incompleteness error*. However, a discussion of that issue would bring us too far from the main topic.

8.5 Summary

Here we collect some of the most useful results for Legendre and Bessel functions in order to provide a convenient reference for later work on boundary-value problems. Additional formulas are found in the problems, the chapter, and standard compendia. Unless stated otherwise, we assume $\{n, l, m\}$ are nonnegative integers, $\{x, t\}$ are real, and $\{\nu, z\}$ are arbitrary (possibly complex). We also assume that $\{a, b\}$ are constant coefficients independent of the degree or order of any functions. Note that some of these results are proven in the exercises or are generalizations of results obtained in the text. When working problems, do not quote without proof any results beyond those presented in the text itself.

8.5.1 Legendre Functions

8.5.1.1 Generating Function

$$\frac{1}{\sqrt{1-2xt+t^2}} = \sum_{n=0}^{\infty} P_n[x] t^n \tag{8.316}$$

$$\frac{1}{|\vec{r}-\vec{r}'|} = \sum_{n=0}^{\infty} \frac{r_<^n}{r_>^{n+1}} P_n[x] \tag{8.317}$$

$$P_n[z] = \frac{1}{2^n n!} \left(\frac{d}{dz}\right)^n (z^2-1)^n \tag{8.318}$$

8.5.1.2 Orthonormality

$$\int_{-1}^{1} P_n[x] P_m[x]\, dx = \frac{2}{2n+1} \delta_{n,m} \tag{8.319}$$

8.5.1.3 Differential Equation

$$(1-z^2) f''[z] - 2z f'[z] + \nu(\nu+1) f[z] == 0 \Longrightarrow f_\nu[z] = a P_\nu[z] + b Q_\nu[z] \tag{8.320}$$

8.5.1.4 Recursion Relations

$$(\nu+1) f_{\nu+1} - (2\nu+1) z f_\nu + \nu f_{\nu-1} = 0 \tag{8.321}$$

$$f'_{\nu+1} - 2z f'_\nu + f'_{\nu-1} = f_\nu \tag{8.322}$$

$$f'_{\nu+1} - f'_{\nu-1} = (2\nu+1) f_\nu \tag{8.323}$$

$$(z^2-1) f'_\nu = \nu z f_\nu - \nu f_{\nu-1} \tag{8.324}$$

8.5.1.5 Special Values

$$P_n[1] = 1 \tag{8.325}$$

$$P_n[-x] = (-)^n P_n[x] \tag{8.326}$$

$$P_{2n}[0] = (-)^n \frac{(2n)!}{2^{2n}(n!)^2}, \quad P_{2n+1}[0] = 0 \tag{8.327}$$

8.5.1.6 Integral Representations

$$P_\nu[z] = \frac{2^{-\nu}}{2\pi i} \oint \frac{(t^2-1)^\nu}{(t-z)^{\nu+1}}\, dt \tag{8.328}$$

$$Q_\nu[z] = \frac{2^{-\nu}}{4i\,\mathrm{Sin}[\nu\pi]} \oint \frac{(t^2-1)^\nu}{(z-t)^{\nu+1}}\, dt \tag{8.329}$$

See Fig. 8.3 for contours.

8.5.2 Associated Legendre Functions

8.5.2.1 Differential Equation

$$(1-z^2)f''[z] - 2zf'[z] + \left(\nu(\nu+1) - \frac{\mu^2}{1-z^2}\right) f[z] == 0 \Longrightarrow f_{\nu,\mu}[z] = aP_{\nu,\mu}[z] + bQ_{\nu,\mu}[z] \tag{8.330}$$

8.5.2.2 Orthonormality

$$\int_{-1}^{1} P_{l,m}[x]P_{l',m}[x]\, dx = \frac{2}{(2l+1)}\frac{(l+m)!}{(l-m)!}\delta_{l,l'} \tag{8.331}$$

8.5.2.3 Rodrigues' Formula

$$P_{l,m}[x] = (-)^m(1-x^2)^{m/2}\left(\frac{d}{dx}\right)^m P_l[x] \tag{8.332}$$

$$P_{l,-m}[x] = (-)^m \frac{(l-m)!}{(l+m)!} P_{l,m}[x] \tag{8.333}$$

8.5.2.4 Special Values

$$P_{l,m}[\pm 1] = P_l[\pm 1]\delta_{m,0} = (\pm 1)^l \delta_{m,0} \tag{8.334}$$

$$P_{2n+m,m}[0] = (-)^{m+n}\frac{(2m)!}{2^{m+n}m!n!}\prod_{k=1}^{n}(2m+2k-1) \tag{8.335}$$

$$P_{2n+m+1,m}[0] = 0 \tag{8.336}$$

8.5.2.5 Recursion Relations

$$P_{l,m+2}[x] + (m+1)\frac{2x}{1-x^2}P_{l,m+1}[x] + \big(l(l+1) - m(m+1)\big)P_{l,m}[x] == 0 \tag{8.337}$$

$$xP_{l,m} + (l-m+1)(1-x^2)^{1/2}P_{l,m-1} - P_{l-1,m} == 0 \tag{8.338}$$

$$P_{l-1,m+1} - (2l+1)(1-x^2)^{1/2}P_{l,m} - P_{l+1,m+1} == 0 \tag{8.339}$$

$$(l-m+1)P_{l+1,m} - (2l+1)xP_{l,m} + (l+m)P_{l-1,m} == 0 \tag{8.340}$$

8.5.3 Spherical Harmonics

8.5.3.1 Definition

$$Y_{l,m}[\theta,\phi] = \left(\frac{2l+1}{4\pi}\frac{(l-m)!}{(l+m)!}\right)^{1/2} P_{l,m}[\cos\theta]e^{im\phi} \tag{8.341}$$

8.5.3.2 Symmetries

$$Y^*_{l,m}[\theta,\phi] = (-)^m Y_{l,-m}[\theta,\phi] \tag{8.342}$$

$$Y_{l,m}[-\hat{r}] = (-)^l Y_{l,m}[\hat{r}] \tag{8.343}$$

8.5.3.3 Special Values

$$Y_{l,m}[\theta,0] = \left(\frac{2l+1}{4\pi}\right)^{1/2} P_l[\cos\theta]\delta_{m,0} \tag{8.344}$$

$$Y_{l,m}[\hat{z}] = Y_{l,m}[0,0] = \left(\frac{2l+1}{4\pi}\right)^{1/2} \delta_{m,0} \tag{8.345}$$

8.5.3.4 Orthonormality

$$\int Y_{l,m}[\hat{r}]Y^*_{l',m'}[\hat{r}]\,d\Omega = \delta_{l,l'}\delta_{m,m'} \tag{8.346}$$

8.5.3.5 Addition Theorem

$$P_l[\hat{r}\cdot\hat{r}'] = \frac{4\pi}{2l+1}\sum_{m=-l}^{l} Y_{l,m}[\hat{r}]Y_{l,m}[\hat{r}']^* = \frac{4\pi}{2l+1}Y_l[\hat{r}]\cdot Y_l[\hat{r}'] \tag{8.347}$$

$$\frac{1}{|\vec{r}-\vec{r}'|} = \sum_{l=0}^{\infty}\frac{4\pi}{2l+1}\frac{r_<^l}{r_>^{l+1}}Y_l[\hat{r}]\cdot Y_l[\hat{r}'] \tag{8.348}$$

8.5.4 Cylindrical Bessel Functions

8.5.4.1 Differential Equation

$$z^2 f''[z] + zf'[z] + (z^2-\nu^2)f[z] == 0 \Longrightarrow f_\nu[z] = aJ_\nu[z] + bN_\nu[z] \tag{8.349}$$

8.5.4.2 Series

$$J_\nu[z] = \left(\frac{z}{2}\right)^\nu \sum_{m=0}^{\infty}\frac{1}{\Gamma[\nu+m+1]\Gamma[m+1]}\left(-\frac{z^2}{4}\right)^m \tag{8.350}$$

$$N_\nu[z] = \text{Cot}[\nu\pi]J_\nu[z] - \text{Csc}[\nu\pi]J_{-\nu}[z] \tag{8.351}$$

8.5.4.3 Hankel Functions

$$H_\nu^{(1)}[z] = J_\nu[z] + iN_\nu[z] \tag{8.352}$$

$$H_\nu^{(2)}[z] = J_\nu[z] - iN_\nu[z] \tag{8.353}$$

8.5.4.4 Asymptotic Forms

$$J_\nu[z] \simeq \sqrt{\frac{2}{\pi}} z^{-1/2} \operatorname{Cos}\left[z - (2\nu + 1)\frac{\pi}{4}\right] \tag{8.354}$$

$$N_\nu[z] \simeq \sqrt{\frac{2}{\pi}} z^{-1/2} \operatorname{Sin}\left[z - (2\nu + 1)\frac{\pi}{4}\right] \tag{8.355}$$

$$H_\nu^{(1)}[z] \simeq \sqrt{\frac{2}{\pi}} z^{-1/2} \operatorname{Exp}\left[i\left(z - (2\nu + 1)\frac{\pi}{4}\right)\right] \tag{8.356}$$

$$H_\nu^{(2)}[z] \simeq \sqrt{\frac{2}{\pi}} z^{-1/2} \operatorname{Exp}\left[-i\left(z - (2\nu + 1)\frac{\pi}{4}\right)\right] \tag{8.357}$$

8.5.4.5 Recursion Relations

$$f_{\nu-1}[z] + f_{\nu+1}[z] = \frac{2\nu}{z} f_\nu[z] \tag{8.358}$$

$$f_{\nu-1}[z] - f_{\nu+1}[z] = 2f_\nu'[z] \tag{8.359}$$

$$\frac{d}{dz}(z^{\pm\nu} f_\nu[z]) = \pm z^{\pm\nu} f_{\nu\mp1}[z] \tag{8.360}$$

8.5.4.6 Orthonormality

$$\int_0^R J_\nu[k_1\xi]J_\nu[k_2\xi]\xi\, d\xi = R\frac{k_2 J_\nu[k_1R]J_\nu'[k_2R] - k_1 J_\nu[k_2R]J_\nu'[k_1R]}{k_1^2 - k_2^2}$$

$$\int_0^\infty J_\nu[k\xi]J_\nu[k'\xi]\xi\, d\xi = \frac{\delta[k - k']}{\sqrt{kk'}} \tag{8.361}$$

$$\int_0^\infty J_\nu[k\xi]J_\nu[k\xi']k\, dk = \frac{\delta[\xi - \xi']}{\sqrt{\xi\xi'}} \tag{8.362}$$

8.5.5 Spherical Bessel Functions

8.5.5.1 Differential Equation

$$f''[z] + \frac{2}{z}f'[z] + \left(1 - \frac{\nu(\nu+1)}{z^2}\right)f[z] == 0 \Longrightarrow f_\nu[z] = a j_\nu[z] + b n_\nu[z] \tag{8.363}$$

$$j_\nu[z] = \sqrt{\frac{\pi}{2}} z^{-1/2} J_{\nu+\frac{1}{2}}[z] \tag{8.364}$$

$$n_\nu[z] = \sqrt{\frac{\pi}{2}} z^{-1/2} N_{\nu+\frac{1}{2}}[z] \tag{8.365}$$

$$h_\nu^{(1)}[z] = j_\nu[z] + i n_\nu[z] = \sqrt{\frac{2}{\pi}} z^{-1/2} H_{\nu+1/2}^{(1)}[z] \tag{8.366}$$

$$h_\nu^{(2)}[z] = j_\nu[z] - i n_\nu[z] = \sqrt{\frac{2}{\pi}} z^{-1/2} H_{\nu+1/2}^{(2)}[z] \tag{8.367}$$

8.5.5.2 Recursion Relations

$$f_{\nu-1}[z] + f_{\nu+1}[z] = \frac{2\nu+1}{z} f_\nu[z] \tag{8.368}$$

$$\nu f_{\nu-1}[z] - (\nu+1) f_{\nu+1}[z] = (2\nu+1) f_\nu'[z] \tag{8.369}$$

$$\frac{d}{dz}\left(z^{\nu+1} f_\nu[z]\right) = z^{\nu+1} f_{\nu-1}[z] \tag{8.370}$$

$$\frac{d}{dz}\left(z^{-\nu} f_\nu[z]\right) = -z^{-\nu} f_{\nu+1}[z] \tag{8.371}$$

8.5.5.3 Orthonormality

$$\int_0^R j_l[k_1 r] j_l[k_2 r] r^2\, dr = R^2 \frac{k_2 j_l[k_1 R] j_l'[k_2 R] - k_1 j_l[k_2 R] j_l'[k_1 R]}{k_1^2 - k_2^2} \tag{8.372}$$

8.5.5.4 Series

$$j_\nu[z] = \frac{\sqrt{\pi}}{2}\left(\frac{z}{2}\right)^\nu \sum_{m=0}^{\infty} \frac{1}{\Gamma[\nu+m+\frac{3}{2}]\Gamma[m+1]}\left(-\frac{z^2}{4}\right)^m \tag{8.373}$$

8.5.5.5 Asymptotic Forms

$$z j_\nu[z] \simeq \mathrm{Sin}\left[z - \frac{\nu\pi}{2}\right] \tag{8.374}$$

$$z n_\nu[z] \simeq -\mathrm{Cos}\left[z - \frac{\nu\pi}{2}\right] \tag{8.375}$$

$$z h_\nu^{(1)}[z] \simeq \mathrm{Exp}\left[i\left(z - \frac{(\nu+1)\pi}{2}\right)\right] \tag{8.376}$$

$$z h_\nu^{(2)}[z] \simeq \mathrm{Exp}\left[-i\left(z - \frac{(\nu+1)\pi}{2}\right)\right] \tag{8.377}$$

8.5.6 Fourier–Bessel Expansions

$$\text{Exp}[i\vec{k}\cdot\vec{r}] = 4\pi\sum_{l=0}^{\infty} i^l j_l[kr] Y_l[\hat{k}]\cdot Y_l[\hat{r}] \tag{8.378}$$

$$\text{Exp}[ik\xi\,\text{Sin}[\phi]] = \sum_{m=-\infty}^{\infty} J_m[k\xi] e^{im\phi} \tag{8.379}$$

Problems for Chapter 8

1. Rearrangement formulas for doubly infinite sums

a) Prove that the rearrangement formulas

$$\sum_{m=0}^{\infty}\sum_{n=0}^{\infty} a_{n,m} = \sum_{p=0}^{\infty}\sum_{q=0}^{p} a_{q,p-q} = \sum_{r=0}^{\infty}\sum_{s=0}^{\lfloor r/2\rfloor} a_{s,r-2s} \tag{8.380}$$

where

$$r \text{ even} \implies \lfloor r/2\rfloor = \frac{r}{2} \tag{8.381}$$

$$r \text{ odd} \implies \lfloor r/2\rfloor = \frac{r-1}{2} \tag{8.382}$$

is the floor function, are equivalent provided that the series is absolutely convergent.

b) Use this result to supply the missing steps in the derivation of the Legendre series.

2. Leibnitz's formula

Prove Leibnitz's formula

$$\left(\frac{d}{dx}\right)^n (f[x]g[x]) = \sum_{m=0}^{n}\binom{n}{m} f^{(n-m)}[x] g^{(m)}[x] \tag{8.383}$$

for multiple derivatives of a product.

3. Special values for Legendre functions

a) Evaluate $P_n[0]$.

b) Evaluate $P_n'[1]$.

c) Use the Schläfli integral representation to evaluate $P_\nu[1]$ for arbitrary ν.

4. Legendre integrals

a) Use a recursion relation to evaluate

$$\int_0^1 P_n[x]\,dx \tag{8.384}$$

b) Evaluate

$$\int_0^1 x^m P_n[x]\, dx \tag{8.385}$$

for positive integer m.

5. Legendre series

a) Evaluate

$$\sum_{n=0}^{\infty} \frac{x^{n+1}}{n+1} P_n[x] \tag{8.386}$$

b) Evaluate

$$\int_{-1}^{1} \frac{P_n[x]}{\sqrt{1-x^2}}\, dx \tag{8.387}$$

c) Produce Legendre expansions for $\delta[1 \pm x]$.

6. Parseval relation for Legendre polynomials

Suppose that $f[x]$ is expanded according to

$$f[x] = \sum_{n=0}^{\infty} a_n P_n[x] \tag{8.388}$$

Express $\int_{-1}^{1} |f[x]|^2\, dx$ in terms of the a_n coefficients.

7. Legendre expansion of powers

Develop a Legendre expansion for powers, such that

$$x^m = \sum_{n=0}^{m} A_{m,n} P_n[x] \tag{8.389}$$

by expressing the Legendre polynomials in terms of Rodrigues' formula and using partial integration to eliminate derivatives. You may find the beta function

$$B[p,q] = \int_0^1 t^{p-1}(1-t)^{q-1}\, dt = \frac{\Gamma[p]\Gamma[q]}{\Gamma[p+q]} \tag{8.390}$$

useful, but a proof is not required.

8. Legendre function $P_\nu[z]$ for arbitrary ν

Show that Schläfli's integral representation for the Legendre function of the first-kind satisfies Legendre's differential equation for arbitrary ν provided that the contour and cuts are defined properly.

9. Laplace's integral representation for Legendre functions

a) Suppose that the contour used in Schläfli's integral representation is a circle around z with radius $|\sqrt{z^2-1}|$. Show that

$$P_n[z] = \frac{1}{\pi}\int_0^{\pi}\left(z+\sqrt{z^2-1}\,\mathrm{Cos}[\phi]\right)^n d\phi \tag{8.391}$$

b) Use Laplace's integral representation to evaluate the Legendre generating function

$$g[t,z] = \sum_{n=0}^{\infty} t^n P_n[z] \tag{8.392}$$

Thus, one obtains a generalization for complex z.

10. Generating function for associated Legendre functions

Derive the generating function for associated Legendre functions with $m \geq 0$

$$g_m[t,x] = (-)^m \frac{(2m)!}{2^m m!}\frac{(1-x^2)^{m/2}}{\left(1-2tx+t^2\right)^{m+1/2}} = \sum_{l=0}^{\infty} P_{l+m,m}[x]t^l \tag{8.393}$$

from the generating function for Legendre polynomials. It is not shown often because it appears cumbersome, but it can be useful in deriving relationships or performing integrals involving associated Legendre functions.

11. An integral used in the normalization of associated Legendre functions

There are several methods for obtaining

$$\int_{-1}^{1}(1-x^2)^l\,dx = \sqrt{\pi}\frac{\Gamma[l+1]}{\Gamma\left[l+\frac{3}{2}\right]} = \frac{2^{2l+1}(l!)^2}{(2l+1)!} \tag{8.394}$$

for $l \geq 0$ (other than looking it up, of course). A relatively painless method that also provides many related integral is based upon the *beta function*

$$B[p,q] = \frac{\Gamma[p]\Gamma[q]}{\Gamma[p+q]} \tag{8.395}$$

where p, q are generally complex. For our purposes it will be sufficient to assume that p, q are nonnegative real numbers so that we can employ the simplest integral representation

$$\Gamma[p] = \int_0^{\infty} e^{-u}u^{p-1}\,du \tag{8.396}$$

Show that the substitution $u \to x^2$ and a transformation from Cartesian to polar coordinates facilitates evaluation of the numerator as an integral over the first quadrant, such that

$$p,q \geq 0 \Longrightarrow B[p,q] = 4\int_0^{\pi/2}\mathrm{Cos}[\theta]^{2p-1}\,\mathrm{Sin}[\theta]^{2q-1}\,d\theta \tag{8.397}$$

It should now be a simple matter to obtain the desired integral using appropriate choices of p, q and an obvious change of variable.

12. Recursion relations for associated Legendre functions

Many recursion relations for associated Legendre functions can be developed by differentiating either the differential equation or a recursion relation for Legendre polynomials m times. Use this technique to derive the following relations. For simplicity, assume that $m \geq 0$ for any $P_{l,m}$. Observe that a) varies the order, d) the degree, while b) and c) vary both. It is probably easiest to derive these relations in the order listed.

a) $$P_{l,m+2}[x] + (m+1)\frac{2x}{1-x^2}P_{l,m+1}[x] + \big(l(l+1) - m(m+1)\big)P_{l,m}[x] == 0 \quad (8.398)$$

b) $$xP_{l,m} + (l-m+1)(1-x^2)^{1/2}P_{l,m-1} - P_{l-1,m} == 0 \quad (8.399)$$

c) $$P_{l-1,m+1} - (2l+1)(1-x^2)^{1/2}P_{l,m} - P_{l+1,m+1} == 0 \quad (8.400)$$

d) $$(l-m+1)P_{l+1,m} - (2l+1)xP_{l,m} + (l+m)P_{l-1,m} == 0 \quad (8.401)$$

13. m-raising and lowering operators for $P_{l,m}$

a) Use the Rodrigues formula to derive the m-raising relation

$$P_{l,m+1} = -(1-x^2)^{1/2}P'_{l,m} - m\frac{x}{(1-x^2)^{1/2}}P_{l,m} \quad (8.402)$$

b) Then use a recursion relation to deduce the corresponding m-lowering formula.

14. Special values for associated Legendre functions

Use the generating function to evaluate the following special values or limiting cases for associated Legendre functions. Assume that $m \geq 0$.

a) $P_{l,m}[\pm 1]$

b) $P_{l,m}[0]$

c) $P_{m,m}[x]$ for $x = \mathrm{Cos}[\theta]$. Notice that this result can be used with the m-lowering operator to generate the entire set of $P_{l,m}[x]$; this is a common algorithm.

15. Orthogonality of $P_{l,m}$ with respect to m

The most useful orthogonality for $P_{l,m}$ concerns differing l but common m. Alternatively, one can show that

$$\int_{-1}^{1} \frac{P_{l,m}[x]P_{l,m'}[x]}{1-x^2}\,dx = \frac{1}{m}\frac{(l+m)!}{(l-m)!}\delta_{m,m'}, \quad \big(m, m' > 0\big) \quad (8.403)$$

expresses orthogonality between associated Legendre functions differing in order. This result is generally less useful because orthogonality between azimuthal eigenfunctions usually ensures matching order anyway.

a) Show that the associated Legendre equation is a Sturm–Liouville equation with eigenvalue m and weight function $w[x] = (1-x^2)^{-1/2}$ and, hence, provides this orthogonality integral between eigenfunctions differing in order.

b) Obtain the normalization for $P_{l,m=l}$ and use a recursion relation to step down m.

16. Recursion relations for Bessel functions
Suppose that $f_\nu[z] = aJ_\nu[z] + bN_\nu[z]$ where the coefficients a, b are independent of ν or z. Verify the following recursion relations.

a) $$f_{\nu\pm 1}[z] = \frac{\nu}{z} f_\nu[z] \mp f'_\nu[z] \quad (8.404)$$

b) $$\frac{d}{dz}\left(z^{\pm\nu} f_\nu[z]\right) = \pm z^{\pm\nu} f_{\nu\mp 1}[z] \quad (8.405)$$

c) $$\left(\frac{1}{z}\frac{d}{dz}\right)^k \left(z^\nu f_\nu[z]\right) = z^{\nu-k} f_{\nu-k}[z] \quad (8.406)$$

d) $$\left(\frac{1}{z}\frac{d}{dz}\right)^k \left(z^{-\nu} f_\nu[z]\right) = (-)^k z^{-\nu-k} f_{\nu+k}[z] \quad (8.407)$$

17. Interleaving of roots for Bessel functions
Use the recursion relation

$$\frac{d}{dx}\left(x^{\pm n} f_n[x]\right) = \pm x^{\pm n} f_{n\mp 1}[x] \quad (8.408)$$

where $f_n[x] = aJ_n[x] + bN_n[x]$ is a linear combination of cylindrical Bessel functions with constant coefficients to prove that there is one root of f_{n+1} between successive roots of f_n and vice versa.

18. Rodrigues formula for Bessel functions
Derive a Rodrigues formula of the form

$$z^{-n} f_n[z] = \mathcal{D}^n f_0[z] \quad (8.409)$$

where f_n is a solution to Bessel's equation (J_n, N_n, H_n, etc.) and $\mathcal{D}$ is a differential operator applied n times to f_0.

19. Orthonormality relations for Bessel functions and eigenfunction expansions
Piecewise continuous functions on a finite interval that satisfy linear homogeneous boundary conditions can be expanded in terms of eigenfunctions for Bessel's equation. Here we develop the required orthonormality relations for simple boundary conditions and formulate the corresponding expansions.

a) Use the differential equations satisfied by $J_\nu[k_1\xi]$ and $J_\nu[k_2\xi]$ for arbitrary k_1 and k_2 to show

$$R(k_2 J_\nu[k_1 R] J'_\nu[k_2 R] - k_1 J_\nu[k_2 R] J'_\nu[k_1 R]) = (k_1^2 - k_2^2) \int_0^R J_\nu[k_1\xi] J_\nu[k_2\xi] \xi \, d\xi \quad (8.410)$$

for $\nu \geq 0$. Differentiation of this result and application of recursion relations or the differential equations can then help with specific boundary conditions.

b) Derive the orthonormality relation

$$\int_0^R J_\nu[k_{\nu,m}\xi] J_\nu[k_{\nu,n}\xi]\xi \, d\xi = \frac{R^2}{2} J_{\nu+1}[k_{\nu,n}R]^2 \delta_{m,n} \tag{8.411}$$

where $J_\nu[k_{\nu,n}R] == 0$ defines the eigenvalues $k_{\nu,n}$ for Dirichlet boundary conditions. Provide an explicit expression for the Bessel expansion of an arbitrary function satisfying Dirichlet boundary conditions.

c) Derive the orthonormality relation

$$\int_0^R J_\nu[k_{\nu,m}\xi] J_\nu[k_{\nu,n}\xi]\xi \, d\xi = \frac{R^2}{2}\left(1 - \left(\frac{\nu}{k_{\nu,n}R}\right)^2\right) J_\nu[k_{\nu,n}R]^2 \delta_{m,n} \tag{8.412}$$

where $J'_\nu[k_{\nu,n}R] == 0$ defines the eigenvalues $k_{\nu,n}$ for Neumann boundary conditions. Provide an explicit expression for the Bessel expansion of an arbitrary function satisfying Neumann boundary conditions assuming that $\nu \neq 0$.

d) How is the Bessel expansion using $\nu = 0$ for an arbitrary function satisfying Neumann boundary conditions affected by the null eigenvalue $k_0 = 0$? Show that the corresponding eigenfunction is orthogonal to those with $k_n > 0$ and formulate the appropriate expansion explicitly.

20. Continuum orthonormality for Bessel functions

The continuum orthonormality relation

$$\int_0^\infty J_\nu[k\xi] J_\nu[k'\xi]\xi \, d\xi = \frac{\delta[k-k']}{\sqrt{kk'}} \tag{8.413}$$

for Bessel functions with $\nu \geq 0$ on a semi-infinite range can be derived by careful analysis of the limit of the result for a finite interval

$$\int_0^R J_\nu[k_1\xi] J_\nu[k_2\xi]\xi \, d\xi = R\frac{k_2 J_\nu[k_1 R] J'_\nu[k_2 R] - k_1 J_\nu[k_2 R] J'_\nu[k_1 R]}{k_1^2 - k_2^2} \tag{8.414}$$

as $R \to \infty$. This result should have been obtained in the preceding problem. Use a recursion relation to eliminate the derivatives first and then substitute the leading asymptotic behavior for the remaining Bessel functions. (Why eliminate derivatives first?) After some algebraic manipulation, more easily performed with MATHEMATICA® than by hand, one should recognize a familiar nascent delta function.

21. Parseval relation for Bessel functions

Suppose that $f[x]$ is expanded according to

$$f[x] = \sum_{n=1}^{\infty} a_n J_\nu[k_{\nu,n}x], \quad J_\nu[k_{\nu,n}R] = 0 \tag{8.415}$$

in the range $0 \leq x \leq R$. Express $\int_0^R f[x]^2 x \, dx$ in terms of the a_n coefficients.

22. Summation formula for Bessel functions
Use the generating function to evaluate $J_n[x+y]$ for integer n.

23. Laplace transform of Bessel functions
Use Bessel's integral for $J_n[x]$ to evaluate the Laplace transform for Bessel functions of integral order.

24. Bessel's integral for noninteger ν
Show that

$$J_\nu[z] = \frac{1}{\pi}\int_0^\pi \operatorname{Cos}\left[z\operatorname{Sin}[\theta] - \nu\theta\right] d\theta - \frac{\operatorname{Sin}[\nu\pi]}{\pi}\int_1^\infty \operatorname{Exp}\left[-(z\operatorname{Sinh}[s] + \nu s)\right] ds \quad (8.416)$$

for a suitable domain of z.

25. Plane wave in cylindrical Bessel functions
Show that a plane wave can be expanded in terms of cylindrical Bessel functions according to

$$\operatorname{Exp}\left[ik\xi\operatorname{Sin}[\phi]\right] = \sum_{m=-\infty}^{\infty} J_m[k\xi]e^{im\phi} \quad (8.417)$$

26. Leading asymptotic behavior of Hankel functions
Apply the method of steepest descent to deduce the leading asymptotic behavior of Hankel functions from their integral representations. Then deduce the corresponding behavior of Bessel and Neumann functions. For simplicity assume that $|z| \gg |\nu|$ (why?).

27. Power series for Neumann functions
Complete the derivation of the power series for Neumann functions with integer $n \geq 0$ using the fact that gamma functions have simple poles with known residues at negative integers.

28. Wronskians for Bessel functions
Recall that the Wronskian

$$W[f, g] = f[z]g'[z] - g[z]f'[z] \propto \frac{1}{p[z]} \quad (8.418)$$

for two independent solutions f, g to a Sturm–Liouville equation

$$(\mathcal{L} + \lambda w[z])f[z] == 0 \quad (8.419)$$

$$(\mathcal{L} + \lambda w[z])g[z] == 0 \quad (8.420)$$

generalized to complex z where

$$\mathcal{L} = \frac{d}{dz}\left(p[z]\frac{d}{dz}\right) - q[z] \quad (8.421)$$

takes the form

$$W[f, g] \propto \frac{1}{p[z]} \quad (8.422)$$

Confirm the validity of this result for Bessel functions and identify $p[z]$. Then evaluate the following Wronskian relations between various Bessel functions. Also, express these

Wronskians for parts a) and b) as combinations of products of Bessel function of different orders, without explicit derivatives.

a) $W[J_\nu, J_{-\nu}]$. Explain the result for $\nu \to n$, where n is an integer.

b) $W[J_\nu, N_\nu]$

c) $W[H_\nu^{(1)}, H_\nu^{(2)}]$.

29. Recursion relations for spherical Bessel functions

Suppose that $f_\nu[z] = a j_\nu[z] + b n_\nu[z]$ where the coefficients a and b are independent of ν or z.

a) Starting with the corresponding relations for cylindrical Bessel functions, verify the basic recursion relations:

$$f_{\nu-1}[z] + f_{\nu+1}[z] = \frac{2\nu+1}{z} f_\nu[z] \tag{8.423}$$

$$\nu f_{\nu-1}[z] - (\nu+1) f_{\nu+1}[z] = (2\nu+1) f_\nu'[z] \tag{8.424}$$

$$f_{\nu+1}[z] = \frac{\nu}{z} f_\nu[z] - f_\nu'[z] \tag{8.425}$$

$$f_{\nu-1}[z] = \frac{\nu+1}{z} f_\nu[z] + f_\nu'[z] \tag{8.426}$$

Then demonstrate the following variations:

b) $$\frac{d}{dz}(z^{\nu+1} f_\nu[z]) = z^{\nu+1} f_{\nu-1}[z] \tag{8.427}$$

c) $$\frac{d}{dz}(z^{-\nu} f_\nu[z]) = -z^{-\nu} f_{\nu+1}[z] \tag{8.428}$$

d) $$\left(\frac{1}{z}\frac{d}{dz}\right)^k (z^{\nu+1} f_\nu[z]) = z^{\nu-k} f_{\nu-k}[z] \tag{8.429}$$

e) $$\left(\frac{1}{z}\frac{d}{dz}\right)^k (z^{-\nu} f_\nu[z]) = (-)^k z^{-\nu-k} f_{\nu+k}[z] \tag{8.430}$$

30. Poisson integral representation for spherical Bessel functions

Use

$$j_l[kr] = \frac{i^{-l}}{2} \int_{-1}^{1} \mathrm{Exp}[ikrx] P_l[x]\, dx \tag{8.431}$$

derived in the text to obtain the Poisson integral representation

$$j_l[\rho] = \frac{\rho^l}{2^{l+1} l!} \int_0^\pi \mathrm{Cos}\,[\rho\, \mathrm{Cos}[\theta]]\, \mathrm{Sin}[\theta]^{2l+1}\, d\theta \tag{8.432}$$

for real ρ and nonnegative integer l.

31. Orthonormality relations for spherical Bessel functions

a) Use the differential equations satisfied by $j_l[k_1 r]$ and $j_l[k_2 r]$ for arbitrary k_1 and k_2 to show

$$R^2(k_2 j_l[k_1 R] j_l'[k_2 R] - k_1 j_l[k_2 R] j_l'[k_1 R]) = (k_1^2 - k_2^2) \int_0^R j_l[k_1 r] j_l[k_2 r] r^2 \, dr \tag{8.433}$$

for $l \geq 0$.

b) Derive the orthonormality relation

$$j_l\left[k_{l,n} R\right] = 0 \Longrightarrow \int_0^R j_l\left[k_{l,m} r\right] j_l\left[k_{l,n} r\right] r^2 \, dr = \frac{R^3}{2} j_{l+1}\left[k_{l,n} R\right]^2 \delta_{m,n} \tag{8.434}$$

for Dirichlet boundary conditions. Provide an explicit expression for the spherical Bessel expansion of an arbitrary function satisfying Dirichlet boundary conditions.

c) Derive the continuum orthonormality relation

$$\frac{2}{\pi} \int_0^\infty j_l[kr] j_l[k' r] r^2 \, dr = \frac{\delta[k - k']}{kk'} \tag{8.435}$$

9 Boundary-Value Problems

Abstract. Several methods are developed for constructing solutions to partial differential equations that match boundary conditions on a surface. We focus on equations based upon the Laplacian operator, such as the Helmholtz equation, that are expressed in separable orthogonal coordinate systems. Eigenfunction methods can then be used to develop series expansions or integral representations. Green functions provide general solutions for inhomogeneous problems. Examples include electrostatics, magnetostatics, and scattering theory. Exercises at the end of the chapter explore a wider variety of problems.

9.1 Introduction

We are often faced with the problem of solving a partial differential equation within a volume V bounded by a surface S subject to *boundary conditions* that specify the solution and/or some of its derivatives on S. Since neither time nor space nor patience nor expertise permits an exhaustive survey of this extremely broad subject, we shall be content to study equations based upon the Laplacian operator expressed in *separable orthogonal coordinate systems*. We take as a prototype the inhomogeneous *Helmholtz equation*

$$(\nabla^2 + k^2)\psi[\vec{r}, t] == -4\pi\rho[\vec{r}, t] \tag{9.1}$$

which includes among its special cases Laplace's equation ($k = 0, \rho = 0$), Poisson's equation ($k = 0$), the wave equation ($k = \omega/c$), the diffusion equation ($k^2 \to -\alpha$), and the Schrödinger equation ($k^2 = 2mE, -4\pi\rho = 2mV\psi$). In the happy circumstance that the boundary consists of surfaces on which one of the coordinates is constant, such problems are often amenable to the *method of separation of variables*.

Rather than plunging directly into the general theory, we illustrate the basic strategy by solving Poisson's equation

$$\nabla^2\psi[\vec{r}] == -4\pi\rho[\vec{r}] \tag{9.2}$$

within a rectangular box $\{0 \le x \le a,\ 0 \le y \le b,\ 0 \le z \le c\}$ with specified potentials upon its surfaces. We can divide the problem into several parts. First, we determine the potentials $\psi_i[\vec{r}, t]$ that satisfy Laplace's equation

$$\nabla^2\psi_i[\vec{r}] == 0, \quad \vec{r} \in S_i \implies \psi_i == V_i, \quad \vec{r} \in S_{j\neq i} \implies \psi_i == 0 \tag{9.3}$$

where ψ_i is specified by the two-dimensional function V_i on surface S_i and vanishes on the other five faces of the box. Then we determine the Green function that satisfies

$$\nabla^2 G[\vec{r}, \vec{r}\,'] == -4\pi\delta[\vec{r} - \vec{r}\,'] \tag{9.4}$$

Graduate Mathematical Physics. James J. Kelly

ISBN: 3-527-40637-9

with boundary conditions

$$\psi[0, y, z] == \psi[a, y, z] == \psi[x, 0, z] == \psi[x, b, z] == \psi[x, y, 0] == \psi[x, y, c] == 0 \quad (9.5)$$

on the surface of a grounded box. Finally, recognizing that Poisson's equation is linear, we can superimpose these solutions to obtain the general solution

$$\psi[\vec{r}] = \int_V G[\vec{r}, \vec{r}']\rho[\vec{r}']\, dV' + \sum_{i=1}^{6} \psi_i[\vec{r}] \quad (9.6)$$

for an arbitrary internal charge distribution and arbitrary potentials on each of the six sides of the box.

9.1.1 Laplace's Equation in Box with Specified Potential on one Side

Suppose that we wish to solve Laplace's equation $\nabla^2\psi == 0$ in a rectangular volume $\{0 \le x \le a, 0 \le y \le b, 0 \le z \le c\}$ with $\psi = 0$ on all faces except

$$z = c \Longrightarrow \psi[x, y, c] == V[x, y] \quad (9.7)$$

This problem is ideally suited for the method of separation of variables, wherein we propose a factorized solution of the form

$$\psi[x, y, z] = X[x]Y[y]Z[z] \quad (9.8)$$

for which

$$\nabla^2\psi == 0 \Longrightarrow \frac{1}{X}\frac{d^2X}{dx^2} + \frac{1}{Y}\frac{d^2Y}{dy^2} + \frac{1}{Z}\frac{d^2Z}{dz^2} == 0 \quad (9.9)$$

Recognizing that each of the three terms is a function of a different variable, each must separately be constant, such that

$$\frac{1}{X}\frac{d^2X}{dx^2} == -\alpha^2, \quad \frac{1}{Y}\frac{d^2Y}{dy^2} == -\beta^2, \quad \frac{1}{Z}\frac{d^2Z}{dz^2} == \gamma^2 \quad (9.10)$$

where the separation constants $\{-\alpha^2, -\beta^2, \gamma^2\}$ must satisfy

$$\alpha^2 + \beta^2 == \gamma^2 \quad (9.11)$$

Note that the separation constants were chosen with the foresight of experience (disingenuous hindsight), but lacking such foresight we would have used simply $\{A, B, C\}$ and then discovered later their most natural interpretations. The boundary conditions with respect to x, y are satisfied using

$$X[0] == X[a] == 0 \Longrightarrow X_n[x] = \mathrm{Sin}[\alpha_n x], \quad \alpha_n = \frac{n\pi}{a} \quad (9.12)$$

$$Y[0] == Y[b] == 0 \Longrightarrow Y_m[y] = \mathrm{Sin}[\beta_m y], \quad \beta_m = \frac{m\pi}{b} \quad (9.13)$$

with integer m, n, while the lower boundary condition on z requires

$$Z[0] == 0 \Longrightarrow Z[z] = \text{Sinh}\left[\gamma_{nm} z\right], \quad \gamma_{nm} = \sqrt{\alpha_n^2 + \beta_m^2} \tag{9.14}$$

A linear superposition of solutions of this form then gives

$$\psi[x, y, z] = \sum_{n,m} A_{nm} \, \text{Sin}[\alpha_n x] \, \text{Sin}[\beta_m y] \, \text{Sinh}[\gamma_{nm} z] \tag{9.15}$$

where the upper boundary condition on z requires

$$\psi[x, y, c] == V[x, y] \Longrightarrow A_{nm} = \frac{4}{ab \, \text{Sinh}[\gamma_{nm} c]} \int_0^a dx \int_0^b dy \, \text{Sin}[\alpha_n x] \, \text{Sin}[\beta_m y] V[x, y] \tag{9.16}$$

Solutions to more general problems in which nonzero potentials are specified on more than one face can then be constructed simply by adding several solutions of this type.

9.1.2 Green Function for Grounded Box

We can reduce the three-dimensional equation

$$\nabla^2 G[\vec{r}, \vec{r}\,'] == -4\pi \delta[\vec{r} - \vec{r}\,'] \tag{9.17}$$

with boundary conditions

$$\psi[0, y, z] == \psi[a, y, z] == \psi[x, 0, z] == \psi[x, b, z] == \psi[x, y, 0] == \psi[x, y, c] == 0 \tag{9.18}$$

to a one-dimensional equation by using the expansions

$$\delta[\vec{r} - \vec{r}\,'] = \delta[z - z'] \left(\frac{2}{a} \sum_{i=1}^{\infty} \text{Sin}[\alpha_i x] \, \text{Sin}[\alpha_i x'] \right) \left(\frac{2}{b} \sum_{j=1}^{\infty} \text{Sin}[\beta_j y] \, \text{Sin}[\beta_j y'] \right) \tag{9.19}$$

$$G[\vec{r}, \vec{r}\,'] = \sum_{i,j=1}^{\infty} g_{i,j}[z, z'] \, \text{Sin}[\alpha_i x] \, \text{Sin}[\alpha_i x'] \, \text{Sin}[\beta_j y] \, \text{Sin}[\beta_j y'] \tag{9.20}$$

to obtain

$$\left(\gamma_{i,j}^2 - \frac{\partial^2}{\partial z^2} \right) g_{i,j}[z, z'] == \frac{16\pi}{ab} \delta[z - z'], \quad g_{i,j}[0, z'] == g_{i,j}[c, z'] \tag{9.21}$$

where

$$\gamma_{i,j}^2 = \alpha_i^2 + \beta_j^2 \tag{9.22}$$

is the separation constant formed from two eigenvalues for the degrees of freedom that we chose to eliminate. The sine expansions satisfy the boundary conditions on four of the six sides automatically, but we must use the resulting one-dimensional inhomogeneous

Helmholtz equation to reconstruct the delta-function source. We chose to retain the z variable because the nontrivial boundary condition occurs on a side with constant z.

Using solutions to the homogeneous Helmholtz equation

$$0 \leq z \leq z' \Longrightarrow g_{i,j}[z, z'] = A_{i,j}\,\mathrm{Sinh}[\gamma_{i,j} z] \tag{9.23}$$

$$z' \leq z \leq c \Longrightarrow g_{i,j}[z, z'] = B_{i,j}\,\mathrm{Sinh}[\gamma_{i,j}(c - z)] \tag{9.24}$$

that satisfy the boundary conditions one either side of the source and applying the matching conditions

$$\Delta g_{i,j}[z', z'] == 0 \Longrightarrow B_{i,j}\,\mathrm{Sinh}[\gamma_{i,j}(c - z')] - A_{i,j}\,\mathrm{Sinh}[\gamma_{i,j} z'] == 0 \tag{9.25}$$

$$\Delta g'_{i,j}[z', z'] == -\frac{16\pi}{ab} \Longrightarrow B_{i,j}\,\mathrm{Cosh}[\gamma_{i,j}(c - z')] + A_{i,j}\,\mathrm{Cosh}[\gamma_{i,j} z'] == \frac{16\pi}{ab\gamma_{i,j}} \tag{9.26}$$

across the $z = z'$ interface, we find

$$A_{i,j} = \frac{16\pi}{ab\gamma_{i,j}} \frac{\mathrm{Sinh}[\gamma_{i,j}(c - z')]}{\mathrm{Sinh}[\gamma_{i,j} c]}, \quad B_{i,j} = \frac{16\pi}{ab\gamma_{i,j}} \frac{\mathrm{Sinh}[\gamma_{i,j} z']}{\mathrm{Sinh}[\gamma_{i,j} c]} \tag{9.27}$$

Thus, we find

$$g_{i,j}[z, z'] = \frac{16\pi}{ab\gamma_{i,j}} \frac{\mathrm{Sinh}[\gamma_{i,j} z_<]\,\mathrm{Sinh}[\gamma_{i,j}(c - z_>)]}{\mathrm{Sinh}[\gamma_{i,j} c]} \tag{9.28}$$

where $z_<$ is the smaller and $z_>$ is the larger of z and z'. This form should be very familiar, by now, from our work with Sturm–Liouville systems. Therefore, we can assemble the entire result

$$G[\vec{r}, \vec{r}\,'] = \frac{16\pi}{ab} \sum_{i,j=1}^{\infty} \mathrm{Sin}[\alpha_i x]\,\mathrm{Sin}[\alpha_i x']\,\mathrm{Sin}[\beta_j y]\,\mathrm{Sin}[\beta_j y'] \cdot \frac{\mathrm{Sinh}[\gamma_{i,j} z_<]\,\mathrm{Sinh}[\gamma_{i,j}(c - z_>)]}{\gamma_{i,j}\,\mathrm{Sinh}[\gamma_{i,j} c]} \tag{9.29}$$

in the form of a piecewise continuous, two-dimensional Fourier sine expansion. The expressions may be lengthy, but the analysis is hardly more complicated than that for one-dimensional Sturm–Liouville systems. Two other similar representations can be obtained by simply permuting variables and their associated boundary conditions; the most convenient choice may depend upon the representation of the source density, $\rho[\vec{r}]$, in the Poisson equation. Notice that the reciprocity condition

$$G[\vec{r}\,', \vec{r}] = G[\vec{r}, \vec{r}\,'] \tag{9.30}$$

is satisfied automatically.

Alternatively, we can employ Fourier sine expansions in all three dimensions

$$\delta[\vec{r}-\vec{r}'] = \frac{8}{abc}\sum_{i,j,k=1}^{\infty} \mathrm{Sin}[\alpha_i x]\,\mathrm{Sin}[\alpha_i x']\,\mathrm{Sin}[\beta_j y]\,\mathrm{Sin}[\beta_j y']\,\mathrm{Sin}[\gamma_k z]\,\mathrm{Sin}[\gamma_k z'] \tag{9.31}$$

$$G[\vec{r},\vec{r}'] = \sum_{i,j,k=1}^{\infty} g_{i,j,k}\,\mathrm{Sin}[\alpha_i x]\,\mathrm{Sin}[\alpha_i x']\,\mathrm{Sin}[\beta_j y]\,\mathrm{Sin}[\beta_j y']\,\mathrm{Sin}[\gamma_k z]\,\mathrm{Sin}[\gamma_k z'] \tag{9.32}$$

where now we define $\gamma_k = k\pi/c$ with a single index. Substituting into Poisson's equation and using orthogonality to equate coefficients

$$-(\alpha_i^2+\beta_j^2+\gamma_k^2)g_{i,j,k} = -\frac{32\pi}{abc}\,\mathrm{Sin}[\alpha_i x']\,\mathrm{Sin}[\beta_j y']\,\mathrm{Sin}[\gamma_k z'] \tag{9.33}$$

we obtain

$$G[\vec{r},\vec{r}'] = \frac{32\pi}{abc}\sum_{i,j,k=1}^{\infty} \frac{\mathrm{Sin}[\alpha_i x]\,\mathrm{Sin}[\alpha_i x']\,\mathrm{Sin}[\beta_j y]\,\mathrm{Sin}[\beta_j y']\,\mathrm{Sin}[\gamma_k z]\,\mathrm{Sin}[\gamma_k z']}{\alpha_i^2+\beta_j^2+\gamma_k^2} \tag{9.34}$$

Equivalence between Eq. (9.29) and Eq. (9.34) may be verified by performing a Fourier sine expansion of $g_{i,j}[z,z']$, whereby

$$\begin{aligned} g_{i,j,k} &= \frac{2}{c}\int_0^c g_{i,j}[z,z']\,\mathrm{Sin}[\gamma_k z]\,dz \\ &= \frac{32\pi}{abc\gamma_{i,j}}\left(\frac{\mathrm{Sinh}[\gamma_{i,j}(c-z')]}{\mathrm{Sinh}[\gamma_{i,j}c]}\int_0^{z'} \mathrm{Sinh}[\gamma_{i,j}z]\,\mathrm{Sin}[\gamma_k z]\,dz \right. \\ &\quad \left. + \frac{\mathrm{Sinh}[\gamma_{i,j}z']}{\mathrm{Sinh}[\gamma_{i,j}c]}\int_{z'}^{c} \mathrm{Sinh}[\gamma_{i,j}(c-z)]\,\mathrm{Sin}[\gamma_k z]\,dz\right) \end{aligned} \tag{9.35}$$

Although it is not difficult to evaluate these integrals by hand and simplify the result using standard trigonometric identities, we prefer to let *MATHEMATICA®* perform the 'grunt' work to obtain

$$\mathbf{Simplify}\left[\frac{32\pi}{a\,b\,c\,\gamma_{i,j}}\left(\frac{\mathbf{Sinh}[\gamma_{i,j}(c-z')]\int_0^{z'}\mathbf{Sinh}[\gamma_{i,j}z]\,\mathbf{Sin}[\gamma_k z]\,dz}{\mathbf{Sinh}[\gamma_{i,j}c]} + \frac{\mathbf{Sinh}[\gamma_{i,j}z']\int_{z'}^{c}\mathbf{Sinh}[\gamma_{i,j}(c-z)]\,\mathbf{Sin}[\gamma_k z]\,dz}{\mathbf{Sinh}[\gamma_{i,j}c]}\right)\right] /. \{\gamma_{i,j}^2 \to \alpha_i^2+\beta_j^2,\ \mathbf{Sin}[\gamma_k c]\to 0\}\Big) //\mathbf{Simplify}$$

$$\frac{32\pi\,\mathrm{Sin}[\gamma_k z']}{a\,b\,c\left(\alpha_i^2+\beta_j^2\right)+a\,b\,c\,\gamma_k^2}$$

in agreement with our analysis based upon Poisson's equation.

A point charge q located at position $\vec{r}'$ within a grounded box produces an electrostatic potential

$$\rho[\vec{r}] = q\delta[\vec{r} - \vec{r}'] \Longrightarrow \psi[\vec{r}] = qG[\vec{r}, \vec{r}'] \tag{9.36}$$

while a charge density distributed within the box produces

$$\psi[\vec{r}] = \int_V G[\vec{r}, \vec{r}']\rho[\vec{r}']\, dV' \tag{9.37}$$

where V denotes the enclosed volume. The integration can be performed using either form of the Green function, whichever seems simplest for the actual interior charge distribution. Of course, in real life, such integrations usually must be performed numerically outside the classroom.

9.2 Green's Theorem for Electrostatics

In electrostatics or magnetostatics we are often given or seek to define either the potential or the field on some closed surface and then need to determine those quantities everywhere within the volume enclosed by the bounding surface. Although one could, in principle, compute the potentials by adding the contributions of all charges and currents, we often do not know the detailed distributions of charge or current outside the volume of interest. For example, if we use a battery to maintain a constant potential on an electrode, we probably cannot guess the distribution of charge on its surface except in the simplest of geometries; hence, we are faced with a boundary-value problem. Once we know the Green function for a point charge within the volume of interest subject to the specified boundary conditions, we could compute the distribution of surface charge induced upon the electrode by the interior charge density. *Dirichlet* boundary conditions specify the value of a scalar potential ψ on the surface S while *Neumann* boundary conditions specify its normal derivative. Sometimes one encounters *mixed boundary conditions* for which the value is specified on some portions and the normal derivative on others. *Cauchy* boundary conditions specifying both the potential and its normal derivative are too restrictive for Poisson's equation, generally precluding existence of a solution, but can be useful for other subjects.

In this section we use Green's theorem to construct formal solutions to either Dirichlet or Neumann boundary value problems for Poisson's equation. First we review the derivation of Green's identities based upon the Gauss divergence theorem. Let $\vec{A}$ represent a vector field within a volume V bounded by a surface S. The divergence theorem then states

$$\int \vec{\nabla} \cdot \vec{A}\, dV = \oint \vec{A} \cdot d\vec{S} \tag{9.38}$$

where $d\vec{S} = \hat{n}\, dS$ is the directed element of surface area with $\hat{n}$ being the outward normal. If we choose $\vec{A} = \psi\vec{\nabla}\phi$ where ϕ and ψ are differentiable scalar fields, substitution of

$$\vec{A} = \psi\vec{\nabla}\phi \Longrightarrow \vec{\nabla} \cdot \vec{A} = \psi\nabla^2\phi + \vec{\nabla}\psi \cdot \vec{\nabla}\phi \tag{9.39}$$

into Gauss' theorem immediately yields *Green's first identity*

$$\int (\psi \nabla^2 \phi + \vec{\nabla}\psi \cdot \vec{\nabla}\phi)\, dV = \oint \psi \vec{\nabla}\phi \cdot d\vec{S} = \oint \psi \frac{\partial \phi}{\partial n}\, dS \tag{9.40}$$

where

$$\frac{\partial \phi}{\partial n} = \hat{n} \cdot \vec{\nabla}\phi \tag{9.41}$$

is a convenient shorthand notation for the normal derivative, the component of the gradient in the direction of the outward normal to a surface. Interchanging ϕ and ψ and subtracting the new form of the first identity then provides the more symmetric *Green's second identity*

$$\int (\psi \nabla^2 \phi - \phi \nabla^2 \psi)\, dV = \oint (\psi \vec{\nabla}\phi - \phi \vec{\nabla}\psi) \cdot d\vec{S} = \oint (\psi \frac{\partial \phi}{\partial n} - \phi \frac{\partial \psi}{\partial n})\, dS \tag{9.42}$$

Poisson's equation for the electrostatic potential ψ takes the form

$$\nabla^2 \psi[\vec{r}] == -4\pi \rho[\vec{r}] \tag{9.43}$$

where ρ is the charge density found within volume V. The contributions of charges that might be found outside the volume of interest are represented by the boundary conditions, either the potential or its normal derivative on the bounding surface. It is useful to define the Green function $G[\vec{r}, \vec{r}']$ to be the potential at $\vec{r}$ produced by a unit point charge at $\vec{r}'$ as the solution to

$$\nabla^2 G[\vec{r}, \vec{r}'] == -4\pi \delta[\vec{r} - \vec{r}'] \tag{9.44}$$

subject to appropriate boundary conditions to be specified later. Next we apply Green's second identity by choosing ψ to be the electrostatic potential and ϕ to be the Green function, such that

$$\int (\psi[\vec{r}']\nabla'^2 G[\vec{r}, \vec{r}'] - G[\vec{r}, \vec{r}']\nabla'^2 \psi[\vec{r}'])\, dV' = \oint (\psi[\vec{r}'] \frac{\partial G[\vec{r}, \vec{r}']}{\partial n'} - G[\vec{r}, \vec{r}'] \frac{\partial \psi[\vec{r}']}{\partial n'})\, dS' \tag{9.45}$$

Using Poisson's equation and the definition of the Green function this becomes

$$-4\pi \int (\psi[\vec{r}']\delta[\vec{r} - \vec{r}'] - G[\vec{r}, \vec{r}']\rho[\vec{r}'])\, dV' = \oint (\psi[\vec{r}'] \frac{\partial G[\vec{r}, \vec{r}']}{\partial n'} - G[\vec{r}, \vec{r}'] \frac{\partial \psi[\vec{r}']}{\partial n'})\, dS' \tag{9.46}$$

Thus, we obtain a formal solution to Poisson's equation in the form of *Green's electrostatic theorem*

$$\psi[\vec{r}] = \int G[\vec{r}, \vec{r}']\rho[\vec{r}']\, dV' + \frac{1}{4\pi} \oint (G[\vec{r}, \vec{r}'] \frac{\partial \psi[\vec{r}']}{\partial n'} - \psi[\vec{r}'] \frac{\partial G[\vec{r}, \vec{r}']}{\partial n'})\, dS' \tag{9.47}$$

If there were no bounding surfaces, the first term would represent the familiar electrostatic potential produced by a specified charge distribution. The second term then represents the contribution made by external charges that determine the conditions on the bounding surface.

For Dirichlet boundary conditions specifying ψ on S, it is convenient to require that the Dirichlet Green function, G_{D}, vanish on S such that

$$\vec{r}' \in S' \Longrightarrow G_{\mathrm{D}}[\vec{r},\vec{r}'] == 0 \Longrightarrow$$
$$\psi[\vec{r}] = \int G_{\mathrm{D}}[\vec{r},\vec{r}']\rho[\vec{r}']\,dV' - \frac{1}{4\pi}\oint \psi[\vec{r}']\frac{\partial G_{\mathrm{D}}[\vec{r},\vec{r}']}{\partial n'}\,dS' \tag{9.48}$$

The corresponding result for Neumann boundary conditions, specifying $\partial\psi/\partial n'$ on S', is slightly more complicated because we cannot simply require $\partial G_{\mathrm{N}}/\partial n'$ to vanish on S'. The problem is that the average value of the normal derivative on the bounding surface is constrained by Poisson's equation and Gauss' theorem such that

$$\int \vec{\nabla}' \cdot \vec{\nabla}' G[\vec{r},\vec{r}']\,dV' = \int -4\pi\delta[\vec{r}-\vec{r}']\,dV' = \oint \vec{\nabla}' G[\vec{r},\vec{r}'] \cdot d\vec{S}' \tag{9.49}$$

requires

$$\oint \frac{\partial G[\vec{r},\vec{r}']}{\partial n'}\,dS' = -4\pi \tag{9.50}$$

Hence, the simplest boundary condition we can impose upon G_{N} is

$$\frac{\partial G_{\mathrm{N}}[\vec{r},\vec{r}']}{\partial n'} = -\frac{4\pi}{S} \tag{9.51}$$

where here S is the area of the bounding surface. Thus, a formal solution to the Neumann boundary-value problem can be expressed as

$$\begin{aligned} &\frac{\partial G_{\mathrm{N}}[\vec{r},\vec{r}']}{\partial n'} = -\frac{4\pi}{S} \\ &\Longrightarrow \psi[\vec{r}] = \langle\psi\rangle_S + \int G_{\mathrm{N}}[\vec{r},\vec{r}']\rho[\vec{r}']\,dV' + \frac{1}{4\pi}\oint G_{\mathrm{N}}[\vec{r},\vec{r}']\frac{\partial\psi[\vec{r}']}{\partial n'}\,dS' \end{aligned} \tag{9.52}$$

where $\langle\psi\rangle_S$ denotes the average potential on S. Often Neumann problems appear in exterior form where the region of interest is outside an inner boundary and can be interpreted as within an outer boundary that is taken to ∞, such that

$$S \to \infty \Longrightarrow \langle\psi\rangle_S \to 0, \quad \frac{\partial G_{\mathrm{N}}[\vec{r},\vec{r}']}{\partial n'} \to 0 \Longrightarrow \tag{9.53}$$

$$\psi[\vec{r}] = \int G_{\mathrm{N}}[\vec{r},\vec{r}']\rho[\vec{r}']\,dV' + \frac{1}{4\pi}\oint G_{\mathrm{N}}[\vec{r},\vec{r}']\frac{\partial\psi[\vec{r}']}{\partial n'}\,dS' \tag{9.54}$$

where only the inner surface provides a finite contribution.

9.3 Separable Coordinate Systems

The Laplacian operator can be expressed in a curvilinear coordinate system with $\vec{r} = (\xi_1, \xi_2, \xi_3)$ as

$$\nabla^2\psi = \frac{1}{h_1 h_2 h_3} \sum_{i=1}^{3} \frac{\partial}{\partial \xi_i} \left(\frac{h_1 h_2 h_3}{h_i^2} \frac{\partial \psi}{\partial \xi_i} \right) \tag{9.55}$$

where

$$h_i^2 = \sum_{j=1}^{3} \left(\frac{\partial x_j}{\partial \xi_i} \right)^2 \tag{9.56}$$

are diagonal elements of the metric tensor

$$g_{i,j} = \sum_{k=1}^{3} \frac{\partial x_i}{\partial \xi_k} \frac{\partial x_j}{\partial \xi_k} \tag{9.57}$$

relating the cartesian coordinates $(x_1, x_2, x_3) = (x, y, z)$ to the curvilinear coordinates (ξ_1, ξ_2, ξ_3). We consider a coordinate system $\vec{r} = (\xi_1, \xi_2, \xi_3)$ separable when a product function of the form

$$\psi[\xi_1, \xi_2, \xi_3] = X_1[\xi_1] X_2[\xi_2] X_3[\xi_3] \tag{9.58}$$

permits the Laplacian to be expressed in the form

$$\frac{\nabla^2\psi}{\psi} = f_1[\xi_1] \left(\frac{1}{X_1[\xi_1]} \frac{\partial}{\partial \xi_1} \left(g[\xi_1] \frac{\partial X_1[\xi_1]}{\partial \xi_1} \right) + \frac{f_2[\xi_2]}{X_2[\xi_2]} \frac{\partial}{\partial \xi_2} \left(g[\xi_2] \frac{\partial X_2[\xi_2]}{\partial \xi_2} \right) + \frac{f_{3,2}[\xi_2] f_{3,3}[\xi_3]}{X_3[\xi_3]} \frac{\partial}{\partial \xi_3} \left(g[\xi_3] \frac{\partial X_3[\xi_3]}{\partial \xi_3} \right) \right) \tag{9.59}$$

One can then separate the homogeneous Helmholtz equation

$$\nabla^2\psi + k^2\psi == 0 \tag{9.60}$$

in a sequence of steps

$$\frac{f_{3,3}[\xi_3]}{X_3[\xi_3]} \frac{d}{d\xi_3} \left(g[\xi_3] \frac{dX_3[\xi_3]}{d\xi_3} \right) == \lambda_3 \tag{9.61}$$

$$\frac{f_2[\xi_2]}{X_2[\xi_2]} \frac{d}{d\xi_2} \left(g[\xi_2] \frac{dX_2[\xi_2]}{d\xi_2} \right) + \lambda_3 f_{3,2}[\xi_2] == \lambda_2 \tag{9.62}$$

$$f_1[\xi_1] \left(\frac{1}{X_1[\xi_1]} \frac{d}{d\xi_1} \left(g[\xi_1] \frac{dX_1[\xi_1]}{d\xi_1} \right) + \lambda_2 \right) == -k^2 \tag{9.63}$$

that produces three second-order ordinary differential equations coupled through the two separation constants λ_3 and λ_2.

It turns out that there are 11 orthogonal coordinate systems that are separable in this manner. A comprehensive analysis can be found in the treatise by Morse and Feshbach. Here we will be content to illustrate the general method using just the most common of these systems: rectangular, spherical polar, and cylindrical. We have already discussed the rectangular case and now proceed to spherical and cylindrical coordinates.

9.3.1 Spherical Polar Coordinates

Using the familiar spherical polar coordinates

$$x = r\,\mathrm{Sin}[\theta]\,\mathrm{Cos}[\phi] \tag{9.64}$$

$$y = r\,\mathrm{Sin}[\theta]\,\mathrm{Sin}[\phi] \tag{9.65}$$

$$z = r\,\mathrm{Cos}[\theta] \tag{9.66}$$

the Laplacian takes the form

$$\nabla^2\psi = \frac{1}{r^2}\frac{\partial}{\partial r}\left(r^2\frac{\partial\psi}{\partial r}\right) + \frac{1}{r^2\,\mathrm{Sin}[\theta]}\frac{\partial}{\partial\theta}\left(\mathrm{Sin}[\theta]\frac{\partial\psi}{\partial\theta}\right) + \frac{1}{r^2\,\mathrm{Sin}[\theta]^2}\frac{\partial^2\psi}{\partial\phi^2} \tag{9.67}$$

We propose a separable solution of the form

$$\psi[\vec{r}] = R[r]\Theta[\theta]\Phi[\phi] \tag{9.68}$$

for which the Helmholtz equation becomes

$$\frac{1}{R}\frac{\partial}{\partial r}\left(r^2\frac{\partial R}{\partial r}\right) + \frac{1}{\mathrm{Sin}[\theta]}\frac{1}{\Theta}\frac{\partial}{\partial\theta}\left(\mathrm{Sin}[\theta]\frac{\partial\Theta}{\partial\theta}\right) + \frac{1}{\mathrm{Sin}[\theta]^2}\frac{1}{\Phi}\frac{\partial^2\Phi}{\partial\phi^2} + k^2r^2 == 0 \tag{9.69}$$

Recognizing that only one term depends upon ϕ, we define a separation constant m using

$$\frac{1}{\Phi}\frac{d^2\Phi}{d\phi^2} == -m^2 \Longrightarrow \Phi = e^{\pm im\phi} \tag{9.70}$$

leaving

$$\frac{1}{R}\frac{\partial}{\partial r}\left(r^2\frac{\partial R}{\partial r}\right) + \frac{1}{\mathrm{Sin}[\theta]}\frac{1}{\Theta}\frac{\partial}{\partial\theta}\left(\mathrm{Sin}[\theta]\frac{\partial\Theta}{\partial\theta}\right) - \frac{m^2}{\mathrm{Sin}[\theta]^2} + k^2r^2 == 0 \tag{9.71}$$

Isolating the θ dependence using a foresightful separation constant

$$\mathrm{Sin}[\theta]\frac{d}{d\theta}\left(\mathrm{Sin}[\theta]\frac{d\Theta}{d\theta}\right) - m^2\Theta == -l(l+1)\,\mathrm{Sin}[\theta]^2\Theta \tag{9.72}$$

and employing the transformation

$$x = \mathrm{Cos}[\theta] \Longrightarrow \mathrm{Sin}[\theta]\frac{d\Theta}{d\theta} = -\left(1-x^2\right)\frac{d\Theta}{dx} \tag{9.73}$$

we recognize the associated Legendre equation

$$\left(1-x^2\right)\frac{d^2\Theta}{dx^2} - 2x\frac{d\Theta}{dx} + \left(l(l+1) - \frac{m^2}{1-x^2}\right)\Theta == 0 \Longrightarrow \Theta[x] = P_{l,m}[x], Q_{l,m}[x] \tag{9.74}$$

whose solutions are combinations of regular and irregular associated Legendre polynomials. Finally, using

$$R[r] = \frac{u[r]}{r} \tag{9.75}$$

the radial equation reduces to the spherical form of Bessel's equation

$$u'' + \left(k^2 - \frac{l(l+1)}{r^2}\right) u == 0 \Longrightarrow \frac{u[r]}{kr} = j_l[kr], n_l[kr] \tag{9.76}$$

whose solutions for $k > 0$ are composed of regular and irregular spherical Bessel functions. Therefore, general solutions can be constructed from appropriate linear combinations of the schematic form

$$\psi\left[\vec{r}\right] = \sum_{l,m} \psi_{l,m} \begin{pmatrix} j_l[kr] \\ n_l[kr] \end{pmatrix} \begin{pmatrix} P_{l,m}\left[\text{Cos}[\theta]\right] \\ Q_{l,m}\left[\text{Cos}[\theta]\right] \end{pmatrix} \begin{pmatrix} e^{im\phi} \\ e^{-im\phi} \end{pmatrix} \tag{9.77}$$

where $\psi_{l,m}$ is a constant coefficient that multiplies terms obtained from the appropriate choices from each of the three columns. Often the boundary conditions will eliminate one of the choices from each column, but otherwise we need to superimpose all possible combinations for the most general conditions. Similarly, depending upon the nature of k it might be more appropriate to employ modified Bessel functions or

$$k \to 0 \Longrightarrow j_l[kr] \to r^l, \quad n_l[kr] \to r^{-l-1} \tag{9.78}$$

Remember, this method provides a useful strategy, but it still must be adapted to the particular features of the problem at hand.

Often it is much more convenient to combine the angular functions into spherical harmonics and to use the identity

$$\nabla^2 \frac{u[r]}{r} Y_{l,m}[\theta, \phi] = \frac{1}{r}\left(u''[r] - \frac{l(l+1)}{r^2} u[r]\right) Y_{l,m}[\theta, \phi] \tag{9.79}$$

The Helmholtz equation then reduces to the radial equation

$$u''[r] + \left(k^2 - \frac{l(l+1)}{r^2}\right) u[r] == 0 \tag{9.80}$$

such that $\psi[\vec{r}]$ can be represented as a *multipole expansion* of the form

$$\psi[\vec{r}] = \sum_{l,m} \psi_{l,m} \frac{u_l[kr]}{kr} Y_{l,m}[\theta, \phi] \tag{9.81}$$

where the $\psi_{l,m}$ are expansion coefficients and $u_l[kr]$ is a conveniently normalized solution to the radial equation and its boundary conditions. Note that we have assumed, for simplicity, that the irregular Legendre functions can be discarded, but generalization should not be difficult.

It is also useful to commit to memory some of the standard Green functions for open boundary conditions. Namely, for Poisson's equation we find

$$\nabla^2 G[\vec{r}, \vec{r}\,'] == -4\pi\delta[\vec{r}, \vec{r}\,'] \Longrightarrow G[\vec{r}, \vec{r}\,'] = \frac{1}{|\vec{r} - \vec{r}\,'|} = \sum_l \frac{4\pi}{2l+1} \frac{r_<^l}{r_>^{l+1}} Y_l[\hat{r}] \cdot Y_l[\hat{r}\,'] \tag{9.82}$$

where we remind the reader of the convenient notation

$$\begin{aligned} Y_l[\hat{r}] \cdot Y_l[\hat{r}'] &= \sum_{m=-l}^{l} Y_{l,m}[\theta, \phi] Y_{l,m}^*[\theta', \phi'] = \sum_{m=-l}^{l} Y_{l,m}^*[\theta, \phi] Y_{l,m}[\theta', \phi'] \\ &= \sum_{m=-l}^{l} (-)^m Y_{l,-m}[\theta, \phi] Y_{l,m}[\theta', \phi'] \end{aligned} \tag{9.83}$$

Similarly, for the Helmholtz equation one finds

$$\begin{aligned} &(\nabla^2 + k^2) G[\vec{r}, \vec{r}'] == -4\pi\delta[\vec{r}, \vec{r}'] \\ &\Longrightarrow G^{(\pm)}[\vec{r}, \vec{r}'] = \frac{\mathrm{Exp}[\pm ik|\vec{r} - \vec{r}'|]}{|\vec{r} - \vec{r}'|} = \pm 4\pi ik \sum_l j_l[kr_<] h_l^{(\pm)}[kr_>] Y_l[\hat{r}] \cdot Y_l[\hat{r}'] \end{aligned} \tag{9.84}$$

where $h_l^{(\pm)} = j_l \pm i n_l$ are spherical Hankel functions for outgoing (+) or incoming (−) boundary conditions. The radial factors are obtained by interface matching with the aid of the Wronskian for the radial differential equation; we leave the algebra as an exercise for the reader.

9.3.2 Cylindrical Coordinates

Using the cylindrical coordinates

$$x = \xi \,\mathrm{Cos}[\phi] \tag{9.85}$$

$$y = \xi \,\mathrm{Sin}[\phi] \tag{9.86}$$

$$z = z \tag{9.87}$$

the Laplacian takes the form

$$\nabla^2 \psi = \frac{1}{\xi} \frac{\partial}{\partial \xi} \left(\xi \frac{\partial \psi}{\partial \xi} \right) + \frac{1}{\xi^2} \frac{\partial^2 \psi}{\partial \phi^2} + \frac{\partial^2 \psi}{\partial z^2} \tag{9.88}$$

We propose a separable solution of the form

$$\psi\left[\vec{r}\right] = R[\xi] \Phi[\phi] Z[z] \tag{9.89}$$

for which the Helmholtz equation becomes

$$\frac{1}{R} \frac{1}{\xi} \frac{\partial}{\partial \xi} \left(\xi \frac{\partial R}{\partial \xi} \right) + \frac{1}{\Phi} \frac{1}{\xi^2} \frac{\partial^2 \Phi}{\partial \phi^2} + \frac{1}{Z} \frac{\partial^2 Z}{\partial z^2} + k^2 == 0 \tag{9.90}$$

Separating first the ϕ dependence using trigonometric functions with periodic boundary conditions and then the z dependence using hyperbolic functions assuming finite range, we obtain

$$\frac{1}{\Phi}\frac{\partial^2\Phi}{\partial\phi^2} == -m^2 \Longrightarrow \Phi[z] \in \{e^{im\phi}, e^{-im\phi}\} \tag{9.91}$$

$$\frac{1}{Z}\frac{\partial^2 Z}{\partial z^2} == \alpha_n^2 \Longrightarrow Z[z] \in \{\mathrm{Sinh}[\alpha_n z], \mathrm{Cosh}[\alpha_n z]\} \tag{9.92}$$

$$\frac{1}{\xi}\frac{\partial}{\partial\xi}\left(\xi\frac{\partial R}{\partial\xi}\right) + \left(k^2 + \alpha^2 - \frac{m^2}{\xi^2}\right)R == 0 \Longrightarrow R[\xi] \in \{J_m[\beta_n\xi], N_m[\beta_n\xi]\} \tag{9.93}$$

$$\text{with} \quad \beta_n^2 = k^2 + \alpha_n^2 \tag{9.94}$$

where we have assumed, somewhat arbitrarily, that the separation constants $\{m, \alpha, \beta\}$ are real; if not, we replace a trigonometric function by an exponential function or a Bessel function by a modified Bessel function. The schematic notation $f \in \{f_1, f_2\}$ indicates a suitable choice or linear combination of solutions that satisfies the appropriate boundary conditions. Therefore, complete solutions can be constructed from linear superpositions of the schematic form

$$\psi[\vec{r}] = \sum_{m,n} \psi_{m,n} \begin{pmatrix} J_m[\beta_n\xi] \\ N_m[\beta_n\xi] \end{pmatrix} \begin{pmatrix} \mathrm{Sinh}[\alpha_n z] \\ \mathrm{Cosh}[\alpha_n z] \end{pmatrix} \begin{pmatrix} e^{im\phi} \\ e^{-im\phi} \end{pmatrix} \tag{9.95}$$

Depending upon the boundary conditions, it might be more convenient to employ trigonometric functions for Z and then modified Bessel functions for R. Alternatively, sometimes it is more convenient to separate the R dependence first and then match across an interface with respect to Z. If the range of the angular variable is restricted, different boundary conditions and different angular functions could be needed. It takes experience to anticipate the optimum choices for a particular problem, and some of that experience can be acquired by solving the problems at the end of the chapter! In the remainder of this chapter we will concentrate upon spherical geometries, but many problems with cylindrical geometries are provided. Rather than attempt to apply the results derived here directly, it is usually better to perform the separation of variables for each problem anew in order to make informed decisions regarding the order of separation and the nature of separation constants.

9.4 Spherical Expansion of Dirichlet Green Function for Poisson's Equation

We seek to construct a Green function for Poisson's equation that vanishes on the concentric spheres $r = a$ and $r = b$ and satisfies

$$\nabla^2 G[\vec{r}, \vec{r}'] == -4\pi\delta[\vec{r} - \vec{r}'], \quad G == 0 \text{ on } r = a, b \tag{9.96}$$

in the enclosed volume. It is natural to employ spherical polar coordinates and to make a multipole expansion of the form

$$G[\vec{r},\vec{r}'] = \sum_l g_l[r,r'] Y_l[\hat{r}] \cdot Y_l[\hat{r}'] \tag{9.97}$$

$$\delta[\vec{r}-\vec{r}'] = \frac{\delta[r-r']}{r^2} \sum_l Y_l[\hat{r}] \cdot Y_l[\hat{r}'] \tag{9.98}$$

where g_l satisfies the radial equation

$$\frac{1}{r}\frac{d^2(rg_l)}{dr^2} - \frac{l(l+1)}{r^2} g_l[r,r'] == -\frac{4\pi}{r^2}\delta[r-r'] \tag{9.99}$$

Solutions to the homogeneous equation that satisfy either the inner or the outer boundary condition take the form

$$r < r' \Longrightarrow g_l = A\left(\left(\frac{r}{a}\right)^l - \left(\frac{r}{a}\right)^{-l-1}\right) \tag{9.100}$$

$$r > r' \Longrightarrow g_l = B\left(\left(\frac{r}{b}\right)^l - \left(\frac{r}{b}\right)^{-l-1}\right) \tag{9.101}$$

and are subject to matching conditions

$$g_> == g_< \Longrightarrow B\left(\left(\frac{r'}{b}\right)^l - \left(\frac{r'}{b}\right)^{-l-1}\right) == A\left(\left(\frac{r'}{a}\right)^l - \left(\frac{r'}{a}\right)^{-l-1}\right) \tag{9.102}$$

$$\Delta g' == -\frac{4\pi}{(r')^2} \Longrightarrow \frac{B}{r'}\left(l\left(\frac{r'}{b}\right)^l + (l+1)\left(\frac{r'}{b}\right)^{-l-1}\right) - \frac{A}{r'}\left(l\left(\frac{r'}{a}\right)^l + (l+1)\left(\frac{r'}{a}\right)^{-l-1}\right) == -\frac{4\pi}{(r')^2} \tag{9.103}$$

across the interface. It is convenient to rewrite these equations as

$$B(1-\beta) == A(1-\alpha) \tag{9.104}$$

$$B(l+\beta(l+1)) - A(l+\alpha(l+1)) == -4\pi(r')^{-l-1} \tag{9.105}$$

where

$$\alpha = \left(\frac{a}{r'}\right)^{2l+1} \quad \beta = \left(\frac{b}{r'}\right)^{2l+1} \tag{9.106}$$

It is now easy, albeit tedious, to solve these equations

```
sol = Solve[{B(1 - β) == A(1 - α), B(l + β(l + 1)) - A(l + α(l + 1)) == -4π (r')^(-l-1)},
{A, B}][[1]]//Simplify
```

$$\left\{A \to \frac{4\pi(-1+\beta)(r')^{-1-l}}{(1+2l)(-\alpha+\beta)},\ B \to -\frac{4\pi(-1+\alpha)(r')^{-1-l}}{(1+2l)(\alpha-\beta)}\right\}$$

and to construct the two pieces

```
{A((r/a)^l - (r/a)^(-l-1)), B((r/b)^l - (r/b)^(-l-1))} /. sol /.
{α → (a/r')^(2l+1), β → (b/r')^(2l+1)} // Simplify
```

$$\left\{\frac{4\pi\left(-\left(\frac{r}{a}\right)^{-1-l}+\left(\frac{r}{a}\right)^{l}\right)\left(-1+\left(\frac{b}{r'}\right)^{1+2l}\right)(r')^{-1-l}}{(1+2l)\left(-\left(\frac{a}{r'}\right)^{1+2l}+\left(\frac{b}{r'}\right)^{1+2l}\right)},\right.$$
$$\left.-\frac{4\pi\left(-\left(\frac{r}{b}\right)^{-1-l}+\left(\frac{r}{b}\right)^{l}\right)\left(-1+\left(\frac{a}{r'}\right)^{1+2l}\right)(r')^{-1-l}}{(1+2l)\left(\left(\frac{a}{r'}\right)^{1+2l}-\left(\frac{b}{r'}\right)^{1+2l}\right)}\right\}$$

Therefore, combining the two pieces, we obtain the Dirichlet Green function for the volume between concentric grounded spheres as

$$g_l[r, r'] = \frac{4\pi}{2l+1}\frac{1}{(r_< r_>)^{l+1}}\frac{(b^{2l+1} - r_>^{2l+1})(r_<^{2l+1} - a^{2l+1})}{b^{2l+1} - a^{2l+1}} \tag{9.107}$$

where $r_<$ is the smaller and $r_>$ is the larger of r and r'. Notice that this function is simply the product of the inner and outer solutions with a normalization based upon their Wronskian; that general form should be familiar from our work on Sturm–Liouville systems and appears often in various guises.

Several important special cases

$$a \to 0 \Longrightarrow g_l[r, r'] = \frac{4\pi}{2l+1}\frac{r_<^l}{r_>^{l+1}}\left(1 - \left(\frac{r_>}{b}\right)^{2l+1}\right) \tag{9.108}$$

$$b \to \infty \Longrightarrow g_l[r, r'] = \frac{4\pi}{2l+1}\frac{r_<^l}{r_>^{l+1}}\left(1 - \left(\frac{a}{r_<}\right)^{2l+1}\right) \tag{9.109}$$

$$a \to 0, b \to \infty \Longrightarrow g_l[r, r'] = \frac{4\pi}{2l+1}\frac{r_<^l}{r_>^{l+1}} \tag{9.110}$$

can be obtained from the limiting behaviors when one or both of the radii assume extreme values. The first case listed above describes the interior of a sphere of radius b, the second describes the exterior of a sphere of radius a, while the third describes an open geometry without bounding surfaces. The result for open geometry should already be familiar from the multipole expansion

$$\frac{1}{|\vec{r} - \vec{r}'|} = \sum_l \frac{4\pi}{2l+1}\frac{r_<^l}{r_>^{l+1}} Y_l[\hat{r}] \cdot Y_l[\hat{r}'] \tag{9.111}$$

for the potential of a unit point charge in empty space.

Given an arbitrary charge distribution $\rho[\vec{r}']$ in the volume between concentric spheres with specified potentials $\psi_a = \psi[a, \theta, \phi]$ and $\psi_b = \psi[b, \theta, \phi]$, we can now compute the electrostatic potential anywhere within the enclosed volume by integrating

$$\psi[\vec{r}] = \int G[\vec{r}, \vec{r}']\rho[\vec{r}']\, dV' - \frac{1}{4\pi}\oint \psi[\vec{r}']\vec{\nabla}' G[\vec{r}, \vec{r}'] \cdot d\vec{S}' \tag{9.112}$$

taking care to interpret the directed element of surface area as outward with respect to the enclosed volume such that

$$r' = a \Longrightarrow d\vec{S}' = -a^2\, d\Omega' \hat{r}' \tag{9.113}$$

$$r' = b \Longrightarrow d\vec{S}' = +b^2\, d\Omega' \hat{r}' \tag{9.114}$$

where $\hat{r}'$ is the radial unit vector. Thus, the second contribution to the potential produced by a point charge in the presence of a grounded sphere must represent the potential due to charges on the surface of the conductor. Several examples of the use of this Green function are given below. Some of these examples are simple enough to evaluate symbolically, but generally numerical integration would be required for nontrivial distributions. Nevertheless, a mathematician would consider the problem solved because numerical integration can be performed by computer in a straightforward manner. A physicist, on the other hand, would probably require a more practical demonstration that the numerical method works!

9.4.1 Example: Multipole Expansion for Localized Charge Distribution

Suppose that we seek the electrostatic potential

$$\psi[\vec{r}] = \int G[\vec{r}, \vec{r}']\rho[\vec{r}']\, dV' \tag{9.115}$$

for distances $r \gg R$ where R represents the radius containing all of the charge, such that

$$G[\vec{r}, \vec{r}'] = \sum_l g_l[r, r'] Y_l[\hat{r}] \cdot Y_l[\hat{r}'] \tag{9.116}$$

$$g_l[r, r'] = \frac{4\pi}{2l+1} \frac{r_<^l}{r_>^{l+1}} \longrightarrow \frac{4\pi}{2l+1} \frac{r'^l}{r^{l+1}} \tag{9.117}$$

The potential then takes the form

$$\psi[\vec{r}] = \sum_{l,m} \frac{q_{l,m}}{r^{l+1}} Y_{l,m}[\hat{r}] \tag{9.118}$$

where the multipole amplitudes are

$$q_{l,m} = \frac{4\pi}{2l+1} \int r^l Y_{l,m}^*[\theta, \phi] \rho[\vec{r}]\, dV \tag{9.119}$$

The lowest nonvanishing multipoles tend to dominate for $r \gg R$.

9.4.2 Example: Point Charge Near Grounded Conducting Sphere

The potential due to a point charge near a grounded conducting sphere has two contributions, one from the point charge itself and another from the distribution of surface charge needed to maintain the conductor at constant potential. One can show that the contribution of the induced surface charge density is equivalent to the potential produced by an *image charge*. Suppose first that the point charge is inside the conducting sphere, such that $r, r' < b$. By writing

$$\frac{r_<^l}{r_>^{l+1}}\left(\frac{r_>}{b}\right)^{2l+1} = \frac{1}{b}\left(\frac{r_< r_>}{b^2}\right)^l = \frac{1}{b}\left(\frac{rr'}{b^2}\right)^l = \frac{r_b}{b}\frac{r^l}{r_b^{l+1}} \tag{9.120}$$

we obtain

$$\sum_l \frac{4\pi}{2l+1}\frac{r_<^l}{r_>^{l+1}}\left(\frac{r_>}{b}\right)^{2l+1} Y_l[\hat{r}]\cdot Y_l[\hat{r}'] = \frac{r_b}{b}\frac{1}{|\vec{r}-\vec{r}_b|} \tag{9.121}$$

where $\vec{r}_b = \vec{r}'(b/r')^2$ is the location of an image charge outside a sphere of radius b induced by a point charge inside the sphere at $\vec{r}'$. Similarly, by writing

$$\frac{r_<^l}{r_>^{l+1}}\left(\frac{a}{r_<}\right)^{2l+1} = \frac{1}{a}\left(\frac{a^2}{r_< r_>}\right)^{l+1} = \frac{1}{a}\left(\frac{a^2}{rr'}\right)^l = \frac{r_a}{a}\frac{r_a^l}{r^{l+1}} \tag{9.122}$$

we obtain

$$\sum_l \frac{4\pi}{2l+1}\frac{r_<^l}{r_>^{l+1}}\left(\frac{a}{r_<}\right)^{2l+1} Y_l[\hat{r}]\cdot Y_l\left[\hat{r}'\right] = \frac{r_a}{a}\frac{1}{|\vec{r}-\vec{r}_a|} \tag{9.123}$$

where $\vec{r}_a = \vec{r}'(a/r')^2$ is the location of an image charge inside a sphere of radius a induced by a point charge outside the sphere at $\vec{r}'$. Therefore, the potential for a point charge at $\vec{r}'$ in the presence of a ground conducting sphere of radius R becomes

$$\psi[\vec{r}] = \frac{q}{|\vec{r}-\vec{r}'|} + \frac{q_i}{|\vec{r}-\vec{r}_i|} \quad \text{with} \quad q_i = -q\frac{R}{r'}, \quad \vec{r}_i = \left(\frac{R}{r'}\right)^2 \vec{r}' \tag{9.124}$$

where q_i is the magnitude and $\vec{r}_i$ is the location of the image charge. The same expression is obtained whether the charge is inside or outside the sphere but, of course, the potential vanishes in the region on the opposite side of the conducting surface from the point charge.

The actual surface charge density is determined by the normal derivative of the electrostatic potential according to

$$\sigma = -\frac{1}{4\pi}\left(\frac{\partial\psi}{\partial r}\right)_{r=R} \tag{9.125}$$

which, after simple but tedious algebra, reduces to

$$\sigma = -\frac{q}{4\pi R}\frac{|r'^2 - R^2|}{\left(R^2 - 2r'R\,\text{Cos}[\theta] + r'^2\right)^{3/2}} \tag{9.126}$$

where $\text{Cos}[\theta] = \hat{r}\cdot\hat{r}'$ is the polar angle on the sphere relative to the line between its center and the point charge. A simple calculation shows that the total induced charge

$$2\pi R^2 \int_{-1}^{1} \sigma\, d\text{Cos}[\theta] = \begin{cases} -q, & r' < R \\ -qR/r', & r' > R \end{cases} \tag{9.127}$$

is not necessarily equal to the image charge. If the point charge is inside the grounded sphere the induced charge is equal to $-q$ so that the net charge and the potential outside

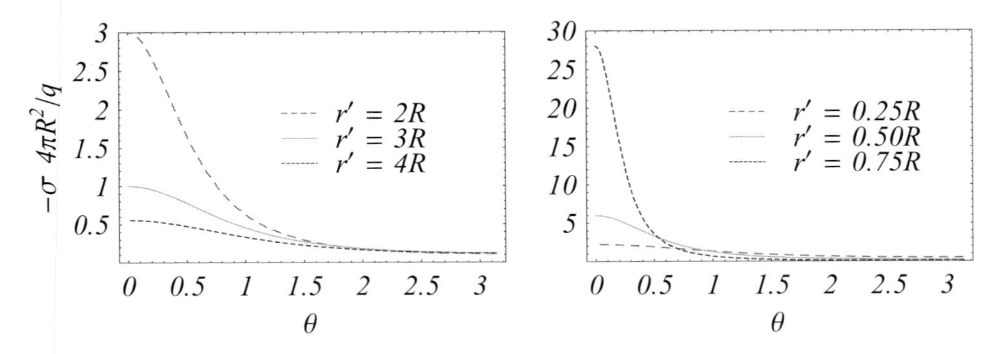

Figure 9.1. Surface charge density induced upon a grounded sphere of radius R by a point charge at r'.

the sphere vanish. On the other hand, if the point charge is outside the sphere the total induced charge is inversely proportional to the distance from the center of the sphere. In either case, the induced charge is concentrated nearby when the point charge is near the surface and is spread more uniformly as the distance between the point charge and the surface increases. Figure 9.1 show the angular distribution of induced surface charge density for several values of r'/R.

Although it is probably easier to obtain the Green function for a single grounded sphere using the method of images than the expansion in spherical harmonics, the problem of two concentric spheres would be rather difficult to solve using images because an infinite set of images is required. The expansion in spherical harmonics, by contrast, is no more difficult for a spherical shell than for a sphere and contains the latter as a special case. The Green function method is much more versatile than the method of images.

9.4.3 Example: Specified Potential on Surface of Empty Sphere

Suppose that $\psi == f[\theta, \phi]$ is specified on the surface of an empty sphere of radius R. The interior potential is then determined by

$$\psi[\vec{r}] = -\frac{R^2}{4\pi} \oint f[\theta, \phi] \left(\frac{\partial G[\vec{r}, \vec{r}']}{\partial r'} \right)_{r'=R} d\Omega' \tag{9.128}$$

where $d\vec{S}' = \hat{r}' R^2\, d\Omega'$ is the directed differential area and

$$G[\vec{r}, \vec{r}'] = \sum_l g_l[r, r'] Y_l[\hat{r}] \cdot Y_l[\hat{r}'] \tag{9.129}$$

$$g_l[r, r'] = \frac{4\pi}{2l+1} \frac{r_<^l}{r_>^{l+1}} \left(1 - \left(\frac{r_>}{R} \right)^{2l+1} \right) \tag{9.130}$$

is the Green function for the interior of a sphere. The radial derivative is simply

$$\left(\frac{\partial g_l[r, r']}{\partial r'} \right)_{r'=R} = -\frac{4\pi}{R^2} \left(\frac{r}{R} \right)^l \tag{9.131}$$

such that

$$r < R \Longrightarrow \psi[r,\theta,\phi] = \sum_{l,m}\left(\left(\frac{r}{R}\right)^l Y_{l,m}[\theta,\phi]\int Y_{l,m}^*\left[\theta',\phi'\right]f\left[\theta',\phi'\right]d\Omega'\right) \tag{9.132}$$

can be evaluated numerically for any surface potential. Similarly, the exterior potential is determined by

$$g_l[r,r'] = \frac{4\pi}{2l+1}\frac{r_<^l}{r_>^{l+1}}\left(1-\left(\frac{R}{r_<}\right)^{2l+1}\right) \Longrightarrow \left(\frac{\partial g_l[r,r']}{\partial r'}\right)_{r'=R} = \frac{4\pi}{R^2}\left(\frac{R}{r}\right)^{l+1} \tag{9.133}$$

Then, using $d\vec{S}' = -\hat{r}' R^2\, d\Omega'$ as the outward normal to the boundary of the exterior region, we obtain

$$r > R \Longrightarrow \psi[r,\theta,\phi] = \sum_{l,m}\left(\left(\frac{R}{r}\right)^{l+1} Y_{l,m}[\theta,\phi]\int Y_{l,m}^*\left[\theta',\phi'\right]f\left[\theta',\phi'\right]d\Omega'\right) \tag{9.134}$$

These results can now be combined in the form

$$\psi[r,\theta,\phi] = \sum_{l,m}\frac{r_<^l}{r_>^{l+1}}\psi_{l,m}Y_{l,m}[\theta,\phi] \tag{9.135}$$

where now $r_<$ is interpreted as the smaller and $r_>$ as the larger of r and R and where the multipole amplitudes are given by

$$\psi_{l,m} = \int Y_{l,m}^*[\theta,\phi]f[\theta,\phi]\,d\Omega \tag{9.136}$$

The same results could have been obtained by writing the eigenfunction expansions for ψ in the interior and exterior regions and matching the boundary condition at the surface without using the Green function. (Try it!)

Suppose that a point charge is found near a sphere of specified potential. According to the superposition principle, the electrostatic potential can be constructed by adding the potential for a point charge near a grounded sphere to the contribution by a sphere with specified surface potential in otherwise empty space. More generally, if there is a charge distribution near a sphere with specified potential, one adds the contribution for each element of charge, $dq' = \rho[\vec{r}']\,dV'$ for a grounded sphere to the contribution of the surface potential, as represented by the formula

$$\psi[\vec{r}] = \int G[\vec{r},\vec{r}']\rho[\vec{r}']\,dV' - \frac{1}{4\pi}\oint \psi[\vec{r}']\vec{\nabla}' G[\vec{r},\vec{r}']\cdot d\vec{S}' \tag{9.137}$$

where the appropriate Green function is used in the interior and exterior regions and where $d\vec{S}'$ is radially outward for the interior or inward for the exterior regions. The surface contribution can be represented by a multipole expansion while the net contribution of the charge density and its corresponding induced surface charges can be represented either by multipoles or by images.

9.4.4 Example: Charged Ring at Center of Grounded Conducting Sphere

Suppose that a charged ring of radius a is centered within a grounded conducting sphere of radius b. It is convenient to align the $\hat{z}$ axis with the normal to the ring, such that

$$\rho[\vec{r}] = \frac{q}{2\pi a^2}\delta[r-a]\delta\big[\mathrm{Cos}[\theta]\big] \tag{9.138}$$

represents the charge density. The electrostatic potential is determined by

$$\psi[\vec{r}] = \int G[\vec{r},\vec{r}\,']\rho[\vec{r}\,']\,dV' \tag{9.139}$$

with

$$G[\vec{r},\vec{r}\,'] = \sum_l g_l[r,r']Y_l[\hat{r}]\cdot Y_l[\hat{r}'] \tag{9.140}$$

$$g_l[r,r'] = \frac{4\pi}{2l+1}\frac{r_<^l}{r_>^{l+1}}\left(1-\left(\frac{r_>}{b}\right)^{2l+1}\right) \tag{9.141}$$

Azimuthal symmetry with respect to ϕ' limits the expansion to terms with $m = 0$, for which

$$Y_{l,0}[\theta,\phi] = \left(\frac{2l+1}{4\pi}\right)^{1/2} P_l\big[\mathrm{Cos}[\theta]\big] \tag{9.142}$$

Therefore, we obtain

$$\psi[\vec{r}] = q\sum_l \frac{r_<^l}{r_>^{l+1}}\left(1-\left(\frac{r_>}{b}\right)^{2l+1}\right)P_l[0]P_l\big[\mathrm{Cos}[\theta]\big] \tag{9.143}$$

where $r_<$ is the smaller and $r_>$ is the larger of r and a. Although $P_l[0]$ can be evaluated in closed form, the series is not simplified thereby. Nevertheless, the potential can be evaluated numerically using a simple program; however, it is often necessary to include many terms in order to achieve the desired accuracy. An example is provided in Fig. 9.2 for a/b and summation up to $l \leq 50$.

9.5 Magnetic Field of Current Loop

The current density for a circular loop of radius a in the xy plane can be expressed as

$$\vec{J} = J_\phi\hat{\phi} \tag{9.144}$$

$$J_\phi = I\delta\big[\mathrm{Cos}[\theta]\big]\frac{\delta[r-a]}{a} \tag{9.145}$$

where the azimuthal unit vector can be expressed in either Cartesian or spherical bases as

$$\hat{\phi} = -\mathrm{Sin}[\phi]\hat{x} + \mathrm{Cos}[\phi]\hat{y} = \frac{i}{\sqrt{2}}\left(e^{-i\phi}\hat{e}_+ + e^{i\phi}\hat{e}_-\right) \tag{9.146}$$

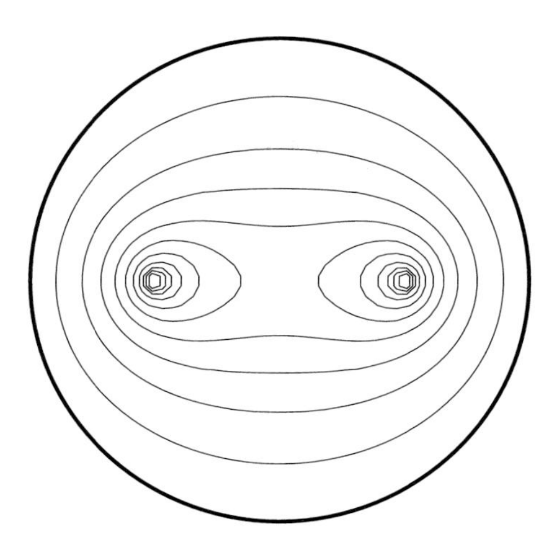

Figure 9.2. Equipotentials for a charged ring of radius a centered within a grounded sphere of radius b, where $a/b = 0.5$.

where

$$\hat{e}_+ = -\frac{\hat{e}_x + i\hat{e}_y}{\sqrt{2}}, \quad \hat{e}_- = \frac{\hat{e}_x - i\hat{e}_y}{\sqrt{2}}, \quad \hat{e}_0 = \hat{e}_z \tag{9.147}$$

The multipole expansion of the magnetic vector potential

$$\begin{aligned}\vec{A}[\vec{r}] &= \frac{1}{c}\int dV' \frac{\vec{J}[\vec{r}']}{|\vec{r} - \vec{r}'|} \\ &= \frac{1}{c}\sum_{l=0}^{\infty}\sum_{m=-l}^{l} \frac{4\pi}{2l+1} Y_{l,m}[\theta, \phi] \int d\Omega'\, dr' r'^2 \vec{J}[\vec{r}'] \frac{r_<^l}{r_>^{l+1}} Y_{l,m}^*[\theta', \phi']\end{aligned} \tag{9.148}$$

then becomes

$$\vec{A}[\vec{r}] = \frac{Ia}{c}\sum_{l=0}^{\infty} \frac{4\pi}{2l+1}\frac{r_<^l}{r_>^{l+1}} \sum_{m=-l}^{l} Y_{l,m}[\theta, \phi] \int d\phi' Y_{l,m}^*\left[\frac{\pi}{2}, \phi'\right] \frac{i}{\sqrt{2}}(e^{-i\phi'}\hat{e}_+ + e^{i\phi'}\hat{e}_-) \tag{9.149}$$

where $r_<$ is the smaller and $r_>$ is the larger of r and a. By writing

$$Y_{l,m}^*[\theta, \phi] = Y_{l,m}[\theta, 0]e^{-im\phi} \Longrightarrow \int d\phi Y_{l,m}^*[\theta, \phi]e^{\pm i\phi} = 2\pi\delta_{m,\pm 1} Y_{l,m}[\theta, 0] \tag{9.150}$$

we obtain

$$\begin{aligned}\vec{A}[\vec{r}] = 2\pi\frac{Ia}{c}&\sum_{l=0}^{\infty} \frac{4\pi}{2l+1}\frac{r_<^l}{r_>^{l+1}}\frac{i}{\sqrt{2}} \\ &\left(Y_{l,-1}[\theta, 0]Y_{l,-1}\left[\frac{\pi}{2}, 0\right]e^{-i\phi}\hat{e}_+ + Y_{l,1}[\theta, 0]Y_{l,1}\left[\frac{\pi}{2}, 0\right]e^{i\phi}\hat{e}_-\right)\end{aligned} \tag{9.151}$$

Figure 9.3. A_ϕ for current loop.

Next we use

$$Y_{l,-m} = (-)^m Y^*_{l,m}, \qquad Y^*_{l,m}[\theta, 0] = Y_{l,m}[\theta, 0] \tag{9.152}$$

to obtain

$$A_\phi[\vec{r}] = 2\pi \frac{Ia}{c} \sum_{l=0}^{\infty} \frac{4\pi}{2l+1} \frac{r_<^l}{r_>^{l+1}} Y_{l,1}[\theta, 0] Y_{l,1}\left[\frac{\pi}{2}, 0\right] \tag{9.153}$$

Finally, we use

$$Y_{l,1}\left[\frac{\pi}{2}, 0\right] = \begin{cases} 0, & l = 2n \\ (-)^{n+1}\left(\frac{4\pi l(l+1)}{2l+1}\right)^{-1/2} \frac{(2n+1)!!}{(2n)!!}, & l = 2n+1 \end{cases} \tag{9.154}$$

and

$$Y_{l,1}[\theta, 0] = -\left(\frac{4\pi l(l+1)}{2l+1}\right)^{-1/2} P_{l,1}[\mathrm{Cos}[\theta]] \tag{9.155}$$

to obtain

$$A_\phi[\vec{r}] = 2\pi \frac{Ia}{c} \sum_{n=0}^{\infty} \frac{(-)^n}{2^n} \frac{(2n-1)!!}{(n+1)!} \frac{r_<^{2n+1}}{r_>^{2n+2}} P_{2n+1,1}\left[\mathrm{Cos}[\theta]\right] \tag{9.156}$$

in terms of associated Legendre polynomials. The appearance of associated Legendre polynomials in the vector potential, instead of ordinary Legendre polynomials in the scalar potential for a ring, reflects the different symmetry properties of vector and scalar fields. Contours of A_ϕ are sketched in Fig. 9.3.

The magnetic field is obtained using

$$\vec{A} = A_\phi \hat{\phi} \Longrightarrow B_r = \frac{1}{r\,\mathrm{Sin}[\theta]}\frac{\partial}{\partial\theta}(\mathrm{Sin}[\theta]A_\phi), \quad B_\theta = -\frac{1}{r}\frac{\partial}{\partial r}(rA_\phi), \quad B_\phi = 0 \tag{9.157}$$

The angular derivative can also be expressed as

$$\frac{1}{\mathrm{Sin}[\theta]}\frac{\partial}{\partial\theta} = -\frac{\partial}{\partial x} \Longrightarrow \frac{1}{\mathrm{Sin}[\theta]}\frac{\partial}{\partial\theta}(\mathrm{Sin}[\theta]A_\phi) = -\frac{\partial}{\partial x}\left((1-x^2)^{1/2}A_\phi[x]\right) \tag{9.158}$$

where $x = \mathrm{Cos}[\theta]$. Expressing the associated Legendre polynomial as

$$P_{l,1}[x] = \left(1-x^2\right)^{1/2} P_l'[x] \Longrightarrow \frac{\partial}{\partial x}\left(\left(1-x^2\right)^{1/2} P_{l,1}[x]\right) = \left(1-x^2\right)P_l''[x] - 2xP_l'[x] \tag{9.159}$$

and using the Legendre differential equation, we find

$$\frac{1}{\mathrm{Sin}[\theta]}\frac{\partial}{\partial\theta}\left(\mathrm{Sin}[\theta]P_{l,1}\left[\mathrm{Cos}[\theta]\right]\right) = l(l+1)P_l\left[\mathrm{Cos}[\theta]\right] \tag{9.160}$$

and obtain

$$B_r = 2\pi\frac{Ia}{cr}\sum_{n=0}^{\infty}\frac{(-)^n}{2^n}\frac{(2n+1)!!}{n!}\frac{r_<^{2n+1}}{r_>^{2n+2}}P_{2n+1}\left[\mathrm{Cos}[\theta]\right] \tag{9.161}$$

The B_θ component is easily evaluated but must be separated into inner and outer regions, whereby

$$r < a \Longrightarrow B_\theta = 4\pi\frac{Ia}{cr}\sum_{n=0}^{\infty}\frac{(-)^n}{2^n}\frac{(2n-1)!!}{n!}\frac{r^{2n+1}}{a^{2n+2}}P_{2n+1,1}\left[\mathrm{Cos}[\theta]\right] \tag{9.162}$$

$$r > a \Longrightarrow B_\theta = -2\pi\frac{Ia}{cr}\sum_{n=0}^{\infty}\frac{(-)^n}{2^n}\frac{(2n+1)!!}{(n+1)!}\frac{a^{2n+1}}{r^{2n+2}}P_{2n+1,1}\left[\mathrm{Cos}[\theta]\right] \tag{9.163}$$

9.6 Inhomogeneous Wave Equation

9.6.1 Spatial Representation of Time-Independent Green Function

Consider an inhomogeneous scalar wave equation of the form

$$\left(\frac{1}{c^2}\frac{\partial^2}{\partial t^2} - \nabla^2\right)\psi[\vec{r},t] == 4\pi\rho[\vec{r},t] \tag{9.164}$$

where ψ represents the wave amplitude, c is the phase velocity, and ρ represents a source that we assume is localized in both space and time. It is useful to perform a Fourier analysis of the time dependence

$$\rho[\vec{r},t] = \int_{-\infty}^{\infty}\frac{d\omega}{2\pi}e^{-i\omega t}\rho_\omega[\vec{r}] \tag{9.165}$$

$$\psi[\vec{r},t] = \int_{-\infty}^{\infty}\frac{d\omega}{2\pi}e^{-i\omega t}\psi_k[\vec{r}] \tag{9.166}$$

such that the spatial dependence is described by the inhomogeneous Helmholtz equation

$$\left(\nabla^2 + k^2\right)\psi_k[\vec{r}] == -4\pi\rho_\omega[\vec{r}] \tag{9.167}$$

where $k = \omega/c$ is the wave number. One could, of course, perform a Fourier analysis of the spatial dependence also, but we prefer to begin with a more traditional partial-wave expansion of the Green function, which satisfies an equation of the form

$$\nabla^2 G + k^2 G == -4\pi\delta[\vec{r} - \vec{r}\,'] \tag{9.168}$$

with boundary conditions to be specified later. Once we have the Green function, the solution to the original wave equation becomes

$$\psi_k[\vec{r}] = \phi_k[\vec{r}] + \int G_k[\vec{r}, \vec{r}\,']\rho_\omega[\vec{r}\,']\, dV' \tag{9.169}$$

where ϕ_k is a solution to the homogeneous equation

$$(\nabla^2 + k^2)\phi_k[\vec{r}] == 0 \tag{9.170}$$

that is determined by matching the boundary conditions.

For example, in the time-independent formalism for scattering by a localized distribution, one specifies that the asymptotic wave function takes the form of an incident plane wave plus an outgoing spherical wave of the form

$$r \gg r' \Longrightarrow \psi_k[\vec{r}] \simeq \text{Exp}[i\vec{k}\cdot\vec{r}] + f[\vec{k}', \vec{k}]\frac{\text{Exp}[ik'r]}{r} \tag{9.171}$$

where the scattering amplitude f is the amplitude of the spherical wave. The incident wave vector $\vec{k} = \frac{\omega}{c}\hat{k}$ specifies both the frequency ω and direction $\hat{k}$ for the incident wave while the direction of the outgoing wave is specified by $\hat{k}' = \hat{r}$. The magnitude of the scattered and incident wave vectors are equal for elastic scattering, such that $k' = k = \omega/c$, but differ for inelastic scattering. If we choose the $\hat{z}$ axis along the incident direction, we may write the asymptotic wave function for elastic scattering as

$$r \gg r' \Longrightarrow \psi_k[\vec{r}] \simeq \text{Exp}[ikz] + f_k[\theta, \phi]\frac{\text{Exp}[ikr]}{r} \tag{9.172}$$

Including the time dependence, the planes of constant phase in the incident component

$$\phi[\vec{r}, t] = \text{Exp}[i(kz - \omega t)] \tag{9.173}$$

are clearly seen to travel in the direction of increasing z with phase velocity c. Similarly, spheres of constant phase in the scattered wave

$$\psi_{\text{sc}}[\vec{r}, t] \propto \frac{\text{Exp}[i(kr - \omega t)]}{r} \tag{9.174}$$

move radially outward with the same phase velocity. Thus, *outgoing boundary conditions* require

$$r \gg r' \Longrightarrow G_k^{(+)}[\vec{r}, \vec{r}\,'] \simeq \frac{\text{Exp}[ikr]}{r} \tag{9.175}$$

However, a more rigorous treatment would represent the incoming wave by a localized wave packet of finite extent, rather than a plane wave.

Recognizing that $k \to 0 \Longrightarrow G \to |\vec{r} - \vec{r}'|^{-1}$, we seek a solution of the form

$$G = \frac{g[R]}{R} \tag{9.176}$$

where $R = |\vec{r} - \vec{r}'|$. It is convenient to temporarily shift the coordinate system to make $\vec{r}' \to 0$ and to employ

$$\nabla^2(fg) = f\nabla^2 g + g\nabla^2 f + 2\nabla f \cdot \nabla g \Longrightarrow \nabla^2 \frac{g}{r} = \frac{1}{r}\nabla^2 g - \frac{2}{r^2}\frac{\partial g}{\partial r} - 4\pi\delta[\vec{r}] \tag{9.177}$$

Using the spherical symmetry of $g[r]$ to write

$$\nabla^2 g[r] = \frac{1}{r}\frac{\partial^2}{\partial r^2}(rg[r]) = rg''[r] + 2g'[r] \tag{9.178}$$

the Helmholtz equation becomes

$$g''[r] + k^2 g[r] == -4\pi(1 - g[r])\delta[r] \tag{9.179}$$

Therefore, by requiring $g[0] == 1$ we immediately obtain $g = a\,\text{Exp}[ikr] + b\,\text{Exp}[-ikr]$ with $a + b = 1$. Outgoing boundary conditions require $G \propto e^{ikR}/R$ for $R \to \infty$, such that

$$G_k^{(+)}[\vec{r}, \vec{r}'] = \frac{\text{Exp}[ik|\vec{r} - \vec{r}'|]}{|\vec{r} - \vec{r}'|} \tag{9.180}$$

represents an outgoing spherical wave produced by a point source at $\vec{r}'$. Similarly, the Green function for *incoming* boundary conditions becomes

$$G_k^{(-)}[\vec{r}, \vec{r}'] = \frac{\text{Exp}[-ik|\vec{r} - \vec{r}'|]}{|\vec{r} - \vec{r}'|} = (G_k^{(+)}[\vec{r}, \vec{r}'])^* \tag{9.181}$$

It is also useful to express these functions in terms of spherical Hankel functions

$$G_k^{(\pm)}[\vec{r}, \vec{r}'] = \frac{\text{Exp}[\pm ik|\vec{r} - \vec{r}'|]}{|\vec{r} - \vec{r}'|} = \pm ikh_0^{(\pm)}[k|\vec{r} - \vec{r}'|] \tag{9.182}$$

where only $l = 0$ is needed because the Green function is spherically symmetric with respect to the distance from the source.

Suppose that ρ_ω represents a localized source, such as an antenna, that radiates outgoing waves and that there is no incident wave. We can then drop the ϕ_k contribution and employ the Green function for outgoing boundary conditions, such that the solution to the inhomogeneous Helmholtz equation takes the form

$$\psi_k[\vec{r}] = \int \frac{\text{Exp}[ik|\vec{r} - \vec{r}'|]}{|\vec{r} - \vec{r}'|}\rho_\omega[\vec{r}']\,dV' \tag{9.183}$$

In the far field, where r is much greater than any r' for which ρ_ω has appreciable strength, we may use

$$r \gg r' \Longrightarrow |\vec{r} - \vec{r}'| \approx r\left(1 - \frac{\hat{r} \cdot \vec{r}'}{r}\right) \Longrightarrow \frac{\text{Exp}[ik|\vec{r} - \vec{r}'|]}{|\vec{r} - \vec{r}'|} \approx \frac{\text{Exp}[ikr]}{r}\text{Exp}[-i\vec{k} \cdot \vec{r}'] \tag{9.184}$$

where $\vec{k} = k\hat{r}$ to approximate the Green function. Note that it is important to include the dependence of the phase of the exponential upon r' because cancellations between various parts of the source are crucial to the amplitude and angular distribution of the scattered wave, but the variation of the denominator is much less important and can be neglected for a localized source. Thus, in the far field we obtain an asymptotic approximation

$$r \gg r' \Longrightarrow \psi_k[\vec{r}] \simeq f[\vec{k}]\frac{\mathrm{Exp}[ikr]}{r} \tag{9.185}$$

that has the form of an outgoing spherical wave whose angular distribution

$$f[\vec{k}] = \int \mathrm{Exp}[-i\vec{k}\cdot\vec{r}']\rho_\omega[\vec{r}']\,dV' \tag{9.186}$$

is governed by the Fourier transform of the source. The quantity $f[\vec{k}]$ is known as the *form factor*. Similar form factors appear throughout theories of radiation or scattering. The interpretation of the form factor should now be clear. When the field point is sufficiently far from the source to treat the rays received from various parts of the source as parallel, the total amplitude is given by the sum of amplitudes from each part of the source with a phase, relative to the center, given by $-\vec{k}\cdot\vec{r}'$.

It is often useful to expand the form factor in terms of multipoles of the source. Expanding the plane wave in spherical harmonics

$$\mathrm{Exp}[-i\vec{k}\cdot\vec{r}'] = 4\pi\sum_l i^{-l} j_l[kr']Y_l[\hat{k}]\cdot Y_l[\hat{r}'] \tag{9.187}$$

the angular distribution can be expressed in the form

$$f[\vec{k}] = \sum_l C_l\cdot Y_l[\hat{k}] \Longrightarrow f[\theta,\phi] = \sum_{l,m} C^*_{l,m}[k]Y_{l,m}[\theta,\phi] \tag{9.188}$$

where the coefficients

$$C_{l,m}[k] = i^l\int j_l[kr]Y_{l,m}[\hat{r}]\rho_\omega[\vec{r}]\,dV \tag{9.189}$$

are Fourier–Bessel multipole amplitudes for the source.

9.6.2 Partial-Wave Expansion

It will often be useful to have a partial-wave expansion of the Helmholtz Green function. Let

$$G_k[\vec{r},\vec{r}'] = \sum_l g_l[r,r']Y_l[\hat{r}]\cdot Y_l[\hat{r}'] \tag{9.190}$$

$$\delta[\vec{r}-\vec{r}'] = \frac{\delta[r-r']}{rr'}\sum_l Y_l[\hat{r}]\cdot Y_l[\hat{r}'] \tag{9.191}$$

where the boundary conditions will be specified later and where the k dependence in g_l is left implicit. Using

$$\nabla^2 g_l[r]Y_{l,m}[\theta,\phi] = \left(\frac{1}{r}\frac{\partial^2\left(rg_l[r]\right)}{\partial r^2} - \frac{l(l+1)}{r^2}g_l[r]\right)Y_{l,m}[\theta,\phi] \tag{9.192}$$

the radial equation becomes

$$\frac{\partial^2 g_l[r,r']}{\partial r^2} + \frac{2}{r}\frac{\partial g_l[r,r']}{\partial r} + \left(k^2 - \frac{l(l+1)}{r^2}\right) g_l[r,r'] == -4\pi \frac{\delta[r-r']}{rr'} \tag{9.193}$$

The homogeneous equation is recognized as the spherical Bessel equation. If we require g_l to be regular for $r < r'$ and to obey outgoing boundary conditions for $r > r'$, then we expect g_l to take the form

$$g_l[r,r'] = A j_l\left[kr_<\right] h_l^{(+)}\left[kr_>\right] \tag{9.194}$$

where A is a constant, j_l is the regular spherical Bessel function, and $h_l^{(+)}$ is the outgoing spherical Hankel function. The constant is determined by the discontinuity in slope

$$\Delta\left(\frac{\partial g_l[r,r']}{\partial r}\right)_{r=r'} == -\frac{4\pi}{r^2} \Longrightarrow A\left(j_l[kr]\frac{\partial h_l^{(+)}[kr]}{\partial r} - h_l^{(+)}[kr]\frac{\partial j_l[kr]}{\partial r}\right) == -\frac{4\pi}{r^2} \tag{9.195}$$

where the term in the final parentheses is the Wronskian between the solutions in the inner and outer regions. Using

$$h_l^{(+)}[x] = j_l[x] + i n_l[x] \Longrightarrow j_l[x]\frac{\partial h_l^{(+)}[x]}{\partial x} - h_l^{(+)}[x]\frac{\partial j_l[x]}{\partial x} = j_l[x]n_l'[x] - n_l[x]j_l'[x] \tag{9.196}$$

and the Wronskian

$$j_l[x]n_l'[x] - n_l[x]j_l'[x] = \frac{i}{x^2} \Longrightarrow A = 4\pi i k \tag{9.197}$$

we finally obtain the partial-wave expansion

$$G_k^{(+)}[\vec{r},\vec{r}'] = \frac{\mathrm{Exp}[ik|\vec{r}-\vec{r}'|]}{|\vec{r}-\vec{r}'|} = 4\pi i k \sum_l j_l[kr_<]h_l^{(+)}[kr_>]Y_l[\hat{r}]\cdot Y_l[\hat{r}'] \tag{9.198}$$

The asymptotic form of the wave function for $r \gg r'$, where r' is limited by the source ρ, can now be written as

$$r \gg r' \Longrightarrow \psi_k[\vec{r}] \simeq \phi_k[\vec{r}] + ik\sum_{l,m} h_l^{(+)}[kr]Y_{l,m}[\hat{r}]\int j_l\left[kr'\right]Y_{l,m}^*[\hat{r}']\rho_\omega[\vec{r}']\,dV' \tag{9.199}$$

The angular dependence of the scattered wave is coupled to the angular properties of its source. Using the asymptotic behavior of the Hankel function

$$h_l^{(+)}[kr] \simeq \frac{\mathrm{Exp}[i(kr - \frac{l+1}{2}\pi)]}{kr} \tag{9.200}$$

we find

$$\psi_k[\vec{r}] \simeq \phi_k[\vec{r}] + \frac{e^{ikr}}{r}\sum_{l,m} C_{l,m}^*[k]Y_{l,m}[\hat{r}] \tag{9.201}$$

$$C_{l,m} = i^l \int j_l[kr']Y_{l,m}[\hat{r}']\rho_\omega[\vec{r}']\,dV' \tag{9.202}$$

Thus, the scattered wave is expanded in outgoing waves with definite orbital angular momentum whose amplitudes are determined by a multipole expansion of the Fourier–Bessel transform of the source.

9.6.3 Momentum Representation of Time-Independent Green Function

Another useful representation of the Helmholtz Green function is obtained by expanding

$$G[\vec{r},\vec{r}'] = \int \frac{d^3k'}{(2\pi)^3}\beta[\vec{k}',\vec{r}']\phi[\vec{k}',\vec{r}] \tag{9.203}$$

where $\phi[\vec{k},\vec{r}]$ is an eigenfunction of the homogeneous equation

$$(\nabla^2 + k^2)\phi[\vec{k},\vec{r}] == 0 \tag{9.204}$$

and where

$$(\nabla^2 + k^2)G[\vec{r},\vec{r}'] == -4\pi\delta[\vec{r}-\vec{r}'] \tag{9.205}$$

with the appropriate boundary conditions. Using $\phi[\vec{k},\vec{r}] = \mathrm{Exp}[i\vec{k}\cdot\vec{r}]$ and applying the differential operator $\nabla^2 + k^2$ we obtain

$$\begin{aligned}(\nabla^2 + k^2)G[\vec{r},\vec{r}'] &= \int \frac{d^3k'}{(2\pi)^3}(k^2 - k'^2)\beta[\vec{k}',\vec{r}']\,\mathrm{Exp}[i\vec{k}'\cdot\vec{r}]\\ &= -4\pi\int\frac{d^3k'}{(2\pi)^3}\mathrm{Exp}[i\vec{k}'\cdot(\vec{r}-\vec{r}')]\end{aligned} \tag{9.206}$$

where in the last step we employ the Fourier representation of the delta function. Using the orthogonality of plane waves, we deduce

$$\beta[\vec{k}',\vec{r}'] = 4\pi\frac{\mathrm{Exp}[-i\vec{k}'\cdot\vec{r}']}{k'^2 - k^2} \Longrightarrow G[\vec{r},\vec{r}'] = 4\pi\int\frac{d^3k'}{(2\pi)^3}\frac{\mathrm{Exp}[i\vec{k}'\cdot(\vec{r}-\vec{r}')]}{k'^2-k^2} \tag{9.207}$$

The angular integral can be done by expanding the plane wave, such that

$$\begin{aligned}G[\vec{r},\vec{r}'] &= \frac{2}{\pi}\int_0^\infty dk'k'^2\frac{j_0[k'R]}{k'^2-k^2} = \frac{2}{\pi R}\int_0^\infty dk'k'\frac{\mathrm{Sin}[k'R]}{k'^2-k^2}\\ &= \frac{1}{i\pi R}\int_{-\infty}^{\infty} dk'k'\frac{\mathrm{Exp}[ik'R]}{k'^2-k^2}\end{aligned} \tag{9.208}$$

where $R = |\vec{r}-\vec{r}'|$. Ordinarily we would exploit the symmetry of the integrand to replace the sine function by an exponential and to extend the range of integration with respect to k' to $\pm\infty$ so that we can use a great semicircular contour closed in the upper half-plane, but the integrand has singularities at $k' = \pm k$ that are on the contour. Fortunately, we still have some unused information, namely the boundary conditions, which can be used to determine the proper handling of these singularities. Imposing outgoing boundary conditions in the far field, $R \to \infty$, suggests that we should include the positive pole and exclude the negative pole. With this choice we obtain the same outgoing Green function

$$G_k^{(+)}[\vec{r},\vec{r}'] = \frac{\mathrm{Exp}[ik|\vec{r}-\vec{r}'|]}{|\vec{r}-\vec{r}'|} \tag{9.209}$$

as before. Alternatively, incoming boundary conditions are satisfied by including the negative and excluding the positive pole. The appropriate contours for these conditions are sketched in Fig. 9.4.

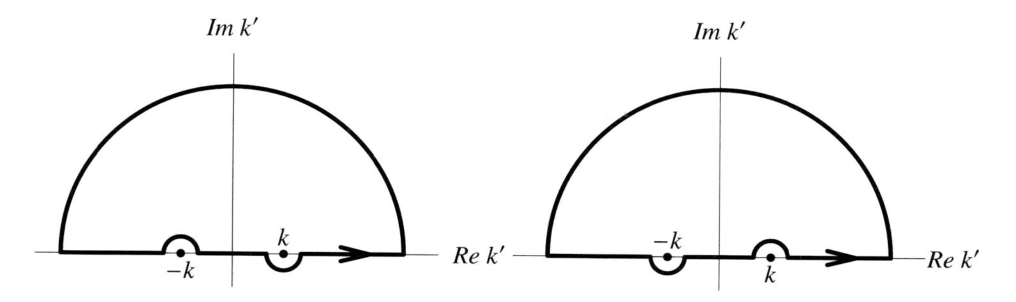

Figure 9.4. left: outgoing bc, right: incoming bc.

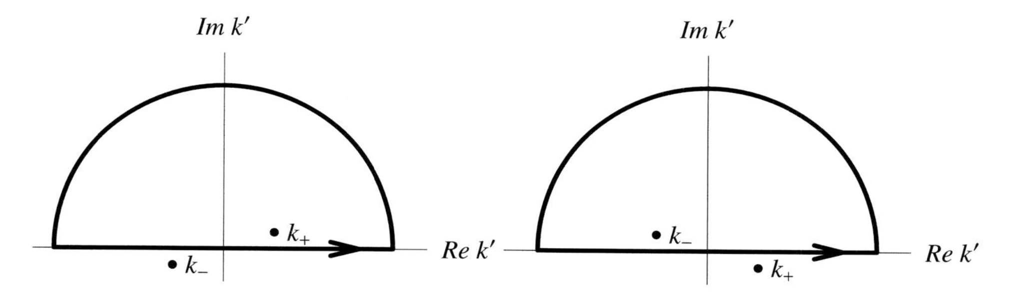

Figure 9.5. left: outgoing bc, right: incoming bc.

Another method for achieving the same result is to modify the Helmholtz equation by adding an infinitesimal δ

$$k \to k + i\frac{\delta}{2} \Longrightarrow k'_{\pm} = k \pm i\frac{\delta}{2} \tag{9.210}$$

such that the poles at $k'_{\pm}$ are shifted off the real axis asymmetrically, as sketched in Fig. 9.5. If $\delta \to 0^+$, we obtain outgoing boundary conditions while if $\delta \to 0^-$ we obtain incoming boundary conditions. Therefore, the momentum representation of the Green function can be expressed as

$$\tilde{G}^{(\pm)}[k, k'] = \frac{-4\pi}{k^2 - k'^2 \pm i\delta} \tag{9.211}$$

where δ is a positive infinitesimal and where the spatial representation is obtained using the Fourier transform

$$G_k^{(\pm)}[\vec{r}, \vec{r}'] = -4\pi \int \frac{d^3k'}{(2\pi)^3} \frac{\text{Exp}[i\vec{k}' \cdot (\vec{r} - \vec{r}')]}{k^2 - k'^2 \pm i\delta} \tag{9.212}$$

9.6.4 Retarded Green Function

If one performs a Fourier analysis of both the time and spatial dependencies for

$$\rho[\vec{r},t] = \int \frac{d^3k}{(2\pi)^3} \int_{-\infty}^{\infty} \frac{d\omega}{2\pi} e^{i(\vec{k}\cdot\vec{r}-\omega t)} \tilde{\rho}_\omega[\vec{k}] \tag{9.213}$$

$$\psi[\vec{r},t] = \int \frac{d^3k}{(2\pi)^3} \int_{-\infty}^{\infty} \frac{d\omega}{2\pi} e^{i(\vec{k}\cdot\vec{r}-\omega t)} \tilde{\psi}_\omega[\vec{k}] \tag{9.214}$$

the inhomogeneous wave equation takes the form

$$\left(k^2 - \left(\frac{\omega}{c}\right)^2\right) \tilde{\psi}_\omega[\vec{k}] == \tilde{\rho}_\omega[\vec{k}] \Longrightarrow \tilde{\psi}_\omega[\vec{k}] = \frac{\tilde{\rho}_\omega\left[\vec{k}\right]}{k^2 - \left(\frac{\omega}{c}\right)^2} \tag{9.215}$$

where, in this approach, k is not restricted to ω/c. The spatial wave function is obtained by inverting the Fourier transform

$$\psi[\vec{r},t] = -c^2 \int \frac{d^3k}{(2\pi)^3} \int_{-\infty}^{\infty} \frac{d\omega}{2\pi} e^{i\left(\vec{k}\cdot\vec{r}-\omega t\right)} \frac{\tilde{\rho}_\omega[\vec{k}]}{\omega^2 - k^2c^2} \tag{9.216}$$

but care must be exercised in handling the singularity. It is simplest to consider the Green function for a point source in both space and time, which satisfies an equation of the form

$$\left(\frac{1}{c^2}\frac{\partial^2}{\partial t^2} - \nabla^2\right) G\left[\vec{r},t;\vec{r}',t'\right] == 4\pi\delta[\vec{r}-\vec{r}']\delta\left[t-t'\right] \tag{9.217}$$

with boundary conditions to be specified later. Thus, using

$$\rho[\vec{r},t] = 4\pi\delta[\vec{r}-\vec{r}']\delta[t-t'] \Longrightarrow \tilde{\rho}_\omega[\vec{k}] = 4\pi\,\mathrm{Exp}[-i(\vec{k}\cdot\vec{r}' - \omega t')] \tag{9.218}$$

we find that

$$G[\vec{r},t;\vec{r}',t'] = 4\pi c^2 \int \frac{d^3k}{(2\pi)^3} \int_{-\infty}^{\infty} \frac{d\omega}{2\pi} \frac{\mathrm{Exp}[i(\vec{k}\cdot(\vec{r}-\vec{r}') - \omega(t-t'))]}{\omega^2 - k^2c^2} \tag{9.219}$$

represents a wave produced by a point source at position $\vec{r}'$ at time t'.

The angular part of the integral can be performed easily using

$$\mathrm{Exp}[i\vec{k}\cdot\vec{R}] = 4\pi \sum_l i^l j_l[kR] Y_l[\hat{k}] \cdot Y_l[\hat{R}] \tag{9.220}$$

where $\vec{R} = \vec{r} - \vec{r}'$ and $R = |\vec{R}|$, such that

$$G[\vec{r},t;\vec{r}',t'] = \frac{2c^2}{\pi} \int_0^{\infty} dk k^2 j_0[kR] \int_{-\infty}^{\infty} \frac{d\omega}{2\pi} \frac{\mathrm{Exp}[-i\omega(t-t')]}{\omega^2 - k^2c^2} \tag{9.221}$$

is limited by spherical symmetry to $l = 0$. The integrand exhibits poles on the real axis of the complex frequency plane at $\omega = \pm kc$. The integral does not have a unique value

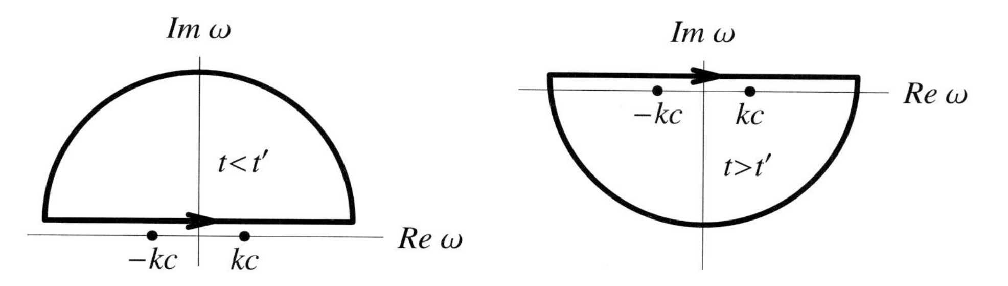

Figure 9.6. Contours for Retarded Green Function.

because the poles lie on the integration path, but we can appeal to causality to determine boundary conditions that specify the appropriate contour. Thus, we require G to vanish for $t < t'$ such that causes precede effects. This is accomplished by employing a contour that is slightly above the real axis, such that $\omega \to \omega + i\epsilon$ where ϵ is a positive infinitesimal, and closing in the contour in the upper half-plane for $t < t'$ or the lower half-plane for $t > t'$, as shown in Fig. 9.6. Alternatively, one could imagine shifting the poles slightly below the real axis, such that $\omega_{\pm} = \pm kc - i\epsilon$ and taking the limit $\epsilon \to 0^+$ after evaluating the integral. Either way we obtain

$$\int_{-\infty}^{\infty} \frac{d\omega}{2\pi} \frac{\text{Exp}[-i\omega(t-t')]}{(\omega + i\epsilon)^2 - k^2c^2} = \frac{\text{Sin}[kc(t-t')]}{kc}\Theta[t-t'] \tag{9.222}$$

where Θ is the unit step function.

The retarded Green function then takes the form

$$\begin{aligned} G^{(+)}[\vec{r},t;\vec{r}',t'] &= \frac{2c}{\pi}\frac{\Theta[t-t']}{R}\int_0^{\infty} dk\,\text{Sin}[kR]\,\text{Sin}[kc(t-t')] \\ &= \frac{c}{\pi}\frac{\Theta[t-t']}{R}\int_0^{\infty} dk\big(\text{Cos}\big[k\big(R - c(t-t')\big)\big] \\ &\quad - \text{Cos}\big[k\big(R + c(t-t')\big)\big]\big) \\ &= \frac{c}{2\pi}\frac{\Theta[t-t']}{R}\int_{-\infty}^{\infty} dk\big(\text{Exp}[ik(R - c(t-t'))]. \\ &\quad - \text{Exp}\big[ik\big(R + c(t-t')\big)\big]\big) \end{aligned} \tag{9.223}$$

Therefore, we finally obtain a retarded Green function

$$G^{(+)}[\vec{r},t;\vec{r}',t'] = c\frac{\delta[|\vec{r}-\vec{r}'| - c(t-t')]\Theta[t-t']}{|\vec{r}-\vec{r}'|} \tag{9.224}$$

describing a spherical delta-function shell propagating outwards with velocity c. The contribution that propagates inward for earlier times is suppressed by the step function. However, for some problems it might be appropriate to consider either the advanced Green function, $G^{(-)}$, or a standing wave obtained using different boundary conditions.

The wave function for a variable source can now be computed using

$$\psi[\vec{r},t] = c\int \frac{\delta[|\vec{r}-\vec{r}'| - c(t-t')]\Theta[t-t']}{|\vec{r}-\vec{r}'|}\rho[\vec{r}',t']\,dV'\,dt' \tag{9.225}$$

such that

$$\psi[\vec{r},t] = \int \frac{\rho\left[\vec{r}', t - \frac{|\vec{r}-\vec{r}'|}{c}\right]}{|\vec{r}-\vec{r}'|} dV' \tag{9.226}$$

resembles the solution to the static Poisson equation except that the potential at position $\vec{r}$ contributed by a source at $\vec{r}'$ is determined by an earlier, or retarded, time $t - |\vec{r} - \vec{r}'|/c$ as expected for a finite propagation velocity.

9.6.5 Lippmann–Schwinger Equation

The time-independent Schrödinger equation

$$\left(-\frac{\hbar^2}{2m}\nabla^2 + V[\vec{r}]\right)\psi[\vec{r}] == E\psi[\vec{r}] \tag{9.227}$$

for positive energies $E = \hbar^2 k^2/2m$ can be expressed in a form

$$(\nabla^2 + k^2)\psi[\vec{r}] == v[\vec{r}]\psi[\vec{r}] \tag{9.228}$$

that strongly resembles the inhomogeneous Helmholtz equation except that the source term is proportional to the wave function itself. Here we define $v = 2mV/\hbar^2$ for convenience. Thus, a formal solution can be expressed in the form of the *Lippmann–Schwinger equation*

$$\psi[\vec{r}] = \phi[\vec{r}] + \int G[\vec{r},\vec{r}']v[\vec{r}']\psi[\vec{r}']\, d^3r' \tag{9.229}$$

where ϕ is a solution to the homogeneous equation

$$(\nabla^2 + k^2)\phi[\vec{r}] == 0 \tag{9.230}$$

and where the Green function satisfies

$$(\nabla^2 + k^2)G[\vec{r},\vec{r}'] == \delta[\vec{r}-\vec{r}'] \tag{9.231}$$

and with outgoing boundary conditions becomes

$$G_k^{(+)}[\vec{r},\vec{r}'] = -\frac{1}{4\pi}\frac{\text{Exp}[ik|\vec{r}-\vec{r}'|]}{|\vec{r}-\vec{r}'|} \tag{9.232}$$

Note that one must always be aware that the normalization of the Green function depends upon the form chosen for its point source and tends to vary with the application.

With the unknown wavefunction ψ appearing on both sides of this integral equation, one may question whether anything has been accomplished. Nevertheless, the Lippmann–Schwinger equation has proven to be very valuable in developing approximations and analyzing the general properties of scattering solutions. For example, if the potential is sufficiently weak, we can assume that its effect is relatively small and use

$$\psi \approx \phi \Longrightarrow \psi[\vec{r}] \approx \phi[\vec{r}] + \int G[\vec{r},\vec{r}']V[\vec{r}']\phi[\vec{r}']\, d^3r' \tag{9.233}$$

to obtain the *Born approximation*. This approximation can be improved by substituting the first-order approximation back into the Lippmann–Schwinger equation to obtain

$$\psi[\vec{r}] \approx \phi[\vec{r}] + \int d^3r' G[\vec{r}, \vec{r}'] V[\vec{r}'] \phi[\vec{r}'] + \int d^3r' \, d^3r'' G[\vec{r}, \vec{r}'] V[\vec{r}'] G[\vec{r}', \vec{r}''] V[\vec{r}''] \phi[\vec{r}''] + \cdots \quad (9.234)$$

Although it is often difficult to prove that the series converges, iteration of this successive-approximation procedure provides a multiple-scattering series that is formally exact. The notation for this series can be simplified considerably by using an operator representation of the Lippmann–Schwinger equation

$$\psi == \phi + Gv\psi \quad (9.235)$$

Next we define the *transition operator* T by the operator equation

$$T\phi == v\psi \quad (9.236)$$

and multiply both sides of the Lippmann–Schwinger equation by v, such that

$$v\psi == v\phi + vGv\psi \Longrightarrow T\phi == (v + vGT)\psi \quad (9.237)$$

and conclude that the transition operator satisfies

$$T == v + vGT \Longrightarrow T = v + vGv + vGvGv + \cdots \quad (9.238)$$

For elastic scattering by a short-ranged potential, one naturally chooses a plane wave for ϕ and uses the outgoing Green function such that

$$\psi[\vec{r}] = \text{Exp}[i\vec{k} \cdot \vec{r}] - \frac{m}{2\pi\hbar^2} \int \frac{\text{Exp}[ik|\vec{r} - \vec{r}'|]}{|\vec{r} - \vec{r}'|} V[\vec{r}'] \psi[\vec{r}'] \, d^3r' \quad (9.239)$$

For distances much larger than the size of the source, we can approximate the phase of the Green function using

$$r \gg r' \Longrightarrow |\vec{r} - \vec{r}'| \approx r\left(1 - \frac{\hat{r} \cdot \vec{r}'}{r}\right) \Longrightarrow \frac{\text{Exp}[ik|\vec{r} - \vec{r}'|]}{|\vec{r} - \vec{r}'|} \approx \frac{\text{Exp}[ikr]}{r} \text{Exp}[-i\vec{k}' \cdot \vec{r}'] \quad (9.240)$$

where $\vec{k}' = k\hat{r}$ is the wave vector for the scattered wave. Note that we must treat the phase carefully because it varies rapidly over the size of the source, but need not be overly concerned with the denominator because the magnitude of the Green function varies slowly when $r \gg r'$. Thus, the asymptotic wavefunction takes the form

$$r \gg r' \Longrightarrow \psi[\vec{r}] \simeq \text{Exp}[i\vec{k} \cdot \vec{r}] + f[\vec{k}', \vec{k}] \frac{\text{Exp}[ikr]}{r} \quad (9.241)$$

where the scattering amplitude is identified as

$$f[\vec{k}', \vec{k}] = -\frac{m}{2\pi\hbar^2} \int \text{Exp}[-i\vec{k}' \cdot \vec{r}'] V[\vec{r}'] \psi[\vec{r}'] \, d^3r' \quad (9.242)$$

Although this result is exact, it is not especially useful without a suitable approximation for ψ. Provided that the potential is short-ranged and sufficiently weak, we can use the Born approximation to replace ψ with ϕ and obtain

$$f[\vec{q}] \approx -\frac{m}{2\pi\hbar^2} \int \mathrm{Exp}[i\vec{q} \cdot \vec{r}'] V[\vec{r}']\, d^3r' \tag{9.243}$$

where $\vec{q} = \vec{k} - \vec{k}'$ is the momentum transfer to the target.

Problems for Chapter 9

1. Between the sheets

Two infinite parallel conducting planes at $z = 0, L$ are held at $\psi == 0$.

a) Using cylindrical coordinates, $\vec{r} = (\xi, \phi, z)$, demonstrate that the Dirichlet Green function takes the form

$$G[\vec{r}, \vec{r}'] = \frac{4}{L} \sum_{m=-\infty}^{\infty} \left(\mathrm{Exp}[im(\phi - \phi')] \sum_{n=1}^{\infty} \mathrm{Sin}[k_n z]\, \mathrm{Sin}[k_n z'] I_m[k_n \xi_<] K_m[k_n \xi_>] \right) \tag{9.244}$$

where $k_n = \frac{n\pi}{L}$ and where I_m and K_m are modified Bessel functions of the first and second kinds, respectively.

b) Show that this Green function can also be expressed in the form

$$G[\vec{r}, \vec{r}'] = 2 \sum_{m=-\infty}^{\infty} \left(\mathrm{Exp}[im(\phi - \phi')] \int_0^{\infty} dk J_m[k\xi] J_m[k\xi'] \frac{\mathrm{Sinh}[kz_<]\, \mathrm{Sinh}[k(L - z_>)]}{\mathrm{Sinh}[kL]} \right) \tag{9.245}$$

c) Use either of these representations to evaluate the electrostatic potential produced by a point charge on the z axis at height h between grounded planes.

d) Find expressions for the surface charge densities, σ_0 and σ_L, on the conducting planes and compute the total charge on each.

2. Polar caps

Suppose that a portion of a sphere of radius R contained within $\theta < \theta_0$, described as a north polar cap, is maintained at potential $+V_0$ and that the corresponding south polar cap with $\theta > \pi - \theta_0$ is held at potential $-V_0$ while the remainder of the sphere is grounded.

a) Develop an expansion for the electrostatic potential within the sphere. The coefficients may involve Legendre polynomials with fixed argument. Which terms contribute?

b) Write the corresponding expansion for $r > R$. Describe the asymptotic behavior of the potential for $r \gg R$ and find an explicit expression for the effective dipole moment.

3. Dirichlet Green function for two-dimensional semicircle

Consider a two-dimensional semicircular region defined by $(0 \le \xi \le a, 0 \le \phi \le \pi)$. Use the eigenfunctions of

$$\left(\nabla^2 + \lambda_{n,m}^2\right)\psi[\vec{r}] == 0, \quad \psi[a,\phi] == \psi[\xi,0] == \psi[\xi,\pi] == 0 \tag{9.246}$$

to expand the two-dimensional Dirichlet Green function for the Poisson equation

$$\nabla^2 G[\vec{r},\vec{r}\,'] == -4\pi\delta[\vec{r}-\vec{r}\,'] \tag{9.247}$$

Although numerical values for the eigenvalues are not required, you must provide the equations that determine them.

4. Piece of pie

Suppose that an electrode is shaped like a piece of pie – a wedge of radius a and opening angle α. Evaluate the two-dimensional Dirichlet Green function using separation of variables in polar coordinates, (ξ, ϕ).

5. Cylinder with grounded endcaps

An empty cylinder of radius a with its axis along the $\hat{z}$ axis has grounded endcaps (ψ ==0) at $z = 0$ and $z = L$ while its curved surface is held at potential $\psi[a,\phi,z] == V[\phi,z]$.

a) Develop an expansion for the electrostatic potential $\psi[\xi,\phi,z]$ within the cylinder and express the coefficients in terms of the appropriate integral over $V[\phi,z]$.

b) Determine the coefficients for the simple case $V[\phi,z] = V_0(1 - 2\Theta[\phi-\pi])$ where Θ is the unit step function.

6. Cylinder with opposite potentials on its endcaps

The curved surface of a cylinder of radius a is grounded while the endcaps at $z = \pm L/2$ are maintained at opposite potentials $\psi[\xi,\phi,\pm L/2] = \pm V[\xi,\phi]$.

a) Develop an expansion for the electrostatic potential $\psi[\xi,\phi,z]$ within the cylinder and express the coefficients in terms of the appropriate integral over $V[\xi,\phi]$.

b) Determine the coefficients for the simple case $V[\xi,\phi] = V_0$ where V_0 is constant.

7. Cylinder with diametrically opposed electrodes

Suppose that a long conducting cylinder of radius a has edge electrodes at potentials $\psi = V$ for $|\theta| \le \alpha$ and $\psi = -V$ for $|\theta - \pi| \le \alpha$. The remaining portions of the cylinder are held at $\psi = 0$. Evaluate the electrostatic potential in both interior and exterior regions.

8. Split-sphere acoustic antenna

Sound waves produced by a split-sphere antenna with radius R satisfy

$$\frac{1}{c^2}\frac{\partial^2\psi}{\partial t^2} - \nabla^2\psi == 0, \quad \psi[R,\theta,\phi,t] = \begin{cases} A e^{-i\omega t} & 0 < \theta < \frac{\pi}{2} \\ -A e^{-i\omega t} & \frac{\pi}{2} < \theta < \pi \end{cases} \tag{9.248}$$

Evaluate $\psi[\vec{r},t]$ for $r > R$ with outgoing boundary conditions.

9. Acoustical waves from vibrating cylinder
The surface of an infinite cylinder of radius R vibrates harmonically such that the pressure at the surface is given by

$$\psi[R,\theta,t] = f[\theta]e^{-i\omega t} \tag{9.249}$$

where the angular function $f[\theta]$ is prescribed. The pressure for $r > R$ satisfies the wave equation

$$\left(\frac{1}{c^2}\frac{\partial^2}{\partial t^2} - \nabla^2\right)\psi == 0 \tag{9.250}$$

with outgoing boundary conditions. Construct a formal solution for the field.

10. Acoustical wave guides
Density displacements $\psi[\vec{r},t]$ for a gas in a confined volume satisfy a wave equation of the form

$$\frac{1}{c^2}\frac{\partial^2\psi}{\partial t^2} - \nabla^2\psi == 0, \qquad \frac{\partial\psi}{\partial n} == 0 \tag{9.251}$$

where c is the sound velocity and $\partial\psi/\partial n = \hat{n}\cdot\vec{\nabla}\psi$ is the normal derivative at a boundary surface.

a) Evaluate the normal modes for a long rectangular wave guide with cross-sectional dimensions $a \times b$. Show that for most modes there is a minimum frequency for transmission. Compare the phase and group velocities.

b) Evaluate the normal modes for a long cylindrical wave guide with cross-sectional radius R. Show that for most modes there is a minimum frequency for transmission. Compare the phase and group velocities.

11. Acoustic modes
Sound waves in a confined volume satisfy a wave equation of the form

$$\frac{1}{v^2}\frac{\partial^2\psi}{\partial t^2} - \nabla^2\psi == 0, \qquad \frac{\partial\psi}{\partial n} == 0 \tag{9.252}$$

where c is the sound velocity and $\partial\psi/\partial n = \hat{n}\cdot\vec{\nabla}\psi$ is the normal derivative at a boundary surface. The normal modes take the form

$$\psi_i[\vec{r},t] = \psi_i[\vec{r}]e^{-i\omega_i t} \tag{9.253}$$

where the eigenfrequencies ω_i may be degenerate, requiring additional indices to distinguish between degenerate modes.

a) Evaluate the normal modes and eigenfrequencies in a box with dimensions $a \times b \times c$.

b) Evaluate the normal modes and eigenfrequencies for a sphere of radius R.

c) Evaluate the normal modes and eigenfrequencies for a cylinder of radius R and length L.

12. Organ pipe

The pressure fluctuation ψ in an organ pipe of length L satisfies the wave equation

$$\frac{1}{c^2}\frac{\partial^2\psi[x,t]}{\partial t^2} - \frac{\partial^2\psi[x,t]}{\partial x^2} == 0 \tag{9.254}$$

where c is the speed of sound. The end at $x = 0$ is open while the end at $x = L$ is partly closed such that the boundary conditions are

$$\psi[0,t] == 0\,, \quad \left(\psi + \alpha\frac{\partial\psi}{\partial x}\right)_{x=L} == 0 \tag{9.255}$$

where α is a constant.

a) Construct the normal modes and derive the equation that determines the characteristic frequencies. Describe a graphical solution to this equation.
b) Demonstrate that the normal modes are orthogonal.
c) Suppose that the pressure in the pipe is in equilibrium for $t < 0$ and that constant pressure ψ_0 is applied to the open end for $t \geq 0$. Find a series representation for $\psi[x,t]$.

13. Vibrations of a membrane wedge

Suppose that a taut membrane is stretched across the area described by $a \leq \xi \leq b$, $\theta_1 \leq \theta \leq \theta_2$. Develop the vibrational eigenfunctions and the condition satisfied by the eigenfrequencies. Demonstrate that the eigenfunctions are orthogonal and provide a formal expansion for the initial-value problem (specified initial displacement $\Psi[\xi,\theta,t=0]$). It is not necessary to obtain an explicit formula for the radial normalization integral.

14. Current flow in a disk with diametrically opposed electrodes

Suppose that a disk of radius a has uniform conductivity σ and thickness τ. Current I enters through an edge electrode at $|\theta| \leq \alpha$ and leaves through an electrode at $|\theta - \pi| \leq \alpha$. Show that the electric potential satisfies Laplace's equation and evaluate the current density $\vec{j}$ in the steady state. Assume that the radial current is independent of angle at the electrodes.

15. Critical mass

Suppose that the neutron density ψ in a fissionable material satisfies an inhomogeneous diffusion equation of the form

$$\frac{1}{\kappa}\frac{\partial\psi}{\partial t} == \nabla^2\psi + \lambda\psi \tag{9.256}$$

where κ is the diffusion constant and λ depends upon the fission probability per neutron. For simplicity assume that $\psi = 0$ on the surface of the material (neutrons escape rapidly at the surface or are absorbed by the surrounding medium).

a) Suppose that the material forms a sphere of radius R. Determine the critical radius R_c, beyond which the neutron density is unstable (exponentially increasing) and produces an explosion.

b) Determine the critical radius R_1, for a hemisphere of the same material. If two hemispheres that are barely stable are brought together as a sphere, the final configuration will be unstable. Express the explosive time constant τ in terms of κ and λ, such that $\psi \sim e^{t/\tau}$.

16. Heating simple solids

Suppose that a simple solid (brick, sphere, cylinder, etc.) with uniform initial temperature is immersed at time $t = 0$ in a heat bath. The temperature $\psi[\vec{r}, t]$ within the material satisfies

$$\frac{1}{\kappa}\frac{\partial \psi}{\partial t} == \nabla^2 \psi, \quad \psi[\vec{r}, 0] == \psi_i, \quad \psi[\vec{R}, t \geq 0] == \psi_f \tag{9.257}$$

where $\vec{R}$ is on the surface.

a) Determine the temperature distribution for positive times within a brick with $0 \leq x \leq a$, $0 \leq y \leq b$, and $0 \leq z \leq c$.

b) Determine the temperature distribution for positive times within a sphere with $r \leq R$. What is the asymptotic time dependence of the central temperature?

c) Determine the temperature distribution for positive times within a cylinder with $\xi \leq R$ and $0 \leq z \leq L$.

17. Cooling sphere

A solid sphere of radius R has a spherically symmetric initial temperature distribution $\psi_0[r]$ and cools at its surface according to Newton's law of cooling

$$r = R \Longrightarrow \psi + \alpha R \frac{\partial \psi}{\partial r} == 0 \tag{9.258}$$

If one can neglect the radial variation of the diffusion constant, this problem provides a simple model for the cooling of a planet.

a) Develop a series representation for $\psi[r, t]$ and express the coefficients in terms of ψ_0.

b) Evaluate the special case where ψ_0 is constant and α is negligible. Plot the temperature dependence for several radii.

18. Diffusion in a cylinder with constant surface temperature

A long solid cylinder of radius a has initial temperature distribution $\psi_0[\xi, \phi]$ and constant surface temperature $\psi[a, \phi] = 0$. The temperature satisfies the diffusion equation

$$\frac{1}{\kappa}\frac{\partial \psi}{\partial t} == \nabla^2 \psi \tag{9.259}$$

where the variation with height is ignored. Develop a series expansion for $\psi[\xi, \phi, t]$ and express the coefficients in terms of ψ_0.

19. Surface waves on ideal fluid

The velocity field $\vec{v} = \vec{\nabla}\psi$ for the flow of an ideal fluid can be expressed in terms of a velocity potential ψ that satisfies Laplace's equation, $\nabla^2 \psi == 0$, subject to appropriate boundary conditions. Suppose that a fluid with equilibrium depth h is essentially infinite in two dimensions, x and y, but has a flat floor. The boundary conditions are then

$$\left(\frac{\partial\psi}{\partial z}+\frac{1}{g}\frac{\partial^2\psi}{\partial t^2}\right)_{z=0} == 0, \quad \left(\frac{\partial\psi}{\partial z}\right)_{z=-h} == 0 \tag{9.260}$$

a) Find a general expression $\psi_\omega[x, y, z, t]$ for a plane wave with frequency ω that propagates in the x direction.

b) Determine and sketch the phase and group velocities as functions of kh, where k is the wave number.

20. Standing waves on ideal fluid in rectangular tank

The velocity field $\vec{v} = \vec{\nabla}\psi$ for the flow of an ideal fluid can be expressed in terms of a velocity potential ψ that satisfies Laplace's equation, $\nabla^2\psi == 0$, subject to appropriate boundary conditions. Suppose that a fluid with equilibrium depth h is in a rectangular tank with dimensions $a \times b$ and a flat floor. The boundary conditions are then

$$\left(\frac{\partial\psi}{\partial z}+\frac{1}{g}\frac{\partial^2\psi}{\partial t^2}\right)_{z=0} == 0, \quad \left(\frac{\partial\psi}{\partial z}\right)_{z=-h} == 0, \quad \left(\frac{\partial\psi}{\partial y}\right)_{x=0,a} == 0, \quad \left(\frac{\partial\psi}{\partial x}\right)_{y=0,b} == 0 \tag{9.261}$$

a) Construct standing-wave solutions $\psi_{n,m}[x, y, z, t]$ and determine their frequencies.

b) Construct general solutions that satisfy the initial condition

$$v_z[x, y, 0, 0] = v_0[x, y] \tag{9.262}$$

21. Seasonal variation of ground temperature

In the flat-earth approximation, the temperature ψ at depth z can be described by a one-dimensional diffusion equation of the form

$$\frac{\partial^2\psi}{\partial z^2} == \frac{1}{\kappa}\frac{\partial\psi}{\partial t} \tag{9.263}$$

where κ is the thermal diffusion constant. Suppose that the external temperature can be approximated by a sinusoidal variation of the form

$$\psi[0, t] = \psi_0 + \psi_1 \operatorname{Sin}[\omega t] \tag{9.264}$$

where $\omega = 2\pi/T$ for period T.

a) Solve for $\psi[z, t]$ and determine the *penetration depth* d and the *phase delay* for propagation of thermal waves into the ground.

b) The penetration depth for annual variations is approximately 3 m. Determine the phase delay for annual variations. Also determine the corresponding penetration depth and phase delay for daily variations. Discuss the separability of these periods.

c) Obtain $\psi[z, t]$ for combined annual and diurnal variations.

22. Current distribution in wire

One can show that the current density $\vec{J}$ satisfies an equation of the form

$$\left(\nabla^2 - \frac{\epsilon\mu}{c^2}\frac{\partial^2}{\partial t^2} - \frac{4\pi\sigma\mu}{c^2}\frac{\partial}{\partial t}\right)\vec{J} == 0 \tag{9.265}$$

when the permittivity ϵ, permeability μ, and conductivity σ are constant. The radial distribution of current flowing with sinusoidal time dependence along a long straight wire with circular cross-section of radius a can be represented as

$$J_z[\xi, t] = \mathrm{Re}\left[\psi[\xi]\,\mathrm{Exp}[-i\omega t]\right] \tag{9.266}$$

a) Show that $\psi[\xi]$ is related to a Bessel function with complex argument and normalize the solution to total current I.

b) Express the solution for a good conductor with $4\pi\sigma/\epsilon\omega \gg 1$ as a function of γ a where $\gamma = \sqrt{4\pi\sigma\mu\omega/c^2}$. Plot and discuss $\left|\psi[\xi]/\psi[a]\right|^2$ for both large and small values of γa.

c) Evaluate the impedance $Z = V/I$ where V is the voltage drop over a length l measured at the surface of the wire. The impedance can be expressed in the form $Z = R - i\omega L$ where R is the resistance and L is the inductance for frequency ω. Evaluate R and L for a good conductor in both small and large ω limits. Sketch and interpret $R[\omega]$ and $L[\omega]$ for a good conductor. It is useful to express these quantities in terms of the *skin thickness* $\delta = \sqrt{2}/\gamma$.

d) Express the solution for a good conductor in terms of the Kelvin functions, $\mathrm{ber}_\nu[x]$ and $\mathrm{bei}_\nu[x]$, defined by

$$J_\nu[e^{-i\pi/4}x] \equiv \mathrm{ber}_\nu[x] + i\,\mathrm{bei}_\nu[x] \tag{9.267}$$

for positive real x where J_ν is the regular Bessel function of order ν. (Hint: compare the differential equations for $J_\nu[e^{\pm i\pi/4}x]$ with the present equation.) This representation is sometimes seen in electrical engineering.

23. Evolution of line vortex

The vorticity distribution $\omega[x, y, t]$ in a viscous fluid satisfies a diffusion equation of the form

$$\frac{\partial\omega}{\partial t} == \nu\nabla^2\omega \tag{9.268}$$

where the kinematic viscosity ν is a positive constant. Suppose that at $t = 0$ there is a line vortex

$$\omega[x, y, 0] = \Gamma\delta[x]\delta[y] \tag{9.269}$$

where Γ is constant and assume that $\omega \to 0$ when $r \to \infty$. Determine the subsequent behavior of ω using each of the following methods.

a) Use a Laplace transform with respect to time and look up the required inverse transform.

b) Use a two-dimensional Fourier transform with respect to position.

c) Assume a similarity solution of the form $\omega = \frac{\Gamma}{\nu t} f\left[\frac{r^2}{\nu t}\right]$ and determine f. (Hint: examine the behavior of the differential equation for $f[\xi]$ to show that the substitution $f[\xi] = g[\xi]e^{-\xi/4}$ provides an integrable equation for $g[\xi]$.)

24. Scattering by separable *s*-wave potential

Often the interaction between two particles is nonlocal, for which the time-independent Schrödinger equation takes the form

$$\left(k^2 + \nabla^2\right)\psi\left[\vec{k}, \vec{r}\right] - 2\mu \int V[\vec{r}, \vec{r}\,']\psi\left[\vec{k}, \vec{r}\,'\right] d^3r' == 0 \tag{9.270}$$

where k is the wave number in the center of mass and μ is the reduced mass.

a) Develop an integral equation for scattering solutions and identify the scattering amplitude.

b) A separable *s*-wave potential is defined by

$$2\mu V[\vec{r}, \vec{r}\,'] = \lambda v[r] v\left[r'\right] \tag{9.271}$$

where λ is a strength parameter and $v[r]$ depends only upon the distance $r = |\vec{r}|$. Obtain an exact solution for the scattering amplitude in terms of $\tilde{v}[k]$, where

$$\tilde{v}[k] = \int_0^\infty v[r] j_0[kr] r^2\, dr \tag{9.272}$$

Under what conditions does the Born series converge? (Hint: solve first for the quantity $A[\vec{k}] = \int v[r]\psi[\vec{k}, \vec{r}]\, d^3r$.)

c) Evaluate the scattering amplitude for the specific case of a Yamaguchi potential with

$$\tilde{v}[k] = \frac{\Lambda^2}{\Lambda^2 + k^2} \tag{9.273}$$

where Λ is a positive range parameter. Express the phase of the scattering amplitude in terms of the dimensionless parameters $\gamma = -2\pi\lambda\Lambda$ and $\kappa = k/\Lambda$. Under what conditions does the potential support a bound state? For neutron–proton scattering there is a low-energy bound state in the $^3S_1$ channel but the potential for the $^1S_0$ channel is not quite strong enough to produce a bound state. Compare the phase shifts for these channels as functions of κ assuming that γ is either 5% stronger or 5% weaker than the minimum strength for a bound state. (Note: for the phase shift it is best to use the quadrant-sensitive version of the inverse tangent function.)

d) Express $k\,\mathrm{Cot}\left[\delta_0[k]\right]$, where $\delta_0[k]$ is the phase shift, in terms of $\tilde{v}[k]$. Obtain general expressions for the scattering length and effective range assuming that $\tilde{v}[k]$ can be expanded in even powers of k. Then evaluate the scattering length and effective range for the specific case of a Yamaguchi potential. Describe the behavior of the scattering length for γ near the critical value for a bound state.

25. Scattering by a soft cylinder

Plane waves with wave number k are incident upon an infinitely long cylinder of radius R with the direction of propagation perpendicular to the axis of the cylinder. The surface of the cylinder is soft, such that the total wave (incident plus scattered) vanishes at R. Construct a formal expansion for the scattered field and then examine the leading term in the long-wavelength limit $kR \ll 1$.

10 Group Theory

Abstract. Group theory provides powerful methods for analyzing the consequences of symmetries. We begin with the basic concepts for finite groups and their matrix representations, including orthogonality theorems and methods for constructing and analyzing irreducible representations. These concepts are then extended to continuous Lie groups. The irreducible representations of the quantum mechanical rotation group and their direct products are analyzed in detail. Finally, the application of unitary groups to particle theory is discussed briefly.

10.1 Introduction

Many of the general features of the dynamics of physical systems are determined or largely constrained by their symmetries, and symmetries are often associated with conserved quantities. For example, momentum conservation is a consequence of translational symmetry, angular momentum conservation originates in rotational symmetry, and parity embodies reflection symmetry. By classifying the states of a system according to irreducible representations of its symmetry group, one can deduce general selection rules that forbid certain transitions or can deduce patterns among the allowed transitions. These and other implications of symmetries can be studied using the mathematical theory of groups in which the symmetry operators are treated as elements of an algebraic system. For example, the symmetries of an equilateral triangle in a plane consist of an identity element, two rotations, and three reflections. These operators are described as elements of a group that obeys algebraic rules similar to those of ordinary algebra, except that the group elements are more abstract than the natural numbers. Thus, one can often deduce important properties of the physical system by applying the theorems of group theory to its symmetry operators instead of by solving the differential equations that describe its dynamics. The early development of group theory was largely motivated by crystallography, but its most important applications are now in quantum mechanics, field theory, and particle physics. The symmetries of global gauge transformations determine the conserved quantities while the symmetries of local gauge transformations of quantum fields determine, up to coupling constants, the types of interactions among the particles represented by those fields. In fact, so important has group theory become to particle theory that an early theory of grand unification using supersymmetry is generally referenced by its symmetry group, SU(5), rather than by either its dynamical principles or its authors' names (Georgi and Glashow).

The theory of groups is a very large and highly developed subject in both mathematics and physics. A comprehensive treatment would require at least one semester and a dedicated textbook. Nevertheless, this topic is sufficiently important to modern physics that we include a synopsis even though it diverges from the main topic of this book. We begin with

Graduate Mathematical Physics. James J. Kelly

ISBN: 3-527-40637-9

the theory of finite groups and their matrix representations. We include proofs of most of the theorems; these proofs presume familiarity with linear algebra. Our primary examples study small amplitude vibrations of highly symmetric configurations of masses. We then generalize these results to continuous groups, especially Lie groups, but do not always provide proofs – many of the results are analogous to those for finite groups but the proofs are often more technical. We discuss several types of representations, including matrices, differential operators, and eigenfunction multiplets, emphasizing that general results depend upon the abstract structure of the associated Lie algebra instead of the particular representation at hand. These techniques are then used to analyze the quantum mechanical rotation group in considerable detail. Finally, we give a brief discussion of the application of unitary groups in particle physics.

10.2 Finite Groups

10.2.1 Definitions

Consider a finite set of objects, $S_a = \{a_i, i = 1, N\}$. A *transformation* M is a one-to-one mapping of $S_a \to S_a$ described by $Ma_i = a_j$. For example, the permutation

$$Ma_1 = a_2, \quad Ma_2 = a_3, \quad Ma_3 = a_1 \tag{10.1}$$

is one example of a one-to-one mapping of $\{a_1, a_2, a_3\}$ upon itself. A set of such transformations, $G = \{M_i, i = 1, m\}$, is described as a *group* with respect to a law of composition, figuratively described as multiplication, if it satisfies the following conditions.

1. If M_i and M_j are elements of the group, then their product M_iM_j is also a member of the group.
2. Multiplication of group elements is associative, such that $M_i(M_jM_k) = (M_iM_j)M_k$.
3. The group must contain an identity element, labeled I, such that $IM_i = M_iI = M_i$ for every $M_i \in G$.
4. For every $M_i \in G$, there exists an inverse element $M_i^{-1} \in G$ such that $M_i^{-1}M_i = M_iM_i^{-1} = I$.

The number of elements in a group is called its *order*. Groups may be finite, countably infinite, or continuous in order. The present section develops some of the general properties of finite groups but many will apply, with obvious generalizations, to infinite groups also.

Note that "multiplication" means composition of successive transformations, which is not necessarily multiplication in the conventional sense. For example, the set of integers constitutes a countably-infinite group with respect to addition: the set is closed with respect to addition (even though the number of elements is infinite); addition of integers is associative; the identity element is 0 because $0 + x = x + 0 = x$; and for each x the inverse $-x$ belongs to the group. Similarly, the integers $\{0, 1, \ldots, m-1\}$ constitute a finite group of order m under addition modulo m. Ordinary multiplication of integers, on the other hand, does not constitute a group because there is no inverse element for 0.

The structure of a finite group can be exhibited by its multiplication table. Consider, for example, addition of integers modulo three. There are then three elements that we will label abstractly by $I = 0$, $a = 1$, $b = 2$ such that $ab = \mathrm{Mod}[a + b, 3] = 0 = I$. The complete "multiplication" table is shown in Table 10.1.

Table 10.1. Group multiplication table for cyclic group of order 3.

C_3	I	a	b
I	I	a	b
a	a	b	I
b	b	I	a

Notice that each element occurs exactly once in every row and every column and that each element has an inverse. One often omits the labels from a group multiplication table because they simply repeat the first row and column when the identity element is placed in the upper left interior corner. This particular group has order 3 and is identified as C_3, the cyclic group of order 3. Observe that each row or column is a cyclic permutation of the group elements. Later we will prove that any group of prime order is cyclic. Two elements *commute* if $M_i M_j = M_j M_i$. A group is described as *abelian* if its multiplication law is commutative, such that $M_i M_j = M_j M_i$, for any i, j and as *nonabelian* otherwise. The group multiplication table for an abelian group is symmetric about its diagonal. The addition of integers is abelian, but most groups are nonabelian.

Two groups that can be arranged to have the same multiplication table are *isomorphic*. For example, the symmetric group S_n consists of all permutations of n distinguishable objects. Thus, S_3 is of order 6 and its elements

$$P_{1,2,3}[\{A, B, C\}] = \{A, B, C\} \tag{10.2}$$

$$P_{1,3,2}[\{A, B, C\}] = \{A, C, B\} \tag{10.3}$$

$$P_{2,1,3}[\{A, B, C\}] = \{B, A, C\} \tag{10.4}$$

$$P_{2,3,1}[\{A, B, C\}] = \{B, C, A\} \tag{10.5}$$

$$P_{3,1,2}[\{A, B, C\}] = \{C, A, B\} \tag{10.6}$$

$$P_{3,2,1}[\{A, B, C\}] = \{C, B, A\} \tag{10.7}$$

can be enumerated by considering their transformation of a generic set of three objects, denoted $\{A, B, C\}$. Alternatively, there are six symmetry operations that leave an equilateral triangle invariant, consisting of rotations through 120°, 240°, 360°, and three reflections about bisectors. The entire set of symmetries can be constructed from the identity element I, the basic 120° counterclockwise rotation R, and the reflection P (for parity) across the vertical bisector. The elements of these two groups can be placed in a one-to-one correspondence that makes their multiplication tables identical. Therefore, these groups are isomorphic and their formal properties are identical. In Fig. 10.1 we compare the symmetries of an equilateral triangle with the elements of S_3 and assign generic labels for use in composing the multiplication Table 10.2.

The permutation notation may be transparent, but the group properties would be the same whatever the context in which S_3 emerges. Therefore, it is customary to employ

Table 10.2. Isomorphism between symmetries of an equilateral triangle and permutations of three objects (vertices).

Generic	Triangle	Permutation	Order
I	I	[123]	1
a	R	[312]	3
b	R^2	[231]	3
c	P	[132]	2
d	PR	[321]	2
e	PR^2	[213]	2

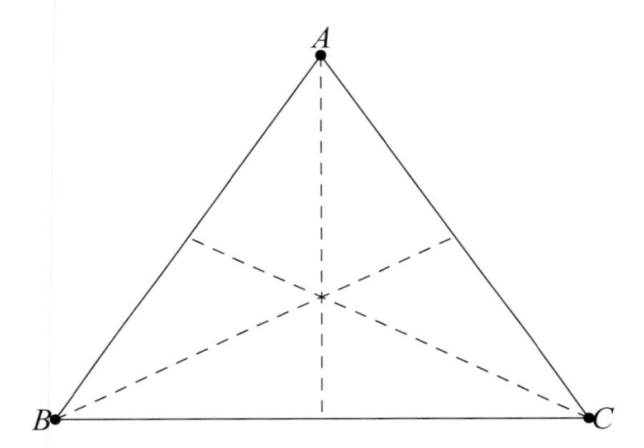

Figure 10.1. The symmetries of an equilateral triangle include the identity I, 120° counterclockwise rotation R, reflection across the vertical bisector P, and the combinations R^2, PR, and PR^2.

more abstract labels, such as $\{I, a, b, c, d, e\}$, and to display the group multiplication table in the form in Table 10.3. Notice that this group is nonabelian. Some find assembly of the multiplication table easier using permutations and others using geometrical reasoning. (Make sure that you can reproduce this table using both methods!)

Table 10.3. Group multiplication table for S_3.

S_3	I	a	b	c	d	e
I	I	a	b	c	d	e
a	a	b	I	e	c	d
b	b	I	a	d	e	c
c	c	d	e	I	a	b
d	d	e	c	b	I	a
e	e	c	d	a	b	I

Successive applications of the same transformation is represented by a power of the group element, such that

$$M_i^n = M_i M_i^{n-1}, \quad M_i^0 = I \tag{10.8}$$

where I is the identity element. A transformation is described as *order* k if each M_i^n is distinct for $0 \le n < k$. In the case of S_3, we find that I is order 1, $\{P, PR, PR^2\}$ are order 2, and that $\{R, R^2\}$ are order 3. The powers of an element of order k form a *cyclic group* of order k, defined by

$$CC_i = C_{i+1}, \quad C_k = C_0 = I \tag{10.9}$$

Cyclic groups are abelian.

Finally, we mention a trivial result that is so useful that it is often called the *rearrangement theorem.* Suppose that g is any element of the group $G = \{G_i\}$. The product $gG = \{(gG_i), i = 1, n\}$ contains every element of G exactly once but probably in a new order. Thus, any complete sum over functions of a group element can then be rearranged by multiplying the argument by an arbitrary group element according to

$$\sum_{h \in G} f[gh] = \sum_{h \in G} f[h] \tag{10.10}$$

without affecting its value.

10.2.2 Equivalence Classes

An element b is described as *conjugate* to element a if there exists an element $u \in G$ such that $uau^{-1} = b$. In other contexts this relationship might be described as similarity, so we will employ the same symbol $a \sim b$. Obviously, every element is conjugate to itself, $a \sim a$, and this relationship is commutative, $a \sim b \Longrightarrow b \sim a$. It is also transitive: $a \sim b, b \sim c \Longrightarrow a \sim c$. An *equivalence class* consists of the set of all elements that are conjugate to each other. The identity element forms a class unto itself and each element of an abelian group also constitutes its own class; hence, the notion of classes is most useful for nonabelian groups. Note that all elements of a class must have the same order. If we construct b^n using

$$b = uau^{-1} \Longrightarrow b^n = (uau^{-1}) \cdots (uau^{-1}) = ua^n u^{-1} \tag{10.11}$$

we find $a^k = I \Longrightarrow b^k = I$.

Let us return to S_3. Recognizing that there are three elements of order 2, consisting of $\{c, d, e\}$ or $\{P, PR, PR^2\}$, we wonder whether they constitute an equivalence class. By direct calculation using Table 10.3, we find

$$aca^{-1} = acb = eb = d \Longrightarrow c \sim d \tag{10.12}$$

$$beb^{-1} = bea = ca = d \Longrightarrow e \sim d \tag{10.13}$$

and use transitivity to conclude that $c \sim d \sim e$ forms an equivalence class. Similarly, $a \sim b$ or $R \sim R^2$ also form a class.

10.2.3 Subgroups

A subset H of the elements of G that forms a group with respect to the same composition law is described as a *subgroup* of G. This relationship is denoted using the notation of sets, $H \subset G$. Every group G contains two trivial subgroups, one consisting of just the identity element $\{I\}$ and the other consisting of G itself; these are described as *improper*. One of the main problems of group theory is the construction and classification of all proper subgroups of a specified group. The relevant criteria are that H be closed with respect to multiplication and that inverses are present for each element; the associative property is satisfied automatically and every subgroup must contain the identity element.

If K is a subgroup of H and H is a subgroup of G, then K is a subgroup of G. Because every group must contain its subgroups, one can construct the sequence $G \supset H \supset K \cdots \supset I$. We now derive *Lagrange's divisor theorem*, which states that the order of a subgroup must be a divisor of the order of its parent. This theorem is obviously true for improper subgroups containing either one or all elements. Suppose that H is a proper subgroup of G and label its elements $\{H_i, i = 1, h\}$ where h is the order of H and $H_1 = I$ is the identity element. Consider an element $a \in G$ that does not belong to H and construct the *left coset* $aH = \{aH_i, i = 1, h\}$ containing the products of a with each H_i. This set is not a group because it does not contain the identity element. Each element of aH is distinct and none are included in H. (Note that $aH_i = H_j \Longrightarrow a = H_j H_i^{-1} \in H$, contrary to the assumption that $a \notin H$.) If the union of H and aH, both of order h, is not G itself, we select another element b and construct a new left coset bH of order h that has no overlap with either H or aH. (If $bH_i = aH_j$ then $b = aH_j H_i^{-1} \in aH$ because $H_j H_i^{-1} \in H$.) Thus, we can divide G into a subgroup H and its cosets according to

$$G = H + aH + bH + \cdots \tag{10.14}$$

where each subset contains h elements. If g is the order of G and h is the order of H, then $g = mh$ where m is the *index* of the subgroup and equal to the number of independent cosets plus one. Therefore, h is a divisor of g, as stipulated by Lagrange's theorem.

Once again consider S_3, where 3 and 2 are divisors of its order. There is one subgroup of order 3, namely $\{I, a, b\}$, that is isomorphic to the group of invariant proper rotations of an equilateral triangle, and there are three subgroups of order 2, namely $\{I, c\}$, $\{I, d\}$, $\{I, e\}$ based upon reflections across a bisector.

If $a \in G$ is an element of order k, we can construct a cyclic group $\{I, a, a^2, \ldots, a^{k-1}\}$ where $a^k = I$. This cyclic group, called the *period* of a, is the smallest subgroup of G that contains a and its order must be a divisor of the order of G. Therefore, all groups with prime order must be cyclic and can be generated from one of its elements other than the identity.

Although our derivation of the divisor theorem employed left cosets aH, we could just as easily have chosen right cosets Ha. Left and right cosets need not be the same, but the argument still works. In the special case that $aH = Ha \Longrightarrow H = aHa^{-1}$ for any $a \in G$, we describe H as an *invariant subgroup* of G. Invariant subgroups clearly must contain one or more complete equivalence classes. More generally, if H is a subgroup of G, and a is any element of G, then aHa^{-1} is also a subgroup of G described as conjugate (or similar) to H. Thus, invariant subgroups are self-conjugate with respect to any element of its parent group. A group that contains no proper invariant subgroups is described as *simple*. Cyclic groups are simple by that criterion.

Invariant subgroups have some very important properties. Consider the multiplication of two cosets, aH and bH, of an invariant group H. Such a product is defined as the set of all products of pairs of elements from the two sets, taking care not to assume that multiplication commutes. We find

$$(aH)(bH) = a(Hb)H = a(bH)H = (ab)(HH) = (ab)H \tag{10.15}$$

where $HH = H$ because H must be closed to form a subgroup. Thus, the product of two cosets of H is itself a coset of H. In fact, we can identify the cosets constructed from an invariant subgroup H as the elements of a new *factor group*, denoted $F = G/H$, such that

$$F = \{H, g_1 H, g_2 H, \dots\} \tag{10.16}$$

where the g_i are elements of G that are not contained in H. (If g_i were a member of H, the coset would simply be a re-ordering of H itself.) Proving that F actually is a group is an instructive exercise. First, we test for closure. Assuming that both g_i and g_j belong to G but not to H, we find that the product

$$F_i F_j = (g_i H)(g_j H) = (g_i g_j)H \Longrightarrow F_i F_j \in F \tag{10.17}$$

is indeed an element of F because the product $g_i g_j$ belongs to G but not H. Second, we verify that the associative property

$$F_i(F_j F_k) = g_i H\big((g_j H)(g_k H)\big) = \big((g_i H)(g_j H)\big)g_k H = (F_i F_j)F_k \tag{10.18}$$

is simply inherited from G. Third, we identify H as the identity element of F because

$$HF_i = H(g_i H) = g_i(HH) = g_i H = F_i \tag{10.19}$$

$$F_i H = (g_i H)H = g_i H = F_i \tag{10.20}$$

using associativity and $H^2 = H$. Finally, we verify that F contains an inverse

$$F_i^{-1} = (g_i H)^{-1} \Longrightarrow F_i F_i^{-1} = g_i H g_i^{-1} H^{-1} = g_i g_i^{-1} H H^{-1} = I \tag{10.21}$$

for every element because the inverses for each g_i belong to G and not H while H^{-1} is the list of inverses for each member of H, all of which belong to H. Therefore, F is indeed a group.

10.2.4 Homomorphism

Consider a mapping $G \to G'$ where a in G is mapped onto the image a' in G'. A mapping is described as *homomorphic* when: (1) each element of group G is mapped onto an element of G'; (2) the mapping preserves multiplication rules such that $ab = c \Longrightarrow a'b' = c'$;

and (3) some of the mappings are several-to-one. As a trivial example, suppose that all elements of the cyclic group C_n are mapped onto $G' = \{I\}$ containing only the identity element. G' obviously meets the requirements of a group and the multiplication rule for C is preserved by the mapping because the $C_k = C_i C_j \to I$. Therefore, this mapping satisfies the requirements of homomorphism.

Suppose that $G \to G'$ is a homomorphism. The image of the identity element I in G must be the identity element I' in G', such that $I \to I'$. If any element $g_1 \to I'$ is mapped onto I', then $g_1^{-1} \to I'$ is also needed to preserve the multiplication rules. Furthermore, if $g_1 \to I'$ then all members of its class must also be mapped onto I'. Thus, all elements $H = \{g_1, g_2, \ldots, g_m\}$ of G that are mapped onto I' constitute an invariant subgroup of G. All elements of the coset aH, where $a \notin H$, are mapped onto the same image $a \to a' \neq I'$ in G'. Therefore, G' is isomorphic to the factor group G/H.

10.2.5 Direct Products

A group $G = H_1 \otimes H_2 \otimes \cdots \otimes H_n$ is described as a *direct product* of its subgroups if: (1) elements in different subgroups commute; and (2) each $g \in G$ is uniquely expressible in the form $g = h_1 h_2 \cdots h_n$ where $h_i \in H_i$. These conditions require that each H_i be an invariant subgroup and that these subgroups, described as *direct factors*, share only the identity element.

Similarly, the direct product $G \otimes G'$ of two different groups is formed by taking all pairs (g_i, g'_j), where $g_i \in G$ and $g'_j \in G'$, and evaluating the products

$$(g_i, g'_j)(g_k, g'_l) = (g_i g_k, g'_j g'_l) \tag{10.22}$$

It is then a simple matter to verify that $G \otimes G'$ satisfies all the requirements of a group and that its order is the product of the orders of its factors.

10.3 Representations

10.3.1 Definitions

Suppose that

$$\psi = \sum_{i=1}^{n} \psi_i u_i, \quad \phi = \sum_{i=1}^{n} \phi_i u_i \tag{10.23}$$

are vectors in an n-dimensional linear vector space where the basis vectors $\{u_i\}$ are orthonormal with respect to the scalar product

$$\langle \psi | \phi \rangle = \sum_{i=1}^{n} \psi_i^* \phi_i, \quad \langle u_i \mid u_j \rangle = \delta_{i,j} \tag{10.24}$$

and suppose that L is a linear operator satisfying

$$L(a\psi + b\phi) = aL\psi + bL\phi \tag{10.25}$$

where a, b are complex numbers. The complex coefficients $\{\psi_i\}$ or $\{\phi_i\}$ are described as *coordinates* with respect to basis u and are obtained using

$$\psi_i = \langle u_i \mid \psi \rangle \tag{10.26}$$

Further suppose that T is a linear transformation that effects an invertible one-to-one mapping between old basis vectors u_i and new basis vectors v_i, such that

$$v_i = Tu_i = \sum_{j=1}^{n} T_{i,j} u_j \tag{10.27}$$

Such a transformation is described as *unitary* if it preserves scalar products, whereby

$$\langle T\psi \mid T\phi \rangle = \langle \psi \mid \phi \rangle \Longrightarrow T^{\dagger}T = I \tag{10.28}$$

for any ψ, ϕ. Let

$$\psi' = T\psi = \sum_{i=1}^{n} \psi_i T u_i = \sum_{i,j=1}^{n} \psi_i T_{i,j} v_j = \sum_{j=1}^{n} \psi'_j v_j \tag{10.29}$$

represent the vector ψ in terms of the new basis v, whereby

$$\psi'_j = \sum_{i=1}^{n} \psi_i T_{i,j} \tag{10.30}$$

It is important to recognize that the coordinates and the basis vectors transform differently under T – notice the ordering of the indices. These transformations are sometimes described as *cogredient* for basis vectors and *contragedient* for coordinates. The vector ψ is physically the same whatever basis (or coordinate system) we choose. Therefore, matrix elements of the linear operator L should be independent of basis, whereby

$$\langle \psi \mid L \mid \phi \rangle = \langle \psi \mid T^{\dagger}TLT^{\dagger}T \mid \phi \rangle = \langle \psi' \mid L' \mid \phi' \rangle \tag{10.31}$$

shows that matrix representations of operators in the two bases are related by the unitary similarity transformation

$$L' = TLT^{\dagger} \tag{10.32}$$

An operator L is described as *invariant* with respect to the transformation T if $TLT^{\dagger} = L$. Assuming that T is unitary, the condition $TLT^{-1} = L$ implies that L and T commute, such that $LT - TL = 0$. Thus, it is useful to define the *commutator* of two operators by

$$[L, T] = LT - TL \tag{10.33}$$

The symmetry group for L consists of the set of all unitary transformations that leave L invariant.

The effect of a symmetry operation R upon the basis vectors is described by a unitary matrix $D[R]$ defined by

$$Ru_i = \sum_{j=1}^{n} u_j D_{j,i}[R] \tag{10.34}$$

where the ordering of indices is designed to preserve the correspondence between matrix representations and group elements. Thus, successive application of transformations gives

$$R_1 R_2 u_i = R_1 \sum_{j=1}^{n} u_j D_{j,i}[R_2] = \sum_{k=1}^{n} \sum_{j=1}^{n} u_k D_{k,j}[R_1] D_{j,i}[R_2] = \sum_{k=1}^{n} u_k D_{k,i}[R_1 R_2] \tag{10.35}$$

such that

$$R_1 R_2 = R_3 \Longrightarrow D[R_1] D[R_2] = D[R_3] \tag{10.36}$$

Therefore, group multiplication is homomorphic to multiplication of matrices and the corresponding matrices are described as a *representation* of the group and the underlying basis vectors are described as *belonging* to that representation. Naturally we require the representation to obey the same properties as the group with respect to closure, existence of an identity element, associativity, and existence of inverse transformations. Both the symmetry group R and the representation $D[R]$ are groups, but the former is defined in abstract terms while the latter is a realization that applies to a specific physical system. Even if the symmetry of the physical system is only approximate, the analysis of its dynamical properties can often be simplified by use of group theoretical techniques. These methods also provide insight – sometimes very different physical systems are described by the same abstract symmetry group.

A set of nonsingular square matrices $\{M_i, i = 1, N_G\}$ is a representation of the group $G = \{g_i, i = 1, N_G\}$ if

$$g_i g_j = g_k \Longrightarrow M_i M_j = M_k \tag{10.37}$$

for all $i, j, k \leq N_G$. Clearly any representation is itself a group under matrix multiplication. A representation that maps each element of G onto a distinct matrix is isomorphic to G and is described as a *faithful* representation, while homomorphic several-to-one mappings onto matrices are described as *unfaithful*. Every group has a trivial one-dimensional representation with all $M_i = 1$ and a trivial n-dimensional representation with all $M_i = I_n$ where I_n is the n-dimensional unit matrix. Let $D[g]$ denote the matrix associated with element g in representation D. Using any nonsingular matrix of the same dimensionality, the similarity transformation

$$D'[g] = S D[g] S^{-1} \tag{10.38}$$

produces a new representation D' that is *equivalent* to D. Clearly there are an infinite number of equivalent representations of G. We can also form new representations simply by arranging two known representations, D_1 and D_2, in block-diagonal form

$$D = D_1 \oplus D_2 = \begin{pmatrix} D_1 & 0 \\ 0 & D_2 \end{pmatrix} \tag{10.39}$$

where the diagonal blocks usually have different dimensions and the off-diagonal blocks indicate null matrices with the appropriate dimensions. This construction obviously satisfies the group multiplication rules of G required of a representation and the other properties required of a group. *Reducible representations* can be cast in block-diagonal form by a well-chosen similarity transformation and each block is itself a representation of G. Any block that cannot be reduced further is described as an *irreducible representation.* Irreducible representations are so important, and the term is used so frequently, that the abbreviated neologism *irrep* finds common usage. The construction of irreps of specified dimension is one of the central problems of group theory.

Before we tackle general theorems perhaps it would help to exhibit a couple of the irreps of S_3, our prototypical group. A trivial irrep assigns -1 to the reflections $\{P, PR, PR^2\}$ and $+1$ to the rotations $\{I, R, R^2\}$. Direct calculation would show that this representation is consistent with the multiplication table for S_3, although the representation is unfaithful because it uses only two elements to represent six. A faithful two-dimensional irreducible representation is provided below. Notice that the determinants are $+1$ for $\{I, a, b\} \equiv \{I, R, R^2\}$ and -1 for $\{c, d, e\} \equiv \{P, PR, PR^2\}$.

$$\begin{aligned}
&I \to \begin{pmatrix} 1 & 0 \\ 0 & 1 \end{pmatrix}, \quad
a \to \frac{1}{2}\begin{pmatrix} -1 & -\sqrt{3} \\ \sqrt{3} & -1 \end{pmatrix}, \quad
b \to \frac{1}{2}\begin{pmatrix} -1 & \sqrt{3} \\ -\sqrt{3} & -1 \end{pmatrix}, \\
&c \to \begin{pmatrix} -1 & 0 \\ 0 & 1 \end{pmatrix}, \quad
d \to \frac{1}{2}\begin{pmatrix} 1 & \sqrt{3} \\ \sqrt{3} & -1 \end{pmatrix}, \quad
e \to \frac{1}{2}\begin{pmatrix} 1 & -\sqrt{3} \\ -\sqrt{3} & -1 \end{pmatrix}
\end{aligned} \tag{10.40}$$

If you take the time to spot-check some of the products, you should find that they conform with the S_3 multiplication table. Alternatively, we can use MATHEMATICA® to check the entire set as follows. We construct a list of matrices and copy the multiplication table into a matrix whose elements are representations of the group elements. The function **Outer** forms all pairs of elements of G and uses **Dot** to multiply the matrices. The final argument of **Outer** instructs **Dot** to use objects at level 1, a structural device for handling lists of matrices. Finally, an equation is formed and simplified to verify that the left- and right-hand sides are equal, element by element. It is certainly a lot easier to let a machine perform these tedious calculations using instructions formulated at a more conceptual level!

```
i = (1  0
     0  1);   a = 1/2 (-1   -√3
                        √3   -1);   b = 1/2 (-1    √3
                                              -√3   -1);

c = (-1  0
      0  1);   d = 1/2 (1    √3
                         √3   -1);   e = 1/2 (1    -√3
                                               -√3   -1);

G = {i, a, b, c, d, e};

           (i  a  b  c  d  e
            a  b  i  e  c  d
S3table =   b  i  a  d  e  c  ;
            c  d  e  i  a  b
            d  e  c  b  i  a
            e  c  d  a  b  i)

Simplify[Outer[Dot, G, G, 1] == S3table]

True
```

10.3.2 Example: Vibrating triangle

Suppose that, in equilibrium, three equal masses are found at the vertices of an equilateral triangle and that those masses are coupled by three identical springs k along the sides of the triangle. For small displacements, changes in the lengths of the springs depend upon displacements parallel to the sides of the triangle; transverse motions contribute only in second order. Thus, we can approximate the energy for small-amplitude motions as

$$T = \frac{m}{2}\sum_{i=1}^{3}\left(\dot{x}_i^2 + \dot{y}_i^2\right) \tag{10.41}$$

$$\begin{aligned} V = \frac{k}{2}\Bigg(&(x_1 - x_2)^2 + \left(\left(\frac{1}{2}x_1 + \frac{\sqrt{3}}{2}y_1\right) - \left(\frac{1}{2}x_3 + \frac{\sqrt{3}}{2}y_3\right)\right)^2 \\ &+ \left(\left(\frac{1}{2}x_2 - \frac{\sqrt{3}}{2}y_2\right) - \left(\frac{1}{2}x_3 - \frac{\sqrt{3}}{2}y_3\right)\right)^2\Bigg) \end{aligned} \tag{10.42}$$

The kinetic energy is obviously invariant with respect to permutation of the mass indices or any orthogonal transformation of the coordinates. The potential energy can be expressed in the form

$$V = \frac{k}{2}\sum_{i,j} v_{i,j}\xi_i\xi_j \tag{10.43}$$

where $\xi = \{x_1, y_1, x_2, y_2, x_3, y_3\}$ is the coordinate vector and

$$v = \frac{1}{4}\begin{pmatrix} 5 & \sqrt{3} & -4 & 0 & -1 & -\sqrt{3} \\ \sqrt{3} & 3 & 0 & 0 & -\sqrt{3} & -3 \\ -4 & 0 & 5 & -\sqrt{3} & -1 & \sqrt{3} \\ 0 & 0 & -\sqrt{3} & 3 & \sqrt{3} & -3 \\ -1 & -\sqrt{3} & -1 & \sqrt{3} & 2 & 0 \\ -\sqrt{3} & -3 & \sqrt{3} & -3 & 0 & 6 \end{pmatrix} \tag{10.44}$$

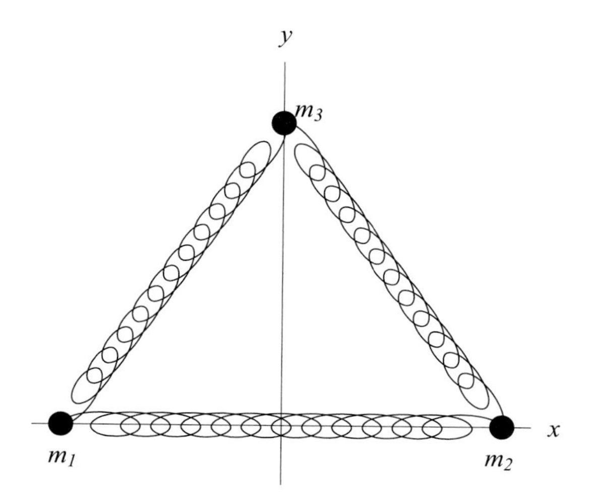

Figure 10.2. Vibrating equilateral triangle.

is the coupling matrix. We would like to prove that the Hamiltonian for this system is invariant with respect to the symmetry group $\{I, R, R^2, P, PR, PR^2\}$, where R denotes rotation by 120° and P is reflection across the y-axis, by constructing explicit matrix representations for each of these symmetry operations and evaluating the similarity transforms GvG^{-1} where G is any of those matrices.

The basic symmetry operations combine a permutation of the mass indices with either reflection or rotation of the coordinates at each vertex. Thus, reflection across the y-axis is represented by

$$D[P] = \begin{pmatrix} 0 & p & 0 \\ p & 0 & 0 \\ 0 & 0 & p \end{pmatrix} \tag{10.45}$$

where

$$p = \begin{pmatrix} -1 & 0 \\ 0 & 1 \end{pmatrix} \tag{10.46}$$

is a 2×2 matrix that performs the reflection $\{x \to -x, y \to y\}$ at a single vertex. Similarly, 120° counterclockwise rotation is represented by

$$D[R] = \begin{pmatrix} 0 & 0 & r \\ 0 & r & 0 \\ r & 0 & 0 \end{pmatrix} \tag{10.47}$$

where

$$r = \begin{pmatrix} -\frac{1}{2} & -\frac{\sqrt{3}}{2} \\ \frac{\sqrt{3}}{2} & -\frac{1}{2} \end{pmatrix} \tag{10.48}$$

rotates the coordinates at a particular vertex. Thus, the reducible six-dimensional representations of the primitive symmetry operators take the form

$$D[P] = \begin{pmatrix} 0 & 0 & -1 & 0 & 0 & 0 \\ 0 & 0 & 0 & 1 & 0 & 0 \\ -1 & 0 & 0 & 0 & 0 & 0 \\ 0 & 1 & 0 & 0 & 0 & 0 \\ 0 & 0 & 0 & 0 & -1 & 0 \\ 0 & 0 & 0 & 0 & 0 & 1 \end{pmatrix} \tag{10.49}$$

$$D[R] = \begin{pmatrix} 0 & 0 & 0 & 0 & -\frac{1}{2} & -\frac{\sqrt{3}}{2} \\ 0 & 0 & 0 & 0 & \frac{\sqrt{3}}{2} & -\frac{1}{2} \\ -\frac{1}{2} & -\frac{\sqrt{3}}{2} & 0 & 0 & 0 & 0 \\ \frac{\sqrt{3}}{2} & -\frac{1}{2} & 0 & 0 & 0 & 0 \\ 0 & 0 & -\frac{1}{2} & -\frac{\sqrt{3}}{2} & 0 & 0 \\ 0 & 0 & \frac{\sqrt{3}}{2} & -\frac{1}{2} & 0 & 0 \end{pmatrix} \tag{10.50}$$

and the remaining group elements can be constructed from these using matrix multiplication.

At this point we would rather use *MATHEMATICA®* to verify the symmetry relations than perform tedious matrix multiplications by hand. Let

$$\xi = \{x_1, y_1, x_2, y_2, x_3, y_3\};$$

$$V = \frac{k}{2}\left((x_1 - x_2)^2 + \left(\left(\frac{1}{2}x_1 + \frac{\sqrt{3}}{2}y_1\right) - \left(\frac{1}{2}x_3 + \frac{\sqrt{3}}{2}y_3\right)\right)^2 + \left(\left(\frac{1}{2}x_2 - \frac{\sqrt{3}}{2}y_2\right) - \left(\frac{1}{2}x_3 - \frac{\sqrt{3}}{2}y_3\right)\right)^2\right);$$

and verify that the matrix

$$\nu = \frac{1}{4}\begin{pmatrix} 5 & \sqrt{3} & -4 & 0 & -1 & -\sqrt{3} \\ \sqrt{3} & 3 & 0 & 0 & -\sqrt{3} & -3 \\ -4 & 0 & 5 & -\sqrt{3} & -1 & \sqrt{3} \\ 0 & 0 & -\sqrt{3} & 3 & \sqrt{3} & -3 \\ -1 & -\sqrt{3} & -1 & \sqrt{3} & 2 & 0 \\ -\sqrt{3} & -3 & \sqrt{3} & -3 & 0 & 6 \end{pmatrix};$$

reproduces the potential energy.

$$V == \frac{k}{2}\,\xi.\nu.\xi\ //\ \mathrm{Simplify}$$

```
True
```

Next define the primitive symmetry operators

$$P = \begin{pmatrix} 0 & 0 & -1 & 0 & 0 & 0 \\ 0 & 0 & 0 & 1 & 0 & 0 \\ -1 & 0 & 0 & 0 & 0 & 0 \\ 0 & 1 & 0 & 0 & 0 & 0 \\ 0 & 0 & 0 & 0 & -1 & 0 \\ 0 & 0 & 0 & 0 & 0 & 1 \end{pmatrix};$$

$$R = \begin{pmatrix} 0 & 0 & 0 & 0 & -\frac{1}{2} & -\frac{\sqrt{3}}{2} \\ 0 & 0 & 0 & 0 & \frac{\sqrt{3}}{2} & -\frac{1}{2} \\ -\frac{1}{2} & -\frac{\sqrt{3}}{2} & 0 & 0 & 0 & 0 \\ \frac{\sqrt{3}}{2} & -\frac{1}{2} & 0 & 0 & 0 & 0 \\ 0 & 0 & -\frac{1}{2} & -\frac{\sqrt{3}}{2} & 0 & 0 \\ 0 & 0 & \frac{\sqrt{3}}{2} & -\frac{1}{2} & 0 & 0 \end{pmatrix};$$

and construct the entire group as a list of matrices.

```
G = {IdentityMatrix[6], R, R.R, P, P.R, P.R.R};
```

Finally, we check that the potential energy is invariant with respect to a similarity transformation using any of the symmetry operations in the group.

```
Table[ν == G[[i]].ν.Inverse[G[[i]]], {i, 1, Length[G]}]
```

```
{True, True, True, True, True, True}
```

Therefore, we do indeed find that ν is invariant with respect to G, as expected. Furthermore, this G is isomorphic to a reducible representation of S_3. Later we will learn how to use the properties of S_3 to determine the eigenvalues of this physical system without solving a sixth-order secular equation.

10.3.3 Orthogonality Theorem

In this section we prove an orthogonality theorem that is central to the theory of group representations. Although we use different notation, the methods are based upon those of Tinkham whose proofs, in turn, were based upon those of Wigner. It is necessary to develop a few preliminary lemmas first before finally arriving at the grand result. Those without the patience to wade through the linear algebra can skip to the finale.

Unitarity

First we demonstrate that any representation using matrices with nonvanishing determinants is equivalent through a similarity transformation to a representation in terms of unitary matrices. Consider the hermitian matrix

$$H = \sum_{g \in G} D[g] D[g]^{\dagger} \tag{10.51}$$

composed of manifestly hermitian terms summed over all group elements. A fundamental theorem of linear algebra tells us that any hermitian matrix is diagonalized by the similarity transformation

$$\mathcal{D} = U^{-1} H U \tag{10.52}$$

using a unitary matrix U whose columns are the orthonormal eigenvectors of H and that the resulting diagonal matrix $\mathcal{D}$ has positive eigenvalues on its diagonal. Thus,

$$\mathcal{D} = \sum_{g \in G} (U^{-1} D[g] U)(U^{-1} D[g]^\dagger U) = \sum_{g \in G} D'[g] D'[g]^\dagger \tag{10.53}$$

where the matrices

$$D'[g] = U^{-1} D[g] U \tag{10.54}$$

represent G in the representation for which H is diagonal. Recognizing that $\mathcal{D}$ is positive-definite, we can construct a useful representation of the unit matrix for this representation according to

$$I = \mathcal{D}^{-1/2} \mathcal{D} \mathcal{D}^{-1/2} = \mathcal{D}^{-1/2} \left(\sum_{g \in G} D'[g] D'[g]^\dagger \right) \mathcal{D}^{-1/2} = \sum_{g \in G} D''[g] D''[g]^\dagger \tag{10.55}$$

where $\mathcal{D}^\alpha$ is the diagonal matrix containing the elements of $\mathcal{D}$ raised to power α and where the doubly-primed representation

$$D''[g] = \mathcal{D}^{-1/2} D'[g] \mathcal{D}^{1/2} \tag{10.56}$$

results from a similarity transformation based upon $\mathcal{D}^{1/2}$. Finally, we demonstrate that the $D''[g]$ matrices are unitary by evaluating the products

$$\begin{aligned} D''[g] D''[g]^\dagger &= D''[g] I D''[g]^\dagger \\ &= (\mathcal{D}^{-1/2} D'[g] \mathcal{D}^{1/2}) \mathcal{D}^{-1/2} (\sum_{h \in G} D'[h] D'[h]^\dagger) \\ &\quad \times \mathcal{D}^{-1/2} (\mathcal{D}^{-1/2} D'[g]^\dagger \mathcal{D}^{1/2}) \end{aligned} \tag{10.57}$$

with the aid of the aforementioned representation of the unit matrix. Note that we must use a distinct summation index. Using the commutation properties of diagonal matrices, the representation property $D[g]D[h] = D[gh]$, and the hermitian conjugation property $(D[g]D[h])^\dagger = D[h]^\dagger D[g]^\dagger$, we find that

$$\begin{aligned} D''[g] D''[g]^\dagger &= \mathcal{D}^{-1} \sum_{h \in G} D'[g] D'[h] D'[h]^\dagger D'[g]^\dagger \\ &= \mathcal{D}^{-1} \sum_{h \in G} D'[gh] D'[gh]^\dagger = \mathcal{D}^{-1} \mathcal{D} = I \end{aligned} \tag{10.58}$$

is indeed unitary. Note that summation of gh over all h is equivalent to summation over all $g \in G$ in a different order. Therefore, one can always construct a unitary representation using suitable similarity transformations and we henceforth assume, without loss of generality, that all representations employed are unitary.

Schur's Lemma

Next we demonstrate that any matrix that commutes with all elements of an irreducible representation must be a multiple of the unit matrix. Suppose that M is a matrix that commutes with all elements of unitary representation D, such that

$$MD[g] = D[g]M \Longrightarrow D[g]^\dagger M^\dagger = M^\dagger D[g]^\dagger \tag{10.59}$$

Multiplying by $D[g]$ on both left and right and using its unitarity,

$$M^\dagger D[g] = D[g]M^\dagger \tag{10.60}$$

we find that the hermitian conjugate $M^\dagger$ also commutes with every $D[g]$. Thus, there exist hermitian matrices

$$H_+ = M + M^\dagger, \quad H_- = i\left(M - M^\dagger\right) \tag{10.61}$$

that also commute with all $D[g]$. If we show that any commuting hermitian matrix is constant, than any $M = (H_+ - iH_-)/2$ must also be constant. Of course, any hermitian matrix can be reduced to the diagonal form $\mathcal{D} = U^{-1}HU$ by a similarity transformation, so we use the corresponding representation

$$D'[g] = U^{-1}D[g]U \tag{10.62}$$

for group elements to write the commutation condition as

$$\mathcal{D}D'[g] = D'[g]\mathcal{D} \tag{10.63}$$

and must prove that the elements of $\mathcal{D}$ are identical. Using the fact that $\mathcal{D}_{i,j} = \mathcal{D}_i\delta_{i,j}$, each element of this matrix equation takes the form

$$\left(\mathcal{D}_i - \mathcal{D}_j\right)D'_{i,j}[g] == 0 \tag{10.64}$$

If $\mathcal{D}_i \neq \mathcal{D}_j$, then all $D'_{i,j}[g] = 0$ for any g such that the similarity transformation U reduces representation D to block diagonal form, contradicting the stipulation that D is irreducible. Therefore, if D is irreducible then $\mathcal{D}_i = \mathcal{D}_j$ for any i, j is a constant diagonal matrix, thereby proving Schur's lemma.

Equivalence of Irreducible Representations of Common Dimension

The next intermediate result takes as much care to state as to prove. Suppose that $D^{(1)}$ is an irreducible representation of dimension n_1 and $D^{(2)}$ is another irreducible representation of dimension n_2. Further, suppose that M is a rectangular matrix of dimensions $n_1 \times n_2$ such that

$$D^{(1)}[g]M = MD^{(2)}[g] \tag{10.65}$$

for any $g \in G$. Then, $n_1 \neq n_2$ requires $M = 0$ while for $n_1 = n_2$ either $M = 0$ or $\mathrm{Det}[M] \neq 0$. When $\mathrm{Det}[M] \neq 0$, M is invertible and $D^{(1)}$ and $D^{(2)}$ are equivalent representations that are related by the similarity transformation effected by M.

We may again assume that both $D^{(1)}$ and $D^{(2)}$ are unitary and that $n_1 \leq n_2$. Using the fact that the representations are both unitary and irreducible, the hermitian adjoint takes the form

$$M^\dagger D^{(1)}[g]^\dagger = D^{(2)}[g]^\dagger M^\dagger \Longrightarrow M^\dagger D^{(1)}[g^{-1}] = D^{(2)}[g^{-1}]M^\dagger \tag{10.66}$$

Multiplying both sides by M and using the stipulated commutation property, we find

$$MM^\dagger D^{(1)}[g^{-1}] = MD^{(2)}[g^{-1}]M^\dagger \Longrightarrow MM^\dagger D^{(1)}[g^{-1}] = D^{(1)}[g^{-1}]MM^\dagger \tag{10.67}$$

Therefore, according to Schur's lemma, $MM^\dagger$ must be a multiple of the unit matrix. First suppose that $n_1 = n_2$ such that M is a square matrix and $MM^\dagger = cI$ where c is a number. Then

$$\mathrm{Det}\left[MM^\dagger\right] = \mathrm{Det}[M]\,\mathrm{Det}[M]^* = |\mathrm{Det}[M]|^2 = c \tag{10.68}$$

requires that c be nonnegative. If $M \neq 0$, then $\mathrm{Det}[M] \neq 0$ and M is invertible. Next suppose that $n_1 < n_2$ and consider the $n_2 \times n_2$ matrix N formed by appending $n_2 - n_1$ columns of zeros to M. Matrix multiplication

$$\left(NN^\dagger\right)_{i,j} = \sum_{k=1}^{n_2} N_{i,k}N^*_{j,k} = \sum_{k=1}^{n_1} M_{i,k}M^*_{j,k} = \left(MM^\dagger\right)_{i,j} \tag{10.69}$$

shows that $NN^\dagger = MM^\dagger$ because the appended columns do not contribute. However, $\mathrm{Det}[N] = 0$ by construction, such that $|\mathrm{Det}[M]|^2 = |\mathrm{Det}[N]|^2 = 0 \Longrightarrow c = 0$. Therefore, if $n_1 \neq n_2$, then $M = 0$.

Orthogonality Theorem

We are finally ready to derive the orthogonality theorem for irreducible representations. Let $D^{(i)}$ and $D^{(j)}$ represent irreducible representations of group G of order N_G in terms of nonsingular matrices of dimension n_i and n_j that are either identical or inequivalent; equivalent representations of the same dimensions are excluded. Then

$$\sum_{g\in G} D^{(i)}_{\mu,\nu}[g]^* D^{(j)}_{\mu',\nu'}[g] = \frac{N_G}{n_i}\delta_{i,j}\delta_{\mu,\mu'}\delta_{\nu,\nu'} \tag{10.70}$$

where the summation runs over all group elements. This theorem will play a central role in determining the number of inequivalent irreducible representations that exist with specified dimensions.

Consider

$$M = \sum_{g\in G} D^{(j)}[g] X D^{(i)}\left[g^{-1}\right] \tag{10.71}$$

where X is an arbitrary matrix and both M and X have dimensions $n_j \times n_i$. We can demonstrate that M satisfies the condition of the third lemma as follows.

$$\begin{aligned} D^{(j)}[g]M &= \sum_{h\in G} D^{(j)}[gh]XD^{(i)}\left[h^{-1}\right] \\ &= \sum_{h\in G} D^{(j)}[gh]XD^{(i)}\left[h^{-1}g^{-1}g\right] \\ &= \sum_{a\in G} D^{(j)}[a]XD^{(i)}\left[a^{-1}\right]D^{(i)}[g] \\ &= MD^{(i)}[g] \end{aligned} \tag{10.72}$$

Any particular matrix element takes the form

$$M_{\alpha\mu} = \sum_{g\in G}\sum_{\beta',\gamma} D^{(j)}_{\alpha,\beta'}[g]X_{\beta',\gamma}D^{(i)}_{\gamma,\mu}\left[g^{-1}\right] \tag{10.73}$$

First suppose that $n_i \neq n_j \implies M = 0$. Recognizing that the matrix elements $X_{\beta',\gamma}$ are arbitrary, we are free to choose $X_{\beta',\gamma} = \delta_{\beta,\beta'}\delta_{\gamma,\nu}$, such that

$$0 = \sum_{g\in G} D^{(j)}_{\alpha,\beta}[g]D^{(i)}_{\nu,\mu}\left[g^{-1}\right] \tag{10.74}$$

The unitarity of $D^{(1)}$ then requires

$$0 = \sum_{g\in G} D^{(i)}_{\mu,\nu}[g]^* D^{(j)}_{\alpha,\beta}[g] \tag{10.75}$$

for $n_i \neq n_j$.

Next suppose that $n_i = n_j$, such that either $M = 0$ or $D^{(i)} \sim D^{(j)}$. The former shows that inequivalent representations are orthogonal. Different equivalent representations are excluded by the conditions of the theorem, leaving us only to deduce the normalization of the sum over products of matrix elements for the same representation. Again using the fact that M is a multiple of the unit matrix gives

$$\sum_{g\in G}\sum_{\gamma,\kappa} D^{(i)}_{\mu,\gamma}[g]^* D^{(i)}_{\alpha,\kappa}[g]X_{\kappa,\gamma} = c\delta_{\alpha,\mu} \tag{10.76}$$

for any X. Making an inspired choice

$$X_{\kappa,\gamma} = \delta_{\kappa,\beta}\delta_{\gamma,\nu} \implies \sum_{g\in G} D^{(i)}_{\mu,\nu}[g]^* D^{(i)}_{\alpha,\beta}[g] = c\delta_{\alpha,\mu} \tag{10.77}$$

we can use the unitarity of $D^{(i)}$ to perform the sum over the n_i identical diagonal matrix elements to obtain

$$\sum_{\mu}\sum_{g\in G} D^{(i)}_{\mu,\nu}[g]^* D^{(i)}_{\mu,\beta}[g] = cn_i = N_G\delta_{\beta,\nu} \implies c = \frac{N_G}{n_i}\delta_{\beta,\nu} \tag{10.78}$$

Therefore, we finally obtain the *orthogonality theorem*

$$\sum_{g \in G} D^{(i)}_{\mu,\nu}[g]^* D^{(j)}_{\alpha,\beta}[g] = \frac{N_G}{n_i} \delta_{\alpha,\mu} \delta_{\beta,\nu} \delta_{i,j} \tag{10.79}$$

An important special case is that the normalization of matrix elements

$$\sum_{g \in G} \left| D^{(i)}_{\mu,\nu}[g] \right|^2 = \frac{N_G}{n_i} \tag{10.80}$$

summed over group elements is independent of the indices of the matrices.

10.3.4 Character

Simple Characters

In the face of an unlimited number of equivalent irreducible representations that can be produced by similarity transformations, it is desirable to describe irreps in terms of invariant properties. An obvious candidate is the trace of the matrices, which is invariant under similarity transformations. Furthermore, because all members of a class are related by similarity transformations, the trace of representation matrices must be the same for every member of a class. Therefore, the *character* $\chi^{(j)}[C_k]$ of class C_k in representation $D^{(j)}$ is defined as $\mathrm{Tr}\left[D^{(j)}[g]\right]$ where $g \in C_k$. The characters for irreducible representations are described as *simple* while those for reducible representations are *composite*. In this section we use the orthogonality theorem to develop several valuable relationships for simple characters and in the next we consider composite characters.

Specializing the orthogonality theorem for inequivalent irreducible representations to diagonal elements

$$\sum_{g \in G} D^{(i)}_{\mu,\mu}[g]^* D^{(j)}_{\alpha,\alpha}[g] = \frac{N_G}{n_i} \delta_{i,j} \delta_{\alpha,\mu} \tag{10.81}$$

and summing over indices, we find

$$\sum_{g \in G} \chi^{(i)}[g]^* \chi^{(j)}[g] = N_G \delta_{i,j} \tag{10.82}$$

or

$$\sum_k N_k \chi^{(i)}\left[C_k\right]^* \chi^{(j)}\left[C_k\right] = N_G \delta_{i,j} \tag{10.83}$$

where N_k is the number of elements in class C_k. Evidently, simple characters also satisfy an orthogonality theorem. When the two irreps are the same, we obtain a formula

$$\sum_g \left| \chi^{(i)}[g] \right|^2 = \sum_k N_k \left| \chi^{(i)}\left[C_k\right] \right|^2 = N_G \tag{10.84}$$

that severely constrains the total number of inequivalent irreps that exist for a finite group.

For example, we already know that S_3 contains three classes: $\{I\}$, $\{R, R^2\}$, and $\{P, PR, P, R^2\}$. The character table for S_3 is given in Table 10.4. Each column is labeled by class and the number of elements in that class. The rows list the characters for representative matrices in each irreducible representation. Because the identity element is in its own class, the first column also gives the dimensionality of each irrep. The first row is the trivial irrep in which each element is represented by the integer 1. For S_3 there is another one-dimensional representation in terms of parity. These rows are orthogonal, weighted by class size, and satisfy the normalization condition. No other one-dimensional irreps can exist. Using the orthogonality conditions, Eq. (10.83), a two-dimensional representation must satisfy the equations

$$2 + 2\chi_2^{(3)} + 3\chi_3^{(3)} == 0 \tag{10.85}$$

$$2 + 2\chi_2^{(3)} - 3\chi_3^{(3)} == 0 \tag{10.86}$$

Thus, we find that there is one two-dimensional irrep and no higher-dimensional irreps are possible because the system of orthogonality equations would be overdetermined. We also observe that these results are consistent with the normalization condition. Therefore, these simple considerations are sufficient to construct the character table without actually producing representation matrices. This exercise suggests that *the number of irreducible representations is equal to the number of classes*. This is a general result, but we will forgo the formal proof.

Table 10.4. Character table for S_3.

$\chi[S_3]$	C_1	$2C_2$	$3C_3$
$D^{(1)}$	1	1	1
$D^{(2)}$	1	1	−1
$D^{(3)}$	2	−1	0

Suppose that we define a square matrix

$$M_{i,j} = \chi^{(j)}[C_i] \tag{10.87}$$

from the body of the character table. The inverse of this matrix is found using the orthogonality formula

$$M^{-1}_{i,j} = \frac{N_j}{N_G}\chi^{(i)}[C_j]^* \Longrightarrow (M^{-1}M)_{i,j} = \sum_k \frac{N_k}{N_G}\chi^{(i)}[C_k]^*\chi^{(j)}[C_k] = \delta_{i,j} \tag{10.88}$$

where the summation ranges over classes. However, because a matrix commutes with its inverse, we can also express orthogonality in the form

$$(MM^{-1})_{i,j} = \sum_k \frac{N_j}{N_G}\chi^{(k)}[C_i]\chi^{(k)}[C_j]^* = \delta_{i,j} \tag{10.89}$$

where the summation ranges over irreps.

Therefore, orthogonality of character

$$\sum_i N_i \chi^{(\mu)}[C_i]^* \chi^{(\nu)}[C_i] = N_G \delta_{\mu,\nu} \tag{10.90}$$

$$\sum_\mu \chi^{(\mu)}[C_i]^* \chi^{(\mu)}[C_j] = \frac{N_G}{N_i} \delta_{i,j} \tag{10.91}$$

can be expressed in terms of either class or irrep and both can be useful in construction of character tables. Here we chose Greek indices for irreps and Latin indices for class.

Composite Character

Recall that a reducible representation can be cast in block diagonal form with a suitable similarity transformation. Upon reducing such a representation as fully as possible, such that each block is irreducible, we write

$$D[g] = \sum_\mu a_\mu D^{(\mu)}[g] \tag{10.92}$$

where the summation ranges over irreducible representations and the nonnegative integers a_μ count the number of times each irrep $D^{(\mu)}$ appears in D. The same irrep may appear in several equivalent forms within D and a_μ counts each equivalent form equally. The summation is schematic – it enumerates the structure along the diagonal of the block diagonalized D. The trace of D then becomes

$$\chi[g] = \sum_\mu a_\mu \chi^{(\mu)}[g] \Longrightarrow \chi[C_i] = \sum_\mu a_\mu \chi^{(\mu)}[C_i] \tag{10.93}$$

where C_i is the class that contains g. Thus, the character for each class in a reducible representation is the sum over the simple characters for irreducible representations of the same class weighted by the number of appearances of equivalent forms of those irreps. The orthogonality theorems for simple characters facilitate projection of a_μ

$$\sum_i N_i \chi^{(\nu)}[C_i]^* \chi[C_i] = \sum_i \sum_\mu a_\mu N_i \chi^{(\nu)}[C_i]^* \chi^{(\mu)}[C_i] = N_G a_\nu \tag{10.94}$$

whereby

$$a_\mu = \frac{1}{N_G} \sum_i N_i \chi^{(\mu)}[C_i]^* \chi[C_i] = \frac{1}{N_G} \sum_{g \in G} \chi^{(\mu)}[g]^* \chi[g] \tag{10.95}$$

Suppose that $D = D^{(\nu)}$ is irreducible, such that $a_\mu = \delta_{\mu,\nu}$. Then

$$\sum_i N_i \left|\chi^{(\mu)}[C_i]\right|^2 = N_G \tag{10.96}$$

is consistent with previous results for the character normalization summed over classes of a particular irrep. In fact, this condition provides a test for irreducibility. For an arbitrary representation, one finds

$$\sum_i N_i \left|\chi\left[C_i\right]\right|^2 \geq N_G \tag{10.97}$$

with equality if and only if χ is simple and D is irreducible. Expanding χ again gives

$$\sum_i N_i \sum_{\mu,\nu} a_\mu a_\nu^* \chi^{(\mu)}\left[C_i\right] \chi^{(\nu)}\left[C_i\right]^* = N_G \sum_\mu \left|a_\mu\right|^2 \tag{10.98}$$

such that

$$\frac{1}{N_G} \sum_i N_i \left|\chi\left[C_i\right]\right|^2 = \sum_\mu \left|a_\mu\right|^2 \geq 1 \tag{10.99}$$

If computation of the quantity on the left gives 1, then the representation is irreducible. If it happens to be 2, then D contains two different irreps, with each occurring just once, because the a_μ are nonnegative integers. If it happens to be 4, then D contains either one irrep occurring twice or four different irreps all occurring once.

Dimensionality Theorem

Table 10.5. Multiplication table for S_3 with elements arranged to place I on the main diagonal.

S_3	I	a	b	c	d	e
I	I	a	b	c	d	e
a^{-1}	b	I	a	d	e	c
b^{-1}	a	b	I	e	c	d
c^{-1}	c	d	e	I	a	b
d^{-1}	d	e	c	b	I	a
e^{-1}	e	c	d	a	b	I

The *regular representation* of any finite group is constructed by arranging the group elements to place the identity element along the main diagonal of the group multiplication table. The matrices $D^{(\mathrm{reg})}[g]$ are then obtained by replacing g in the multiplication table by 1 and all other elements by 0. Thus, if we label columns by g_i we can label rows by g_i^{-1}. We illustrate this construction using S_3 again. Using the multiplication table in the form shown in Table 10.5, we obtain the matrices

$$D[I] = \begin{pmatrix} 1 & 0 & 0 & 0 & 0 & 0 \\ 0 & 1 & 0 & 0 & 0 & 0 \\ 0 & 0 & 1 & 0 & 0 & 0 \\ 0 & 0 & 0 & 1 & 0 & 0 \\ 0 & 0 & 0 & 0 & 1 & 0 \\ 0 & 0 & 0 & 0 & 0 & 1 \end{pmatrix}, \tag{10.100}$$

$$D[a] = \begin{pmatrix} 0 & 1 & 0 & 0 & 0 & 0 \\ 0 & 0 & 1 & 0 & 0 & 0 \\ 1 & 0 & 0 & 0 & 0 & 0 \\ 0 & 0 & 0 & 0 & 1 & 0 \\ 0 & 0 & 0 & 0 & 0 & 1 \\ 0 & 0 & 0 & 1 & 0 & 0 \end{pmatrix}, \tag{10.101}$$

$$D[b] = \begin{pmatrix} 0 & 0 & 1 & 0 & 0 & 0 \\ 1 & 0 & 0 & 0 & 0 & 0 \\ 0 & 1 & 0 & 0 & 0 & 0 \\ 0 & 0 & 0 & 0 & 0 & 1 \\ 0 & 0 & 0 & 1 & 0 & 0 \\ 0 & 0 & 0 & 0 & 1 & 0 \end{pmatrix}, \tag{10.102}$$

$$D[c] = \begin{pmatrix} 0 & 0 & 0 & 1 & 0 & 0 \\ 0 & 0 & 0 & 0 & 0 & 1 \\ 0 & 0 & 0 & 0 & 1 & 0 \\ 1 & 0 & 0 & 0 & 0 & 0 \\ 0 & 0 & 1 & 0 & 0 & 0 \\ 0 & 1 & 0 & 0 & 0 & 0 \end{pmatrix}, \tag{10.103}$$

$$D[d] = \begin{pmatrix} 0 & 0 & 0 & 0 & 1 & 0 \\ 0 & 0 & 0 & 1 & 0 & 0 \\ 0 & 0 & 0 & 0 & 0 & 1 \\ 0 & 1 & 0 & 0 & 0 & 0 \\ 1 & 0 & 0 & 0 & 0 & 0 \\ 0 & 0 & 1 & 0 & 0 & 0 \end{pmatrix}, \tag{10.104}$$

$$D[e] = \begin{pmatrix} 0 & 0 & 0 & 0 & 0 & 1 \\ 0 & 0 & 0 & 0 & 1 & 0 \\ 0 & 0 & 0 & 1 & 0 & 0 \\ 0 & 0 & 1 & 0 & 0 & 0 \\ 0 & 1 & 0 & 0 & 0 & 0 \\ 1 & 0 & 0 & 0 & 0 & 0 \end{pmatrix} \tag{10.105}$$

and can easily verify that these obey the group multiplication table.

The character of the identity element of the regular representation is N_G while the characters of all other classes are zero by construction. Furthermore, the simple character $\chi^{(\mu)}[I] = N_\mu$ for the identity element of an irreducible representation is just the dimensionality of irrep μ. Application of the projection theorem, Eq. (10.95),

$$a_\mu^{(\text{reg})} = \frac{1}{N_G}\sum_i N_i \chi^{(\mu)}[C_i]^* \chi^{(\text{reg})}[C_i] = \frac{1}{N_G} N_\mu N_G \Longrightarrow a_\mu^{(\text{reg})} = N_\mu \tag{10.106}$$

then shows that the number of times each irrep appears in the regular representation is equal to its dimensionality. Finally, the composite character of the identity element of the regular representation

$$\chi^{(\text{reg})}[I] = \sum_\mu a_\mu \chi^{(\mu)}[I] \Longrightarrow N_G = \sum_\mu N_\mu^2 \tag{10.107}$$

provides a simple relationship between the dimensionality of irreps and the order of the group. This result is general because neither of these quantities depends upon the representation; the fact that we obtained it with the aid of the regular representation does not limit its generality. Therefore, we obtain the *dimensionality theorem*

$$\sum_\mu N_\mu^2 = N_G \tag{10.108}$$

For example, the fact that $N_G = 6$ for S_3 is sufficient to deduce that S_3 supports exactly two different one-dimensional irreps and one two-dimensional irrep because $1^2 + 1^2 + 2^2 = 6$ is the only solution to the dimensionality equation in terms of positive integers N_μ. Furthermore, the regular representation contains both one-dimensional irreps once and the two-dimensional irrep twice.

10.3.5 Example: Character table for symmetries of a square

In the preceding few subsections we derived many abstract theorems and we need a non-trivial example to illustrate their use. Consider the symmetries of a square, excluding operations that twist its sides. These symmetries are among the permutations of the four vertices and can be labeled as

$$I = [1234] \tag{10.109}$$
$$R = [2341] \tag{10.110}$$
$$R^2 = [3412] \tag{10.111}$$
$$R^3 = [4123] \tag{10.112}$$
$$P = [4321] \tag{10.113}$$
$$PR = [1432] \tag{10.114}$$
$$PR^2 = [2143] \tag{10.115}$$
$$PR^3 = [3214] \tag{10.116}$$

where R indicates a rotation (cyclic permutation) and P a reflection across the x-axis (parity operation). The combinations of reflection and rotations then fill out the list of permissible permutations. Table 10.6 provides the group multiplication table with an extra column for the order of each element. This table shows that multiplication is associative, closed, and

Table 10.6. Group multiplication table for symmetries of a square.

Square	I	R	R^2	R^3	P	PR	PR^2	PR^3	Order
I	I	R	R^2	R^3	P	PR	PR^2	PR^3	1
R	R	R^2	R^3	I	PR^3	P	PR	PR^2	4
R^2	R^2	R^3	I	R	PR^2	PR^3	P	PR	2
R^3	R^3	I	R	R^2	PR	PR^2	PR^3	P	4
P	P	PR	PR^2	PR^3	I	R	R^2	R^3	2
PR	PR	PR^2	PR^3	P	R^3	I	PR	R^2	2
PR^2	PR^2	PR^3	P	PR	R^2	R^3	I	R	2
PR^3	PR^3	P	PR	PR^2	R	R^2	R^3	I	2

that there is an inverse for each element. Therefore, this set of operations forms a group of order 8 and, because each element is a member of S_4, we conclude that the symmetries of a square are a subgroup of S_4.

The calculations below establish class membership.

$$PRP^{-1} = PRP = PPR^3 = R^3 \Longrightarrow R \sim R^3 \tag{10.117}$$

$$RPR^{-1} = PR^3R^3 = PR^2 \Longrightarrow P \sim PR^2 \tag{10.118}$$

$$RPRR^{-1} = PR^3 \Longrightarrow PR \sim PR^3 \tag{10.119}$$

Thus, the five classes consist of

$$C_1 = \{I\} \tag{10.120}$$

$$C_2 = \{R, R^3\} \tag{10.121}$$

$$C_3 = \{R^2\} \tag{10.122}$$

$$C_4 = \{P, PR^2\} \tag{10.123}$$

$$C_5 = \{PR, PR^3\} \tag{10.124}$$

and there are also five irreps with dimensions $d = \{1, 1, 1, 1, 2\}$, the only solution to $d{\cdot}d = 8$ among positive integers. Placing the trivial one-dimensional representation $D^{(1)}[g] = I$ on the first row and the dimensions d on the first column gives us the start of a character table (10.3.5). We can also use the parity of the permutations as a second one-dimensional representation giving row 2. By inspection we verify that these rows are orthogonal and normalized properly when weighted by class size.

Table 10.7. First steps in construction of the character table for symmetries of a square.

χ	C_1	$2C_2$	C_3	$2C_4$	$2C_5$
$D^{(1)}$	1	1	1	1	1
$D^{(2)}$	1	1	1	−1	−1
$D^{(3)}$	1				
$D^{(4)}$	1				
$D^{(5)}$	2				

Next, we identify $A = \{I, R^2\}$ as an invariant subgroup by examining the multiplication table and observing that A commutes with all other elements of G. Thus, its factor group is

$F = G/A = \{A, RA, PA, PRA\}$ and an irrep of F must be an irrep of G by homomorphism. The multiplication table for F is given in Table 10.8.

Table 10.8.

F	A	RA	PA	PRA
A	A	RA	PA	PRA
RA	RA	A	PRA	PA
PA	PA	PRA	A	RA
PRA	PRA	PA	RA	A

Since F is abelian each element is in its own class, such that F supports four one-dimensional irreps. Every element other than the identity (A) is order 2, such that irreps must assign them values of ± 1. Thus, the irreps are $\{1, 1, 1, 1\}$, $\{1, 1, -1, -1\}$, $\{1, -1, 1, -1\}$, and $\{1, -1, -1, 1\}$; it is a simple exercise to verify that these representations obey the multiplication table for F. We can now use the homomorphism with G to complete rows three and four of its character table (Tab. 10.3.5). It is comforting to observe that the orthonormality conditions are satisfied at this stage.

Table 10.9.

χ	C_1	$2C_2$	C_3	$2C_4$	$2C_5$
$D^{(1)}$	1	1	1	1	1
$D^{(2)}$	1	1	1	−1	−1
$D^{(3)}$	1	−1	1	1	−1
$D^{(4)}$	1	−1	1	−1	1
$D^{(5)}$	2				

Finally, we use orthogonality relations to produce a system of four linear equations for the four remaining elements of this table.

$$2 + 2\chi_2^{(5)} + \chi_3^{(5)} + 2\chi_4^{(5)} + 2\chi_5^{(5)} == 0 \tag{10.125}$$

$$2 + 2\chi_2^{(5)} + \chi_3^{(5)} - 2\chi_4^{(5)} - 2\chi_5^{(5)} == 0 \tag{10.126}$$

$$2 - 2\chi_2^{(5)} + \chi_3^{(5)} + 2\chi_4^{(5)} - 2\chi_5^{(5)} == 0 \tag{10.127}$$

$$2 - 2\chi_2^{(5)} + \chi_3^{(5)} - 2\chi_4^{(5)} + 2\chi_5^{(5)} == 0 \tag{10.128}$$

After some straightforward algebra, we obtain row five and complete the character table, as shown in Table 10.10. You should verify, for practice, that both the row and column orthonormality relations are satisfied for this character table. The reasoning employed here is typical of that used to construct character tables. There are variations in the sequence of steps, of course, depending upon what information is available. Sometimes one can dispense with construction of factor groups; other times more than one is both available and helpful.

Table 10.10. Completed character table for symmetries of a square.

χ	C_1	$2C_2$	C_3	$2C_4$	$2C_5$
$D^{(1)}$	1	1	1	1	1
$D^{(2)}$	1	1	1	-1	-1
$D^{(3)}$	1	-1	1	1	-1
$D^{(4)}$	1	-1	1	-1	1
$D^{(5)}$	2	0	-2	0	0

Next we construct a reducible dimension-four representation using the permutation matrices for four objects; examples from each class are listed below.

$$D[1,2,3,4] = \begin{pmatrix} 1 & 0 & 0 & 0 \\ 0 & 1 & 0 & 0 \\ 0 & 0 & 1 & 0 \\ 0 & 0 & 0 & 1 \end{pmatrix} \Longrightarrow \chi_1 = 4 \tag{10.129}$$

$$D[2,3,4,1] = \begin{pmatrix} 0 & 1 & 0 & 0 \\ 0 & 0 & 1 & 0 \\ 0 & 0 & 0 & 1 \\ 1 & 0 & 0 & 0 \end{pmatrix} \Longrightarrow \chi_2 = 0 \tag{10.130}$$

$$D[3,4,1,2] = \begin{pmatrix} 0 & 0 & 1 & 0 \\ 0 & 0 & 0 & 1 \\ 1 & 0 & 0 & 0 \\ 0 & 1 & 0 & 0 \end{pmatrix} \Longrightarrow \chi_3 = 0 \tag{10.131}$$

$$D[4,3,2,1] = \begin{pmatrix} 0 & 0 & 0 & 1 \\ 0 & 0 & 1 & 0 \\ 0 & 1 & 0 & 0 \\ 1 & 0 & 0 & 0 \end{pmatrix} \Longrightarrow \chi_4 = 0 \tag{10.132}$$

$$D[3,2,1,4] = \begin{pmatrix} 0 & 0 & 1 & 0 \\ 0 & 1 & 0 & 0 \\ 1 & 0 & 0 & 0 \\ 0 & 0 & 0 & 1 \end{pmatrix} \Longrightarrow \chi_5 = 2 \tag{10.133}$$

Then

$$\frac{1}{N_G} \sum_k c_k \left|\chi_k\right|^2 = \frac{1}{8}\left(1 \times 4^2 + 2 \times 0^2 + 1 \times 0 + 2 \times 0^2 + 2 \times 2^2\right) = 3 \tag{10.134}$$

shows that this representation contains three irreps, once each ($1^2 + 1^2 + 1^2 = 3$). Thus, D contains the two-dimensional irrep plus two of the one-dimensional irreps. We can determine which one-dimensional representations are within D using

$$a_1 = \frac{1}{N_G} \sum_k c_k \chi_k^{(1)*} \chi_k = \frac{1}{8}(1 \times 4 + 2 \times 0 + 1 \times 0 + 2 \times 0 + 2 \times 2) = 1 \tag{10.135}$$

$$a_2 = \frac{1}{N_G} \sum_k c_k \chi_k^{(2)*} \chi_k = \frac{1}{8}(1 \times 4 + 2 \times 0 + 1 \times 0 - 2 \times 0 - 2 \times 2) = 0 \tag{10.136}$$

$$a_3 = \frac{1}{N_G} \sum_k c_k \chi_k^{(3)*} \chi_k = \frac{1}{8}(1 \times 4 - 2 \times 0 + 1 \times 0 + 2 \times 0 - 2 \times 2) = 0 \tag{10.137}$$

$$a_4 = \frac{1}{N_G} \sum_k c_k \chi_k^{(4)*} \chi_k = \frac{1}{8}(1 \times 4 - 2 \times 0 + 1 \times 0 - 2 \times 0 + 2 \times 2) = 1 \tag{10.138}$$

such that

$$D = D^{(1)} \oplus D^{(4)} \oplus D^{(5)} \Longrightarrow \chi_k = 1 + \chi_k^{(4)} + \chi_k^{(5)} \tag{10.139}$$

is also satisfied. If one has suitable reducible representations at hand, relationships of this type can sometimes be used to compute the character table, perhaps instead of using factor groups.

10.3.6 Example: Vibrational eigenvalues of square

Suppose that, at equilibrium, four equal masses m are found at the vertices of a square and that these masses are connected by four equal springs k along its sides and two springs κk along the diagonals. We consider small-amplitude vibrations and expand the potential energy to second order in the spatial coordinates. Let $\xi = \{x_1, y_1, x_2, y_2, x_3, y_3, x_4, y_4\}$ represent a state vector for the system where the masses are labeled sequentially. The kinetic and potential energies are given by

$$T = \frac{1}{2} m \sum_i \dot{\xi}_i^2 \tag{10.140}$$

$$\begin{aligned} V &= \frac{1}{2} k \sum_{i,j} v_{i,j} \xi_i \xi_j \\ &= \frac{k}{2} \left((x_1 - x_2)^2 + (y_2 - y_3)^2 + (x_3 - x_4)^2 + (y_1 - y_4)^2 + \kappa \left(\frac{x_2 + y_2}{\sqrt{2}} - \frac{x_4 + y_4}{\sqrt{2}} \right)^2 \right. \\ &\quad \left. + \kappa \left(\frac{x_1 - y_1}{\sqrt{2}} - \frac{x_3 - y_3}{\sqrt{2}} \right)^2 \right) \end{aligned} \tag{10.141}$$

where the symmetric matrix

$$v = \frac{1}{2}\begin{pmatrix} 2+\kappa & -\kappa & -2 & 0 & -\kappa & \kappa & 0 & 0 \\ -\kappa & 2+\kappa & 0 & 0 & \kappa & -\kappa & 0 & -2 \\ -2 & 0 & 2+\kappa & \kappa & 0 & 0 & -\kappa & -\kappa \\ 0 & 0 & \kappa & 2+\kappa & 0 & -2 & -\kappa & -\kappa \\ -\kappa & \kappa & 0 & 0 & 2+\kappa & -\kappa & -2 & 0 \\ \kappa & -\kappa & 0 & -2 & -\kappa & 2+\kappa & 0 & 0 \\ 0 & 0 & -\kappa & -\kappa & -2 & 0 & 2+\kappa & \kappa \\ 0 & -2 & -\kappa & -\kappa & 0 & 0 & \kappa & 2+\kappa \end{pmatrix} \tag{10.142}$$

describes small-amplitude vibrations. (Please check this matrix!) The equations of motion are then

$$m\ddot{\xi}_i = -\frac{\partial V}{\partial \xi_i} = -k\sum_j v_{i,j}\xi_j \tag{10.143}$$

and normal modes satisfy eigenvalue equations of the form

$$\xi_\mu = \hat{\xi}_\mu \,\mathrm{Exp}\left[-i\omega_\mu t\right] \Longrightarrow \omega_0^2 v\hat{\xi}_\mu = \omega_\mu^2\hat{\xi}_\mu \tag{10.144}$$

where $\omega_0^2 = k/m$. Note that we use Greek indices to label the eigenvectors and Latin indices to label components. In this section we will demonstrate that one can deduce the eigenvalues using properties of the symmetry group for this system without solving a secular equation.

We begin by constructing representation matrices for typical members of each class. Consider first the parity operation

$$D[P] = \begin{pmatrix} 0 & 0 & 0 & p \\ 0 & 0 & p & 0 \\ 0 & p & 0 & 0 \\ p & 0 & 0 & 0 \end{pmatrix} \tag{10.145}$$

where

$$p = \begin{pmatrix} 1 & 0 \\ 0 & -1 \end{pmatrix} \tag{10.146}$$

inverts the y coordinates at each vertex and $D[P]$ reflects vertices across the horizontal midplane of the system. Similarly, the basic 90° rotation of vertices is represented by

$$D[R] = \begin{pmatrix} 0 & r & 0 & 0 \\ 0 & 0 & r & 0 \\ 0 & 0 & 0 & r \\ r & 0 & 0 & 0 \end{pmatrix} \tag{10.147}$$

where

$$r = \begin{pmatrix} 0 & 1 \\ -1 & 0 \end{pmatrix} \tag{10.148}$$

operates on the coordinates at each vertex. You should verify, by explicit calculation, that the potential energy is invariant with respect to every member of the symmetry group for the square, such that $v = D[g]vD[g^{-1}]$.

The transformation matrices for this representation are obviously reducible. To determine their irrep content, we evaluate the composite characters for each class

$$C_1 = \{I\} \Longrightarrow \chi_1 = 8 \tag{10.149}$$
$$C_2 = \{R, R^3\} \Longrightarrow \chi_2 = 0 \tag{10.150}$$
$$C_3 = \{R^2\} \Longrightarrow \chi_3 = 0 \tag{10.151}$$
$$C_4 = \{P, PR^2\} \Longrightarrow \chi_4 = 0 \tag{10.152}$$
$$C_5 = \{PR, PR^3\} \Longrightarrow \chi_5 = 0 \tag{10.153}$$

to obtain

$$\sum_\mu |a_\mu|^2 = \frac{1}{N_G} \sum_i N_i |\chi[C_i]|^2 = 8 \tag{10.154}$$

Thus, this representation turns out to be the regular representation

$$D = D^{(1)} \oplus D^{(2)} \oplus D^{(3)} \oplus D^{(4)} \oplus 2D^{(5)} \tag{10.155}$$

Upon application of the similarity transformation that diagonalizes v, we expect to find

$$v = \mathrm{Diag}[\{\lambda_1, \lambda_2, \lambda_3, \lambda_4, \lambda_{5,1}, \lambda_{5,1}, \lambda_{5,2}, \lambda_{5,2}\}] \tag{10.156}$$

where $\lambda_{5,1}$ and $\lambda_{5,2}$ are the eigenvalues for the two occurrences of $D^{(5)}$.

Recognizing that traces are invariant under similarity transformations, we expand

$$\mathrm{Tr}\, D[g]v = \lambda_1 \chi^{(1)}[g] + \lambda_2 \chi^{(2)}[g] + \lambda_3 \chi^{(3)}[g] + \lambda_4 \chi^{(4)}[g] + (\lambda_{5,1} + \lambda_{5,2}) \chi^{(5)}[g] \tag{10.157}$$

and use the character table to deduce five linear equations for the eigenvalues

$$\mathrm{Tr}\, D[I]v = 8 + 4\kappa = \lambda_1 + \lambda_2 + \lambda_3 + \lambda_4 + 2(\lambda_{5,1} + \lambda_{5,2}) \tag{10.158}$$
$$\mathrm{Tr}\, D[R]v = 0 = \lambda_1 + \lambda_2 - \lambda_3 - \lambda_4 \tag{10.159}$$
$$\mathrm{Tr}\, D[R^2]v = 4\kappa = \lambda_1 + \lambda_2 + \lambda_3 + \lambda_4 - 2(\lambda_{5,1} + \lambda_{5,2}) \tag{10.160}$$
$$\mathrm{Tr}\, D[P]v = 4 = \lambda_1 - \lambda_2 + \lambda_3 - \lambda_4 \tag{10.161}$$
$$\mathrm{Tr}\, D[PR]v = -4\kappa = \lambda_1 - \lambda_2 - \lambda_3 + \lambda_4 \tag{10.162}$$

for our particular representation. Combining the first and third equations gives

$$(\lambda_{5,1} + \lambda_{5,2}) = 2 \tag{10.163}$$

such that

$$\lambda_1 + \lambda_2 + \lambda_3 + \lambda_4 = 4 + 4\kappa \tag{10.164}$$
$$\lambda_1 + \lambda_2 - \lambda_3 - \lambda_4 = 0 \tag{10.165}$$
$$\lambda_1 - \lambda_2 + \lambda_3 - \lambda_4 = 4 \tag{10.166}$$
$$\lambda_1 - \lambda_2 - \lambda_3 + \lambda_4 = -4\kappa \tag{10.167}$$

From the first pair we deduce

$$\lambda_1 + \lambda_2 = \lambda_3 + \lambda_4 = 2 + 2\kappa \tag{10.168}$$

and from the second pair

$$\lambda_1 - \lambda_2 = 2 - 2\kappa \tag{10.169}$$

$$\lambda_3 - \lambda_4 = 2 + 2\kappa \tag{10.170}$$

Thus, we find

$$\lambda_1 = 2, \quad \lambda_2 = 2\kappa, \quad \lambda_3 = 2 + 2\kappa, \quad \lambda_4 = 0, \quad \left(\lambda_{5,1} + \lambda_{5,2}\right) = 2 \tag{10.171}$$

Finally, we require an argument that distinguishes between the two occurrences of $D^{(5)}$. On physical grounds we expect to find three modes with $\lambda = 0$ that correspond to either rigid rotation or to horizontal and vertical displacements of the center of mass. The null eigenvalue for $D^{(4)}$ can be identified with the rotational mode because this representation is one-dimensional, leaving one of the two-dimensional representations to describe translations of the center of mass. Therefore, we conclude that the set of eigenvalues for this system is

$$\lambda = \{2, 2\kappa, 2 + 2\kappa, 0, 2, 2, 0, 0\} \tag{10.172}$$

where we account for degeneracy and choose, arbitrarily, to place the rigid translations last.

10.3.7 Direct-Product Representations

Suppose that $D^{(\mu)}[R]$ and $D^{(\nu)}[R]$ are two representations of the same symmetry operator R that act upon different coordinates and that operators acting on different coordinates commute. Let $\{u_i, i = 1, m\}$ and $\{v_j, j = 1, n\}$ represent the basis vectors for these representations, such that the $m \times n$-dimensional set of basis vectors for the direct-product representation $D^{(\mu\otimes\nu)}$ is given by $\{u_i v_j\}$ with two indices. Thus,

$$Ru_i v_j = \sum_{k=1}^{m} \sum_{l=1}^{n} u_k D_{k,i}^{(\mu)}[R] v_l D_{l,j}^{(\mu)}[R] = \sum_{k=1}^{m} \sum_{l=1}^{n} u_k v_l D_{kl,ij}^{(\mu\otimes\nu)}[R] \tag{10.173}$$

where

$$D_{kl,ij}^{(\mu\otimes\nu)}[R] = D_{k,i}^{(\mu)}[R] D_{l,j}^{(\nu)}[R] \tag{10.174}$$

are the matrix elements of the direct-product representation. Similarly, the character for a direct product

$$\chi^{(\mu\otimes\nu)}[R] = \mathrm{Tr}\, D^{(\mu\otimes\nu)}[R] = \sum_{i,j} D_{ij,ij}^{(\mu\otimes\nu)}[R] = \sum_{i,j} D_{i,i}^{(\mu)}[R] D_{j,j}^{(\nu)}[R] \tag{10.175}$$

is simply the product

$$\chi^{(\mu\otimes\nu)}[R] = \chi^{(\mu)}[R]\chi^{(\nu)}[R] \tag{10.176}$$

of the characters of the two representations. Direct-product representations are usually reducible as a *Clebsch–Gordan expansion*

$$D^{(\mu\otimes\nu)} = \sum_{\gamma}(\mu, \nu \mid \gamma)D^{(\gamma)} \tag{10.177}$$

where the summation over irreducible representations $D^{(\gamma)}$ is interpreted as $\oplus$ and where the *reduction coefficients*

$$(\mu, \nu \mid \gamma) = \frac{1}{N_G}\sum_{g\in G}\chi^{(\mu)}[g]\chi^{(\nu)}[g]\chi^{(\gamma)}[g]^* \tag{10.178}$$

are nonnegative integers.

The unitary matrices $D^{(\mu\otimes\nu)}$ can be reduced to the block diagonal form

$$CD^{(\mu\otimes\nu)}C^{-1} = \begin{pmatrix} D^{(\gamma,1)} & 0 & 0 & 0 \\ 0 & \ddots & 0 & 0 \\ 0 & 0 & D^{(\gamma,p)} & 0 \\ 0 & 0 & 0 & \ddots \end{pmatrix} \tag{10.179}$$

where $p = (\mu, \nu \mid \gamma)$ represents, schematically, the number of occurrences of $D^{(\gamma)}$ in the Clebsch–Gordan expansion and where the pattern is replicated for all irreps γ. The unitary matrix C is the matrix of column vectors

$$w_{\gamma,\zeta,k} = \sum_{i,j}\left\langle\begin{matrix}\mu & \nu \\ i & j\end{matrix}\middle|\begin{matrix}\gamma,\zeta \\ k\end{matrix}\right\rangle u_i v_j \tag{10.180}$$

whose *Clebsch–Gordan coefficients* are arranged in a symbol based upon bra-ket notation. Note that specification of the irreps generally requires two labels because there may be multiple occurrences. Using the orthonormality of the bases and the unitarity of C, one obtains the orthogonality relations

$$\sum_{i,j}\left\langle\begin{matrix}\gamma',\zeta' \\ k'\end{matrix}\middle|\begin{matrix}\mu & \nu \\ i & j\end{matrix}\right\rangle\left\langle\begin{matrix}\mu & \nu \\ i & j\end{matrix}\middle|\begin{matrix}\gamma,\zeta \\ k\end{matrix}\right\rangle = \delta_{\gamma,\gamma'}\delta_{\zeta,\zeta'}\delta_{k,k'} \tag{10.181}$$

$$\sum_{\gamma,\zeta,k}\left\langle\begin{matrix}\mu & \nu \\ i & j\end{matrix}\middle|\begin{matrix}\gamma,\zeta \\ k\end{matrix}\right\rangle\left\langle\begin{matrix}\gamma,\zeta \\ k\end{matrix}\middle|\begin{matrix}\mu & \nu \\ i' & j'\end{matrix}\right\rangle = \delta_{i,i'}\delta_{j,j'} \tag{10.182}$$

such that

$$u_i v_j = \sum_{\gamma,\zeta,k}\left\langle\begin{matrix}\mu & \nu \\ i & j\end{matrix}\middle|\begin{matrix}\gamma,\zeta \\ k\end{matrix}\right\rangle^* w_{\gamma,\zeta,k} \tag{10.183}$$

where we define

$$\left\langle\begin{matrix}\gamma,\zeta \\ k\end{matrix}\middle|\begin{matrix}\mu & \nu \\ i & j\end{matrix}\right\rangle = \left\langle\begin{matrix}\mu & \nu \\ i & j\end{matrix}\middle|\begin{matrix}\gamma,\zeta \\ k\end{matrix}\right\rangle^* \tag{10.184}$$

for the inverse coupling according to the conventions for bra-ket notation.

A product representation is described as *simply reducible* when $(\mu, \nu \mid \gamma)$ is either 0 or 1 for all γ and when each g^{-1} is in the same class as g. The index ζ is then superfluous and one can determine the CG coefficients, up to a phase, relatively simply. If we invert the similarity transformation

$$D^{(\mu\otimes\nu)}_{i'j',ij}[g] = D^{(\mu)}_{i',i}[g] D^{(\nu)}_{j',j}[g] = \sum_{\gamma,k,k'} \left\langle \begin{matrix} \mu & \nu \\ i' & j' \end{matrix} \middle| \begin{matrix} \gamma \\ k' \end{matrix} \right\rangle D^{(\gamma)}_{k',k}[g] \left\langle \begin{matrix} \mu & \nu \\ i & j \end{matrix} \middle| \begin{matrix} \gamma \\ k \end{matrix} \right\rangle^* \tag{10.185}$$

and then multiply by $D^{(\gamma)}[g]^*$, sum over group elements, and apply the orthogonality theorem for irreps, we obtain

$$\sum_{g\in G} D^{(\mu)}_{i',i}[g] D^{(\nu)}_{j',j}[g] D^{(\gamma)}_{k',k}[g]^* = \frac{N_G}{n_\gamma} \left\langle \begin{matrix} \mu & \nu \\ i' & j' \end{matrix} \middle| \begin{matrix} \gamma \\ k' \end{matrix} \right\rangle \left\langle \begin{matrix} \mu & \nu \\ i & j \end{matrix} \middle| \begin{matrix} \gamma \\ k \end{matrix} \right\rangle^* \tag{10.186}$$

where N_G is the number of group elements and n_γ is the dimensionality of $D^{(\gamma)}$. Taking $i' = i$, $j' = j$, $k' = k$ will provide at least one nonzero CG coefficient and we are free to choose it to be positive. Varying i', j', k' will then provide enough information to determine the remaining CG coefficients. You are probably familiar with the Clebsch–Gordan (CG) coefficients for coupling of two angular momenta, but we are using similar notation in a more generic sense applicable to direct-product representations of any group; naturally the CG coefficients depend upon the particular group involved. For many groups it is possible to construct bases that make the Clebsch–Gordan coefficients real.

Example

Let us return once more to the problem of small-amplitude vibrations of an equilateral triangle. The six-dimensional representation we employed is actually a direct product of a three-dimensional representation of S_3 describing the permutations of the mass indices and a two-dimensional representation of a group containing rotations and reflections that is isomorphic to S_3. The two-dimensional representation is irreducible, but the three-dimensional representation is not. Suppose that we arrange the group elements as $\{I, R, R^2, P, PR, PR^2\}$. As worked out in a previous section, there are two one-dimensional irreps with characters $\chi^{(1)} = \{1, 1, 1, 1, 1, 1\}$ and $\chi^{(1')} = \{1, 1, 1, -1, -1, -1\}$ and a two-dimenional irrep with character $\chi^{(2)} = \{2, -1, -1, 0, 0, 0\}$. A faithful three-dimensional representation consists of simple permutations of the three mass indices, and a little thought, or explicit construction, will reveal that its character is $\chi^{(3)} = \{3, 0, 0, 1, 1, 1\}$. Hence, the character of the product representation is $\chi^{(6)} = \{6, 0, 0, 0, 0, 0\}$. The reduction coefficients are

$$(3, 2 \mid 1) = \frac{1}{6} \sum \{3, 0, 0, 1, 1, 1\} * \{2, -1, -1, 0, 0, 0\} * \{1, 1, 1, 1, 1, 1\} = 1 \tag{10.187}$$

$$(3, 2 \mid 1') = \frac{1}{6} \sum \{3, 0, 0, 1, 1, 1\} * \{2, -1, -1, 0, 0, 0\} * \{1, 1, 1, -1, -1, -1\} = 1 \tag{10.188}$$

$$(3, 2 \mid 2) = \frac{1}{6} \sum \{3, 0, 0, 1, 1, 1\} * \{2, -1, -1, 0, 0, 0\} * \{2, -1, -1, 0, 0, 0\} = 1 \tag{10.189}$$

where this notation indicates multiplication of corresponding elements of each list and summation over the list of products. Thus, the Clebsch–Gordan expansion takes the form $D^{(3\times2)} = D^{(1)} \oplus D^{(1')} \oplus D^{(2)}$.

10.3.8 Eigenfunctions

The Hamiltonian H commutes with every operator in its symmetry group $\mathcal{G}$, such that $[H, \mathcal{G}] = 0$. Let j denote the labels for an irreducible representation $D^{(j)}$ of $\mathcal{G}$; then the matrix representation of H will commute with each irreducible $D^{(j)}[R]$, such that $[H, D^j[R]] = 0$ for all $R \in \mathcal{G}$. According to Schur's lemma, the matrix representation of H within the group space of $D^{(j)}$ must be a multiple of the unit matrix. In other words, there must be a set of n linearly independent eigenfunctions $\{\psi_m^{(j)}, m = 1, n\}$ with the same eigenvalue ε_j, such that

$$H\psi_m^{(j)} = \varepsilon_j \psi_m^{(j)} \tag{10.190}$$

for $1 \le m \le n_j$. This set of eigenfunctions is described as a *multiplet* that is degenerate with respect to H with degeneracy n_j. If we start with one eigenfunction in a multiplet, the others may be generated by applying symmetry operations because, according to group closure, $R\psi_m^{(j)}$ produces the linear combination

$$R\psi_m^{(j)} = \sum_{m'=1}^{n_j} \psi_{m'}^{(j)} D_{m',m}^{(j)}[R] \tag{10.191}$$

where the coefficients $D_{m',m}^{(j)}[R]$ form a nonsingular n_j-dimensional matrix representation of operator R within the set of eigenfunctions belonging to eigenvalue ε_j. Thus, if H and R commute, sets of eigenfunctions that are degenerate with respect to H transform among themselves under R. If we now enlarge the basis to include all possible values of j, H is represented by a diagonal matrix with the set of ε_j along the diagonal, each appearing n_j times, while representations of R take the fully reduced block-diagonal form

$$D[R] = \begin{pmatrix} D^{(1)}[R] & 0 & 0 \\ 0 & D^{(2)}[R] & 0 \\ 0 & 0 & \ddots \end{pmatrix} \tag{10.192}$$

where the superscripts enumerate ε_j. Each $\psi_m^{(j)}$ is described as belonging to row m of irrep j.

The energy eigenvalues ε_j for each irrep j are usually distinct but occasionally one finds two or more distinct irreps that are degenerate in energy. These situations generally indicate the existence of an *accidental* or *dynamical* symmetry and not all eigenfunctions with the same energy are connected by the symmetry operators within $\mathcal{G}$. Knowing the degeneracy n_ε of an energy level is not always sufficient to determine the dimensionality n_j of the irreps at that energy; one must also verify that the corresponding $D[R]$ are irreducible. Testing for irreduciblity can be done using character tables for $\mathcal{G}$. On the other hand, accidental degeneracies are often broken by either perturbations or by subtle effects neglected by simpler models. A famous example is the Lamb shift in which the accidental degeneracy between the $2s$ and $2p_{1/2}$ states of the hydrogen atom is broken by the fluctuations of the electromagnetic field in vacuum that are predicted by quantum electrodynamics.

Assuming that our system does not possess accidental symmetries, the orthogonality theorem for irreducible representations provides important selection rules. The unitarity of R requires

$$\langle \psi_{m'}^{(j')} \left| \psi_m^{(j)} \right\rangle = \langle R\psi_{m'}^{(j')} \left| R\psi_m^{(j)} \right\rangle = \sum_{\lambda,\lambda'} \langle \psi_{\lambda'}^{(j')} \left| \psi_\lambda^{(j)} \right\rangle D_{\lambda',m'}^{(j')}[R]^* D_{\lambda,m}^{(j)}[R] \tag{10.193}$$

and averaging over the group gives

$$\left\langle \psi_{m'}^{(j')} \left| \psi_m^{(j)} \right\rangle = \sum_{\lambda,\lambda'} \left\langle \psi_{\lambda'}^{(j')} \left| \psi_\lambda^{(j)} \right\rangle \frac{1}{N_G} \sum_R D_{\lambda',m'}^{(j')}[R]^* D_{\lambda,m}^{(j)}[R] = \frac{\delta_{j,j'}\delta_{m,m'}}{n_j} \sum_{\lambda,\lambda'} \left\langle \psi_{\lambda'}^{(j')} \middle| \psi_\lambda^{(j)} \right\rangle \delta_{\lambda,\lambda'} \tag{10.194}$$

where n_j is the dimensionality of irrep j. Thus, eigenfunctions belonging to different irreps or to different rows of the same irrep are orthogonal. Next, suppose that V is invariant with respect to R, such that $[V, \mathcal{G}] = 0$, for any $R \in \mathcal{G}$, where $\mathcal{G}$ is the symmetry group of H. Then

$$\left\langle \psi_{m'}^{(j')} \left| V \right| \psi_m^{(j)} \right\rangle = \left\langle \psi_{m'}^{(j')} \left| R^\dagger V R \right| \psi_m^{(j)} \right\rangle = \sum_{\lambda,\lambda'} \left\langle \psi_{\lambda'}^{(j')} \left| V \right| \psi_\lambda^{(j)} \right\rangle D_{\lambda',m'}^{(j')}[R]^* D_{\lambda,m}^{(j)}[R] \tag{10.195}$$

can again be averaged over R to obtain

$$\left\langle \psi_{m'}^{(j')} \left| V \right| \psi_m^{(j)} \right\rangle = \frac{\delta_{j,j'}\delta_{m,m'}}{n_j} \sum_\lambda \left\langle \psi_\lambda^{(j')} \left| V \right| \psi_\lambda^{(j)} \right\rangle \tag{10.196}$$

Observe that these matrix elements vanish unless $j = j'$, $m = m'$ and are independent of m for specified j. Therefore, we can write

$$[V, \mathcal{G}] = 0 \Longrightarrow \left\langle \psi_{m'}^{(j')} \left| V \right| \psi_m^{(j)} \right\rangle = V_j \delta_{j,j'} \delta_{m,m'} \tag{10.197}$$

where V_j depends only upon the irrep to which these states belong. This is also a consequence of Schur's lemma, but the foregoing derivation provides some practice using the orthogonality theorem.

10.3.9 Wigner–Eckart Theorem

Although most operators of interest will not be invariant with respect to the symmetry group of a Hamiltonian, it is often possible to express them in terms of operators that transform among themselves according to an irreducible representation of that group. Consider an operator $V_m^{(j)}$ that transforms with irrep j according to

$$RV_M^{(J)}R^\dagger = \sum_{M'} V_{M'}^{(J)} D_{M',M}^{(J)}[R] \tag{10.198}$$

for any $R \in \mathcal{G}$. We assume that $D^{(J\otimes j)}$ is simply reducible. Using the orthogonality theorem to average matrix elements of the form

$$\begin{aligned}\left\langle \psi_{m'}^{(j')} \middle| V_M^{(J)} \middle| \psi_m^{(j)} \right\rangle &= \left\langle \psi_{m'}^{(j')} \middle| R^\dagger R V_M^{(J)} R^\dagger R \middle| \psi_m^{(j)} \right\rangle \\ &= \sum_{\lambda,\lambda',M'} \left\langle \psi_{\lambda'}^{(j')} \middle| V_{M'}^{(J)} \middle| \psi_\lambda^{(j)} \right\rangle D_{\lambda',m'}^{(j')}[R]^* D_{M',M}^{(J)}[R] D_{\lambda,m}^{(j)}[R]\end{aligned} \tag{10.199}$$

over all $R \in \mathcal{G}$ then gives

$$\begin{aligned}\left\langle \psi_{m'}^{(j')} \middle| V_M^{(J)} \middle| \psi_m^{(j)} \right\rangle &= \frac{1}{N_\mathcal{G}} \sum_{\lambda,\lambda',M'} \left\langle \psi_{\lambda'}^{(j')} \middle| V_{M'}^{(J)} \middle| \psi_\lambda^{(j)} \right\rangle D_{\lambda',m'}^{(j')}[R]^* D_{M',M}^{(J)}[R] D_{\lambda,m}^{(j)}[R] \\ &= \left\langle \begin{matrix} j & J \\ m & M \end{matrix} \middle| \begin{matrix} j' \\ m' \end{matrix} \right\rangle^* \frac{1}{n_{j'}} \sum_{\lambda,\lambda',M'} \left\langle \psi_{\lambda'}^{(j')} \middle| V_{M'}^{(J)} \middle| \psi_\lambda^{(j)} \right\rangle \left\langle \begin{matrix} j & J \\ \lambda & M' \end{matrix} \middle| \begin{matrix} j' \\ \lambda' \end{matrix} \right\rangle\end{aligned} \tag{10.200}$$

where $n_{j'}$ is the dimensionality of irrep j'. Notice that the summation on the right-hand side depends upon $\{j, J, j'\}$ but is independent of $\{m, M, m'\}$ – the dependencies upon row indices for each irrep are contained entirely within the Clebsch–Gordan coefficient, which is hopefully known for $\mathcal{G}$. Therefore, we obtain the *Wigner–Eckart theorem*

$$\left\langle \psi_{m'}^{(j')} \middle| V_M^{(J)} \middle| \psi_m^{(j)} \right\rangle = \left\langle \begin{matrix} j' \\ m' \end{matrix} \middle| \begin{matrix} j & J \\ m & M \end{matrix} \right\rangle \left\langle j' \middle\| V^{(J)} \middle\| j \right\rangle \tag{10.201}$$

where the *reduced matrix element* $\langle j' \| V^{(J)} \| j\rangle$ is independent of row and can be evaluated for the most convenient selection of $\{m, M, m'\}$. The reduced matrix element can also be obtained using

$$\left\langle j' \middle\| V^{(J)} \middle\| j \right\rangle = \frac{1}{n_{j'}} \sum_{\lambda,\lambda',M} \left\langle \psi_{\lambda'}^{(j')} \middle| V_M^{(J)} \middle| \psi_\lambda^{(j)} \right\rangle \left\langle \begin{matrix} j & J \\ \lambda & M \end{matrix} \middle| \begin{matrix} j' \\ \lambda' \end{matrix} \right\rangle \tag{10.202}$$

but that method offers no obvious economy of effort. Note that additional phase or normalization factors are sometimes included in the definition of the reduced matrix elements.

The Wigner–Eckart theorem is usually presented in textbooks on quantum mechanics in connection with rotational symmetry and angular momentum eigenstates, but the group theoretical derivation is more general and applies to any group with simply reducible direct-product representations. Nor is it limited to quantum mechanics. Generalizations can also be made for direct-product representations that are not simply reducible.

10.4 Continuous Groups

10.4.1 Definitions

Suppose that R is a finite group containing n elements $\{R_a, a = 1, n\}$. The group multiplication table $R_a R_b = R_c$ can be described as a discrete function $c = \phi[a, b]$ that produces a unique output index c for each pair of input indices a, b. Functions of group elements can be defined on n points by $f[a] = f[R_a]$. For example, the representation $D^{(\mu)}_{i,j}[a]$ is a group function while the character $\chi^{(\mu)}[a]$ is a class function. The set of points is described as the group manifold and is discrete for a finite group.

The elements of a continuous group are labeled by a finite number of continuous parameters instead of a discrete index. We assume, for simplicity, that all parameters are *essential* in the sense that there exists no smaller set of parameters, defined as functions of the original parameters, capable of uniquely identifying every group element. If n parameters are essential, the group is described as an n-parameter continuous group. Let $a = \{a_i, i = 1, n\}$ represent the set of parameters and $R[a]$ represent elements of group R. The continuous parameters a_i may have either finite or infinite ranges depending upon the nature of the group. If the parameters have finite ranges, the group manifold is described as *compact*. We will assume that the group manifold is simply connected but generalizations are possible. Group multiplication is represented by the law of composition

$$R[a]R[b] = R[c] \Longrightarrow c_i = \phi_i[a, b] \tag{10.203}$$

where each c_i falls within the allowed range for parameter i and where each $\phi_i[a, b]$ is a continuous function of $2n$ variables. This requirement is analogous to the closure property of finite groups. Similarly, there must exist an identity element $a^{(0)}$ such that

$$R\left[a^{(0)}\right]R[b] = R[b]R\left[a^{(0)}\right] = R[b] \tag{10.204}$$

and every element a must have an inverse $\bar{a}$ such that

$$R\left[\bar{a}\right]R[a] = R[a]R\left[\bar{a}\right] = R\left[a^{(0)}\right] \tag{10.205}$$

It is usually possible to define the group such that all parameters of the identity element vanish; in other words $\{a_i^{(0)} = 0; i = 1, n\}$. Henceforth we will assume, unless stated otherwise, that the identity element is at the origin of the parameter space. Finally, the associative property

$$R[a](R[b]R[c]) = (R[a]R[b])R[c] \Longrightarrow \phi\big[a, \phi[b, c]\big] = \phi\big[\phi[a, b], c\big] \tag{10.206}$$

is usually the most restrictive.

We will concentrate on *Lie groups* which perform transformations

$$x_i' = \xi_i\left[\left\{x_i, i = 1, n\right\}; \left\{a_k, k = 1, m\right\}\right] \Longleftrightarrow x' = \xi[x; a] \tag{10.207}$$

upon an n-dimensional vector space that are described by n functions of m essential parameters that are analytic in both x and a. For example, consider the two-parameter group of nonsingular linear coordinate transformations on a line according to

$$x' = R[a]x \Longrightarrow x' = a_0 + \left(1 + a_1\right)x \tag{10.208}$$

where $-\infty < a_i < \infty$ and $a_1 \neq -1$. The identity element is $\{a_0, a_1\} = \{0, 0\}$ by construction. Group multiplication is closed according to

$$\begin{aligned} R[a]R[b]x &= R[a]\left(b_0 + \left(1 + b_1\right)x\right) \\ &= a_0 + \left(1 + a_1\right)\left(b_0 + \left(1 + b_1\right)x\right) \\ &= \left(a_0 + b_0\left(1 + a_1\right)\right) + (1 + a_1)(1 + b_1)x \end{aligned} \tag{10.209}$$

such that

$$\phi_0[a, b] = a_0 + b_0\left(1 + a_1\right) \tag{10.210}$$

$$\phi_1[a, b] = a_1 + b_1\left(1 + a_1\right) \tag{10.211}$$

are continuous functions within the specified (infinite) ranges. Each element $a = \{a_0, a_1\}$ has an inverse element $\bar{a} = \{\bar{a}_0, \bar{a}_1\}$ where

$$\bar{a}_0 = -\frac{a_0}{1 + a_1} \tag{10.212}$$

$$\bar{a}_1 = -\frac{a_1}{1 + a_1} \tag{10.213}$$

are analytic functions because $a_1 \neq -1$; furthermore, $\bar{a}_1 \neq -1$ is in the proper range. The associative property requires

$$\phi\left[\phi[a, b], c\right] == \phi\left[a, \phi[b, c]\right] \tag{10.214}$$

Although it might seem obvious that linear transformations are associative, perhaps it is instructive to verify this property explicitly. Expansion of the left-hand side gives

$$\begin{aligned} &\phi_0\left[\left\{a_0 + b_0\left(1 + a_1\right), a_1 + b_1\left(1 + a_1\right)\right\}, \left\{c_0, c_1\right\}\right] \\ &\quad = a_0 + b_0\left(1 + a_1\right) + c_0\left(1 + a_1 + b_1\left(1 + a_1\right)\right) \\ &\quad = a_0 + b_0 + c_0 + a_1 b_0 + a_1 c_0 + b_1 c_0 + a_1 b_1 c_0 \end{aligned} \tag{10.215}$$

$$\begin{aligned} &\phi_1\left[\left\{a_0 + b_0\left(1 + a_1\right), a_1 + b_1\left(1 + a_1\right)\right\}, \left\{c_0, c_1\right\}\right] \\ &\quad = a_1 + b_1\left(1 + a_1\right) + c_1\left(1 + a_1 + b_1\left(1 + a_1\right)\right) \\ &\quad = a_1 + b_1 + c_1 + a_1 b_1 + a_1 c_1 + b_1 c_1 + a_1 b_1 c_1 \end{aligned} \tag{10.216}$$

while expansion of the right-hand side gives

$$\begin{aligned} &\phi_0\left[\left\{a_0, a_1\right\}, \left\{b_0 + c_0\left(1 + b_1\right), b_1 + c_1\left(1 + b_1\right)\right\}\right] \\ &\quad = a_0 + \left(b_0 + c_0\left(1 + b_1\right)\right)\left(1 + a_1\right) \\ &\quad = a_0 + b_0 + c_0 + a_1 b_0 + a_1 c_0 + b_1 c_0 + a_1 b_1 c_0 \end{aligned} \tag{10.217}$$

$$\begin{aligned} &\phi_1\left[\left\{a_0, a_1\right\}, \left\{b_0 + c_0\left(1 + b_1\right), b_1 + c_1\left(1 + b_1\right)\right\}\right] \\ &\quad = a_1 + \left(b_1 + c_1\left(1 + b_1\right)\right)\left(1 + a_1\right) \\ &\quad = a_1 + b_1 + c_1 + a_1 b_1 + a_1 c_1 + b_1 c_1 + a_1 b_1 c_1 \end{aligned} \tag{10.218}$$

Thus, these ϕ_i do satisfy the associative property and R satisfies all the requirements of a continuous group. Notice that this group is nonabelian.

The requirement that every transformation be invertible means that one can always solve for x_i in terms of $\{x'_j\}$, which requires that the Jacobian

$$\mathcal{J} = \begin{vmatrix} \frac{\partial \xi_1}{\partial x_1} & \cdots & \frac{\partial \xi_1}{\partial x_n} \\ \vdots & \vdots & \vdots \\ \frac{\partial \xi_n}{\partial x_1} & \cdots & \frac{\partial \xi_n}{\partial x_n} \end{vmatrix} \neq 0 \tag{10.219}$$

be nonsingular. The completeness condition requires successive transformations

$$x' = \xi[x; a] \tag{10.220}$$

$$x'' = \xi\left[x'; b\right] \tag{10.221}$$

to correspond to

$$x'' = \xi[x; c] \tag{10.222}$$

where c is an allowed parameter set, calculable in terms of analytic functions

$$c_k = \eta_k[a; b] \tag{10.223}$$

of the two constituent sets of transformation parameters.

10.4.2 Transformation of Functions

Suppose that new coordinates x' are obtained by applying group element R to the old coordinates x. This transformation of variables is represented schematically by

$$x' = Rx \tag{10.224}$$

even if R is not simply a matrix. Thus, R is described as an *operator* that acts on the coordinate variables. We wish to construct an operator $\mathcal{R}$ that acts on a function $f[x]$ to produce a new function $\tilde{f}[x']$ that has the same values for the transformed coordinates, such that

$$x' = Rx \Longrightarrow \tilde{f}\left[x'\right] = \mathcal{R}f[x] = f\left[R^{-1}x\right] \tag{10.225}$$

It is important to understand that a forward transformation of a function is produced by an inverse transformation of its arguments. Perhaps the simplest example is rotation within a plane. Think of the function $f[x]$, where $x = \{x_1, x_2\}$, as a surface. Positive rotation of that surface relative to fixed coordinate axes is equivalent to negative rotation of those axes relative to a fixed surface. For example, the surface $f[x, y]$, shown on the left side of Fig. 10.3, has a positive ridge in quadrant 4 with $x > 0$, $y < 0$. Positive rotation of the surface by 90° relative to fixed coordinate axes yields the surface $\tilde{f}[x, y]$ with its positive lobe in quadrant 1. Alternatively, rotation of the coordinate axes relative to a fixed surface,

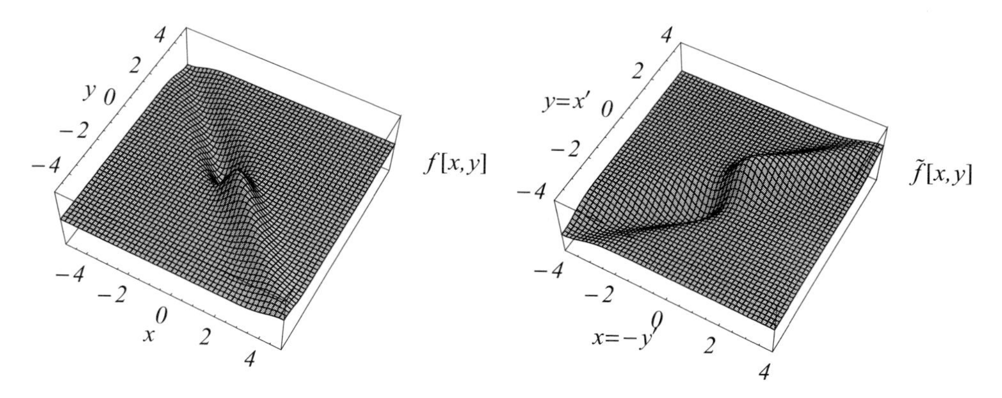

Figure 10.3. Positive rotation of a surface is equivalent to negative rotation of its coordinates.

such that $x \to -y$, $y \to x$, relabels the quadrant 4 as quadrant 1 and produces the same transformation. We describe $\mathcal{R}$ as an *active transformation* of the function and R^{-1} as a *passive transformation* of the coordinate system.

In the next section we will develop a general technique for constructing $\mathcal{R}$ given R, but now we preview this method by analyzing infinitesimal rotations. Suppose that $x = \{x_1, x_2\}$ represents the original coordinates and $x' = R[\varepsilon]x = \{x_1', x_2'\}$ represents coordinates after an infinitesimal rotation

$$R[\varepsilon] = \begin{pmatrix} 1 & -\varepsilon \\ \varepsilon & 1 \end{pmatrix} \Longrightarrow R^{-1} = \begin{pmatrix} 1 & \varepsilon \\ -\varepsilon & 1 \end{pmatrix} \tag{10.226}$$

A first-order Taylor expansion

$$f\left[R^{-1}x\right] \approx f[x] - \frac{\partial f}{\partial x_1}\left(x_1' - x_1\right) - \frac{\partial f}{\partial x_2}\left(x_2' - x_2\right) = f[x] + \varepsilon\left(x_2 \frac{\partial f}{\partial x_1} - x_1 \frac{\partial f}{\partial x_2}\right) \tag{10.227}$$

relates functions at new and old coordinates where negative signs multiply the derivatives because the argument $R^{-1}x$ involves the inverse of R. Therefore, the operator that transforms from $f[x]$ to $\tilde{f}[x]$ is

$$\mathcal{R} = 1 + \varepsilon\left(x_2 \frac{\partial}{\partial x_1} - x_1 \frac{\partial}{\partial x_2}\right) \tag{10.228}$$

for infinitesimal ε. Notice that if we employ polar coordinates with $x = \{r\,\mathrm{Cos}[\theta], r\,\mathrm{Sin}[\theta]\}$, infinitesimal rotation of a function is performed by the differential operator

$$\mathcal{R} = 1 - \varepsilon \frac{\partial}{\partial \theta} \tag{10.229}$$

10.4.3 Generators

The continuity of Lie groups permits a finite transformation to be performed in a sequence of smaller steps. Before confronting the general technique, consider the specific example of coordinate translations. Let $T[a]$ be a translation whose effect upon coordinates is given by

$$T[a]x = x + a \tag{10.230}$$

Translations form an abelian group

$$T[b]T[a]x = T[b](x+a) = x + a + b \Longrightarrow T[b]T[a] = T[a]T[b] = T[a+b] \tag{10.231}$$

with identity element $T[0]$ and inverses $T^{-1}[a] = T[-a]$. The associative property is also satisfied. We seek an operator $\mathcal{T}[a]$ that performs a corresponding translation of an analytic function

$$\mathcal{T}[a]f[x] = f\left[T^{-1}[a]x\right] = f[x-a] \tag{10.232}$$

Any finite translation can be performed in a sequence of infinitesimal steps

$$\mathcal{T}[a] = \lim_{n\to\infty}\left(\mathcal{T}\left[\frac{a}{n}\right]\right)^n \tag{10.233}$$

where an infinitesimal translation is evaluated using the definition of derivative as

$$\mathcal{T}[\varepsilon]f[x] = f[x-\varepsilon] = f[x] - \frac{df[x]}{dx}\varepsilon \Longrightarrow \mathcal{T}[\varepsilon] = \left(1 - \varepsilon\frac{d}{dx}\right)f[x] \tag{10.234}$$

Thus, a finite translation takes the form

$$\mathcal{T}[a] = \lim_{n\to\infty}\left(1 - \frac{a}{n}\frac{d}{dx}\right)^n = \mathrm{Exp}\left[-a\frac{d}{dx}\right] \tag{10.235}$$

where a function $g[A]$ of an operator A is evaluated by substituting that operator into the Maclaurin series for $g[x]$. Although this operator is really little more than a fancy representation for the Taylor expansion

$$f[x-a] = \sum_{n=0}^{\infty}\frac{1}{n!}\left(-a\frac{d}{dx}\right)^n f[x] = \mathrm{Exp}\left[-a\frac{d}{dx}\right]f[x] \tag{10.236}$$

it serves as a prototype for the definition and behavior of generators of Lie groups.

We identify $p_x = -i\partial/\partial x$ as the *generator* of infinitesimal translations along the x-axis, such that arbitrary group elements can be expressed in the form

$$\mathcal{T}_x[a] = \mathrm{Exp}\left[-iap_x\right] \tag{10.237}$$

Translations with respect to the x-axis are obviously a subgroup of translations within three-dimensional space and we can generalize to three dimensions

$$\mathcal{T}\left[\vec{a}\right] = \mathrm{Exp}\left[-\vec{a}\cdot\vec{\nabla}\right] \Longrightarrow \mathcal{T}\left[\vec{a}\right]f\left[\vec{r}\right] = f\left[\vec{r}-\vec{a}\right] \tag{10.238}$$

and obtain an operator representation of the three-dimensional Taylor expansion. In general, an n-parameter Lie group $\mathcal{G}$ is described by n generators

$$G_i = -i\left(\frac{\partial \mathcal{T}}{\partial a_i}\right)_{a=0} \tag{10.239}$$

of infinitesimal transformations. Notice that

$$G^{-1}[a] = G[-a] \tag{10.240}$$

applies to any infinitesimal generator.

More generally, if x is an m-dimensional coordinate vector and a is an n-dimensional group vector whose identity element is $a = 0$, the Taylor expansion

$$T[a]x = \xi[x;a] \Longrightarrow (T[-a]x)_j = \xi_j[x;-a] \approx x_j - \sum_{i=1}^{n} a_i \left(\frac{\partial \xi_j}{\partial a_i}\right)_{a=0} \tag{10.241}$$

can be used to write

$$f\left[T[-a]x\right] \approx f[x] - \sum_{j=1}^{m}\sum_{i=1}^{n} a_i \frac{\partial f}{\partial x_j}\left(\frac{\partial \xi_j}{\partial a_i}\right)_{a=0} \tag{10.242}$$

or

$$\mathcal{T}[a]f[x] \approx \left(1 - i\sum_{i=1}^{n} a_i G_i\right) f[x] \tag{10.243}$$

where the infinitesimal generators

$$G_i = -i\sum_{j=1}^{m}\left(\frac{\partial \xi_j}{\partial a_i}\right)_{a=0}\frac{\partial}{\partial x_j} \tag{10.244}$$

are linear differential operators. Therefore, the operator

$$\mathcal{T}[a] = \text{Exp}[-ia\cdot G] = \text{Exp}\left[-i\sum_i a_i G_i\right] \tag{10.245}$$

performs finite transformations from $f[x]$ to $\mathcal{T}f[x] = f[T^{-1}x]$. The inclusion of the factor of i in the definition of infinitesimal generators ensures that those operators are hermitian when the transformation is unitary. Thus,

$$\mathcal{T}^\dagger[a] = \mathcal{T}^{-1}[a] = \mathcal{T}[-a] \Longrightarrow \text{Exp}\left[ia\cdot G^\dagger\right] = \text{Exp}[ia\cdot G] \Longrightarrow G_i^\dagger = G_i \tag{10.246}$$

for arbitrary real a_i. Although many mathematicians do not include this phase, applications of Lie groups to quantum mechanics benefit from this convention because physical observables are represented by hermitian operators. We will adopt this convention here, but the reader must be aware that it is not universal. Finally, notice that the generators satisfy a product rule

$$G_i f[x] g[x] = g[x] G_i f[x] + f[x] G_i f[x] \tag{10.247}$$

where the differential operator G_i acts on everything to its right within a term. Care must be taken in evaluation of $\mathcal{T}$ because the generators do not necessarily commute with each other.

Next consider an arbitrary Lie group T with parameters $\{a_i\}$ and infinitesimal generators $\{G_i\}$ and assume that a general element can be expressed in a form

$$\mathcal{T}[a] = \text{Exp}[-ia \cdot G] = \text{Exp}\left[-i \sum_i a_i G_i\right] \tag{10.248}$$

that is sometimes called a *Lie series* in analogy with a Taylor series. Closure of the group imposes important conditions upon the generators. Suppose that

$$\mathcal{T}_i = \text{Exp}\left[-i\varepsilon_i G_i\right] \approx 1 - i\varepsilon_i G_i - \tfrac{1}{2}\varepsilon_i^2 G_i^2 \tag{10.249}$$

$$\mathcal{T}_j = \text{Exp}\left[-i\varepsilon_j G_j\right] \approx 1 - i\varepsilon_j G_j - \tfrac{1}{2}\varepsilon_j^2 G_j^2 \tag{10.250}$$

represent two distinct infinitesimal transformations with inverses

$$\mathcal{T}_i^{-1} = \text{Exp}\left[i\varepsilon_i G_i\right] \approx 1 + i\varepsilon_i G_i - \tfrac{1}{2}\varepsilon_i^2 G_i^2 \tag{10.251}$$

$$\mathcal{T}_j^{-1} = \text{Exp}\left[i\varepsilon_j G_j\right] \approx 1 + i\varepsilon_j G_j - \tfrac{1}{2}\varepsilon_j^2 G_j^2 \tag{10.252}$$

such that

$$\begin{aligned}
\mathcal{T}_i^{-1}\mathcal{T}_j^{-1}\mathcal{T}_i\mathcal{T}_j &\approx \left(1 + i\left(\varepsilon_i G_i + \varepsilon_j G_j\right) - \varepsilon_i\varepsilon_j G_i G_j - \tfrac{1}{2}\varepsilon_i^2 G_i^2 - \tfrac{1}{2}\varepsilon_j^2 G_j^2\right) \\
&\quad \times \left(1 - i\left(\varepsilon_i G_i + \varepsilon_j G_j\right) - \varepsilon_i\varepsilon_j G_i G_j - \tfrac{1}{2}\varepsilon_i^2 G_i^2 - \tfrac{1}{2}\varepsilon_j^2 G_j^2\right) \\
&= 1 - i\left(\varepsilon_i G_i + \varepsilon_j G_j\right) - \varepsilon_i\varepsilon_j G_i G_j - \tfrac{1}{2}\varepsilon_i^2 G_i^2 - \tfrac{1}{2}\varepsilon_j^2 G_j^2 + i\left(\varepsilon_i G_i + \varepsilon_j G_j\right) \\
&\quad + \varepsilon_i^2 G_i^2 + \varepsilon_j^2 G_j^2 + \varepsilon_i\varepsilon_j\left(G_i G_j + G_j G_i\right) - \varepsilon_i\varepsilon_j G_i G_j - \tfrac{1}{2}\varepsilon_i^2 G_i^2 - \tfrac{1}{2}\varepsilon_j^2 G_j^2 \\
&= 1 - \varepsilon_i\varepsilon_j\left(G_i G_j - G_j G_i\right)
\end{aligned} \tag{10.253}$$

to second order in ε_i and ε_j. If this product is to represent an element of T, the *commutator* $[G_i, G_j] \equiv G_i G_j - G_j G_i$ must be a linear combination of infinitesimal generators of the form

$$\left[G_i, G_j\right] = i \sum_k c_{i,j}^k G_k \tag{10.254}$$

where the *structure constants* $c_{i,j}^k$ are real numbers which are fundamental properties of the generators that are independent of the choice of representation and are invariant with respect to similarity transformations. The structure constants must be antisymmetric in their lower indices, such that

$$c^k_{j,i} = -c^k_{i,j} \tag{10.255}$$

because the commutator is antisymmetric. Although it should be obvious that the structure constants are determined by the generators, Sophus Lie proved the more powerful converse that groups of this kind are uniquely determined by specification of their structure constants.

Another important relationship among the structure constants uses the double commutator

$$\left[G_i, \left[G_j, G_k\right]\right] = \left[G_i, i\sum_l c^l_{j,k} G_l\right] = \sum_{l,m} c^l_{j,k} c^m_{l,i} G_m \tag{10.256}$$

and the *Jacobi identity*

$$\left[G_i, \left[G_j, G_k\right]\right] + \left[G_j, \left[G_k, G_i\right]\right] + \left[G_k, \left[G_i, G_j\right]\right] = 0 \tag{10.257}$$

to find

$$\sum_{l,m} c^l_{j,k} c^m_{l,i} G_m + \sum_{l,m} c^l_{k,i} c^m_{l,j} G_m + \sum_{l,m} c^l_{i,j} c^m_{l,k} G_m = 0 \tag{10.258}$$

or

$$\sum_m \sum_l \left(c^l_{j,k} c^m_{l,i} + c^l_{k,i} c^m_{l,j} + c^l_{i,j} c^m_{l,k}\right) G_m = 0 \tag{10.259}$$

Thus, assuming that the G_m are linearly independent, the structure constants must satisfy a Jacobi identity of the form

$$\sum_l \left(c^l_{j,k} c^m_{l,i} + c^l_{k,i} c^m_{l,j} + c^l_{i,j} c^m_{l,k}\right) = 0 \tag{10.260}$$

Lie proved that antisymmetry of lower indices and compliance with the Jacobi identity are necessary and sufficient conditions for generation of a continuous group from local structure constants.

10.4.4 Example: Linear coordinate transformations in one dimension

To illustrate these relations, we return to the two-parameter group of linear coordinate transformations

$$T[x]x = a_0 + \left(1 + a_1\right)x \tag{10.261}$$

such that

$$\mathcal{T}[a]f[x] = f\left[T^{-1}[a]x\right] = f\left[\bar{a}_0 + \left(1 + \bar{a}_1\right)x\right] \approx f[x] - a_0\frac{\partial f}{\partial x} - a_1 x\frac{\partial f}{\partial x} + \cdots \tag{10.262}$$

where

$$\bar{a}_0 = -\frac{a_0}{1+a_1} \approx -a_0 \tag{10.263}$$

$$\bar{a}_1 = -\frac{a_1}{1+a_1} \approx -a_1 \tag{10.264}$$

for infinitesimal transformations. Thus, we identify the generators as linear differential operators

$$G_0 = -i\frac{\partial}{\partial x} \tag{10.265}$$

$$G_1 = -ix\frac{\partial}{\partial x} \tag{10.266}$$

The commutator is evaluated using

$$\left(G_0G_1 - G_1G_0\right) f[x] = -\frac{\partial}{\partial x}\left(x\frac{\partial f}{\partial x}\right)+\left(x\frac{\partial}{\partial x}\right)\frac{\partial f}{\partial x} = -x\frac{\partial^2 f}{\partial x^2}-\frac{\partial f}{\partial x}+x\frac{\partial^2 f}{\partial x^2} = -\frac{\partial f}{\partial x} \tag{10.267}$$

to obtain

$$\left[G_0, G_1\right] = -iG_0 \Longrightarrow c^0_{0,1} = -1, \quad c^1_{0,1} = 0 \tag{10.268}$$

The structure constants are indeed real numbers and are antisymmetric wrt lower indices. The only set of indices for which the Jacobi identity does not vanish immediately is

$$\sum_l \left(c^l_{0,1}c^0_{l,k} + c^l_{1,k}c^0_{l,0} + c^l_{k,0}c^0_{l,1}\right) = c^0_{0,1}c^0_{0,k} + c^0_{k,0}c^0_{0,1} = 0 \tag{10.269}$$

such that the identity is indeed satisfied. Therefore, the general transformation operator for functions takes the form

$$\mathcal{T}[a] = \text{Exp}\left[-\left(a_0 + a_1 x\right)\frac{\partial}{\partial x}\right] \tag{10.270}$$

10.4.5 Example: SO(2)

A finite rotation within a plane is performed by the transformation matrix

$$R[\theta] = \begin{pmatrix} \text{Cos}[\theta] & -\text{Sin}[\theta] \\ \text{Sin}[\theta] & \text{Cos}[\theta] \end{pmatrix} \tag{10.271}$$

and one can (and should!) verify that $R[\theta]$ satisfies all of the requirements of a group and that the group multiplication law

$$R\left[\theta_2\right]R\left[\theta_1\right] = R\left[\theta_1 + \theta_2\right] \tag{10.272}$$

is abelian. The inverse matrix

$$R^{-1}[\theta] = R[-\theta] = R^T[\theta] \tag{10.273}$$

is the same as the transpose (indicated by superscript T); hence, rotation matrices are orthogonal. Also note that $\text{Det}[R] = 1$. Thus, rotations in the plane are isomorphic with the group of all two-dimensional special orthogonal matrices, designated SO(2); the term "special" refers to the requirement that the determinant be +1, which excludes reflections.

An infinitesimal rotation is described by

$$R[\varepsilon] \approx \begin{pmatrix} 1 & -\varepsilon \\ \varepsilon & 1 \end{pmatrix} = \mathbb{1} - i\sigma_2\varepsilon \tag{10.274}$$

where $\mathbb{1}$ is the unit matrix, here two-dimensional, and where we identify the generator for coordinate transformations as σ_2 where

$$\sigma_2 = \begin{pmatrix} 0 & -i \\ i & 0 \end{pmatrix} \tag{10.275}$$

happens to be one of the Pauli matrices. A finite rotation is then given by

$$R[\theta] = \lim_{n\to\infty} R\left[\frac{\theta}{n}\right]^n = \lim_{n\to\infty}\left(\mathbb{1} - \frac{i\theta}{n}\sigma_2\right)^n = \mathrm{Exp}\left[-i\theta\sigma_2\right] \tag{10.276}$$

Perhaps it is worthwhile to demonstrate that the original matrix is recovered from the power series. Expanding

$$\mathrm{Exp}\left[-i\theta\sigma_2\right] = \sum_{n=0}^{\infty}\frac{\left(-i\theta\sigma_2\right)^n}{n!} = \sum_{n=0}^{\infty}(-)^n\frac{\left(\theta\sigma_2\right)^{2n}}{(2n)!} - i\sum_{n=0}^{\infty}(-)^n\frac{\left(\theta\sigma_2\right)^{2n+1}}{(2n+1)!} \tag{10.277}$$

and using

$$\sigma_2^2 = 1 \Longrightarrow \sigma_2^{2n} = \mathbb{1}, \quad \sigma_2^{2n+1} = \sigma_2 \tag{10.278}$$

we obtain

$$\mathrm{Exp}\left[-i\theta\sigma_2\right] = \mathrm{Cos}[\theta]\mathbb{1} - i\,\mathrm{Sin}[\theta]\sigma_2 = R[\theta] \tag{10.279}$$

as claimed. This result is a generalization of the Euler identity to two dimensions.

Next consider the transformation operator for functions of these coordinates. Let $x = \{x_1, x_2\}$ represent the original coordinates, $x' = R^{-1}[\varepsilon]x = \{x_1', x_2'\}$ represent coordinates after a passive infinitesimal rotation of the axes, and expand

$$f\left[x'\right] \approx f[x] + \frac{\partial f}{\partial x_1}\left(x_1' - x_1\right) + \frac{\partial f}{\partial x_2}\left(x_2' - x_2\right) = f[x] + \varepsilon\left(x_2\frac{\partial f}{\partial x_1} - x_1\frac{\partial f}{\partial x_2}\right) \tag{10.280}$$

to first order. The final term in parentheses is the infinitesimal generator for transformations of functions from the old coordinates to the new coordinates. Therefore, the operator that transforms from $f[x]$ to $\tilde{f}[x] = f[R^{-1}x]$ is

$$L = -i\left(x_1\frac{\partial}{\partial x_2} - x_2\frac{\partial}{\partial x_1}\right) \Longrightarrow \tilde{f}[x] = \mathrm{Exp}[-i\theta L]f[x] \tag{10.281}$$

Notice that the operator L is antisymmetric, like the matrix R, and that when the derivatives are interpreted as momentum operators, L is the operator for orbital angular momentum perpendicular to this plane.

10.4.6 Example: SU(2)

The states of a spin-$\frac{1}{2}$ particle are described by two-component spinors of the form

$$|\psi\rangle = \begin{pmatrix} \psi_1 \\ \psi_2 \end{pmatrix} \tag{10.282}$$

where the amplitudes ψ_i are expressed relative to a particular orientation of the coordinate system. Rotations of the coordinate axes mix the components of the spinors, preserving its norm, and are described by unitary 2×2 matrices, such that

$$|\psi'\rangle = U|\psi\rangle \Longrightarrow \begin{pmatrix} \psi_1' \\ \psi_2' \end{pmatrix} = \begin{pmatrix} U_{1,1} & U_{1,2} \\ U_{2,1} & U_{2,2} \end{pmatrix} \begin{pmatrix} \psi_1 \\ \psi_2 \end{pmatrix} \tag{10.283}$$

where $UU^\dagger = \mathbb{1}$. The set of all unitary 2×2 matrices with unit determinant forms a group that is designated SU(2), which stands for special unitary group in two dimensions. Any 2×2 matrix contains four complex numbers or, equivalently, eight real parameters. The requirement of unitarity imposes four constraints and the restriction upon the determinant imposes another constraint, leaving three independent real parameters. Thus, an infinitesimal transformation U can be parametrized in the form

$$U = \text{Exp}\left[-i\theta\hat{n} \cdot \vec{S}\right] \tag{10.284}$$

where $\hat{n}$ is an arbitrary unit vector in three dimensions and $\vec{S}$ is a set of three appropriately-chosen 2×2 matrices that represent the infinitesimal generators of this group. Unitarity requires $\vec{S}$ to be hermitian and the condition $\text{Det}[U] = 1$ requires $\text{Tr}[\vec{S}] = 0$. Therefore, we express the generators in the form $\vec{S} = \vec{\sigma}/2$ where the Pauli matrices

$$\sigma_1 = \begin{pmatrix} 0 & 1 \\ 1 & 0 \end{pmatrix}, \quad \sigma_2 = \begin{pmatrix} 0 & -i \\ i & 0 \end{pmatrix}, \quad \sigma_3 = \begin{pmatrix} 1 & 0 \\ 0 & -1 \end{pmatrix} \tag{10.285}$$

constitute a complete set of linearly independent, traceless, hermitian, 2×2 matrices and where the factor of $\frac{1}{2}$ ensures that the eigenfunctions

$$S_3 \begin{pmatrix} 1 \\ 0 \end{pmatrix} = \frac{1}{2} \begin{pmatrix} 1 \\ 0 \end{pmatrix}, \quad S_3 \begin{pmatrix} 0 \\ 1 \end{pmatrix} = -\frac{1}{2} \begin{pmatrix} 0 \\ 1 \end{pmatrix} \tag{10.286}$$

have eigenvalues $\pm\frac{1}{2}$ for S_3. Therefore, if we rotate the coordinate axis through angle θ about axis $\hat{n}$, the components of a spin-$\frac{1}{2}$ wave function transform among themselves according to the unitary matrix

$$U\left[\vec{\theta}\right] = \text{Exp}\left[-i\vec{\theta} \cdot \vec{\sigma}/2\right] \tag{10.287}$$

where $\vec{\theta} = \theta\hat{n}$ encodes the three parameters of SU(2).

Direct calculation will show that the Pauli matrices satisfy commutation relations of the form

$$\left[\sigma_i, \sigma_j\right] = 2i\varepsilon_{ijk}\sigma_k \Longrightarrow \left[S_i, S_j\right] = i\varepsilon_{ijk}S_k \tag{10.288}$$

where, by convention, one sums over a repeated index; hence, the right-hand side is summed over k. Also, recall that the *Levi–Civita symbol* $\varepsilon_{i,j,k}$ takes the values $+1$ for a symmetric permutation of the indices, −1 for an antisymmetric permutation, or 0 whenever two indices are repeated. Thus, the structure constants are antisymmetric and one can show that they satisfy the Jacobi identity, as required for a Lie group. Finally, notice that the commutation relations for SU(2) can be expressed in the form

$$\vec{S}\times\vec{S} = i\vec{S} \Longleftrightarrow \vec{\sigma}\times\vec{\sigma} = 2i\vec{\sigma} \tag{10.289}$$

where the cross-product does not vanish because the components of $\vec{\sigma}$ do not commute with each other. One must be careful when generalizing results from vector algebra to noncommuting operators or matrices.

The group SU(2) also admits matrix representations of higher dimension. Suppose that we transform coordinates according to $\vec{x}' = R\vec{x}$ where the matrices $R_i[\theta_i]$ describe independent rotations about each of three axes. The corresponding infinitesimal generators S_i for transformations of a vector are then obtained by differentiating R_i as follows.

$$R_1[\theta_1] = \begin{pmatrix} 1 & 0 & 0 \\ 0 & \mathrm{Cos}[\theta_1] & \mathrm{Sin}[\theta_1] \\ 0 & -\mathrm{Sin}[\theta_1] & \mathrm{Cos}[\theta_1] \end{pmatrix} \Longrightarrow S_1 = -i\left(\frac{\partial R_1}{\partial\theta_1}\right)_{\theta_1=0} = -i\begin{pmatrix} 0 & 0 & 0 \\ 0 & 0 & 1 \\ 0 & -1 & 0 \end{pmatrix} \tag{10.290}$$

$$R_2[\theta_2] = \begin{pmatrix} \mathrm{Cos}[\theta_2] & 0 & -\mathrm{Sin}[\theta_2] \\ 0 & 1 & 0 \\ \mathrm{Sin}[\theta_2] & 0 & \mathrm{Cos}[\theta_2] \end{pmatrix} \Longrightarrow S_2 = -i\left(\frac{\partial R_2}{\partial\theta_2}\right)_{\theta_2=0} = -i\begin{pmatrix} 0 & 0 & -1 \\ 0 & 0 & 0 \\ 1 & 0 & 0 \end{pmatrix} \tag{10.291}$$

$$R_3[\theta_3] = \begin{pmatrix} \mathrm{Cos}[\theta_3] & \mathrm{Sin}[\theta_3] & 0 \\ -\mathrm{Sin}[\theta_3] & \mathrm{Cos}[\theta_3] & 0 \\ 0 & 0 & 1 \end{pmatrix} \Longrightarrow S_3 = -i\left(\frac{\partial R_3}{\partial\theta_3}\right)_{\theta_3=0} = -i\begin{pmatrix} 0 & 1 & 0 \\ -1 & 0 & 0 \\ 0 & 0 & 0 \end{pmatrix} \tag{10.292}$$

These generators also satisfy the commutation relations of SU(2). More generally, the three operators $\vec{S}$ that describe the *intrinsic spin* are traceless hermitian matrices of dimension $2s + 1$ that satisfy the SU(2) commutation relations

$$\left[S_i, S_j\right] = i\varepsilon_{ijk}S_k \tag{10.293}$$

such that

$$U\left[\vec{\theta}\right] = \mathrm{Exp}\left[-i\vec{\theta}\cdot\vec{S}\right] \tag{10.294}$$

generates rotation of the state vectors for particles with spin s.

10.4.7 Example: SO(3)

Coordinate rotations that leave the Euclidean norm $x_1^2 + x_2^2 + x_3^2$ invariant are described by orthogonal matrices $R[\vec{\theta}]$ where $\vec{\theta}$ stands for the three parameters needed to specify the rotation. Proper rotations that exclude reflections also require $\mathrm{Det}[R] = +1$. Thus, rotations are isomorphic with the group of special orthogonal matrices in three dimensions, namely SO(3). Let

$$x'_i = \xi_i[x;\theta] = R_{i,j}\left[\vec{\theta}\right]x_j \tag{10.295}$$

represent the coordinate transformation. Infinitesimal coordinate rotations are described by

$$\vec{\xi} \approx \vec{x} + \vec{\theta}\times\vec{x} \Longrightarrow \xi_i \approx x_i + \varepsilon_{i,j,k}\theta_j x_k \tag{10.296}$$

where $\vec{\theta} = \{\theta_1, \theta_2, \theta_3\}$ is an infinitesimal vector and where the summation convention is employed. The infinitesimal generators are then

$$L_i = -i\frac{\partial \xi_j}{\partial \theta_i}\frac{\partial}{\partial x_j} = -i\varepsilon_{i,j,k}x_j\frac{\partial}{\partial x_k} = \vec{x}\times\vec{p} \tag{10.297}$$

where $\vec{p} = -i\vec{\nabla}$ is the momentum operator. These generators can be expressed in either cartesian or polar coordinates as

$$L_x = -i\left(y\frac{\partial}{\partial z} - z\frac{\partial}{\partial y}\right) = -i\left(-\mathrm{Sin}[\phi]\frac{\partial}{\partial\theta} - \mathrm{Cos}[\phi]\,\mathrm{Cot}[\theta]\frac{\partial}{\partial\phi}\right) \tag{10.298}$$

$$L_y = -i\left(z\frac{\partial}{\partial x} - x\frac{\partial}{\partial z}\right) = -i\left(\mathrm{Cos}[\phi]\frac{\partial}{\partial\theta} - \mathrm{Sin}[\phi]\,\mathrm{Cot}[\theta]\frac{\partial}{\partial\phi}\right) \tag{10.299}$$

$$L_z = -i\left(x\frac{\partial}{\partial y} - y\frac{\partial}{\partial x}\right) = -i\frac{\partial}{\partial\phi} \tag{10.300}$$

and are easily seen to satisfy commutation relations

$$\left[L_i, L_j\right] = i\varepsilon_{i,j,k}L_k \tag{10.301}$$

that are the same as those of SU(2). Nevertheless, these groups are not identical and we will later exploit the homomorphism between SU(2) and SO(3) to analyze their representations in considerable detail. For now, suffice it to recognize that the operator that transforms functions whose coordinates are rotated through a finite angle is given by

$$\mathcal{R}\left[\vec{\theta}\right] = \mathrm{Exp}\left[-i\vec{\theta}\cdot\vec{L}\right] \tag{10.302}$$

where $\vec{\theta} = \{\theta_x, \theta_y, \theta_z\}$ specifies the axis of rotation and the angle and where $\vec{L}$ are the differential operators defined above.

10.4.8 Total angular momentum

We can now combine spin and orbital angular momentum by considering the transformation properties of a wave function of the form

$$\Psi\left[\vec{x}\right] = \begin{pmatrix} \psi_1\left[\vec{x}\right] \\ \psi_2\left[\vec{x}\right] \\ \vdots \end{pmatrix} \tag{10.303}$$

that consists of $2s + 1$ functions of the three-dimensional vector $\vec{x}$. Rotation of the coordinate system will both change the coordinates in each function and mix the components of the state vector. Thus, we write

$$\Psi'\left[\vec{x}\right] = \text{Exp}\left[-i\theta\hat{n}\cdot\vec{S}\right]\Psi\left[R\left[-\theta\hat{n}\right]\vec{x}\right] \tag{10.304}$$

where $\vec{S}$ mixes the components and R acts upon the coordinates. Expanding the infinitesimal transformations to first order

$$\Psi'\left[\vec{x}\right] \approx \left(\mathbb{1} - i\theta\hat{n}\cdot\vec{S}\right)\Psi\left[1 - \theta\hat{n}\times\vec{x}\right] \approx \left(\mathbb{1} - i\theta\hat{n}\cdot\vec{S}\right)\left(\Psi\left[\vec{x}\right] - \theta\hat{n}\cdot\vec{x}\times\vec{\nabla}\Psi\left[\vec{x}\right]\right) \tag{10.305}$$

and dropping higher-order terms, we can identify the infinitesimal generators as

$$\vec{J} = \vec{L} + \vec{S} \tag{10.306}$$

whereby

$$\frac{\partial\Psi'\left[\vec{x}\right]}{\partial\theta} \approx -i\hat{n}\cdot\vec{J}\Psi\left[\vec{x}\right] \Longrightarrow \Psi'\left[\vec{x}\right] = \text{Exp}\left[-i\vec{\theta}\cdot\vec{J}\right]\Psi\left[\vec{x}\right] \tag{10.307}$$

provides the wave function in the rotated coordinate system.

The spin of an elementary particle is an intrinsic property that should be independent of its position or momentum. Therefore, the $\vec{S}$ operators should be independent of $\vec{x}$ or $\vec{p}$ and, hence, commute with $\vec{L}$; $\vec{S}$ acts upon a nonclassical internal space represented by the $2s + 1$ components of the wave function while $\vec{L}$ acts upon geometrical variables with classical interpretations. Thus, the total angular momentum $\vec{J} = \vec{L} + \vec{S}$ contains orbital and spin contributions that satisfy the commutation relations

$$\left[L_i, L_j\right] = i\varepsilon_{i,j,k}L_k \tag{10.308}$$

$$\left[S_i, S_j\right] = i\varepsilon_{i,j,k}S_k \tag{10.309}$$

$$\left[J_i, J_j\right] = i\varepsilon_{i,j,k}J_k \tag{10.310}$$

$$\left[L_i, S_j\right] = 0 \tag{10.311}$$

such that the transformation operators

$$\mathcal{R}\left[\vec{\theta}\right] = \text{Exp}\left[-i\vec{\theta}\cdot\vec{J}\right] = \text{Exp}\left[-i\vec{\theta}\cdot\vec{L}\right]\text{Exp}\left[-i\vec{\theta}\cdot\vec{S}\right] \tag{10.312}$$

contain two factors that operate on different aspects of the wave function. If the hamiltonian is rotationally invariant, such that $[H, \vec{J}] = 0$, we can construct energy eigenfunctions that are also eigenfunctions of J^2 and J_3. Often $[H, S^2] = 0$ also indicates that S^2 is conserved, but L^2, L_3, and S_3 usually are not when $s > 0$.

10.4.9 Transformation of Operators

Suppose that U is a unitary operator that performs coordinate transformations upon functions and that H is an arbitrary operator on the same Hilbert space. Operators in the new and old coordinate systems are related by the familiar similarity transformation

$$\psi' = U\psi \Longrightarrow H' = UHU^{-1} \tag{10.313}$$

If we now represent $U = \mathrm{Exp}[-i\alpha S]$ in terms of a hermitian infinitesimal generator S and a real parameter α, this similarity transformation can be evaluated as a power series in α with the aid of the *Baker–Haussdorff identity*

$$e^{-i\alpha S} H e^{i\alpha S} = H + (-i\alpha)[S, H] + \frac{(-i\alpha)^2}{2!}\left[S, [S, H]\right] + \frac{(-i\alpha)^3}{3!}\left[S, \left[S, [S, H]\right]\right] + \cdots \tag{10.314}$$

where the number of times S appears in the leading position of each multiple commutator is given by the associated power of $-i\alpha$. Verification of this identity is left as an exercise for the reader. For infinitesimal transformations, we find

$$\left(\frac{\partial H'}{\partial \alpha}\right)_{\alpha=0} = -i[S, H] \tag{10.315}$$

An important consequence is that H is invariant with respect to transformations generated by operators with which it commutes, such that

$$[S, H] = 0 \Longrightarrow H' = H \tag{10.316}$$

10.4.10 Invariant Functions

A function ψ that is unchanged when its coordinates are transformed by any group element is described as an *invariant* of the group and satisfies

$$\psi\left[R^{-1}x\right] = \psi[x] \Longrightarrow \mathcal{R}\psi = \psi \Longrightarrow \mathrm{Exp}[-ia \cdot G]\psi = \psi \tag{10.317}$$

where a is the parameter vector and G is the corresponding vector of generators. For small a_i we can use the first-order expansion

$$(1 - ia \cdot G)\psi[x] = \psi[x] \Longrightarrow G_i\psi = 0 \tag{10.318}$$

to conclude that invariants are annihilated by any of the group generators, labeled by $i = 1, n$. For example, consider a central function of the form $\psi[x, y] = \psi[r]$ where $r = \sqrt{x^2 + y^2}$ and the infinitesimal generator

$$L_z = -i\left(x\frac{\partial}{\partial y} - y\frac{\partial}{\partial x}\right) = -i\frac{\partial}{\partial \phi} \tag{10.319}$$

for rotations in the plane. Obviously, ψ is invariant wrt L_z because $L_z\psi[r] = 0$ for any $\{x, y\}$.

If $G_i\psi = 0$ for some generators but not others, we can describe ψ as invariant with respect to the subgroups generated by each G_i that annihilates ψ. Suppose that ψ is an invariant wrt G_i and ϕ is an arbitrary function. Then

$$G_i\psi\phi = \phi G_i\psi + \psi G_i\phi = \psi G_i\phi \tag{10.320}$$

permits the invariant to be factored out as if it were a constant. Similarly, the transformation

$$\mathcal{R}_i\psi\phi = \mathrm{Exp}\left[-ia_i G_i\right]\psi\phi = \psi\,\mathrm{Exp}\left[-ia_i G_i\right]\phi \tag{10.321}$$

only affects ϕ, the noninvariant (i.e., variable) factor.

A function $\psi[x]$ that satisfies

$$G\psi[x] = \lambda[x]\psi[x] \tag{10.322}$$

where $\lambda[x]$ is an invariant of the infinitesimal generator G is described as an *eigenfunction* of G with *eigenvalue* $\lambda[x]$. Finite transformations based upon G then take the form

$$\text{Exp}[-iaG]\psi[x] = e^{-ia\lambda}\psi[x] \tag{10.323}$$

Suppose that ψ_1 and ψ_2 are two different eigenfunctions of G with eigenvalues λ_1 and λ_2, respectively. Then

$$G\psi_1[x]\psi_2[x] = \psi_2 G\psi_1 + \psi_1 G\psi_2 = \left(\lambda_1[x] + \lambda_2[x]\right)\psi_1[x]\psi_2[x] \tag{10.324}$$

shows that $\psi_1\psi_2$ is also an eigenfunction of G whose eigenvalue is $\lambda_1 + \lambda_2$.

Let $T[a]$ represent an element of a symmetry group with parameter a that acts on coordinate x according to $x' = T[a]x$ and let $\mathcal{T}[a]$ represent the corresponding operator upon functions, such that

$$\mathcal{T}[a]\psi[x] = \psi\left[T^{-1}[a]x\right] \tag{10.325}$$

Suppose that the Hamiltonian operator H commutes with $\mathcal{T}$, such that $[H, \mathcal{T}] = 0$ and that $\{\psi_i, i = 1, n\}$ are a linearly independent set of eigenfunctions for H that share a common eigenvalue E, such that

$$H\psi_i = E\psi_i \Longrightarrow H\mathcal{T}\psi_i = E\mathcal{T}\psi_i \tag{10.326}$$

Thus, $\mathcal{T}\psi_i$ is also an eigenfunction of H with the same eigenvalue and can be expanded as the linear superposition

$$\mathcal{T}[a]\psi_i[x] = \psi_i\left[T^{-1}[a]x\right] = \sum_{j=1}^{n}\psi_j[x]D_{j,i}[a] \tag{10.327}$$

with coefficients that depend upon the group parameter. The ordering of the indices may appear unusual but is needed to ensure that the matrices $D[a]$ constitute a representation of the symmetry group $\mathcal{T}$. Suppose that we combine two symmetry operations

$$\mathcal{T}[b]\mathcal{T}[a]\psi_i = \sum_{j=1}^{n}\mathcal{T}[b]\psi_j D_{j,i}[a] = \sum_{j,k=1}^{n}\psi_k D_{k,j}[b]D_{j,i}[a] \tag{10.328}$$

such that

$$T[b]T[a] = T[c] \Longleftrightarrow D[b]D[a] = D[c] \tag{10.329}$$

Therefore, $D[a]$ is an n-dimensional matrix representation of group element $T[a]$. Notice that this ordering is consistent with the inverse relationship between $\mathcal{T}$ and T, whereby

$$\mathcal{T}[c]\psi[x] = \psi\left[T^{-1}[c]x\right] = \psi\left[(T[b]T[a])^{-1}x\right] = \psi\left[T^{-1}[a]T^{-1}[b]x\right] \tag{10.330}$$

10.5 Lie Algebra

10.5.1 Definitions

A *Lie algebra* is a linear vector space with a law of composition (or multiplication) denoted by $[A_i, A_j]$ that satisfies the following properties:

- closure under multiplication: if A and B are members of the algebra, then the product $[A, B]$ is also a member
- antisymmetry under multiplication: $[B, A] = -[A, B]$
- distributive law: $[A + B, C] = [A, C] + [B, C]$
- Jacobi identity: $\left[A, [B, C]\right] + \left[B, [C, A]\right] + \left[C, [A, B]\right] = 0$.

This abstract notion includes both Lie groups expressed in terms of differential operators that generate infinitesimal coordinate transformations and their representations in terms of matrices. The law of composition could be the commutator

$$[A, B] = AB - BA \tag{10.331}$$

for quantum mechanics or it could be the Poisson bracket

$$[A, B] = \sum_i \left(\frac{\partial A}{\partial q_i} \frac{\partial B}{\partial p_i} - \frac{\partial B}{\partial q_i} \frac{\partial A}{\partial p_i} \right) \tag{10.332}$$

for classical mechanics, where q_i, p_i are pairs of canonically conjugate generalized coordinates and momenta. The fact that general theorems depend upon abstract postulates instead of specific realizations gives this branch of mathematics considerable power and versatility.

The dimension of the algebra is the dimension of its underlying vector space or, equivalently, the number of independent generators. An algebra $\mathcal{A}$ is abelian if $[A_i, A_j] = 0$ for all $\{A_i, A_j\} \in \mathcal{A}$. A *subalgebra* is a subset of the vector space that satisfies all conditions for a Lie algebra. An algebra is described as *simple* if it does not contain an invariant subalgebra or *semisimple* if it does not contain an abelian subalgebra. An algebra is semisimple if and only if it is a direct sum of simple algebras. A compact algebra is necessarily semisimple. It is useful to define a metric tensor, also known as a *Killing form*, as

$$g_{i,j} \propto \sum_{k,l} c^l_{j,k} c^k_{l,i} \tag{10.333}$$

where the proportionality constant can be chosen for convenience. *Cartan's theorem* states that a Lie algebra is semisimple if and only if $\text{Det}[g] \neq 0$. Under those conditions we can also define an inverse $g^{i,j}$ that satisfies

$$\sum_k g_{i,k} g^{k,j} = \delta_{i,j} \tag{10.334}$$

The *rank* of an algebra is the maximum number of linearly independent mutually commuting operators that it supports. Let $\{P_i, i = 1, r\}$ represent a set of mutually commuting operators, such that $[P_i, P_j] = 0$ for all $i, j \le r$ and let $\{Q_j, j = r+1, d\}$ represent the remaining generators. Racah proved that r independent *Casimir operators* $\{C_i, i = 1, r\}$ that commute with all $A_i \in \mathcal{A}$, such that

$$[C_i, P_j] = [C_i, Q_j] = 0 \tag{10.335}$$

can be constructed for any semisimple Lie algebra of rank r from linear combinations of products of the generators. For example,

$$C_1 \propto \sum_{i,j} g^{i,j} A_i A_j \tag{10.336}$$

commutes with all $A_i \in \mathcal{A}$; the proportionality constant can be chosen for convenience. This lowest-order Casimir operator is usually very useful but higher-order Casimir operators are often very complicated. Schur's lemma tells us that any matrix which commutes with all the matrices of an irreducible representation must be a multiple of the unit matrix. Therefore, the irreducible representations of $\mathcal{A}$ can be labeled with the eigenvalues $\{c_i\}$ for the Casimir operators while their members are distinguished by the eigenvalues $\{p_i\}$ for the selected $\{P_i\}$. The basis states for such representations can then be expressed in the form

$$C_i \left|c_1 \cdots c_r; p_1 \cdots p_r\right\rangle = c_i \left|c_1 \cdots c_r; p_1 \cdots p_r\right\rangle \tag{10.337}$$
$$P_i \left|c_1 \cdots c_r; p_1 \cdots p_r\right\rangle = p_i \left|c_1 \cdots c_r; p_1 \cdots p_r\right\rangle \tag{10.338}$$

Furthermore, raising and lowering operators (*ladder operators*) for the p_i can then be constructed from linear combinations of the Q_i.

10.5.2 Example: SU(2)

For example, SU(2) is defined by the commutation relations

$$[S_i, S_j] = i\varepsilon_{i,j,k} S_k \tag{10.339}$$

and has dimension 3 and rank 1. Its metric tensor

$$\sum_{k,l} \varepsilon_{j,k,l} \varepsilon_{l,i,k} = \sum_{k,l} \varepsilon_{j,k,l} \varepsilon_{i,k,l} = 2\delta_{i,j} \tag{10.340}$$

shows that it is semisimple. Because the normalization of $g_{i,j}$ is irrelevant, one normally redefines the metric tensor for this group to be $g_{i,j} = g^{i,j} = \delta_{i,j}$. There is only one Casimir operator

$$C = S_1^2 + S_2^2 + S_3^2 \tag{10.341}$$

and it is proportional to the operator for total spin, S^2. Therefore, if $[H, \vec{S}] = 0$, one may classify eigenstates $|s, m\rangle$ by their eigenvalues for S^2 and one of the S_i, arbitrarily chosen as S_3, such that

$$S_3|s, m\rangle = m|s, m\rangle \tag{10.342}$$

Furthermore,

$$\left\langle s, m \left| S^2 \right| s, m \right\rangle \geq m^2 \tag{10.343}$$

demonstrates that there must be a maximum value of m for any finite s; equality applies only when $s = 0$.

To determine the eigenvalues of S^2 and the allowed values of m for each irrep s, it is useful to define the ladder operators

$$S_{\pm} = S_1 \pm i S_2 \tag{10.344}$$

whose commutation relations

$$[S_3, S_{\pm}] = \pm S_{\pm} \tag{10.345}$$

$$[S^2, S_{\pm}] = 0 \tag{10.346}$$

$$[S_+, S_-] = 2S_3 \tag{10.347}$$

can (and should!) be verified easily. Applying the first set of commutation relations

$$(S_3 S_{\pm} - S_{\pm} S_3)|s, m\rangle = (S_3 - m) S_{\pm}|s, m\rangle = \pm S_{\pm}|s, m\rangle \Longrightarrow S_3 S_{\pm}|s, m\rangle = (m \pm 1) S_{\pm}|s, m\rangle \tag{10.348}$$

demonstrates that $S_{\pm}|s, m\rangle$ is an eigenfunction of S_3 with eigenvalue $m \pm 1$. Thus, the effect of S_+ is to raise m by one unit while S_- lowers m by one unit. We can express S^2 in two alternative forms

$$S^2 = S_- S_+ + S_3(S_3 + 1) = S_+ S_- + S_3(S_3 - 1) \tag{10.349}$$

and choose to label irrep s using the maximum possible value of m, such that

$$S_+|s, s\rangle = 0 \Longrightarrow S^2|s, s\rangle = s(s + 1)|s, s\rangle \tag{10.350}$$

Closure of irrep s with respect to the ladder operators, $S_{\pm}$, requires the allowed values of m to form a sequence $s, s - 1, \ldots, s - n$ where n is a finite nonnegative integer, such that

$$S_-|s, s-n\rangle = 0 \Longrightarrow S^2|s, s-n\rangle = (s-n)(s-n-1)|s, s-n\rangle \Longrightarrow (s-n)(s-n-1) = s(s+1) \tag{10.351}$$

requires $n = 2s$. Therefore, s is either integral or half-integral such that the dimension of irrep s is $2s + 1$ and its eigenvalues are given by

$$S^2\left|s, m\right\rangle = s(s + 1)|s, m\rangle, \qquad s = 0, \tfrac{1}{2}, 1, \tfrac{3}{2}, 2\ldots \tag{10.352}$$

$$S_3\left|s, m\right\rangle = m|s, m\rangle, \qquad m = s, s - 1, \ldots, -s \tag{10.353}$$

It is important to recognize that this analysis was based entirely upon the commutation relations of SU(2) and did not employ any specific representation of the associated Lie algebra. Therefore, it applies equally well to representations of any dimension, not just the two-dimensional representation that originally motivated the correspondence between the generators of SU(2) and the operators for spin. One might expect the same analysis also to apply to the representations of SO(3) that describe orbital angular momentum if one simply replaces $S \to L$ and $s \to l$ everywhere. However, that correspondence does not exclude half-integral orbital angular momentum, which is actually a tricky issue. For classical physics the requirement of single-valuedness excludes half-integral values because $e^{im\phi}$ would change sign under $\phi \to \phi + 2\pi$ if m were half-integral, but quantum mechanics permits this sign change because observables depend upon the absolute magnitude of the wave function. On the other hand, because orbital angular momentum is a classical concept, we might appeal to the correspondence principle to require single-valuedness anyway; no such appeal is possible for intrinsic spin because it is inherently nonclassical. For a more rigorous argument that limits l to integer values, we must return to the differential operators and their eigenfunctions; that argument is left to the exercises. Perhaps we should have anticipated that the algebraic argument based upon commutation relations alone would be inconclusive because the generators of SU(2) obey the same commutation relations as those of SO(3), yet SU(2) was designed to describe spin-$\frac{1}{2}$. Therefore, such an analysis does not distinguish between spin and orbital angular momentum. This is actually a strength of the method! If we define angular momentum by the requirement that eigenstates of a rotationally invariant Hamiltonian transform according to

$$\mathcal{R} = \text{Exp}\left[-i\vec{\theta} \cdot \vec{J}\right] \tag{10.354}$$

where the infinitesimal generators $\vec{J} = \vec{L} + \vec{S}$ obey the commutation relations

$$\left[J_i, J_j\right] = i\varepsilon_{i,j,k} J_k \tag{10.355}$$

then both spin and orbital contributions are included.

10.6 Orthogonality Relations for Lie Groups

Orthogonality relations were crucial to the analysis of finite groups and we expect analogous relations to be central to applications using Lie groups, but we must generalize from summation to integration using weight functions that preserve group properties. Let $a = \{a_i, i = 1, m\}$ represent a point in the m-dimensional parameter space of a Lie group and let da represent an infinitesimal volume surrounding that point. If b is another point in the same parameter space, the group multiplication law $c = \phi[a, b]$ assigns point c to the composition of elements a and b. If a is a unique point and the number of elements in the volume at b is $\rho[b]\,db$ where $\rho[b]$ is a density function, then the number of points

$$\rho[c]\,dc = \rho[b]\,db \tag{10.356}$$

near c must be the same because group multiplication by a provides a one-to-one correspondence between elements near c with those near b. Suppose that we choose a to be the identity element, such that

$$b_i = \phi_i[0, b] \Longrightarrow db_i = \sum_{j=1}^{m} \left(\frac{\partial \phi_i[a, b]}{\partial a_j} \right)_{a=0} da_j \tag{10.357}$$

The volume at b is then

$$db = J[b]\, da \tag{10.358}$$

where

$$J[b] = \left| \mathrm{Det}\left[\left(\frac{\partial \phi_i[a, b]}{\partial a_j} \right)_{a=0} \right] \right| \tag{10.359}$$

is the Jacobian determinant. Let $\rho[0]$ denote the density near the identity element, such that

$$\rho[b]\, db = \rho[0]\, da \Longrightarrow \rho[b] = \frac{\rho[0]}{J[b]} \tag{10.360}$$

defines a suitable density function. The density at the origin is usually assigned $\rho[0] = 1$, somewhat arbitrarily. This version is described as a *right density* because it is based upon the right argument of $c = \phi[a, b]$.

Suppose that $\psi[a]$ is a function of the group parameters. Recall that the rearrangement theorem for finite groups states

$$\sum_g \psi[g] = \sum_g \psi[gh] = \sum_g \psi[hg] \tag{10.361}$$

where h belongs to the group and the summations are taken over all group elements g. Generalizations to continuous groups then take the form

$$\int \psi[a]\rho[a]\, da = \int \psi\big[\phi[a, b]\big]\rho_{\mathrm{R}}[b]\, db = \int db \rho_{\mathrm{L}}[b]\psi\big[\phi[b, a]\big] \tag{10.362}$$

where one must recognize that left and right densities need not be equal if one or more parameters has an infinite range. Thus, depending upon application, we may require both

$$\rho_{\mathrm{R}}[b] = \rho[0] \Bigg/ \left| \mathrm{Det}\left[\left(\frac{\partial \phi_i[a, b]}{\partial a_j} \right)_{a=0} \right] \right| \tag{10.363}$$

$$\rho_{\mathrm{L}}[b] = \rho[0] \Bigg/ \left| \mathrm{Det}\left[\left(\frac{\partial \phi_i[b, a]}{\partial a_j} \right)_{a=0} \right] \right| \tag{10.364}$$

where $\rho[0]$ may be chosen for convenience. Fortunately, one can show that these densities are equal for compact groups and we usually will not have to confront this issue.

Perhaps the most important application of group integration is to the orthogonality properties of irreducible representations whose matrix elements are analytic functions of the continuous group parameters. Recall that the orthogonality theorem for two irreps of a finite group with N_G elements takes the form

$$\sum_{g \in G} D^{(i)}_{\mu,\nu}[g]^* D^{(j)}_{\alpha,\beta}[g] = \frac{N_G}{n_i} \delta_{\alpha,\mu} \delta_{\beta,\nu} \delta_{i,j} \tag{10.365}$$

where n_i is the dimensionality of irrep i. This result can now be translated to the continuum according to

$$\int D^{(i)}_{\mu,\nu}[a]^* D^{(j)}_{\alpha,\beta}[a] \rho[a] \, da = \frac{\delta_{\alpha,\mu} \delta_{\beta,\nu} \delta_{i,j}}{n_i} \int \rho[a] \, da \tag{10.366}$$

where the integral over density is the analog of N_G. Naturally we assume that the integrals exist but will not investigate the necessary conditions; suffice it to say that problems are rarely encountered here. Similarly, orthogonality of characters

$$\sum_k N_k \chi^{(i)} \left[C_k\right]^* \chi^{(j)} \left[C_k\right] = N_G \delta_{i,j} \tag{10.367}$$

takes the continuum form

$$\int \chi^{(i)}[a]^* \chi^{(j)}[a] \, da = \delta_{i,j} \int \rho[a] \, da \tag{10.368}$$

Example

For example, consider

$$x' = a_0 + \left(1 + a_1\right)x, \quad a_1 \neq -1 \tag{10.369}$$

such that

$$\phi_0[a,b] = a_0 + b_0\left(1 + a_1\right) \Longrightarrow \frac{\partial \phi_0}{\partial a_0} = 1, \quad \frac{\partial \phi_0}{\partial a_1} = b_0 \tag{10.370}$$

$$\phi_1[a,b] = a_1 + b_1\left(1 + a_1\right) \Longrightarrow \frac{\partial \phi_1}{\partial a_0} = 0, \quad \frac{\partial \phi_1}{\partial a_1} = 1 + b_1 \tag{10.371}$$

results in

$$J[b] = 1 + b_1 \tag{10.372}$$

such that

$$\rho[b] = \left| \frac{1}{1 + b_1} \right| \tag{10.373}$$

10.7 Quantum Mechanical Representations of the Rotation Group

10.7.1 Generators and Commutation Relations

We have seen in previous sections that the wave functions for spin-$\frac{1}{2}$ transform among themselves according to SU(2) while those for orbital angular momentum are governed by SO(3). Both of those groups have dimension 3, rank 1, and satisfy the same commutation relations. Therefore, we generalize the definition of angular momentum to include both spin and orbital angular momentum by postulating commutation relations of the form

$$\left[J_i, J_j\right] = i\varepsilon_{i,j,k} J_k \tag{10.374}$$

for the generators of rotations of the wave function and define the square of the total angular momentum as the Casimir operator

$$J^2 = J_x^2 + J_y^2 + J_z^2 \Longrightarrow \left[J^2, J_i\right] = 0 \tag{10.375}$$

that commutes with all of the generators. Therefore, we can define basis vectors $\psi_{j,m}$ that are simultaneous eigenfunctions of J^2 and one of its components, arbitrarily chosen as J_z, such that

$$J_z \psi_{j,m} = m\psi_{j,m} \tag{10.376}$$

It is also useful to define raising and lowering operators, also known as *ladder operators* , according to

$$J_\pm = J_x \pm iJ_y \tag{10.377}$$

such that

$$\left[J_z, J_\pm\right] = \pm J_\pm \tag{10.378}$$

$$\left[J_\pm, J^2\right] = 0 \tag{10.379}$$

$$\left[J_+, J_-\right] = 2J_z \tag{10.380}$$

Thus,

$$\left(J_z J_+ - J_+ J_z\right)\psi_{j,m} = +J_+\psi_{j,m} \Longrightarrow J_z J_+ \psi_{j,m} = (m+1) J_+ \psi_{j,m} \tag{10.381}$$

$$\left(J_z J_- - J_- J_z\right)\psi_{j,m} = -J_-\psi_{j,m} \Longrightarrow J_z J_- \psi_{j,m} = (m-1) J_- \psi_{j,m} \tag{10.382}$$

demonstrates that $J_\pm \psi_{j,m}$ is an eigenfunction of J_z with eigenvalue $m \pm 1$ – the effect of J_+ is to raise m by one unit whereas J_- lowers m by the same amount. Suppose that μ is the maximum possible m in the irreducible representation labeled j. Expressing J^2 in the form

$$J^2 = J_- J_+ + J_z\left(J_z + 1\right) \tag{10.383}$$

we obtain

$$\mu = \text{Max}[m] \Longrightarrow J_+ \psi_{j,\mu} = 0 \Longrightarrow J^2 \psi_{j,\mu} = \mu(\mu+1)\psi_{j,\mu} \tag{10.384}$$

and can label the representation by its maximum m such that

$$J^2\psi_{j,m} = j(j+1)\psi_{j,m} \tag{10.385}$$

for any allowed m. Application of $(J_-)^n$ to $\psi_{j,j}$ yields an eigenfunction $\psi_{j,j-n}$ and must terminate at a finite value of n because the representation j is finite. Expressing J^2 in the form

$$J^2 = J_+J_- + J_z(J_z - 1) \tag{10.386}$$

and using $J_-\psi_{j,j-n} = 0$, we obtain

$$J^2\psi_{j,j-n} = (j-n)(j-n-1)\psi_{j,j-n} = j(j+1)\psi_{j,j-n} \tag{10.387}$$

where the final step uses the fact that the eigenvalue of J^2 is independent of m. Thus, we find $j = n/2$ where n is an integer, and conclude that j is either integral or half-integral according to whether n is even or odd. The operators $J_\pm$, J_z transform the eigenfunctions of J^2 with common j among themselves and must span the vector space because the representation we seek is irreducible. Therefore, the irreducible representation labeled j contains $2j+1$ basis vectors that satisfy

$$J^2\psi_{j,m} = j(j+1)\psi_{j,m} \tag{10.388}$$

$$J_z\psi_{j,m} = m\psi_{j,m} \tag{10.389}$$

for $m = -j, -j+1, \ldots, j-1, j$.

All rotations through the same angle, regardless of axis, are similar and belong to the same class. Therefore, there are an infinite number of classes and, correspondingly, an infinite number of irreducible representations. To demonstrate this class structure, consider a rotation $S = R_x[\theta]R_z[\phi]$ that creates a new coordinate system with its polar axis in the direction $\{\theta, \phi\}$ wrt to the original coordinate system. If we follow a rotation about this new axis by S^{-1} to restore the original coordinate system, the net result of $SR_z[\alpha]S^{-1}$ is to perform a rotation through angle α about an axis specified by $\{\theta, \phi\}$. This general rotation is similar to $R_z[\alpha]$ by construction. Thus, all rotations through the same angle around any axis are similar to each other and form a class. The irreducible representations for rotations about the z-axis take the form

$$D^{(j)}\left[R_z[\theta]\right] = \mathrm{Exp}[-im\theta]\mathbb{1} \tag{10.390}$$

with character

$$\chi^{(j)}[\theta] = \mathrm{Tr}\,\mathrm{Exp}[-im\theta]\mathbb{1} = \sum_{m=-j}^{j} \mathrm{Exp}[im\theta] = \frac{\mathrm{Sin}[(2j+1)\theta/2]}{\mathrm{Sin}[\theta/2]} \tag{10.391}$$

where θ labels the class. More generally, an arbitrary rotation is represented by

$$D^{(j)}[\omega] = \mathrm{Exp}[-i\omega\cdot J] \tag{10.392}$$

where $\omega = \{\theta_x, \theta_y, \theta_z\}$. The $D^{(j)}$ matrices are unitary, which requires the J_i operators to be hermitian.

Finally, we evaluate the matrix elements of $J_\pm$. Let

$$J_+\psi_{j,m} = a_{j,m}\psi_{j,m+1} \tag{10.393}$$
$$J_-\psi_{j,m} = b_{j,m}\psi_{j,m-1} \tag{10.394}$$

and observe that these operators are hermitian conjugates of each other

$$J_- = J_+^\dagger, \quad J_+ = J_-^\dagger \tag{10.395}$$

Furthermore, the cross-products take the form

$$J_+J_- = J^2 - J_z\left(J_z - 1\right) \tag{10.396}$$
$$J_-J_+ = J^2 - J_z\left(J_z + 1\right) \tag{10.397}$$

Thus, the norms

$$\left|a_{j,m}\right|^2 = \left\langle J_+\psi_{j,m}\middle|J_+\psi_{j,m}\right\rangle = \left\langle\psi_{j,m}\left|J_-J_+\right|\psi_{j,m}\right\rangle = j(j+1) - m(m+1) \tag{10.398}$$
$$\left|b_{j,m}\right|^2 = \left\langle J_-\psi_{j,m}\middle|J_-\psi_{j,m}\right\rangle = \left\langle\psi_{j,m}\left|J_+J_-\right|\psi_{j,m}\right\rangle = j(j+1) - m(m-1) \tag{10.399}$$

determine the matrix elements up to phase. Choosing the so-called Condon–Shortley conventions

$$J_+\psi_{j,m} = \sqrt{j(j+1) - m(m+1)}\,\psi_{j,m+1} \tag{10.400}$$
$$J_-\psi_{j,m} = \sqrt{j(j+1) - m(m-1)}\,\psi_{j,m-1} \tag{10.401}$$

then determines the relative phases among the $\psi_{j,m}$ basis vectors. Notice that $J_+\psi_{j,j} = J_-\psi_{j,-j} = 0$, as required by the preceding analysis.

10.7.2 Euler Parametrization

The orientation of a three-dimensional object can be specified by three *Euler angles* describing the following sequence of operations:

- rotation about the $\hat{z}$-axis through angle α, producing new axes $\hat{x}', \hat{y}', \hat{z}' = \hat{z}$
- rotation about the $\hat{y}'$-axis through angle β, producing new axes $\hat{x}'', \hat{y}'' = \hat{y}', \hat{z}''$
- rotation about the $\hat{z}''$-axis through angle γ.

Alternatively, the same orientation can be achieved using the following sequence operations with respect to space-fixed axes:

- rotation about the $\hat{z}$-axis through angle γ
- rotation about the $\hat{y}$-axis through angle β
- rotation about the $\hat{z}$-axis through angle α.

Thus, a general rotation matrix takes the form

$$R[\alpha, \beta, \gamma] = R_z[\alpha] R_y[\beta] R_z[\gamma] \tag{10.402}$$

with inverses

$$(R[\alpha, \beta, \gamma])^{-1} = R_z[-\gamma] R_y[-\beta] R_z[-\alpha] = R[-\gamma, -\beta, -\alpha] \tag{10.403}$$

The ranges for these angles are $0 \le \alpha \le 2\pi$, $0 \le \beta \le \pi$, and $0 \le \gamma \le 2\pi$. However, a specified rotation does not uniquely determine the Euler angles – for example, the parameters $\{\alpha, 0, \gamma\}$ and $\{\alpha + \kappa, 0, \gamma - \kappa\}$ produce the same rotation for any κ.

Suppose that $\psi_{j,m}$ is an eigenfunction with total angular momentum j and magnetic quantum number m. The effect of a rotation is then

$$\mathcal{R}[\alpha, \beta, \gamma] \psi_{j,m} = \sum_{m'=-j}^{j} \psi_{j,m'} D^{(j)}_{m'm}[\alpha, \beta, \gamma] \tag{10.404}$$

where $D^{(j)}$ is the rotation matrix for irreducible representation j. Please note once again the ordering of the indices. Having chosen basis functions to be eigenfunctions of J_z, we can write

$$D^{(j)}[\alpha, \beta, \gamma] = \text{Exp}\left[-i\alpha J_z\right] d^{(j)}[\beta] \, \text{Exp}\left[-i\gamma J_z\right] \tag{10.405}$$

where the reduced rotation matrix $d^{(j)}$ depends only upon β while the rotations about the z-axis are diagonal. Thus,

$$\mathcal{R}[\alpha, \beta, \gamma] \psi_{j,m} = \sum_{m'=-j}^{j} \psi_{j,m'} d^{(j)}_{m'm}[\beta] \, \text{Exp}\left[-i\left(m\gamma + m'\alpha\right)\right] \tag{10.406}$$

10.7.3 Homomorphism Between SU(2) and SO(3)

The groups SO(3) and SU(2) both require three real parameters and their generators satisfy the same commutation relations, suggesting that these groups are homomorphic. There are two common parametrizations of SU(2). First, consider the Pauli representation

$$U = \begin{pmatrix} p_0 + i p_3 & p_2 + i p_1 \\ -p_2 + i p_1 & p_0 - i p_3 \end{pmatrix} = p_0 \sigma_0 + i \vec{p} \cdot \vec{\sigma} \tag{10.407}$$

where the parameters p_i are real,

$$\sigma_0 = \begin{pmatrix} 1 & 0 \\ 0 & 1 \end{pmatrix}, \quad \sigma_1 = \begin{pmatrix} 0 & 1 \\ 1 & 0 \end{pmatrix}, \quad \sigma_2 = \begin{pmatrix} 0 & -i \\ i & 0 \end{pmatrix}, \quad \sigma_3 = \begin{pmatrix} 1 & 0 \\ 0 & -1 \end{pmatrix} \tag{10.408}$$

are the basis matrices, and where the constraint

$$p_0^2 = 1 - p_1^2 - p_2^2 - p_3^2 \tag{10.409}$$

ensures that $\mathrm{Det}[U] = +1$. Note that there are only three independent parameters and that we require $|\vec{p}| \leq 1$. Alternatively, the Cayley–Klein parametrization

$$U[\xi,\psi,\zeta] = \begin{pmatrix} e^{i\zeta}\,\mathrm{Cos}[\psi] & e^{i\xi}\,\mathrm{Sin}[\psi] \\ -e^{-i\xi}\,\mathrm{Sin}[\psi] & e^{-i\zeta}\,\mathrm{Cos}[\psi] \end{pmatrix} \tag{10.410}$$

has the advantage that the parameters are unconstrained and U is manifestly unitary.

Consider the matrix

$$\rho = \vec{r}\cdot\vec{\sigma} = \begin{pmatrix} z & x - iy \\ x + iy & -z \end{pmatrix} \tag{10.411}$$

from which one can read the components of $\vec{r}$ in a particular coordinate system using

$$x = \left(\rho_{21} + \rho_{12}\right)/2 \tag{10.412}$$
$$y = \left(\rho_{21} - \rho_{12}\right)/2i \tag{10.413}$$
$$z = \rho_{11} \tag{10.414}$$

Next, suppose that we evaluate

$$\rho' = U\rho U^{-1} \tag{10.415}$$

with a similarity transform and evaluate the new coordinates

$$\vec{r}\,' = A\vec{r} \tag{10.416}$$

by interpreting the elements of ρ' in the same manner. The resulting linear transformation is represented by the 3×3 matrix A. If A is an orthogonal matrix with determinant $+1$, it can be interpreted as a coordinate rotation that establishes a homomorphism between the SU(2) matrix U and the SO(3) matrix A. After some straightforward but tedious algebra (see exercises), one finds

$$A = \begin{pmatrix} 1 - 2\left(p_2^2 + p_3^2\right) & 2\left(p_0p_3 + p_1p_2\right) & 2\left(p_1p_3 - p_0p_2\right) \\ 2\left(p_1p_2 - p_0p_3\right) & 1 - 2\left(p_1^2 + p_3^2\right) & 2\left(p_2p_3 + p_0p_1\right) \\ 2\left(p_0p_2 + p_1p_3\right) & 2\left(p_2p_3 - p_0p_1\right) & 1 - 2\left(p_1^2 + p_2^2\right) \end{pmatrix} \tag{10.417}$$

and can verify that $AA^T = \mathbb{1}$ and $\mathrm{Det}[A] = +1$ using $p_0^2 = 1 - p_1^2 - p_2^2 - p_3^2$. Therefore, $A \in \mathrm{SO}(3)$ is a valid rotation matrix.

Next we seek the relationship between the Euler angles and the Cayley–Klein parameters. Although one can evaluate $\rho' = U\rho U^{-1}$ using the Cayley–Klein parametrization in general, it is easier to choose the following special cases:

$$U[0,0,\zeta] = \begin{pmatrix} e^{i\zeta} & 0 \\ 0 & e^{-i\zeta} \end{pmatrix} \Longrightarrow U\rho U^{-1} = \begin{pmatrix} \mathrm{Cos}[2\zeta] & \mathrm{Sin}[2\zeta] & 0 \\ -\mathrm{Sin}[2\zeta] & \mathrm{Cos}[2\zeta] & 0 \\ 0 & 0 & 1 \end{pmatrix} = R_z[2\zeta] \tag{10.418}$$

$$U[0,\psi,0] = \begin{pmatrix} \mathrm{Cos}[\psi] & \mathrm{Sin}[\psi] \\ -\mathrm{Sin}[\psi] & \mathrm{Cos}[\psi] \end{pmatrix} \Longrightarrow U\rho U^{-1} = \begin{pmatrix} \mathrm{Cos}[2\psi] & 0 & \mathrm{Sin}[2\psi] \\ 0 & 1 & 0 \\ -\mathrm{Sin}[2\psi] & 0 & \mathrm{Cos}[2\psi] \end{pmatrix} = R_y[2\psi] \tag{10.419}$$

Thus, we have expressed the two matrices needed for the Euler parametrization of SO(3) in terms of the Cayley–Klein parametrization of SU(2). A general rotation then takes the form

$$R[\alpha,\beta,\gamma] = R_z[\alpha]R_y[\beta]R_z[\gamma] \Longrightarrow U = \pm U\left[0,0,\frac{\alpha}{2}\right]U\left[0,\frac{\beta}{2},0\right]U\left[0,0,\frac{\gamma}{2}\right] \tag{10.420}$$

where the $\pm$ reflects the sign ambiguity in U that arises because R was derived from $U\rho U^{-1}$, which is second-order in U. This homomorphism is double-valued: two elements of SU(2) map onto the same element of SO(3).

Transformation matrices for state vectors with $j = 1/2$ are, as usual, based upon the inverse of the rotation for coordinates

$$\begin{aligned}(R[\alpha,\beta,\gamma])^{-1} &= R_z[-\gamma]R_y[-\beta]R_z[-\alpha] \\ \Longrightarrow U^{-1} &= \pm U\left[0,0,-\frac{\gamma}{2}\right]U\left[0,-\frac{\beta}{2},0\right]U\left[0,0,-\frac{\alpha}{2}\right]\end{aligned} \tag{10.421}$$

such that

$$D^{(1/2)}[\alpha,\beta,\gamma] = \begin{pmatrix} e^{-i(\alpha+\gamma)/2}\,\mathrm{Cos}\left[\frac{\beta}{2}\right] & -e^{i(\alpha-\gamma)/2}\,\mathrm{Sin}\left[\frac{\beta}{2}\right] \\ e^{i(\gamma-\alpha)/2}\,\mathrm{Sin}\left[\frac{\beta}{2}\right] & e^{i(\alpha+\gamma)/2}\,\mathrm{Cos}\left[\frac{\beta}{2}\right] \end{pmatrix} \tag{10.422}$$

is a two-dimensional irreducible representation of SO(3) that describes states with angular momentum $j = 1/2$. It is also useful to factor the rotation matrix

$$D^{(1/2)}[\alpha,\beta,\gamma] = \mathrm{Exp}\left[-i\gamma\,\sigma_3/2\right]d^{(1/2)}[\beta]\,\mathrm{Exp}\left[-i\alpha\,\sigma_3/2\right] \tag{10.423}$$

where the reduced rotation matrix is simply

$$d^{(1/2)}[\beta] = \begin{pmatrix} \mathrm{Cos}\left[\frac{\beta}{2}\right] & -\mathrm{Sin}\left[\frac{\beta}{2}\right] \\ \mathrm{Sin}\left[\frac{\beta}{2}\right] & \mathrm{Cos}\left[\frac{\beta}{2}\right] \end{pmatrix} \tag{10.424}$$

Notice that a rotation of the coordinate system by 2π around any axis simply inverts the sign of the wave function; a rotation through 4π is needed to recover the original wave function. The half-angles appear here because the homomorphism between SU(2) and SO(3) is double-valued. Despite this minor complication, this representation for $D^{(1/2)}$ provides a useful starting point for construction of irreducible representations of higher dimension.

10.7.4 Irreducible Representations of SU(2)

Consider two variables χ_+ and χ_- and construct linear differential operators

$$J_x = \frac{1}{2}\left(\chi_-\frac{\partial}{\partial\chi_+} + \chi_+\frac{\partial}{\partial\chi_-}\right) \tag{10.425}$$

$$J_y = \frac{i}{2}\left(\chi_+\frac{\partial}{\partial\chi_-} - \chi_-\frac{\partial}{\partial\chi_+}\right) \tag{10.426}$$

$$J_z = \frac{1}{2}\left(\chi_+\frac{\partial}{\partial\chi_+} - \chi_-\frac{\partial}{\partial\chi_-}\right) \tag{10.427}$$

that act on functions of the form $\psi[\chi_+, \chi_-]$. With careful examination of commutators like

$$\begin{aligned}
\left(J_x J_y - J_y J_x\right)\psi &= \frac{i}{4}\left(\left(\chi_- \frac{\partial}{\partial \chi_+} + \chi_+ \frac{\partial}{\partial \chi_-}\right)\left(\chi_+ \frac{\partial \psi}{\partial \chi_-} - \chi_- \frac{\partial \psi}{\partial \chi_+}\right)\right. \\
&\quad \left. - \left(\chi_+ \frac{\partial}{\partial \chi_-} - \chi_- \frac{\partial}{\partial \chi_+}\right)\left(\chi_- \frac{\partial \psi}{\partial \chi_+} + \chi_+ \frac{\partial \psi}{\partial \chi_-}\right)\right) \\
&= \frac{i}{4}\left(\chi_- \frac{\partial \psi}{\partial \chi_-} + \chi_- \chi_+ \frac{\partial^2 \psi}{\partial \chi_+ \partial \chi_-} - \chi_-^2 \frac{\partial^2 \psi}{\partial \chi_+^2} + \chi_+^2 \frac{\partial^2 \psi}{\partial \chi_-^2} - \chi_+ \frac{\partial \psi}{\partial \chi_+} - \chi_+ \chi_- \frac{\partial^2 \psi}{\partial \chi_- \partial \chi_+}\right) \\
&\quad - \frac{i}{4}\left(\chi_+ \frac{\partial \psi}{\partial \chi_+} + \chi_+ \chi_- \frac{\partial^2 \psi}{\partial \chi_- \partial \chi_+} + \chi_+^2 \frac{\partial^2 \psi}{\partial \chi_-^2} - \chi_-^2 \frac{\partial^2 \psi}{\partial \chi_+^2} - \chi_- \frac{\partial \psi}{\partial \chi_-}\right. \\
&\quad \left. - \chi_- \chi_+ \frac{\partial^2 \psi}{\partial \chi_+ \partial \chi_-}\right) \\
&= \frac{i}{2}\left(\chi_- \frac{\partial \psi}{\partial \chi_-} - \chi_+ \frac{\partial \psi}{\partial \chi_+}\right) \\
&= J_z \psi
\end{aligned} \tag{10.428}$$

one soon verifies that these operators satisfy the commutation relations

$$\left[J_i, J_j\right] = i\varepsilon_{ijk} J_k \tag{10.429}$$

required for angular momentum operators. Next, suppose that ψ assumes the specific form

$$\psi_{j,m} = \sqrt{\frac{(2j)!}{(j+m)!(j-m)!}}\, \chi_+^{j+m} \chi_-^{j-m} \tag{10.430}$$

of homogeneous polynomials of order $2j$ where $m = -j, -j+1, \ldots, j-1, j$ and where the normalization factor is chosen to ensure that

$$\sum_{m=-j}^{j} \left|\psi_{j,m}\right|^2 = \sum_{m=-j}^{j} \binom{2j}{j+m} \left|\chi_+^{j+m} \chi_-^{j-m}\right|^2 = \left(\left|\chi_+\right|^2 + \left|\chi_-\right|^2\right)^j = 1 \tag{10.431}$$

when the basis vectors are normalized according to $\left|\chi_+\right|^2 + \left|\chi_-\right|^2 = 1$. Then using

$$\chi_+ \frac{\partial}{\partial \chi_+} \psi_{j,m} = (j+m)\psi_{j,m}, \quad \chi_- \frac{\partial}{\partial \chi_-} \psi_{j,m} = (j-m)\psi_{j,m} \Longrightarrow J_z \psi_{j,m} = m \psi_{j,m} \tag{10.432}$$

we observe that $\psi_{j,m}$ is an eigenfunction of J_z with eigenvalue m. Similarly, using

$$J^2 = J_x^2 + J_y^2 + J_z^2 = \frac{1}{4}\left(\chi_+^2 \frac{\partial^2}{\partial \chi_+^2} + \chi_-^2 \frac{\partial^2}{\partial \chi_-^2} + 2\chi_- \chi_+ \frac{\partial^2}{\partial \chi_+ \partial \chi_-} + 3\chi_+ \frac{\partial}{\partial \chi_+} + 3\chi_- \frac{\partial}{\partial \chi_-}\right) \tag{10.433}$$

such that

$$\begin{aligned} J^2\psi_{j,m} &= \tfrac{1}{4}\big((j+m)(j+m-1)+(j-m)(j-m-1)+2(j+m)(j-m)+3(j+m) \\ &\quad +3(j-m)\big)\psi_{j,m} \\ &= j(j+1)\psi_{j,m} \end{aligned} \tag{10.434}$$

we find that $\psi_{j,m}$ is an eigenfunction of J^2 with eigenvalue $j(j+1)$. Therefore, $\psi_{j,m}$ constitutes a suitable set of basis functions for representations of the rotation group.

We now require the variables χ_+, χ_- to transform under SU(2) according to $D^{(1/2)}[\alpha,\beta,\gamma]$, such that

$$\chi'_m = \sum_{m'=-1/2}^{1/2} \chi_{m'} D^{(1/2)}_{m',m}[\alpha,\beta,\gamma] \tag{10.435}$$

where $\chi_{\pm 1/2}$ is equivalent to $\chi_\pm$. It is sufficient to work out the case $\alpha = \gamma = 0$, for which

$$\begin{aligned} (\chi'_+)^{j+m}(\chi'_-)^{j-m} &= \left(\mathrm{Cos}\left[\frac{\beta}{2}\right]\chi_+ - \mathrm{Sin}\left[\frac{\beta}{2}\right]\chi_-\right)^{j+m}\left(\mathrm{Sin}\left[\frac{\beta}{2}\right]\chi_+ + \mathrm{Cos}\left[\frac{\beta}{2}\right]\chi_-\right)^{j-m} \\ &= \sum_{\mu=0}^{j+m}\sum_{\nu=0}^{j-m}(-)^{\mu}\binom{j+m}{\mu}\binom{j-m}{\nu} \\ &\quad \times \mathrm{Cos}\left[\frac{\beta}{2}\right]^{j+m-\mu+\nu}\mathrm{Sin}\left[\frac{\beta}{2}\right]^{j-m+\mu-\nu}\chi_+^{2j-\mu-\nu}\chi_-^{\mu+\nu} \end{aligned} \tag{10.436}$$

where the second line arises from the binomial theorem. The terms under the summation can be expressed in terms of $\psi_{j,m'}$ using the substitution $\mu+\nu \to j-m'$ such that

$$\begin{aligned} (\chi'_+)^{j+m}(\chi'_-)^{j-m} &\sum_{m'=-j}^{j}\sum_{\nu}(-)^{j-m'-\nu}\binom{j+m}{j-m'-\nu}\binom{j-m}{\nu} \\ &\times \mathrm{Cos}\left[\frac{\beta}{2}\right]^{m+m'+2\nu}\mathrm{Sin}\left[\frac{\beta}{2}\right]^{2j-m-m'-2\nu}\chi_+^{j+m'}\chi_-^{j-m'} \end{aligned} \tag{10.437}$$

where the summation over ν is limited by the binomial coefficients, which vanish when the lower argument is negative. Upon insertion of the normalization factors

$$\begin{aligned} \psi'_{j,m} = &\sum_{m'=-j}^{j}\sum_{\nu}(-)^{j-m'-\nu}\left(\frac{(j+m')!\,(j-m')!}{(j+m)!(j-m)!}\right)^{1/2}\binom{j+m}{j-m'-\nu}\binom{j-m}{\nu} \\ &\times \mathrm{Cos}\left[\frac{\beta}{2}\right]^{m+m'+2\nu}\mathrm{Sin}\left[\frac{\beta}{2}\right]^{2j-m-m'-2\nu}\psi_{j,m'} \end{aligned} \tag{10.438}$$

we identify the rotation matrix as

$$d^{(j)}_{m'm}[\beta] = \left(\frac{(j+m')!\,(j-m')!}{(j+m)!(j-m)!}\right)^{1/2} \sum_{\nu} (-)^{j-m'-\nu} \binom{j+m}{j-m'-\nu} \binom{j-m}{\nu} \\ \times \operatorname{Cos}\left[\frac{\beta}{2}\right]^{m+m'+2\nu} \operatorname{Sin}\left[\frac{\beta}{2}\right]^{2j-m-m'-2\nu} \tag{10.439}$$

or

$$d^{(j)}_{m'm}[\beta] = (-)^{j-m'} \left(\frac{(j+m')!\,(j-m')!}{(j+m)!(j-m)!}\right)^{1/2} \operatorname{Sin}\left[\frac{\beta}{2}\right]^{2j} \operatorname{Cot}\left[\frac{\beta}{2}\right]^{m+m'} \\ \times \sum_{\nu} (-)^{\nu} \binom{j+m}{j-m'-\nu} \binom{j-m}{\nu} \operatorname{Cot}\left[\frac{\beta}{2}\right]^{2\nu} \tag{10.440}$$

The summation over ν is expressible in terms of a hypergeometric function or a Jacobi polynomial in $\operatorname{Cot}[\beta/2]^2$, but that does not really simplify the result. One should verify that our previous result for $d^{(1/2)}$ is contained within this formula. Carrying this one step further, one obtains the rotation matrix for spin-1

$$d^{(1)} = \begin{pmatrix} \operatorname{Cos}\left[\frac{\beta}{2}\right]^2 & -\frac{\operatorname{Sin}[\beta]}{\sqrt{2}} & \operatorname{Sin}\left[\frac{\beta}{2}\right]^2 \\ \frac{\operatorname{Sin}[\beta]}{\sqrt{2}} & \operatorname{Cos}[\beta] & -\frac{\operatorname{Sin}[\beta]}{\sqrt{2}} \\ \operatorname{Sin}\left[\frac{\beta}{2}\right]^2 & \frac{\operatorname{Sin}[\beta]}{\sqrt{2}} & \operatorname{Cos}\left[\frac{\beta}{2}\right]^2 \end{pmatrix} = \begin{pmatrix} \frac{1}{2}(1+\operatorname{Cos}[\beta]) & -\frac{\operatorname{Sin}[\beta]}{\sqrt{2}} & \frac{1}{2}(1-\operatorname{Cos}[\beta]) \\ \frac{\operatorname{Sin}[\beta]}{\sqrt{2}} & \operatorname{Cos}[\beta] & -\frac{\operatorname{Sin}[\beta]}{\sqrt{2}} \\ \frac{1}{2}(1-\operatorname{Cos}[\beta]) & \frac{\operatorname{Sin}[\beta]}{\sqrt{2}} & \frac{1}{2}(1+\operatorname{Cos}[\beta]) \end{pmatrix} \tag{10.441}$$

and observes that half-angles are not required for integral j. Wave functions for half-integral spin change sign upon rotation through 2π, whereas those for integral spin do not.

Below we tabulate some of the symmetries and special values for rotation matrices but do not provide proofs. Note that l is restricted to integers while j may be integral or half-integral. Be aware that some authors employ different phase conventions for both rotation matrices and spherical harmonics.

$$D^{(l)}_{m,0}[\alpha,\beta,\gamma] = \sqrt{\frac{4\pi}{2l+1}}\, Y_{l,-m}[\beta,\alpha] \tag{10.442}$$

$$d^{(l)}_{m,0}[\beta] = (-)^m \left(\frac{(l-m)!}{(l+m)!}\right)^{1/2} P_{l,m}[\cos\beta] \tag{10.443}$$

$$d^{(l)}_{0,0}[\beta] = P_l[\cos\beta] \tag{10.444}$$

$$d^{(j)}_{m',m}[-\beta] = d^{(j)}_{m,m'}[\beta] \tag{10.445}$$

$$d^{(j)}_{m',m}[\pm\pi] = (-)^{j\pm m}\delta_{m',-m} \tag{10.446}$$

$$d^{(j)}_{m',m}[\pi-\beta] = (-)^{j-m'}d^{(j)}_{m,-m'}[\beta] \tag{10.447}$$

$$d^{(j)}_{m',m}[\beta] = (-)^{m'-m}d^{(j)}_{-m',-m}[\beta] = (-)^{m'-m}d^{(j)}_{m,m'}[\beta] \tag{10.448}$$

$$D^{(j)}_{m',m}[\alpha,\beta,\gamma]^* = D^{(j)}_{m',m}[-\alpha,\beta,-\gamma] = (-)^{m'-m}D^{(j)}_{-m',-m}[\alpha,\beta,\gamma] \tag{10.449}$$

10.7.5 Orthogonality Relations for Rotation Matrices

Now that we have an expression for the reduced rotation matrix, the matrices for arbitrary rotations can be evaluated using

$$D^{(j)}[\alpha,\beta,\gamma] = \text{Exp}\left[-i\alpha J_z\right]d^{(j)}[\beta]\,\text{Exp}\left[-i\gamma J_z\right] \tag{10.450}$$

$$\Longrightarrow D^{(j)}_{m'm}[\alpha,\beta,\gamma] = \text{Exp}\left[-i\left(m\gamma+m'\alpha\right)\right]d^{(j)}_{m'm}[\beta] \tag{10.451}$$

Let $\omega = \{\alpha,\beta,\gamma\}$ denote the Euler angles. According to the orthogonality theorem for compact Lie groups, we expect the representation matrices to satisfy an orthogonality relation of the form

$$\int\left(D^{(j)}_{m'm}[\omega]\right)^* D^{(j')}_{n'n}[\omega]\rho[\omega]\,d\omega = \frac{\delta_{j,j'}\delta_{m,n}\delta_{m',n'}}{2j+1}\int\rho[\omega]\,d\omega \tag{10.452}$$

where $\rho[\omega]$ is an appropriate density function and where the integration includes all group elements. It is rather difficult to derive the density function directly because the composition law for Euler angles is complicated, but some physical reasoning will provide the required function. Consider the motion of the z-axis under the rotations described by γ and β. The first acts as an azimuthal angle while β acts as a polar angle, describing any point on the unit sphere. Thus, we expect a uniform density for γ and a Sin[β] density for the polar angle. The final rotation by α is also azimuthal with a uniform density. Therefore, we expect that $\rho[\omega] = \text{Sin}[\beta]$ and write

$$\begin{aligned}&\int\left(D^{(j)}_{m'm}[\omega]\right)^* D^{(j')}_{n'n}[\omega]\rho[\omega]\,d\omega\\ &\quad= \int d^{(j)}_{m'm}[\beta]d^{(j')}_{n'n}[\beta]\,\text{Exp}[i(m-n)\gamma]\,\text{Exp}\left[i\left(m'-n'\right)\alpha\right]\text{Sin}[\beta]\,d\alpha\,d\beta\,d\gamma\end{aligned} \tag{10.453}$$

where $d^{(j)}$ is real. Orthogonality with respect to magnetic quantum numbers must be provided by the integrations over α and γ. If j, j' are either both integral or both half-integral the differences between magnetic quantum numbers are integral and integration over the range $(0, 2\pi)$ ensures orthogonality, but if one is integral and the other half-integral, the difference in magnetic quantum numbers is half-integral and integration over the range $(0, 4\pi)$ is needed for both α and γ. This should not be surprising, at least in retrospect, because the identity element for subgroups consisting of rotations about a fixed axis is a rotation of 4π for a half-integral spin. Therefore, the *covering group* capable of accommodating both integral and half-integral angular momentum features an enlarged parameter space with $0 \le \alpha \le 4\pi$ and $0 \le \gamma \le 4\pi$.

Orthogonality with respect to j is then left to integration over β. Treating $d^{(j)}[\beta]$ as representations of a one-parameter subgroup, we expect

$$\int_0^\pi d^{(j)}_{m'm}[\omega] d^{(j')}_{m'm}[\omega] \,\mathrm{Sin}[\beta]\, d\beta = \frac{\delta_{j,j'}}{2j+1} \int_0^\pi \mathrm{Sin}[\beta]\, d\beta \tag{10.454}$$

or

$$\int_0^\pi d^{(j)}_{m'm}[\omega] d^{(j')}_{m'm}[\omega] \,\mathrm{Sin}[\beta]\, d\beta = \frac{2}{2j+1} \delta_{j,j'} \tag{10.455}$$

It would be rather difficult to obtain this result directly from the power series for $d^{(j)}$ without using the underlying theorems from group theory. We can now put everything together to obtain

$$\int_0^{4\pi} d\alpha \int_0^\pi d\beta \,\mathrm{Sin}[\beta] \int_0^{4\pi} d\gamma D^{(j)}_{m'm}[\alpha,\beta,\gamma] \left(D^{(j')}_{n'n}[\alpha,\beta,\gamma]\right)^* = \frac{32\pi^2}{2j+1} \delta_{j,j'} \delta_{m,n} \delta_{m',n'} \tag{10.456}$$

as the orthogonality theorem for representations of the rotational covering group without performing arduous integrations.

10.7.6 Coupling of Angular Momenta

Direct products of two irreducible representations of the rotation group produce a Clebsch–Gordan expansion

$$D^{(j_1 \otimes j_2)} = \sum_j (j_1 j_2 \mid j) D^{(j)} \tag{10.457}$$

Recall that all rotations through the same angle belong to the same class regardless of axis. Thus, the easiest way to determine the reduction coefficients is to evaluate the character

$$\chi^{(j_1 \otimes j_2)}[\omega] = \chi^{(j_1)}[\omega] \chi^{(j_2)}[\omega] \tag{10.458}$$

for $\omega = \{0, 0, \phi\}$ to obtain

$$\sum_j \sum_{m=-j}^{j} (j_1 j_2 \mid j) \,\mathrm{Exp}[-im\phi] = \sum_{m_1=-j_1}^{j_1} \sum_{m_2=-j_2}^{j_2} \mathrm{Exp}\left[-i\left(m_1+m_2\right)\phi\right] \tag{10.459}$$

Multiplying both sides by $\mathrm{Exp}[in\phi]$ and integrating over $0 \le \phi \le 4\pi$ to account for possible half-integral quantum numbers, we obtain

$$\sum_j \sum_{m=-j}^{j} (j_1 j_2 \mid j) \,\delta_{m,n} = \sum_{m_1=-j_1}^{j_1} \sum_{m_2=-j_2}^{j_2} \delta_{m_1+m_2,n} \tag{10.460}$$

Satisfying these constraints limits j to the range $|j_1 - j_2| \le j \le j_1 + j_2$. The total number of terms on the right-hand side is $(2j_1+1)(2j_2+1)$. If we assume, without loss of generality, that $j_1 \geq j_2$, the total number of terms on the left hand side is

$$\sum_{j=j_1-j_2}^{j_1+j_2} (2j+1) = (2j_1+1)(2j_2+1) \tag{10.461}$$

Therefore, every $(j_1 j_2 | j)$ in the allowed range equals 1, such that

$$D^{(j_1 \otimes j_2)} = D^{(|j_1-j_2|)} \oplus D^{(|j_1-j_2|+1)} \oplus \cdots \oplus D^{(j_1+j_2)} \tag{10.462}$$

shows that the rotation group is simply reducible.

The direct product of two angular momentum eigenstates can now be expanded in terms of irreducible representations according to

$$\psi_{j_1,m_1}\psi_{j_2,m_2} = \sum_{j,m} \left\langle \begin{matrix} j_1 & j_2 \\ m_1 & m_2 \end{matrix} \middle| \begin{matrix} j \\ m \end{matrix} \right\rangle \psi_{j,m} \tag{10.463}$$

where the Clebsch–Gordan coefficients can be obtained by generalizing the relationship

$$\sum_{g\in G} D^{(\mu)}_{i',i}[g] D^{(\nu)}_{j',j}[g] D^{(\gamma)}_{k',k}[g]^* = \frac{N_G}{n_\gamma} \left\langle \begin{matrix} \mu & \nu \\ i' & j' \end{matrix} \middle| \begin{matrix} \gamma \\ k' \end{matrix} \right\rangle \left\langle \begin{matrix} \mu & \nu \\ i & j \end{matrix} \middle| \begin{matrix} \gamma \\ k \end{matrix} \right\rangle^* \tag{10.464}$$

for discrete groups to the continuum according to

$$\int_0^{4\pi} d\alpha \int_0^{\pi} d\beta\, \mathrm{Sin}[\beta] \int_0^{4\pi} d\gamma D^{(j_1)}_{m_1',m_1}[\omega] D^{(j_2)}_{m_2',m_2}[\omega] D^{(j)}_{m',m}[\omega]^*$$
$$= \frac{32\pi^2}{2j+1} \left\langle \begin{matrix} j_1 & j_2 \\ m_1' & m_2' \end{matrix} \middle| \begin{matrix} j \\ m \end{matrix} \right\rangle \left\langle \begin{matrix} j_1 & j_2 \\ m_1 & m_2 \end{matrix} \middle| \begin{matrix} j \\ m \end{matrix} \right\rangle^* \tag{10.465}$$

Notice that for the special cases $m' = \pm j$, the rotation matrices have only a single term:

$$D^{(j)}_{j,m}[\omega] = \left(\frac{(2j)!}{(j+m)!(j-m)!} \right)^{1/2} \mathrm{Sin}\left[\frac{\beta}{2}\right]^{j-m} \mathrm{Cos}\left[\frac{\beta}{2}\right]^{j+m} \mathrm{Exp}[-i(m\gamma + j\alpha)] \tag{10.466}$$

$$D^{(j)}_{-j,m}[\omega] = (-)^{j+m} \left(\frac{(2j)!}{(j+m)!(j-m)!} \right)^{1/2} \mathrm{Sin}\left[\frac{\beta}{2}\right]^{j+m} \mathrm{Cos}\left[\frac{\beta}{2}\right]^{j-m} \mathrm{Exp}[-i(m\gamma - j\alpha)] \tag{10.467}$$

Consider the special case $m_1' = j_1$, $m_2' = -j_2$, for which

$$\int_0^{4\pi} d\alpha \int_0^{\pi} d\beta\, \mathrm{Sin}[\beta] \int_0^{4\pi} d\gamma D^{(j_1)}_{j_1,m_1}[\omega] D^{(j_2)}_{-j_2,m_2}[\omega] D^{(j)}_{j_1-j_2,m}[\omega]^*$$
$$= \frac{32\pi^2}{2j+1} \left\langle \begin{matrix} j_1 & j_2 \\ j_1 & -j_2 \end{matrix} \middle| \begin{matrix} j \\ j_1 - j_2 \end{matrix} \right\rangle \left\langle \begin{matrix} j_1 & j_2 \\ m_1 & m_2 \end{matrix} \middle| \begin{matrix} j \\ m \end{matrix} \right\rangle^* \tag{10.468}$$

and let

$$I = \int_0^{4\pi} d\alpha \int_0^{\pi} d\beta \, \text{Sin}[\beta] \int_0^{4\pi} d\gamma D^{(j_1)}_{j_1,m_1}[\omega] D^{(j_2)}_{-j_2,m_2}[\omega] D^{(j)}_{m',m}[\omega]^*$$
$$= (-)^{j_2+m_2} \left(\frac{(2j_1)!}{(j_1+m_1)!(j_1-m_1)!} \frac{(2j_2)!}{(j_2+m_2)!(j_2-m_2)!} \frac{(j+m')!(j-m')!}{(j+m)!(j-m)!} \right)^{1/2}$$
$$\times \sum_\nu (-)^{j-m'-\nu} \binom{j+m}{j-m'-\nu} \binom{j-m}{\nu}$$
$$\times \int_0^{4\pi} d\alpha \int_0^{\pi} d\beta \, \text{Sin}[\beta] \int_0^{4\pi} d\gamma \, \text{Sin}\left[\frac{\beta}{2}\right]^{j_1-m_1+j_2+m_2+2j-m-m'-2\nu}$$
$$\times \text{Cos}\left[\frac{\beta}{2}\right]^{j_1+m_1+j_2-m_2+m+m'+2\nu} \text{Exp}[-i(m_1+m_2-m)\gamma] \, \text{Exp}[-i(j_1-j_2-m')\alpha] \tag{10.469}$$

where ν is an integer whose range is limited by the binomial coefficients in the expansion. Integration over γ requires $m = m_1 + m_2$ while integration over α requires $m' = j_1 - j_2$, such that

$$I = (4\pi)^2 \delta_{m,m_1+m_2} \delta_{m',j_1-j_2} (-)^{j_2+m_2} \left(\frac{(2j_1)!}{(j_1+m_1)!(j_1-m_1)!} \frac{(2j_2)!}{(j_2+m_2)!(j_2-m_2)!} \right.$$
$$\left. \times \frac{(j-j_1+j_2)!(j+j_1-j_2)!}{(j+m)!(j-m)!} \right)^{1/2}$$
$$\times \sum_\nu (-)^{j-j_1+j_2-\nu} \binom{j+m}{j-j_1+j_2-\nu} \binom{j-m}{\nu} 4 \int_0^{\pi/2} d\eta \, \text{Sin}[\eta]^{2j+2j_2-2m_1-2\nu+1}$$
$$\times \text{Cos}[\eta]^{2j_1+2m_1+2\nu+1} \tag{10.470}$$

Using our old friend the beta function (recall Sec. 2.6.2)

$$\int_0^{\pi/2} \text{Cos}[\theta]^{2p-1} \, \text{Sin}[\theta]^{2q-1} \, d\theta = \frac{\Gamma[p]\Gamma[q]}{2\Gamma[p+q]} = \frac{1}{2} B[p,q] \tag{10.471}$$

the integral reduces to

$$I = 32\pi^2 \delta_{m,m_1+m_2} \delta_{m',j_1-j_2} (-)^{j-j_1+2j_2+m_2}$$
$$\times \left(\frac{(2j_1)!}{(j_1+m_1)!(j_1-m_1)!} \frac{(2j_2)!}{(j_2+m_2)!(j_2-m_2)!} \frac{(j-j_1+j_2)!(j+j_1-j_2)!}{(j+m)!(j-m)!} \right)^{1/2}$$
$$\times \sum_\nu (-)^\nu \binom{j+m}{j-j_1+j_2-\nu} \binom{j-m}{\nu} \frac{\Gamma[j_1+m_1+\nu+1]\,\Gamma[j+j_2-m_1-\nu+1]}{\Gamma[j+j_1+j_2+2]} \tag{10.472}$$

and we obtain

$$\left\langle \begin{matrix} j_1 & j_2 \\ j_1 & -j_2 \end{matrix} \middle| \begin{matrix} j \\ j_1 - j_2 \end{matrix} \right\rangle \left\langle \begin{matrix} j_1 & j_2 \\ m_1 & m_2 \end{matrix} \middle| \begin{matrix} j \\ m \end{matrix} \right\rangle^*$$
$$= \delta_{m,m_1+m_2} \delta_{m',j_1-j_2} (2j+1)(-)^{j-j_1+2j_2+m_2}$$
$$\times \left(\frac{(2j_1)!}{(j_1+m_1)!(j_1-m_1)!} \frac{(2j_2)!}{(j_2+m_2)!(j_2-m_2)!} \frac{(j-j_1+j_2)!(j+j_1-j_2)!}{(j+m)!(j-m)!} \right)^{1/2}$$
$$\times \sum_\nu (-)^\nu \binom{j+m}{j-j_1+j_2-\nu} \binom{j-m}{\nu} \frac{\Gamma[j_1+m_1+\nu+1]\Gamma[j+j_2-m_1-\nu+1]}{\Gamma[j+j_1+j_2+2]} \tag{10.473}$$

as the product of CG coefficients. Substitution of $m_1 = j_1, m_2 = -j_2$ then gives

$$\left| \left\langle \begin{matrix} j_1 & j_2 \\ j_1 & -j_2 \end{matrix} \middle| \begin{matrix} j \\ j_1 - j_2 \end{matrix} \right\rangle \right|^2 = (2j+1)(-)^{j-j_1+j_2} \sum_\nu (-)^\nu \binom{j+j_1-j_2}{j-j_1+j_2-\nu} \binom{j-j_1+j_2}{\nu}$$
$$\times \frac{\Gamma[2j_1+\nu+1]\Gamma[j-j_1+j_2-\nu+1]}{\Gamma[j+j_1+j_2+2]} \tag{10.474}$$

The preceding two equations provide enough information to determine the full set of CG coefficients for any $\{m_1, m_2, m\}$, but we would prefer to perform the summation above in closed form. Here we simply quote the result

$$\left| \left\langle \begin{matrix} j_1 & j_2 \\ j_1 & -j_2 \end{matrix} \middle| \begin{matrix} j \\ j_1 - j_2 \end{matrix} \right\rangle \right|^2 = (2j+1) \frac{(2j_1)!(2j_2)!}{(j_1+j_2-j)!(j_1+j_2+j+1)!} \tag{10.475}$$

because evaluation of the sum is a bit tricky and we are not especially interested, at this point, in techniques for manipulating such expressions; the details are postponed to the exercises. We are free to choose the positive root for this particular CG coefficient. Also note that the fact that $j \pm j_1 \pm j_2$ is an integer helps to simplify the phase. The remaining coefficients then become

$$\left\langle \begin{matrix} j_1 & j_2 \\ m_1 & m_2 \end{matrix} \middle| \begin{matrix} j \\ m \end{matrix} \right\rangle = \delta_{m,m_1+m_2} \Delta[j_1, j_2, j] (-)^{j_1-j+m_2}$$
$$\times \left(\frac{(2j+1)(j_1+j_2-j)!(j+j_1-j_2)!(j-j_1+j_2)!(j+m)!(j-m)!}{(j+j_1+j_2+1)!(j_1+m_1)!(j_1-m_1)!(j_2+m_2)!(j_2-m_2)!} \right)^{1/2}$$
$$\times \sum_\nu (-)^\nu \frac{(j_1+m_1+\nu)!(j+j_2-m_1-\nu)!}{\nu!(j-m-\nu)!(j-j_1+j_2-\nu)!(j_1-j_2+m+\nu)!} \tag{10.476}$$

where the triangle test

$$\Delta[j_, j_2, j] = \begin{cases} 1 & |j_1 - j_2| \le j \le j_1 + j_2 \\ 0 & \text{otherwise} \end{cases} \tag{10.477}$$

denotes the selection rule for j. Notice that all CG coefficients are real.

Explicit formulas without summation can be found for small angular momenta in standard compilations. Those references also provide useful symmetry relations for CG coefficients and related quantities, but we will not pursue those details here.

10.7.7 Spherical Tensors

Operators $T_{j,m}$ that transform among themselves under rotations as components of an irreducible representation of the rotation group, such that

$$RT_{j,m}R^{-1} = \sum_{m'=-j}^{j} T_{j,m'} D^{(j)}_{m',m}[R] \tag{10.478}$$

are known as *spherical tensors* of rank j. We also require phases to ensure that complex conjugation properties are the same as those for spherical harmonics, such that

$$T^*_{j,m} = (-)^m T_{j,-m} \tag{10.479}$$

and we define the scalar product of two spherical tensors according to

$$A_j \cdot B_j = \sum_{m=-j}^{j} A^*_{j,-m} B_{j,m} = \sum_{m=-j}^{j} (-)^* A_{j,-m} B_{j,m} \tag{10.480}$$

Similarly, we define tensor products based upon coupling of angular momentum states according to

$$\left[A_{j_1} \otimes B_{j_2}\right]_{j,m} = \sum_{m_1,m_2} \left\langle \begin{matrix} j_1 & j_2 \\ m_1 & m_2 \end{matrix} \middle| \begin{matrix} j \\ m \end{matrix} \right\rangle A_{j_1 m_1} B_{j_2 m_2} \tag{10.481}$$

Matrix elements of spherical tensors are expressed most simply using the Wigner–Eckart theorem in the form

$$\langle j_f m_f \mid T_{j,m} \mid j_i m_i \rangle = \left\langle \begin{matrix} j_i & j \\ m_i & m \end{matrix} \middle| \begin{matrix} j_f \\ m_f \end{matrix} \right\rangle \langle j_f \parallel T_j \parallel j_i \rangle \tag{10.482}$$

where the reduced matrix element $\langle j_f \| T_j \| j_i \rangle$ is independent of magnetic quantum numbers and can be evaluated using whatever choices for those quantum numbers that prove simplest. Obviously, these matrix elements vanish unless $|j_f - j_i| \le j \le j_f + j_i$ and $m = m_f - m_i$ satisfy the selection rules for angular momentum coupling.

For infinitesimal rotations we expand

$$R \to 1 - i\varepsilon \hat{n} \cdot \vec{J} + O[\varepsilon^2] \Longrightarrow RT_{j,m}R^{-1} \to T_{j,m} - i\varepsilon \hat{n} \cdot [\vec{J}, T_{j,m}] \tag{10.483}$$

where $\hat{n}$ is a unit vector and express the rotation matrix in terms of matrix elements of $\vec{J}$ as

$$D^{(j)}_{m'm}[R] = \langle j, m' \mid \mathrm{Exp}[-i\varepsilon \hat{n} \cdot \vec{J}] \mid j, m\rangle \to \delta_{m,m'} - \langle j, m' \mid i\varepsilon \hat{n} \cdot \vec{J} | j, m\rangle \tag{10.484}$$

to obtain commutation relations

$$[\vec{J}, T_{j,m}] = \sum_{m'} T_{j,m'} \langle j, m' \mid \vec{J} \mid j, m \rangle \tag{10.485}$$

between the components of the angular momentum operator and those of the spherical tensor. Thus, we obtain commutation relations

$$[J_0, T_{j,m}] = m T_{j,m} \tag{10.486}$$

$$[J_\pm, T_{j,m}] = T_{j,m\pm1} \sqrt{j(j+1) - m(m \pm 1)} \tag{10.487}$$

for spherical tensors that are often easier to verify than their transformation properties. Naturally, we find $[\vec{J}, T_{0,0}] = 0$ for scalar operators.

For example, the appropriate linear combinations of a cartesian vector should transform as a spherical tensor of rank 1. To identify those combinations, we express cartesian coordinates in terms of spherical harmonics as

$$\frac{x}{r} = \mathrm{Sin}[\theta]\,\mathrm{Cos}[\phi] = \sqrt{\frac{4\pi}{3}} \frac{1}{\sqrt{2}} (Y_{1,-1}[\theta,\phi] - Y_{1,1}[\theta,\phi]) \tag{10.488}$$

$$\frac{y}{r} = \mathrm{Sin}[\theta]\,\mathrm{Sin}[\phi] = \sqrt{\frac{4\pi}{3}} \frac{i}{\sqrt{2}} (Y_{1,-1}[\theta,\phi] + Y_{1,1}[\theta,\phi]) \tag{10.489}$$

$$\frac{z}{r} = \mathrm{Cos}[\theta] = \sqrt{\frac{4\pi}{3}} Y_{1,0}[\theta,\phi] \tag{10.490}$$

Thus, it is useful to define spherical unit vectors as

$$\hat{e}_{+1} = -\frac{1}{\sqrt{2}} (\hat{x} + i\hat{y}) \tag{10.491}$$

$$\hat{e}_{-1} = \frac{1}{\sqrt{2}} (\hat{x} - i\hat{y}) \tag{10.492}$$

$$\hat{e}_0 = \hat{z} \tag{10.493}$$

with the same conjugation property

$$\hat{e}_m^* = (-)^m \hat{e}_{-m} \tag{10.494}$$

as spherical harmonics and with the orthonormality condition

$$\langle \hat{e}_{m'} \mid \hat{e}_m \rangle = \hat{e}_{m'}^* \cdot \hat{e}_m = \delta_{m,m'} \tag{10.495}$$

Spherical components of a cartesian vector are then obtained using

$$r_m = \langle \hat{e}_m \mid \vec{r} \rangle \tag{10.496}$$

such that

$$r_{+1} = \langle \hat{e}_{+1} \mid \vec{r} \rangle = -x + iy = -\sqrt{\frac{4\pi}{3}}\, rY_{1,-1}[\theta, \phi] \tag{10.497}$$

$$r_{-1} = \langle \hat{e}_{-1} \mid \vec{r} \rangle = x + iy = -\sqrt{\frac{4\pi}{3}}\, rY_{1,1}[\theta, \phi] \tag{10.498}$$

$$r_0 = \langle \hat{e}_0 \mid \vec{r} \rangle = z = \sqrt{\frac{4\pi}{3}}\, rY_{1,0}[\theta, \phi] \tag{10.499}$$

Thus,

$$\vec{r} = r\sqrt{\frac{4\pi}{3}} \sum_{m=-1}^{1} (-)^m Y_{1,-m} \hat{e}_m = r\sqrt{\frac{4\pi}{3}}\, Y_1 \cdot \hat{e}_1 \tag{10.500}$$

expresses the basic cartesian vector in a spherical basis using polar coordinates.

The angular momentum operator $\vec{J}$ is itself a spherical tensor. Thus, we can express its components in the form

$$J_{+1} = \langle \hat{e}_{+1} \mid \vec{J} \rangle = -\frac{1}{\sqrt{2}}(J_x + iJ_y) = -\frac{1}{\sqrt{2}} J_+ \tag{10.501}$$

$$J_{-1} = \langle \hat{e}_{-1} \mid \vec{J} \rangle = \frac{1}{\sqrt{2}}(J_x - iJ_y) = \frac{1}{\sqrt{2}} J_- \tag{10.502}$$

$$J_0 = \langle \hat{e}_0 \mid \vec{J} \rangle = J_z \tag{10.503}$$

where one must not confuse the raising and lowering operators, J_+ and J_-, with the spherical components, J_{+1} and J_{-1}, despite their notational similarities. Matrix elements then take the form

$$\langle j, m' \mid J_{\pm 1} \mid j, m \rangle = \mp\sqrt{(j \mp m)(j \pm m + 1)/2}\, \delta_{m',m\pm 1} \tag{10.504}$$

$$\langle j, m' \mid J_0 \mid j, m \rangle = m\delta_{m',m} \tag{10.505}$$

and we can use the special value

$$\left\langle \begin{matrix} j & 1 \\ m & 0 \end{matrix} \middle| \begin{matrix} j \\ m \end{matrix} \right\rangle = \frac{m}{\sqrt{j(j+1)}} \tag{10.506}$$

to deduce the reduced matrix element

$$\left\langle j' \parallel J_\mu \parallel j \right\rangle = \delta_{j,j'} \sqrt{j(j+1)} \tag{10.507}$$

where $\mu = 0, \pm 1$. Thus, the commutation relations for a spherical tensor can now be expressed in the form

$$[J_\mu, T_{j,m}] = \sqrt{j(j+1)} \left\langle \begin{matrix} j & 1 \\ m & \mu \end{matrix} \middle| \begin{matrix} j \\ m+\mu \end{matrix} \right\rangle T_{j,m+\mu} \tag{10.508}$$

Finally, if $T_{j,m} = V_m$ is a vector operator with $j = 1$, we can express these commutation relations in cartesian form as

$$[J_i, V_j] = i\varepsilon_{i,j,k} V_k \tag{10.509}$$

where the summation convention is employed and $i, j, k \in \{x, y, z\}$. Notice that this relationship contains the SU(2) commutation relations as a special case. Although we leave the algebra to the exercises, this form is very helpful in manipulating operators composed of products of $\vec{r}$, $\vec{p}$, and $\vec{L}$.

10.8 Unitary Symmetries in Nuclear and Particle Physics

No discussion of group theory in physics would be complete without at least mentioning the assignment of elementary particles to multiplets and the dynamical consequences of their associated group symmetries. The most obvious multiplets are those associated with isospin symmetry in nuclear physics. The difference between the masses of the proton and neutron is so small, only 0.14 %, that they can be considered to be two states of the same particle, the *nucleon*, that are degenerate with respect to the strong interaction while the small splitting is attributed to the much weaker electromagnetic interaction. Thus, the nucleon (N) is assigned an internal variable, called *isospin*, that is analogous to spin. We assume that isospin states transform according SU(2) with generators T_i satisfying the commutation relations

$$\left[T_i, T_j\right] = i\varepsilon_{i,j,k} T_k \tag{10.510}$$

and that the proton and nucleon belong to a dimension-two representation with

$$T^2|N\rangle = \tfrac{3}{4}|N\rangle \tag{10.511}$$

$$T_3|p\rangle = +\tfrac{1}{2}|p\rangle \tag{10.512}$$

$$T_3|n\rangle = -\tfrac{1}{2}|n\rangle \tag{10.513}$$

Similarly, the three charge states of the pion (π^+, π^0, π^-) are practically degenerate and are assigned to a dimension-three representation of SU(2) isospin. The most prominent feature of low-energy interactions between pions and nucleons is a strong resonance in the $l = 1$ partial wave with four charge states labeled Δ^{++}, Δ^+, Δ^0, and Δ^- that are assigned isospin $\frac{3}{2}$. Expressing the charge states

$$|\pi^+ p\rangle = \left|\begin{matrix}3/2\\3/2\end{matrix}\right\rangle \tag{10.514}$$

$$|\pi^0 p\rangle = \sqrt{\frac{2}{3}}\left|\begin{matrix}3/2\\1/2\end{matrix}\right\rangle - \sqrt{\frac{1}{3}}\left|\begin{matrix}1/2\\1/2\end{matrix}\right\rangle \tag{10.515}$$

$$|\pi^- p\rangle = \sqrt{\frac{1}{3}}\left|\begin{matrix}3/2\\-1/2\end{matrix}\right\rangle - \sqrt{\frac{2}{3}}\left|\begin{matrix}1/2\\-1/2\end{matrix}\right\rangle \tag{10.516}$$

$$|\pi^+ n\rangle = \sqrt{\frac{1}{3}}\left|\begin{matrix}3/2\\1/2\end{matrix}\right\rangle + \sqrt{\frac{2}{3}}\left|\begin{matrix}1/2\\1/2\end{matrix}\right\rangle \tag{10.517}$$

$$|\pi^0 n\rangle = \sqrt{\frac{2}{3}}\left|\begin{matrix}3/2\\-1/2\end{matrix}\right\rangle + \sqrt{\frac{1}{3}}\left|\begin{matrix}1/2\\-1/2\end{matrix}\right\rangle \tag{10.518}$$

$$|\pi^- n\rangle = \left|\begin{matrix}3/2\\-3/2\end{matrix}\right\rangle \tag{10.519}$$

in terms of isospin eigenstates and assuming that the isospin $\frac{1}{2}$ contributions to elastic πN scattering are negligible at the resonance, isospin symmetry predicts that cross-sections for the charge states will be found in the ratios

$$\sigma\left[\pi^+ p\right] : \sigma\left[\pi^0 p\right] : \sigma\left[\pi^- p\right] = \sigma\left[\pi^- n\right] : \sigma\left[\pi^0 n\right] : \sigma\left[\pi^+ n\right] = 9 : 2 : 1 \tag{10.520}$$

and data for the resonance region are indeed consistent with these ratios. Isospin symmetry also plays a central role in the structure of atomic nuclei.

At a deeper level, isospin symmetry is understood to arise from the *flavor symmetry* of the interaction between quarks and the near degeneracy between the masses of the lightest quarks, called up and down (u or d) based upon the older isospin up or isospin down terminology. It is natural to try to extend this symmetry to include also strange baryons containing one or more strange quarks even though the s quark is significantly more massive than the u or d quarks. The mass degeneracy within the resulting SU(3) multiplets is less precise, but SU(3) flavor symmetry does help to explain many of the properties of baryons with useful accuracy. We will not discuss the elegant methods that have been developed for analyzing the structure of direct-product representations of SU(N), but merely quote a couple of results. The coupling of three quarks, each belonging to SU(3), can be decomposed into irreducible representations according to

$$3 \otimes 3 \otimes 3 = 1 \oplus 8_{M,S} \oplus 8_{M,A} \oplus 10 \tag{10.521}$$

where the singlet is fully antisymmetric, one octet is symmetric and the other antisymmetric with respect to exchange of the first two quarks, and the decuplet is symmetric with respect to the exchange of any pair of quarks. The mixed-symmetry representations (MS and MA) can be constructed by coupling two quarks in either symmetric or antisymmetric configurations and then coupling a third. The singlet is identified with the $\Lambda(1405)$ particle, while the other low-lying baryons are assigned to multiplets represented by the accompanying *weight diagrams*. The hypercharge Y is defined so that the electric charge

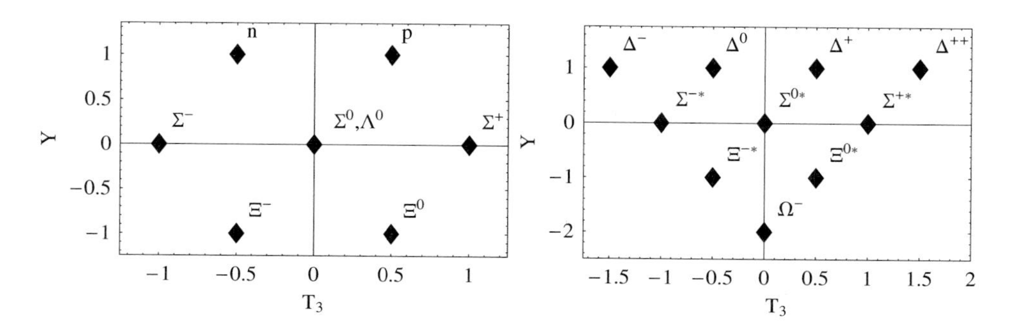

Figure 10.4. Weight diagrams for baryons assigned to symmetric octet (left) and decuplet (right) representations of SU(3) flavor symmetry.

is $Q = T_3 + Y/2$. Each member of $8_{M,S}$ has spin and parity $J^P = \frac{1}{2}^+$ while the decuplet has $J^P = \frac{3}{2}^+$.

This model of the baryon spectrum was called the *eightfold way* by Gell-Mann, co-inventor of the quark model with Zweig, because the SU(3) group has eight generators. The prediction and subsequent discovery of the Ω^- was a major success of the early quark model. Even when symmetries apply only to one sector of a model and are broken by another, group theoretical analyses often explain systematics well without detailed theories or calculations of the underlying dynamics. Subsequent developments in particle theory have relied heavily on such techniques, but are beyond the scope of this text.

Problems for Chapter 10

1. Which are groups?

Which of the following sets constitute groups under the stipulated law of combination? Justify your answers:

a) real numbers under multiplication

b) real numbers under addition

c) all complex numbers except 0 under multiplication

d) rational numbers under multiplication

e) rational numbers under $a * b = a/b$.

2. Groups of order 4

Prove that groups of order 4 must be abelian and that there are only two distinct groups of order 4.

3. Class period

Show that all elements of the same class have the same period.

4. Quaternions

Just as complex numbers are generalizations of real numbers to two dimensions, quaternions are generalizations of complex numbers to four dimensions. Suppose that $q = a + bi + cj + dk$ where a, b, c, d are real numbers and the quaternion basis vectors $1, i, j, k$ satisfy the following multiplication rules

$$i^2 = j^2 = k^2 = -1 \quad ij = k, \quad ji = -k \tag{10.522}$$

a) Show that the set $\{\pm 1, \pm i, \pm j, \pm k\}$ constitute a group, Q, and that the set of all q is an algebra.

b) Determine the classes and construct the character table for this group.

c) Construct a faithful four-dimensional representation and resolve it into irreducible representations.

5. Commutator subgroups

Suppose that a and b are elements of group G. The commutator for $a\,b$ is defined as the element $c \in G$ for which $c(ab) = ba$. Let H consist of all elements of G that can be expressed as products of commutators, such that $h = c_1 c_2 \cdots c_n$. Prove that H is an invariant subgroup of G.

6. Prove that factor groups are abelian

Suppose that H is an invariant subgroup of G. The factor group is defined by $F = G/H = \{H, g_1 H, g_2 H, \dots\}$ where the g_i are all elements of G that are not contained in H. Prove that F is abelian.

7. 3d representation of S_3

Construct and verify a faithful three-dimensional representation of S_3.

8. Group of matrices with unit determinant

Show that the set of all $n \times n$ complex matrices with unit determinant is an invariant subgroup of the nonsingular complex $n \times n$ matrices and that the corresponding factor group is isomorphic to the set of complex numbers excluding 0.

9. Irreps of abelian groups

Show that every irreducible representation of an abelian group is one-dimensional.

10. Symmetries of a square

Show that the symmetries of a square are a subgroup of S_4. Display the group multiplication table and determine the classes and proper subgroups.

11. Rotational symmetries of a tetrahedron

A regular tetrahedron is a solid that contains four sides that are equilateral triangles.

a) Show that the rotational symmetries of a tetrahedron are a subgroup of S_4. (We exclude twists of a single face.)

b) Display the group multiplication table and determine its classes.

c) Construct the character table.

12. Character table for S_4

Construct the character table for S_4. (Hint: if you encounter a situation in which there are two solutions for one of the columns or rows, you should be able to choose the desired solution by considering the two-dimensional representation for elements of order 2.)

13. Vibrating triangle

Three equal masses m are connected by three equal springs k forming an equilateral triangle in its equilibrium configuration.

a) Construct the potential-energy matrix and demonstrate explicitly that it is invariant with respect to S_3.

b) Use group-theoretical arguments to deduce the eigenvalues for small-amplitude vibrations of this system without solving a secular equation.

14. Vibrating square with central mass

Suppose that four masses m occupy the vertices of a square and that another is at the center. The corner masses are connected by four springs k and the corners are connected to the central mass by four springs κk.

a) Construct the potential energy matrix for small-amplitude vibrations and verify that it is symmetric wrt to the symmetry group of a square.

b) Use group theory to obtain the vibrational frequencies. (Hint: use $\mathrm{Tr}[V.V]$ to distinguish between multiple eigenvalues.)

15. Infinite groups are not necessarily reducible

We showed that representations of finite groups are equivalent to unitary representations. As such, if a representation is partially reducible it is fully reducible to block diagonal form. This is not necessarily true for infinite groups.

a) Show that

$$L[l] = \begin{pmatrix} 1 & l \\ 0 & 1 \end{pmatrix} \tag{10.523}$$

for real l is a representation of a Lie group and identify the group.

b) Show that $L[l]$ cannot be reduced and is not unitary.

16. Monomial representations of translation group

Show that the set $P_{n+1} = \{1, x, \cdots, x^n\}$ provides a representation basis of the translation group $T[a]x = x + a$ and determine its transformation matrices. Verify explicitly that these transformation matrices satisfy the group multiplication law $T[a]T[b] = T[a+b]$. Is this representation reducible? Is it unitary; if not, why not?

17. Time evolution operator for two-state system

A two-state system has the Hamiltonian

$$H = H_0 \mathbb{1} - \mu \vec{\sigma} \cdot \vec{B} \tag{10.524}$$

where H_0 is constant, μ is the magnetic moment, and B is a constant magnetic field. The easiest method to evaluate the eigenvalues and eigenvectors is to choose the quantization axis parallel to $\vec{B}$, but the analysis in terms of an arbitrary direction provides valuable practice with matrix exponentiation and the use of rotation matrices.

a) Use matrix exponentiation to evaluate the time evolution operator for arbitrary $\vec{B}$.

b) If the field were along the z-axis, we would write the eigenvectors as (1,0) and (0,1) with respect to that axis. For an arbitrary direction, we should be able to obtain the eigenvectors using the spin-$\frac{1}{2}$ rotation matrix. Verify that states constructed in that manner are indeed eigenstates of H.

18. SO(4)

The group SO(4) consists of all orthogonal matrices in four dimensions that have unit determinant, thereby leaving the Euclidean metric $x_1^2 + x_2^2 + x_3^2 + x_4^2$ invariant.

a) Show that any matrix belonging to SO(4) can be expanded in the form

$$M = \mathrm{Exp}\left[i \sum_{i=1}^{3} \left(a_i A_i + b_i B_i \right) \right] \tag{10.525}$$

where the nonvanishing elements of the basis matrices are

$$\left(A_i\right)_{j,k} = -i\varepsilon_{i,j,k}, \quad 1 \le j, k \le 3 \tag{10.526}$$

$$\left(B_i\right)_{i,4} = -i \tag{10.527}$$

$$\left(B_i\right)_{4,i} = i \tag{10.528}$$

and where the coefficients a_i, b_i are real numbers. Evaluate the commutation relations for this basis.

b) Show that the six generators for the associated Lie group of transformations acting upon these coordinates take the form

$$\{A_1, A_2, A_3, B_1, B_2, B_3\} = \{A_{2,3}, A_{3,1}, A_{1,2}, A_{1,4}, A_{2,4}, A_{3,4}\} \tag{10.529}$$

$$A_{i,j} = -i\left(x_i \frac{\partial}{\partial x_j} - x_j \frac{\partial}{\partial x_i} \right) \tag{10.530}$$

and satisfy the following commutation relations.

$$\left[A_i, A_j\right] = i\varepsilon_{i,j,k} A_k \tag{10.531}$$

$$\left[A_i, B_j\right] = i\varepsilon_{i,j,k} B_k \tag{10.532}$$

$$\left[B_i, B_j\right] = i\varepsilon_{i,j,k} A_k \tag{10.533}$$

c) Show that the linear transformation

$$J_i = \tfrac{1}{2}\left(A_i + B_i\right) \tag{10.534}$$

$$K_i = \tfrac{1}{2}\left(A_i - B_i\right) \tag{10.535}$$

allows one to decompose SO(4) = SO(3) ⊕ SO(3) into two invariant subalgebras and deduce the commutation relations for the new operators. What is the rank of SO(4)? Is it simple? Is it semisimple?

d) Determine the Casimir operators for SO(4). Evaluate the corresponding matrices for coordinate transformations.

19. Lorentz group in one spatial dimension

A Lorentz boost $B[v]$ performs the coordinate transformation

$$t' = \gamma(t + vx) \tag{10.536}$$

$$x' = \gamma(x + vt) \tag{10.537}$$

where $\gamma = (1 - v^2)^{-1/2}$.

a) Show that $B[v]$ is a Lie group and deduce its law of composition. Interpret this result.

b) Deduce the infinitesimal generator K, evaluate the finite transformation $\text{Exp}[i\alpha K]$ in closed form, and determine the relationship between α and v.

c) Construct the operator for transformation of functions $f[t, x]$ and discuss its superficial resemblance to the generator for rotations. How is it related to the Galilean transformation?

20. Lorentz group

a) Construct matrices K_i for infinitesimal boosts and L_i for infinitesimal rotations of a Lorentz four-vector. Then evaluate the structure constants for the Lorentz group and verify compliance with the Jacobi identity. What is the rank of this group? It is important to recognize that these matrices comprise a particular representation of the abstract group that is defined in terms of their commutation relations, which are more general.

b) Evaluate the metric and show that the Lorentz group is semisimple. Express the Casimir operator C_1 in terms of $\vec{L}$ and $\vec{K}$ and obtain the corresponding matrix for this particular representation.

c) Show that there are linear combinations of $\vec{L}$ and $\vec{K}$ that allow the Lie algebra to be separated into two disjoint subalgebras; in other words, show that commutation relations for each subalgebra are closed and do not involve operators from the other subalgebra. Then construct the Casimir operators for each subalgebra.

21. Alternative representation of $\vec{S}$ for $s = 1$

The infinitesimal generators derived in the text for the $s = 1$ representation of SU(2) do not diagonalize S_3. Find a similarity transform that diagonalizes S_3, with decreasing eigenvalues on the diagonal, and makes S_1 real. Then verify that the proper commutation relations are obtained in the new basis.

22. SU(3)

SU(3) matrices are conventionally parametrized in the form

$$U = \text{Exp}\left[-\frac{i}{2}\theta\hat{n}\cdot\vec{\lambda}\right] \tag{10.538}$$

where θ is a real number, $\hat{n}$ is a real unit vector that represents an axis in the internal coordinate space, and

$$\lambda_1 = \begin{pmatrix} 0 & 1 & 0 \\ 1 & 0 & 0 \\ 0 & 0 & 0 \end{pmatrix}, \quad \lambda_2 = \begin{pmatrix} 0 & -i & 0 \\ i & 0 & 0 \\ 0 & 0 & 0 \end{pmatrix}, \quad \lambda_3 = \begin{pmatrix} 1 & 0 & 0 \\ 0 & -1 & 0 \\ 0 & 0 & 0 \end{pmatrix}, \tag{10.539}$$

$$\lambda_4 = \begin{pmatrix} 0 & 0 & 1 \\ 0 & 0 & 0 \\ 1 & 0 & 0 \end{pmatrix}, \quad \lambda_5 = \begin{pmatrix} 0 & 0 & -i \\ 0 & 0 & 0 \\ i & 0 & 0 \end{pmatrix}, \quad \lambda_6 = \begin{pmatrix} 0 & 0 & 0 \\ 0 & 0 & 1 \\ 0 & 1 & 0 \end{pmatrix}, \tag{10.540}$$

$$\lambda_7 = \begin{pmatrix} 0 & 0 & 0 \\ 0 & 0 & -i \\ 0 & i & 0 \end{pmatrix}, \quad \lambda_8 = \frac{1}{\sqrt{3}} \begin{pmatrix} 1 & 0 & 0 \\ 0 & 1 & 0 \\ 0 & 0 & -2 \end{pmatrix} \tag{10.541}$$

are traceless hermitian 3×3 matrices. The appearance of Pauli matrices in the upper-left blocks of $\{\lambda_1, \lambda_2, \lambda_3\}$ reveals an SU(2) subgroup.

a) Tabulate the structure constants for this group and verify that they meet the requirements of a Lie algebra.

b) Show that anticommutators can be expressed in the form

$$\{\lambda_i, \lambda_j\} = \lambda_i \lambda_j + \lambda_j \lambda_i = \delta_{i,j} + d_{i,j}^k \lambda_k \tag{10.542}$$

and tabulate $d_{i,j}^k$.

c) Evaluate Tr $\lambda_i \lambda_j$.

23. Baker–Haussdorff identity

Verify the Baker–Haussdorff identity.

24. Does orbital angular momentum $l = 1/2$ exist?

Show that the ladder operators for orbital angular momentum take the forms

$$L_\pm = \pm \operatorname{Exp}[\pm i\phi] \left(\frac{\partial}{\partial\theta} \pm i \operatorname{Cot}[\theta] \frac{\partial}{\partial\phi} \right) \tag{10.543}$$

in polar coordinates. Let the eigenfunctions for a hypothetical system with $l = \frac{1}{2}$ be represented in the form

$$\psi_{1/2} = u[\theta] e^{i\phi/2} \tag{10.544}$$

$$\psi_{-1/2} = v[\theta] e^{-i\phi/2} \tag{10.545}$$

and use the requirements

$$L_+ \psi_{1/2} = 0, \quad L_- \psi_{-1/2} = 0 \tag{10.546}$$

to obtain first-order differential equations for $u[\theta]$ and $v[\theta]$. Finally, show that

$$L_- u \neq av, \quad L_+ v \neq bu \tag{10.547}$$

where a and b are constants. Therefore, a representation of SO(3) with $l = \frac{1}{2}$ does not exist. More generally, one can show that eigenfunctions of L^2 and L_3 must have integral l.

25. Weights for one-dimensional Lorentz transformations

Recall that the law of composition for the group of one-dimensional Lorentz transformations is expressed either as

$$v_3 = \phi\left[v_1, v_2\right] = \frac{v_1 + v_2}{1 + v_1 v_2} \tag{10.548}$$

for the velocity parametrization or as

$$\eta_3 = \varphi\left[\eta_1, \eta_2\right] = \eta_1 + \eta_2 \tag{10.549}$$

in terms of rapidity $v = \mathrm{Tanh}[\eta]$. Evaluate the density function for group integration using both parametrizations and then re-express those results in terms of dp/E.

26. Compare left and right densities

Suppose that matrices of the form

$$\begin{pmatrix} e^a & b \\ 0 & 1 \end{pmatrix} \tag{10.550}$$

represent a Lie group. Determine the group multiplication laws for $\{a, b\}$ and then evaluate both left and right invariant densities for its parameter space.

27. Density for Euclidean transformations

a) The transformations of the Euclidean plane that preserve distances can be parametrized by the three-parameter group

$$x' = a + x\,\mathrm{Cos}[\theta] + y\,\mathrm{Sin}[\theta] \tag{10.551}$$

$$y' = b - x\,\mathrm{Sin}[\theta] + y\,\mathrm{Cos}[\theta] \tag{10.552}$$

Evaluate the left and right invariant densities.

b) Alternatively, the Euclidean symmetries can be parametrized as

$$x' = \rho\,\mathrm{Cos}[\phi] + x\,\mathrm{Cos}[\theta] + y\,\mathrm{Sin}[\theta] \tag{10.553}$$

$$y' = \rho\,\mathrm{Sin}[\phi] - x\,\mathrm{Sin}[\theta] + y\,\mathrm{Cos}[\theta] \tag{10.554}$$

Determine the density function for this parametrization.

28. Density for SU(2)

a) Show that

$$\begin{pmatrix} p_0 + i p_3 & p_2 + i p_1 \\ -p_2 + i p_1 & p_0 - i p_3 \end{pmatrix} \tag{10.555}$$

with real p_i constrained according to $p_0^2 = 1 - p_1^2 - p_2^2 - p_3^2$ is a parametrization of SU(2) with three parameters. Evaluate the left and right densities.

b) Show that the angular parametrization

$$\begin{pmatrix} e^{i\phi}\,\mathrm{Cos}[\theta] & e^{i\psi}\,\mathrm{Sin}[\theta] \\ -e^{-i\psi}\,\mathrm{Sin}[\theta] & e^{-i\phi}\,\mathrm{Cos}[\theta] \end{pmatrix} \tag{10.556}$$

is also a parametrization of SU(2) and use the preceding result to deduce the corresponding density without working out the composition laws for these parameters.

29. Homomorphism between SU(2) and SO(3)

Here we ask you to use *MATHEMATICA*® or other symbolic manipulation software to complete some of the steps in the demonstration of the homomorphism between SU(2) and SO(3). This can be done by hand, of course, but would be tedious. Let

$$U = \begin{pmatrix} p_0 + i p_3 & p_2 + i p_1 \\ -p_2 + i p_1 & p_0 - i p_3 \end{pmatrix} \tag{10.557}$$

where the parameters are real and where

$$p_0^2 = 1 - p_1^2 - p_2^2 - p_3^2 \tag{10.558}$$

ensures unitarity. Also let

$$\rho = \vec{r}\cdot\vec{\sigma} = \begin{pmatrix} z & x - iy \\ x + iy & -z \end{pmatrix} \tag{10.559}$$

Evaluate $\rho' = U\rho U^{-1}$ and deduce the 3×3 matrix A that performs the rotation $\vec{r}' = A\vec{r}$. Demonstrate that A is orthogonal and that $\mathrm{Det}[A] = +1$.

30. Rotations using space-fixed or body-fixed axes

The text describes two methods for parametrizing rotations in terms of Euler angles. The first,

$$R[\alpha,\beta,\gamma] = R_{z''}[\gamma]R_{y'}[\beta]R_{z}[\alpha], \tag{10.560}$$

employs a sequence of transformations based upon axes embedded in the physical system, described as *body-fixed* axes. Each transformation produces a new set of coordinate axes. The second,

$$R[\alpha,\beta,\gamma] = R_{z}[\alpha]R_{y}[\beta]R_{z}[\gamma], \tag{10.561}$$

describes a sequence of transformations using a common set of axes, described as *space-fixed* axes. Prove that both methods produce the same rotation.

31. Commutation relations for orbital angular momentum

Verify that the differential operators for orbital angular momentum satisfy the commutation relations $[L_i, L_j] = i\varepsilon_{i,j,k}L_k$.

32. Explicit formula for special CG coefficient
We derived the formula

$$\left|\left\langle \begin{matrix} j_1 & j_2 \\ j_1 & -j_2 \end{matrix} \middle| \begin{matrix} j \\ j_1 - j_2 \end{matrix} \right\rangle\right|^2 = (2j+1)(-)^{j-j_1+j_2} \times \sum_\nu (-)^\nu \begin{pmatrix} j+j_1-j_2 \\ j-j_1+j_2-\nu \end{pmatrix} \begin{pmatrix} j-j_1+j_2 \\ \nu \end{pmatrix} \times \frac{\Gamma[2j_1+\nu+1]\Gamma[j-j_1+j_2-\nu+1]}{\Gamma[j+j_1+j_2+2]} \tag{10.562}$$

in the text and would like to perform the summation. Use the identity

$$\sum_\nu (-)^\nu \frac{\Gamma[n+\nu+1]}{\Gamma[m+\nu+1]} \begin{pmatrix} k \\ \nu \end{pmatrix} = \frac{\Gamma[k+m-n]\Gamma[n+1]}{\Gamma[k+m+1]\Gamma[m-n]} \tag{10.563}$$

to obtain

$$\left|\left\langle \begin{matrix} j_1 & j_2 \\ j_1 & -j_2 \end{matrix} \middle| \begin{matrix} j \\ j_1 - j_2 \end{matrix} \right\rangle\right|^2 = (2j+1)\frac{(2j_1)!\,(2j_2)!}{(j_1+j_2-j)!\,(j_1+j_2+j+1)!} \tag{10.564}$$

33. Integration of three spherical harmonics
Evaluation of matrix elements of spherical tensors often requires integration of a product of three spherical harmonics. Evaluate

$$\int Y_{l_1,m_1}[\theta,\phi]Y_{l_2,m_2}[\theta,\phi]Y^*_{l_3,m_3}[\theta,\phi]\,d\Omega \tag{10.565}$$

in terms of Clebsch–Gordan coefficients. Then deduce the reduced matrix element $\langle l_3 \parallel Y_{l_2} \parallel l_1\rangle$.

34. Commutation relations for products
Demonstrate the following commutation relations for products.

$$[A, BC] = B[A, C] + [A, B]C \tag{10.566}$$

$$[AB, C] = A[B, C] + [A, C]B \tag{10.567}$$

35. Commutation relations for vectors

a) Verify the commutation relation

$$[L_i, V_j] = i\varepsilon_{i,j,k}V_k \tag{10.568}$$

where $\vec{L}$ is the orbital angular momentum and $\vec{V}$ is an arbitrary vector operator. Thus, show that $\vec{L}\cdot\vec{V} = \vec{V}\cdot\vec{L}$ for any vector operator $\vec{V}$.

b) Verify $[\vec{L}, \vec{A}\cdot\vec{B}] = [L^2, \vec{A}\cdot\vec{B}] = 0$ for arbitrary vector operators $\vec{A}$ and $\vec{B}$. Do not assume that $\vec{A}$ and $\vec{B}$ commute.

c) Use commutation relations to verify that $\vec{A} \times \vec{B}$ transforms as a vector under rotations. Do not assume that $\vec{A}$ and $\vec{B}$ commute.

d) Next, show that

$$\vec{V} \times \vec{L} + \vec{L} \times \vec{V} = 2i\vec{V} \tag{10.569}$$

for any $\vec{V}$. For example, the identity

$$\hat{r} = \frac{1}{2i}\left(\hat{r} \times \vec{L} + \vec{L} \times \hat{r}\right) \tag{10.570}$$

is often useful.

e) Show that

$$\left[L^2, \vec{V}\right] = i\left(\vec{V} \times \vec{L} - \vec{L} \times \vec{V}\right) = 2\left(\vec{V} + i\vec{V} \times \vec{L}\right) \tag{10.571}$$

for any $\vec{V}$. Thus, $[L^2, \vec{L}] = 0$.

36. Commutation relations involving $\vec{r}$, $\vec{p}$, and $\vec{L}$
Derive each of the commutation relations listed below. Here $\vec{p} = -i\vec{\nabla}$, $\vec{L} = \vec{r} \times \vec{p}$, $\hat{r} = \vec{r}/r$.

a) $\left[p^2, \hat{r}\right] = \frac{2}{r^2}\left(\hat{r} + i\hat{r} \times \vec{L}\right) = \frac{i}{r^2}\left(\hat{r} \times \vec{L} - \vec{L} \times \hat{r}\right)$

b) $\left[\vec{p} \times \vec{L}, \frac{1}{r}\right] = \frac{i}{r^2}\hat{r} \times \vec{L}$
$\left[\vec{L} \times \vec{p}, \frac{1}{r}\right] = \frac{i}{r^2}\hat{L} \times \vec{r}$
$\left[\vec{p} \times \vec{L} - \vec{L} \times \vec{p}, \frac{1}{r}\right] = \frac{i}{r^2}\left(\hat{r} \times \vec{L} - \hat{L} \times \vec{r}\right)$

c) $\vec{p} \times \vec{L} \cdot \vec{p} \times \vec{L} = p^2 L^2$
$\vec{p} \times \vec{L} \cdot \vec{L} \times \vec{p} = -p^2\left(L^2 + 4\right)$
$\vec{L} \times \vec{p} \cdot \vec{p} \times \vec{L} = -p^2 L^2$
$\vec{L} \times \vec{p} \cdot \vec{L} \times \vec{p} = p^2 L^2$

37. Runge–Lenz vector for hydrogenic atoms: An example of dynamical symmetry
In this problem we use group-theoretical methods to analyze the spectrum of hydrogenic atoms without ever solving a differential equation. The method relies upon a generalization of the Runge–Lenz vector which, for the Kepler problem, points in the direction of the major axis of an elliptical orbit. The conservation of the classical Runge–Lenz vector demonstrates that there is a dynamical symmetry for the inverse-square law that leads to closed elliptical orbits; small violations of this symmetry in general relativity are responsible for the famous precession of the perihelion of Mercury. One can show that the n^2 degeneracy of energy level E_n in the Bohr model of hydrogenic atoms arises from a similar dynamical symmetry described by conservation of a quantum mechanical generalization of the Runge–Lenz vector. Here we guide the student through an analysis that is relatively straightforward, even if some of the algebra is tedious.

a) Show that the symmetrized *Runge–Lenz vector*

$$\vec{\Lambda} = \alpha\left(\vec{p} \times \vec{L} - \vec{L} \times \vec{p}\right) - \hat{r} \tag{10.572}$$

is hermitian and that it commutes with the hamiltonian

$$H = \frac{p^2}{2\mu} - \frac{\kappa}{r} \tag{10.573}$$

for a suitable choice of α (real). Thus, there is a dynamical symmetry and the full symmetry group is based upon the six generators $\vec{L}$ and $\vec{\Lambda}$.

b) Show that

$$\Lambda^2 = 1 + \frac{2H}{\mu\kappa^2}\left(L^2 + 1\right) \tag{10.574}$$

c) Show that the symmetry operators satisfy the following commutation relations:

$$\left[L_i, L_j\right] = i\varepsilon_{i,j,k} L_k \tag{10.575}$$

$$\left[L_i, \Lambda_j\right] = i\varepsilon_{i,j,k} \Lambda_k \tag{10.576}$$

$$\left[\Lambda_i, \Lambda_j\right] = -\frac{2H}{\mu\kappa^2} i\varepsilon_{i,j,k} L_k \tag{10.577}$$

d) The presence of H in the preceding commutation relations shows that $\{\vec{\Lambda}, \vec{L}\}$ do not form a Lie group with respect to the full Hilbert space. However, if we define new Runge–Lenz operators according to

$$\vec{\Lambda} = \sqrt{-\frac{2H}{\mu k^2}}\, \vec{\lambda} \tag{10.578}$$

where $H \to E$ within any subspace with common energy $E < 0$, the commutators for $\{\vec{\lambda}, \vec{L}\}$ do form a closed algebraic system. Demonstrate that the new system is closed. Then evaluate the commutation relations for the linear combinations

$$\vec{J} = \tfrac{1}{2}\left(\vec{L} + \vec{\lambda}\right) \tag{10.579}$$

$$\vec{K} = \tfrac{1}{2}\left(\vec{L} - \vec{\lambda}\right) \tag{10.580}$$

and prove that their structure constants satisfy the conditions required for a Lie group. Construct the Casimir operators, identify a complete set of quantum numbers, and specify their allowed values. Note that for this system one of the Casimir operators is actually redundant.

e) Use J^2 to compute the energy eigenvalues and determine the degeneracy of multiplets.

38. Landé formula and deuteron magnetic moment

In this problem you will derive a formula that is often useful in nuclear or atomic spectroscopy and then will apply that formula to analyze the magnetic moment of the deuteron.

a) Suppose that $\vec{V}$ is a vector operator. Show that

$$\langle j \parallel V \parallel j \rangle = \frac{\langle j, m \mid \vec{J} \cdot \vec{V} \mid j, m \rangle}{\sqrt{j(j+1)}} \tag{10.581}$$

for $j > 0$.

b) The magnetic dipole operator for the deuterium nucleus takes the form

$$\vec{\mu} = g_p \vec{S}_p + g_n \vec{S}_n + \tfrac{1}{2} \vec{L} \tag{10.582}$$

where $\vec{S}_p$ and $\vec{S}_n$ are proton and neutron spins and where the factor of $\frac{1}{2}$ for the orbital angular momentum $\vec{L}$ arises because only the proton carries charge and it carries half the total orbital angular momentum. The ground-state wave function can be expressed as

$$|\psi_d\rangle = a \left| {}^3S_1 \right\rangle + b \left| {}^3D_1 \right\rangle \tag{10.583}$$

where $a^2 + b^2 = 1$ and where the standard spectroscopic notation specifies ${}^{2S+1}L_J$. Evaluate the magnetic moment, defined by the matrix element

$$\mu = \langle J, M \mid \mu_0 \mid J, M \rangle \tag{10.584}$$

for the aligned substate with maximum magnetic quantum number. Given that $g_p = 5.586\mu_N$, $g_n = -3.826\mu_N$, and $\mu_d = 0.857\mu_N$ where μ_N is the nuclear magneton, estimate the D-state probability, b^2. Note that this model omits the contributions due to meson exchange between nucleons, but still gives a good approximation to the D-state admixture in the ground state.

Bibliography

Although these sources are not available to the reader, I feel obliged to acknowledge that my most important references were the notes and homework that I saved from similar courses, AMa95 by Prof. Saffman and PH129 by Prof. Peck, at CalTech in the mid-1970s. Below I have compiled a brief bibliography that might be more useful to the reader, with some personal comments. No attempt has been made to quote original sources because I did not use them and they probably would be less useful to most readers than these secondary texts, anyway.

General

1. G. B. Arfken and H. J. Weber, *Mathematical Methods for Physicists*, 6th edition, (Elsevier, Amsterdam, 2005)
   An encyclopedic work that covers a broader range of topics but often at a somewhat lower level than the current text.

2. E. Butkov, *Mathematical Physics*, (Addison-Wesley, Reading MA, 1968)
   Particularly good treatment of generalized functions and the theory of distributions.

3. F. W. Byron and R. W. Fuller, *Mathematics of Classical and Quantum Physics*, (Dover, N.Y., 1969)

4. R. V. Churchill, J. W. Brown, and R. F. Verhey, *Complex Variables and Applications*, 3rd edition, (McGraw-Hill, NY, 1974)
   Clear and concise development of the theory of analytic functions.

5. A. L. Fetter and J. D. Walecka, *Theoretical Mechanics of Particles and Continua*, (McGraw-Hill, NY, 1980)
   The sections on the general string equation, Sturm–Liouville problems, and solitons are the most relevant to the present text.

6. J. D. Jackson, *Classical Electrodynamics*, 2nd edition (Wiley, NY, 1975)
   Extensive treatments of boundary-value problems and dispersion theory.

7. S. M. Lea, *Mathematics for Physicists*, (Thomson Brooks-Cole, Belmont CA, 2004)

8. J. Mathews and R. L. Walker, *Mathematical Methods of Physics*, (Benjamin, Menlo Park, 1964)
   Insightful general text at a slightly higher level.

9. P. M. Morse and H. Feshbach, *Methods of Theoretical Physics*, (McGraw-Hill, NY, 1953)

Graduate Mathematical Physics. James J. Kelly

ISBN: 3-527-40637-9

A massive two-volume text at a rather advanced level. This is an invaluable reference but is not suitable as a textbook, at least for recent generations of students.

Group Theory

1. M. Hamermesh, *Group Theory and Its Application to Physical Problems*, (Addison-Wesley, Reading MA, 1962)
2. W. Ludwig and C. Falter, *Symmetries in Physics*, (Springer-Verlag, Berlin, 1988)
3. L. I. Schiff, *Quantum Mechanics*, 3rd edition, (McGraw-Hill, NY, 1968)
4. M. Tinkham, *Group Theory and Quantum Mechanics*, (McGraw-Hill, NY, 1964)
5. E. P. Wigner, *Group Theory and Its Application to the Quantum Mechanics of Atomic Spectra*, (Academic Press, NY, 1959)

Numerical Methods

1. W. H. Press, B. P. Flannery, S. A. Teukolsky, and W. T. Vetterling, *Numerical Recipes*, (Cambridge University Press, Cambridge, 1986)

Reference Books

1. M. Abramowitz and I. A. Stegun, *Handbook of Mathematical Functions*, (National Bureau of Standards, Washington DC, 1970)
2. I. S. Gradshteyn and I. M. Ryzkhik, *Tables of Integrals, Series, and Products*, 4th edition, (Academic Press, NY, 1965)

Index